21.93

The Engineering Design Process

The Engineering Design Process

Atila Ertas
Jesse C. Jones
Department of Mechanical Engineering
Texas Tech University

John Wiley & Sons, Inc.
New York   Chichester   Brisbane   Toronto   Singapore

Acquisitions Editor	Cliff Robichaud
Marketing Manager	Debra Riegert
Copyediting Supervisor	Deborah Herbert
Production Manager	Lucille Buonocore
Senior Production Supervisor	Savoula Amanatidis
Designer	Laura Nicholls
Manufacturing Manager	Andrea Price

This book was set in Garamond by Publication Services and printed and bound by Hamilton Printing Company. The cover was printed by Phoenix Color.

Library of Congress Catalogibng in Publication Data:

Ertas, Atila, 1944-
    The engineering design process / Atila Ertas. Jesse C. Jones.
      p.   cm.
    Includes Index.
    ISBN 0-471-51796-8 (alk. paper)
    1. Engineering design.   I. Jones, Jesse C.   II. Title.
  TA174.E78   1993   92-34615
  620′.0042–dc20   CIP

Printed in the United States of America

10 9 8 7 6 5 4 3 2 1

Preface

The Engineering Design Process has been written to fulfill the need for a textbook appropriate for use in the capstone senior design project course. It is intended to foster a thorough understanding of the engineering design process, from the recognition of a need and the definition of design objectives through product certification and manufacture of the prototype. The discussion encompasses all of the elements, features, and constraints specified by the Accreditation Board for Engineering and Technology for engineering curricula. Hence, Chapter 1 describes the engineering design process, including elements such as establishing objectives and criteria, synthesis, analysis, testing, evaluation, and manufacturing. Many of the techniques and tools available to assist in the management of design efforts are discussed in Chapter 2. The chapters that follow include material on modeling, economics, optimization, material selection, reliability, safety and health, environmental considerations, ethics, and communications. Many open-ended miniprojects and examples are included to illustrate the discussion.

This book is presently being used in conjunction with a two-semester engineering design course sequence that includes both lecture and laboratory sessions. Engineering Design I, the first course, uses a combination of lectures, case studies, and small design projects to prepare students for undertaking comprehensive, open-ended development projects. Primary emphasis is placed on design efforts that broaden the student's concept of engineering problems and on offering variety in both the nature of the effort (designs, feasibility studies, failure analyses, etc.) and in the disciplines involved (thermal science, fluids engineering, materials science, and mechanical systems).

During the second semester in Engineering Design II, students complete their senior design project. This represents a comprehensive effort involving most of the elements of real-world design projects, including assessment of need, proposal preparation, establishing objectives and criteria, preliminary design, detail design, status briefings, fabrication, customer coordination, budgeting, scheduling, and final written report preparation and oral presentation. Throughout, the text of The Engineering Design Process complements this course sequence and provides lecture and reference material for these subjects.

Each Chapter is self-contained. Thus, individual chapters appropriate for a particular application can be selected, which allows use of the text in a wide variety of classroom and laboratory situations, including graduate courses and disciplines other than mechanical engineering. Although necessarily brief, the content of individual chapter is adequate to provide a grasp of the theory and application of the subject matter covered in the book.

Atila Ertas
Jesse C. Jones

Acknowledgments

We express our appreciation to the Texas Tech University students who assisted in the preparation of the original manuscript, including S. Ekwaro-Osire, G. Mustafa, O. Cuvalci, G. Santos, S. Kavasogullari, I. Kiris, B. Everett, A. K. M. Azizul Hoque, and L. Ordner. We are indebted to Dr. J. Rasty, Texas Tech University, for writing the "Residual Stress Considerations" section of Chapter 4; to Professor T. Floyd, Texas Tech University School of Law, for writing the "Legal Responsibilities of Engineers" section of Chapter 10; and to Dr. M. M. Tanik, Southern Methodist University, for his work on the preparation of the material on "Knowledge-Based Systems in the Design Process." We also thank the individuals who read, critiqued, and made helpful suggestions for improving the material included in the book. They include T. Hembre (ret.), Murrietta, California; J. Sills of General Dynamics Corp., R. J. Bozeman and B. G. Morris of the NASA Johnson Space Center; D. Barnes and E. Morse, Geoscience Consultants Ltd., NASA White Sands Test Facility; R. Soyer of George Washington University; T. J. Kozik of Texas A&M University; M. A. Zahraee of Purdue University; J. Ishii of The Ohio State University; J. Colton of Georgia Tech University; R. C. Progelhof of the University of South Carolina; B. Gilmore of The Pennsylvania State University; J. S. Jones of Michigan State University; and L. A. Reis, B. Walker, J. H. Smith, E. A. O'Hair, C. L. Burford, M. L. Smith, and Ms. J. A. Cantore of Texas Tech University. We are particularly grateful to Dr. T. T. Maxwell, Texas Tech University, for his suggestions about the manuscript and his generous support and help in resolving computer software problems during its preparation.

A. E.
J. C. J.

Contents

Chapter 1

The Engineering Design Process

It is beautiful for an engineer to shape and design the same way that an artist shapes and designs. But whether the whole process makes any sense, whether men become happier-that I can no longer decide. **Rudolf Christian Karl Diesel**

1.1 THE DEFINITION OF DESIGN

Like many expressions in the English language, the term *engineering design* has undergone a significant metamorphosis during the past 20 years. For the engineering graduate of the 1960s the term *design* conjured up visions of long hours at a drafting table using the tools of a draftsman in the process of making drawings, a task not often sought after by most graduates. Most new engineering graduates would invariably select *analysis* in preference to *design*. The idea of working at the "board" was considered to be undesirable because it was thought to be a draftsman's task, not suitably challenging to a young engineer and not likely to lead to job growth. In actuality the situation was just the reverse, the design engineering position was key and the analyst provided support, as requested. The design engineer made the technical decisions applicable to his or her design and coordinated with other appropriate individuals both within and outside his or her own organization in bringing the design to fruition. The design engineer thus performed engineering work as well as low-level management effort and was in a position to gain familiarity with the organization and the key people. The erroneous view that "board work" was not suitably challenging nor rewarding has probably contributed to the negative attitude toward design often observed in present-day students. Universities may have contributed to this attitude by the gradual elimination of courses that teach students the skills and techniques of preparing and using engineering drawings. Elimination of these courses was brought on by pressure to

1

include newly developing technology in the curriculum as well as to add courses in the humanities and other nontechnical fields, while holding the same number of total hours for the degree requirement. With the advent of computer-aided-design capability, the need to include engineering drawing courses in the curriculum was further diminished. Unfortunately, this lack of training in the rudiments of engineering drawing preparation and use has led to the situation that exists at the present time, one in which a significant percentage of graduating engineering students have little or no understanding about engineering drawings or drawing systems. Since the engineering drawing is the principal means of communication (language) for the design engineer, this is a serious gap in the educational process, one that could be likened to poor command of the English language, both written and spoken, by university graduates in general.

Although the meaning of the term *design* has been significantly expanded since the 1960s, many students still retain the relatively narrow interpretation described above. Possibly the most definitive and certainly the most applicable interpretation of the meaning of design for engineering students is that promulgated by the Accreditation Board for Engineering and Technology (ABET),[1] quoted below.

> Engineering design is the process of devising a system, component, or process to meet desired needs. It is a decision making process (often iterative), in which the basic sciences, mathematics, and engineering sciences are applied to convert resources optimally to meet a stated objective. Among the fundamental elements of the design process are the establishment of objectives and criteria, synthesis, analysis, construction, testing and evaluation. The engineering design component of a curriculum must include some of the following features: development of student creativity, use of open ended problems, development and use of design methodology, formation of design problem statements and specifications, consideration of alternative solutions, feasibility considerations, and detailed system descriptions. Further, it is essential to include a variety of realistic constraints such as economic factors, safety, reliability, aesthetics, ethics, and social impact. Courses that contain engineering design normally are taught at the upper division level of the engineering program. Some portion of this requirement must be satisfied by at least one course which is primarily design, preferably at the senior level, and draws upon previous coursework in the relevant discipline.

The primary purpose of this textbook is to provide material suitable for instruction in engineering design in accordance with, and pursuant to, this definition.

1.2 THE DESIGN PROCESS

According to the definition of design given above, the design process begins with an identified need and is completed upon satisfactory qualification testing and acceptance testing of the prototype. In this sense the prototype is the first product completed in the production process using all of the developed manufacturing processes, and checkout and test procedures. The prototype is in every sense identical to the products that follow it in the production process.

[1]Accreditation Board for Engineering and Technology, Inc. Annual Report for the year ending September 30, 1988, New York, 1988.

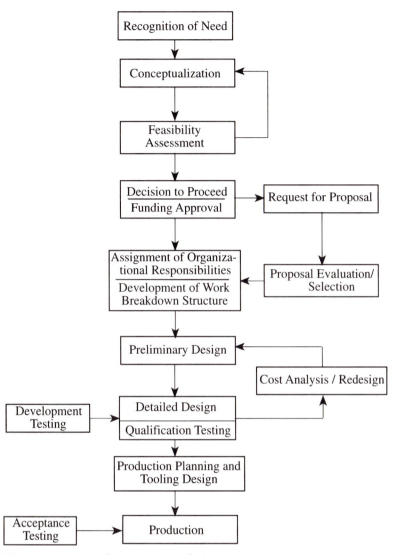

Figure 1.1 Steps in the engineering design process.

Fig. 1.1 depicts the typical steps in the engineering design process for a project of some magnitude. The process shown is considered to be generally applicable to most design efforts but the reader should recognize that individual projects will often require variation of the process, including the elimination of some steps completely.

1.2.1 Recognition of Need

The design process begins with an identified need that can be satisfied by the product of engineering effort. The scale and variety of needs are vast, but any listing would have to include formal requests, as in the case of "requests for proposals" (RFP's); informal requests, in which the potential customer suggests or implies that a proposal in a particular area of interest would be favorably received; unsolicited proposals, which can be based on the ideas of individuals or groups pursuant to a need they believe exists; assignments from superiors; an idea that meets some need formulated during the routine or nonroutine performance of an assigned task or in an assigned area of responsibility; needs pursuant to new legislation; and generation of ideas unrelated to any assigned area of responsibility that satisfy an existing need or for which a need can be developed. It is apparent from even this short listing that the design process can be initiated based on an idea for a solution to an existing or identified need or from an idea for a product or process for which it is thought a need can be generated. The hula hoop is a good example of a product for which the idea preceded the identified need.

In many projects the "need" is identified by an organization other than the one that will eventually accomplish the effort. This is true for most projects sponsored by government organizations and, in these situations, the conceptualization and feasibility assessment will normally be completed prior to the issuance of an RFP for the design, development, and production phases. The NASA Space Shuttle contract with North American/Rockwell is an example of this type of arrangement. Prior to the award of this contract, study contracts were consummated with several aerospace contractors for the definition and feasibility assessment of this significant project. Pursuant to the output of these study contracts, a decision to proceed was made and funding was subsequently approved by the U.S. Congress. The RFP was then issued based on the best ideas presented in the study contracts and the North American/Rockwell proposal was selected as presenting the best overall solution considering both technical content, management, and cost.

It is at the beginning of the design process, during the *conceptualization* phase, that it is most important to consider alternative solutions. Once the *decision to proceed* has been made and *funding approval* has been granted it becomes increasingly costly and difficult to make changes. Human beings tend to lose their objectivity once they have identified with a particular scheme and have contributed to its formulation and initial planning. Thus, it becomes increasingly difficult for the people involved in the process to recognize the value of, or justification for, making significant changes. The project takes on a life of its own and it usually takes some traumatic event to cause a change in approach. This group dynamics syndrome applies to ongoing industrial programs as well as to newly initiated design processes. The changes that took place in the U.S. automobile industry in the 1980s would not have occurred without the combined traumatic events of the 1972 oil embargo and the subsequent increased share of the U.S. automobile market claimed by the Japanese automobile companies. Kodak's lack of interest in the polaroid process when initially approached is well known. Kodak's management apparently had very closed minds when it came to accepting new ideas; the company was profitable and they undoubtedly thought that they had the market captured for both the long and short term. The U.S. Navy's reluctance in accepting adjustable gun-pointing systems on ships provides an interesting study of the resistance to change. This frailty of human nature, the tendency of a group to become cohesive to the point of resisting outside suggestions for change or improvement to the group's solution (close ranks against outside enemy attack), is well recognized and has been the subject of demonstration in university business psychology and management programs.[2] The increasing cost of making changes as programs develop is, of course, due to the sunk costs of manpower and materials as well as to the difficulty in reassigning or terminating an increasingly large number of employees.

[2]H. J. Leavitt, *Managerial Psychology,* Second Edition, The University of Chicago Press, Chicago, 1964.

Recognizing the difficulty of making changes once a course of action has been adopted, what are the important considerations that should be applied to ensure that the selected approach is optimum, given the information and data available? Probably the single most important consideration at this juncture is a thorough understanding of the need. Without a complete understanding of the problem it is impossible to identify the correct solution or solutions. Sanford Cluett, who invented the *sanforizing* process, was able to identify a solution for the shrinkage of clothing only after recognizing that the fibers used in the fabric were distorted by the manufacturing process and subsequent washing of the garment made from the fabric merely restored the individual fibers in the fabric to their original condition. The solution was obvious once the problem was correctly defined: restore the fabric to its natural condition before manufacturing the garment. The personality trait that engineering students need to nurture to enhance their problem analysis capability is curiousity. A curious person will continue to dig until all of the pertinent facts are disclosed and a thorough understanding of the problem is gained. The nature of a curious person is such that they cannot rest until the problem has been thoroughly dissected and understood. Engineers involved in the design process must possess this personality characteristic and apply it with effectiveness during the initial stages of the process, during the need-definition stage.

1.2.2 Conceptualization and Creativity

After the problem has been completely defined, viable solutions need to be identified from which the optimum approach can be selected. Brainstorming, in which a group of participants suggests as many ideas for solutions to the problem as possible in a set period of time, is one technique that has been used to assist in identifying alternate solutions to problems. When several people are involved in the process, this approach may have some merit, since the ideas suggested by one person may trigger ideas for better solutions on the part of another. Success in the development of problem solutions, whether as a member of a group or individually, relates to individual creativity. The question as to whether creativity is an inherent personality trait or can be taught has been debated for many years. A conference sponsored by the American Association for Engineering Education (ASEE), the National Science Foundation (NSF), and the U.S. Department of Commerce held in the mid-1960s, concluded that the creative requisites of invention and innovation could be taught, but that engineering schools were not doing an adequate job.[3] Most parents who have raised more than one child would probably agree that children differ significantly in their inherent creative ability, even within the same household and with the same parents, giving credence to the assumption that this is an inborn characteristic. It is suggested that creativity is a characteristic that is both inherited and capable of being learned. Some people are born with a high degree of potential for creativity and, with the right training, become exceptional. Several great composers and scientists of the past come to mind. People with average creative ability can sharpen their capabilities with appropriate training. The following list includes several personality characteristics thought to correlate with creativity:[4]

---

[3]J. P. Romaldi "The Teaching of Creative Engineering Design," *CIT Engineering News*, Vol. 3, No. 2, Fall 1983.

[4] Communication with colleagues during presentation on "Creativity in Education," Texas Tech University, Lubbock, TX, 1983.

1. Good guessing.
2. Risk taking.
3. Challenging authority and procedures.
4. Preferring the complex and difficult.
5. Being sensitive emotionally, having a sense of beauty.
6. Having a vivid imagination.
7. Desiring honesty and frankness.
8. Being curious.
9. Having high self-esteem.

Those individuals who believe that they have very little inherent creative ability need only remember the comment by Thomas Alva Edison that invention is 99 percent perspiration and 1 percent inspiration. If an individual is committed to achieving a goal, believes that he/she has it within themselves to be successful, and is willing to *perspire*, almost any vision can be realized.

EXERCISE 1.1

Automobile Passive Restraint Design Concept

Recognition of Need

A number of active and passive restraints are available in automobiles today. Current active restraints (manual seat belts) are very effective, but only when used. Passive restraint concepts that have been studied include the air bag, the automatic safety belt, the force-limiting belt, friendly interiors, energy-absorbing steering wheels, knee restraints, and plastic-coated antilacerative windshields. Two of these systems are currently in use, namely, the automatic seat belt and the Air Cushion Restraint System (ACRS), or inflatable air bag. The ACRS seems to be favored by the public;[5] 38 percent of the people surveyed in this study preferred this option. However, only 7 percent reported knowing anything about air bags. On being advised of the additional cost involved to incorporate the air bag into the automobile, 13 percent changed their preference to seat belt restraint.

In addition to being effective in frontal collisions only, the air bag has a number of other drawbacks including initial and replacement cost, lack of protection against secondary impacts, and reduced driver visibility. Accidental deployment can cause serious injuries if the driver's visibility is obscured or he or she reacts to the loud noise of deployment by losing control of the vehicle. Furthermore, the air bag does not satisfy all of the requirements of federal regulations, since it is ineffective against injuries from rollover or side impacts.

The automatic seat belt is not without problems in that ingress/egress difficulties arise with present systems, and a substantial number of consumers appear to be discomfited by belts. A Department of Transportation survey showed that only 38 percent of drivers favored them and that 32 percent would actually go to the extent of disconnecting them. Automatic seat belts

---

[5]J. Kelderman, "Seat Belts Win Approval in Survey," *Auto News*, February 11, 1985, p. 34.

also have the same problems associated with manual seat belts: rubbing the occupant's face and neck; irritation by web-body contact; sliding beneath webbing on impact; and questionable protection in impacts other than frontal. These problems obviously must be addressed for an automatic seat belt system to be completely effective.

General Motors Corporation in conjunction with the American Society for Engineering Education (ASEE) established a program entitled "University Design Competition on Passive Restraints for Automobiles" to address the problem of developing new concepts in passive restraint design. Specifically, the purpose of this program was "to develop alternative passive restraints that are acceptable and functional and to expose university students and faculty to the automotive design process." The DOT requirement[6] for passive restraints specifies that they must protect occupants from serious injury as a result of the vehicular incidents outlined as follows:

- Impact of a vehicle traveling longitudinally forward at any speed, up to and including 30 mph, into a fixed collision barrier that is perpendicular to the line of travel of the vehicle, or at any angle up to 30 degrees in either direction from the perpendicular to the line of travel of the vehicle.

- Impact of a vehicle laterally on either side by a barrier moving at 20 mph.

- Rollover of a vehicle in either lateral direction at 30 mph.

In addressing the problem of providing an acceptable and effective passive restraint system, an appropriate design philosophy must be adopted. The driver/passenger can be restrained in the seat or can be allowed to leave the seat but have his/her resulting motion arrested prior to impacting the steering wheel, dashboard, windshield, or other internal structure. The provision of friendly interiors can significantly mitigate the probability of an unrestrained occupant receiving fatal injury, but the elimination of other restraint devices does not appear warranted. Research indicates that personal safety is greatly enhanced when the occupant is restrained in the seat during an accident. It has been documented by the National Highway Traffic Safety Administration that the fatality rate increases by 500 percent when a person is ejected from his/her seat.

For the purposes of this exercise the body should be considered as two main masses, upper and lower, whose actions are nearly mutually exclusive. The upper mass includes the shoulder and chest region, whereas the lower mass includes the abdomen, pelvic, and leg (knee) regions. The head/neck region should be considered separately, since its full mass moves independently from the upper mass. Consequently, three separate restraint provisions are necessary to effectively prevent injury, namely, head/neck restraint, upper mass restraint, and lower mass restraint.

Head/Neck Region   The human skull can withstand high impact forces without bone damage when the force is evenly distributed over a large contact area. A good mitigation method will alleviate many of the severe/fatal head injuries that occur from accidents of minor/moderate severity. A good mitigation

---

[6]Code of Federal Regulations 49, Parts 400–999, Department of Transportation.

method must include even and maximal surface distribution of head contact, and must consider the force penetration characteristics of the padded materials used.

Upper Mass Restraint In considering the standard shoulder harness with respect to chest restraint, a load of 3000 pounds, distributed over approximately 40 square inches, can be imposed on the occupant's rib cage. Whereas shoulder harnesses are infrequently the cause of internal injuries, clavicle fractures are fairly common, since the load occurs on ribs that are not well anchored. The SAE supports the contention that injuries sustained are usually due to the pressure exerted by the shoulder harness on the "soft spots." By distributing the force over a large surface area, the chance of internal injuries and clavicle fractures can be further reduced.[7] Although the shoulder harness is relatively effective in restraining motion in the sagittal plane, it is not very effective in preventing ejection or injury in a lateral impact. The impact region of the passenger compartment is important to the magnitude of injury risk whereas the shoulder harness merely mitigates the degree of injury. Modifications of the lateral structure of the passenger compartment can reduce injury risk by intrusion. This can be accomplished by installing reinforcing or energy-absorbing elements in the lateral car structure. The SAE also suggests energy-absorbing bolsters that fit into the sides of the compartment in impact-exposed locations.

Lower Mass Restraint The standard seat belt with a lap belt and shoulder harness does not provide satisfactory abdominal protection. The most prevalent and severe type of impact is the head-on collision. The lap belt alone can be potentially more dangerous than no safety belt at all. The lap belt can cause a person to be thrust at an acceleration far exceeding the acceleration the person would have reached had he or she not been wearing a safety belt. This motion is labeled "jackknifing." "Jackknifing" causes the chest and face to strike the steering assembly, causing numerous injuries. It is considered as one of the most critical defects of the lap belt during accidents, since many people have been severely injured or killed by this phenomenon. One study showed that 16.1 percent of all injuries occur in the abdomen area when a seat belt was not worn, while 14.9 percent of all injuries occur in the abdomen area when a seat belt was worn. Abdominal injury frequency is nearly equivalent whether a lap belt is worn or not.[7] It cannot be said that the lap belt does not restrain; however, it does not reduce abdominal injury risk significantly. The optimal seat belt includes a shoulder harness and a thigh restraint rather than the traditional shoulder harness and lap belt.

Development of a Design Concept

Based on the above description of the need for a passive restraint, students should be organized into small groups (four to six per group is recommended) and be asked to schedule a brainstorming session to identify design concepts that will satisfy the need constraints and criteria outlined above. The best concepts can then be selected by each group and be presented to the class along with a rationale as to why that

---

[7]"Restraint Technologies: Front Seat Occupant Protection," Report: (SP-690), International Congress and Exposition, Society of Automotive Engineers, 1987.

particular concept(s) is considered optimum. The solution recommended to General Motors and ASEE by a team of students from Texas Tech University is described in the solution manual and may be useful for comparison purposes.

In conducting the brainstorming sessions the students should try to think of as many ideas as possible in an agreed to period of time. Studies on brainstorming have shown that those sessions that produced twice as many ideas concluded with more than twice as many good solutions in the same period of time. To be successful in brainstorming, criticism of any idea must be absolutely avoided and judgment of any idea must be postponed until after the brainstorming session. Brainstorming is successful only if a goal is set to develop a certain number of ideas during the session or to develop a certain number of ideas in a certain time period. Ideas should be generated with an open mind, with inquisitiveness and with fluent and flexible thinking. The following points should be kept in mind during the brainstorming session:[8]

1. Quantity first.
2. No criticism.
3. No judgment.
4. No evaluation.
5. Encourage uninhibited thinking.
6. Build on the ideas of others.
7. Do not try to make decisions.
8. Set goals.

1.2.3 Feasibility Assessment

Assessment of the feasibility of the selected concept(s) is often accomplished as part of the conceptualization task on relatively small projects but is usually a major element of the overall program on larger projects, sometimes taking several years to complete. The feasibility studies for the Space Shuttle contract mentioned previously covered a period of over two years during which time NASA, aerospace companies, and academic groups considered various vehicle and engine configurations and overall system capabilities. The types of questions that were resolved during that period included:

1. Fully or partially reusable booster.
2. Piggyback or integrated orbiter, launched vertically with horizontal landing.
3. Number and location of crew.
4. Size of cargo bay.
5. Level of engine technology.
6. Use of solid- versus liquid-fueled engines.
7. Orbiter configuration, straight wing, or delta.
8. Projected cost.

[8]J. Ramsey, informal communication, Texas Tech University, Lubbock, TX, 1988.

The purpose of assessing the feasibility of the concept is basically to ensure that the project proceeds into the design phase on the basis of a concept that is achieveable, both technically and within cost constraints, and one that requires new technology only in those areas that have been thoroughly examined and agreed to. The manner in which this is accomplished depends on the size and complexity of the project, but the feasibility assessment period is the time for defining the concept to such a degree that design can proceed with confidence that the end product will accomplish the intended purpose within the available resources. For many projects this effort is largely directed toward refining the cost estimate, since cost is often the limiting constraint. For example, a feasibility study on an aerospace test complex constructed during the early 1960s had a construction budget of slightly over $5.0M. The initial cost estimate presented during the feasibility study first quarterly review was more than $50.0M. Needless to say, the effort to reduce the cost to within the available funding had a significant effect on the design concept. By the midterm review the estimate had been reduced to around $25.0M and, after further consolidation and concept modification, the cost estimate was reduced to approximately $6.0M at the final review. This was considered to be close enough to the available funding (at this stage in the effort) since, by identifying several design features as *options* in the construction bid package, the estimated base cost could be brought to within authorized funding. This example is not considered to be unusual for high technology design projects and, although the focus and financial considerations may be different, is probably reasonably typical for projects involving other products and processes.

It is of paramount importance to have people with broad experience and good judgment involved in this phase of the design process. It is especially important to have people in charge of the feasibility study effort who are directly responsible for the overall (cradle to the grave) performance and functionability of the product, process, or facility—people who have a *work ownership* mentality. During the above-mentioned feasibility study a disagreement arose between the people directing the study and some of the consultants who were providing guidance in one of the principal functional test disciplines within the facility complex. If the consultants' recommendation (which had personnel safety implications) had been followed, the overall test complex capability would have had to be severely compromised or additional funding would have had to be requested, a process that could have severely jeopardized the entire project. In this case the consultants were overruled and after construction, the test complex subsequently proved to be highly functional and safe for operating personnel using the design and operational approach adopted.

1.2.4 Establishing the Objectives and Criteria

Establishing the requirements is one of the most important elements in the overall design process. This is a task that must be accomplished prior to initiating the design and after the concept has been selected. It is often accomplished during the feasibility study after the design concept has been defined, or it may be accomplished early during the design effort. The design requirements both drive and control the design throughout the process, but they are especially important during the early phases of the effort because they are the principal guidelines for the project team at this point in the process. Although it should be self-evident, it is absolutely essential that all project team participants have a complete and agreed-to understanding of the requirements. A good example of what can happen as a result of a misunderstanding of the design requirements occurred during a large (and costly) missile program in the late 1950s. In this program there was confusion between disciplinary organizations as to the meaning of one of the major system design requirements. Unfortunately, this confusion did not become apparent until late in the design process when the overall systems checkout and test procedure was being prepared. The confusion was related to the meaning of the terms *automatic, semiautomatic,* and *manual.* At significant cost, the mechanical design group had provided the capability for all three modes of system operation based on its understanding of these terms, but the control system design group had a different interpretation of these terms and thus did not provide this capability. By the time this misunderstanding was discovered it was impossible to correct the design, and one of the major design requirements was not satisfied.

Establishing the design requirements is one of the most difficult activities in the design process. The requirements are so critical to the ultimate product capability and its cost and they must be established so early in the process that the designer almost needs prescience to be able to specify them accurately. Since engineers are mere mortals and do not have foreknowledge of things to come, the design requirements have to be established by using excellent judgment, with wide and in-depth coordination among key participants in the process, and with test and analysis support, when appropriate. The design requirements need to be specific and to as low a system and component level as possible. Although requirements are established to be permanent and inviolate, nevertheless they should be continuously reviewed and revalidated during the design process, at least, until the design is frozen, to ensure that they continue to reflect the goals and objectives of the project.

To ensure that a common understanding of the design requirements exists, it is essential that adequate coordination be accomplished. Some sort of interdisciplinary coordination group needs to be established to discuss the design effort ongoing in each discipline and how the design requirements are being interpreted and applied. One technique that may be beneficial in this regard is to pick a key individual from each appropriate discipline to start the preparation of system (or product) checkout and test procedures early in the design process. In this way interdisciplinary coordination is forced to focus on a task that ensures proper subsystem integration and uniform understanding of design requirements. Adequate coordination within individual disciplines must be ensured by supervision and management within the discipline. Intradisciplinary coordination is, to a great extent, enhanced by the fact that the personnel involved work together on a day-to-day basis and should normally be aware of each other's actions and plans.

1.2.5 Synthesis and Analysis in the Design Process

The words *synthesis* and *analysis* are antonyms, words that have opposite meaning. To synthesize means to combine the parts or elements of an object into a complex whole whereas to analyze means to separate the whole into its constituent elements. In the design process these activities are totally interrelated and continuously ongoing. At the beginning of the process, when possible solutions are being sought and identified, reasoning proceeds directly from the statement of need and principles established or assumed to a conclusion or possible answer. Ideas tend to be formulated as complete (whole) solutions to the stated need. Analysis begins almost immediately thereafter, if only in a very gross manner at this time in the process, to begin to understand the elements required for the whole and to determine whether the proposed solution is actually feasible.

After the concept has been approved and the design process has been initiated, design requirements must be established. Again, the process involves a combination of synthesis and analysis in order to establish requirements at the component, subsystem, and system level. One of the essential tools that must be developed at this point in the design process is the system schematic, layout, or other engineering documentation (overall product specification), which can used by the various technical organizations to establish and informally maintain system configuration identification and control. All components and subsystems are normally identified on the system schematic; thus, it provides a means for visually depicting the manner in which all of the elements are assembled into a system whole (synthesis/analysis).

Detail design is initiated after the design requirements have been adopted, and the overall system configuration has been established. The design requirements constitute the component specifications that determine the characteristics and configuration of the individual elements. The processes of synthesis and analysis are repeated continuously over the life of the design process as the elements and the total system continue to be defined, tested, evaluated, produced, and finally assembled to form a system. The drawing system that is used to document and define the system and all its components is a visual representation of the synthesis/analysis process. Every manufactured piecepart and component is depicted on its own drawing with its own part number. These detail drawings reflect the number of the *next assembly* drawing, which depicts the relation of the detail part and the part or parts to which it is attached or installed. The resulting subassembly drawing will also identify the drawing number of its *next assembly* , which will show the subassembly installation or attachment. This hierchical process of lower level drawings referring to the *next assembly* drawing continues to the top assembly drawing, thus, providing a completely integrated (synthesized) drawing system.

1.2.6   The Organizational/Work Breakdown Structure

Regardless of the type of organization, industry, or government, once the project becomes viable, a systematic structural breakdown of the effort must be made. For most government projects the preliminary work breakdown structure is incorporated into the RFP and is finalized when the contract is consummated. The work breakdown structure (WBS) is basically a family tree subdivision of the effort that is used to provide a means for management of the various work elements. It relates the tasks to each other and to the end product and provides a baseline against which the technical, schedule, cost, and manpower reporting can be accounted. The WBS is developed by starting with the end objective and subdividing it into manageable elements in terms of size and complexity with each successive level reducing the scope, complexity, and dollar value of the WBS element. For example, the WBS shown in Fig. 1.2 starts with the Space Shuttle Program, which is divided into two phases, Shuttle System Development and Production and Shuttle Operations. The next level is the project level, which includes such elements as the Orbiter Vehicle, the Solid Rocket Motor, External Tank, and Flight Operations. Level 4 corresponds to the systems level and includes Structures, Propulsion, Power, and Avionics. Finally, Level 5 corresponds to the subsystem level, which includes a further breakdown of Level 4 systems. Using the Propulsion System as an example, Level 5 divides this system into subsystems such as the Main, Reaction Control, and Orbital Manuevering

CWBS Level 1 (Total Contract)	CWBS Level 2 (Total Contract)	CWBS Level 3	CWBS Level 4	CWBS Level 5
Space Shuttle Program	1.0 Shuttle System Development and Production Phase	1.1 System Management		
		1.2 System Engineering and Intergration		
		1.3 Orbiter Vehicle	1.3.1 Structures	
			1.3.2 Propulsion	1.3.2.1 Main Propulsion
				1.3.2.2 Reaction Control
			1.3.3 Power	1.3.2.3 Orbital Maneuvers Propulsion
			1.3.4 Avionics	1.3.2.4 Airbreathing Propulsion
			1.3.5 Environmental Control / Life Support	1.3.2.5 Solid Rocket Propulsion
		1.4 Solid Rocket Motor (GFE)	1.3.6 Crew Station and Equipment	
		1.5 Main Engine (GFE)	1.3.7 External Tank Subsystem	
		1.6 External Tank (GFE)	1.3.8 Payload Accommodation	
		1.7 Airbreathing Engine (GFE)	1.3.9 Installation, Assembly, and Checkout	
		1.8 Flight Test Support	1.3.10 Major Ground Testing	
		1.9 System Support		
	2.0 Shuttle Operational Phase	2.1 Flight Operations		
		2.2 Ground Operations		
		2.3 GFE Production		
		2.4 Spares, Repair, and Overhaul		

Figure 1.2 Space shuttle program work breakdown structure (from *Handbook for Preparation of Work Breakdown Structures*, NHB 5610, p. 23, 1975).

Propulsion subsystems. Further breakdown of effort below the subsystem level is often referred to as the component, task, subtask, and work element levels.

The WBS must be compatible with the organizational structure of the work performing entity to ensure that management accountability is maintained and that diffusion of responsibility across organizational lines is minimized. The WBS must reflect and represent the way in which the work is organized, managed, and accounted for so that there is a clear relationship (interface) between each organizational element and the WBS element(s), for which it is responsible. [9]

The degree and level to which any design process is subdivided for work assignment and management responsibility is a function of the complexity and duration of the effort, the overall cost of the project, the organizational structures of the contractor and customer, the number of contracts and their relationships, and the needs of contractor and customer management. Small consumer products and similar design and development efforts obviously do not require the WBS complexity of a project such as the NASA Space Station, for which the preliminary WBS had 80 major systems. In fact, some small projects may not even warrant the preparation of a formal WBS. The point to remember in planning and organizing the work is to subdivide the project into as many tasks and levels as is considered necessary for good management control and organizational element responsibility, taking into account the above considerations.

1.2.7 Preliminary Design

This is the phase of the design process that bridges the gap between the design concept and the detailed design phase of the effort. This phase of the effort is referred to as embodiment design in some textbooks, especially those published in Europe. The design concept is further defined during preliminary design and, if more than one concept is involved, an evaluation leading to selection of the best overall solution must be conducted. As the design concept continues to be refined during the preliminary design phase, the overall project cost estimate will also become more realistic. During preliminary design the overall system configuration is defined and a schematic, layout, definition drawing, or other engineering documentation (depending on the project) will be developed to provide early project configuration control and to assist in ensuring interdisciplinary and intradisciplinary integration and coordination during the detail design phase. System-level and, to the extent possible, component-level design requirements should be established during this phase of the effort. Establishing requirements is a function that involves computation and analysis, literature search, vendor equipment evaluation, evaluation of previous experience, discussion with experts in the particular field, good judgment, and testing (when absolutely necessary). In general, it is costly to run test programs and is thus desirable to minimize testing during the preliminary design phase of the project. The importance of establishing valid requirements cannot be overemphasized, however, and if limited testing will provide the information necessary for this purpose, it may be justified.

The requirements established during the preliminary design phase of the project form the basis for the component specifications that are developed during the detail design phase. If the requirements are too stringent, the project cost will escalate and (possibly) no supplier will be found that is willing to bid on the contract to provide

---

[9] *Handbook for Preparation of Work Breakdown Structures,* National Aeronautics and Space Administration, NHB 5610.1, February 1975.

the item in question. If the requirements are too lax, the overall system requirements may not be met, which could lead to dire consequences for the overall project. An additional problem with loose requirements is that they end up being tightened up, sometimes after the initial contract for the component has been consummated, with greatly increased cost, difficulty, and ill will between the supplier and the customer. The importance of establishing valid design requirements at the outset is thus apparent. A technique that may be beneficial in this regard is to initiate the preparation of the system test, checkout, and maintenance procedures at an early stage in the design. The process of thinking these procedures through may help in quantifying the various operating parameters thus, providing a more valid basis for the component design.

Preliminary design is often accomplished by a small nucleus of personnel who are eventually involved in the detailed design process. Another approach often used is for the preliminary design to be completed by a separate advanced design group prior to turning the project over to the individual detailed design groups, which are usually organized along disciplinary lines. There are basically two reasons for using a separate group at this stage of the design process. A small group, including individuals representing all the necessary technical disciplines, can maintain close coordination during this phase of the effort and can work as an integrated element greatly simplifying management control and technical direction. Also, the costs incurred during this phase of the project can be controlled to a much greater degree by using a small integrated group rather than distributing the task throughout the organization.

1.2.8 Detailed Design

The intent of the detailed design phase of the project is to develop a system of drawings that completely describes a proven and tested design so that it can be man-ufactured. It is during the detailed design phase that the project gets fully underway. At this stage all the various disciplinary organizations are actively involved in the synthesis/analysis process, resolving the system design concept into its component parts, evaluating components to validate previously established requirements and specifying those design requirements left undefined, and assessing the effect of the component requirements on the overall system requirements. As component requirements are finalized, specifications can be prepared for those components that will be procured from outside vendors rather than manufactured internally. These specifications are often drawings (usually A size, 8 1/2 × 11 in.) and, as with internally manufactured parts, the drawing number becomes the part number. Specifications must include all of the requirements for the component including the operating parameters, operating and nonoperating environmental stimuli, test requirements, external dimensions and interfaces that must be controlled, mainte-nance provisions, materials requirements, external surface treatment (if any), design life, packaging requirements, external marking, and any other special requirements such as special lubricants, and the like. A good specification will minimize problems of interpretation that could surface later and result in disagreement with the supplier, possibly with negative impact on the entire project. More importantly in this regard, it is essential that suppliers be selected that are ethical and reliable, as well as capable. A recent trend in U.S. business contracting is the establishment of long-term relations with suppliers based on evidence that the supplier has an effective statistical quality control program in place rather than just being the low bidder.

Under this arrangement a supplier is certified only after an in-depth review of the manufacturing operation, quality control program, and management philosophy.

For components and other system elements that are manufactured internally, detail drawings are prepared that specify the necessary dimensions, the materials of construction, the surface treatments, assembly techniques, and any other information necessary for fabrication. Often, specification requirements for these components are included on the face of the drawing as *Notes*, but also may be tabulated on a separate sheet referred to as a specification sheet. The detail drawing must include all of the information necessary to produce, test, and package the part. The detail drawing must also include adequate orthographic views of the part so that a complete description of its size and shape is furnished to the manufacturing group (see Fig. 1.3).

In the title block of the detail drawing the *next assembly* will be identified. This *next assembly* drawing shows the detail part or component as it is interconnected to other parts and subassemblies. Assembly drawings provide the necessary information for the individual detail parts and components shown on the drawing to be interconnected. Dimensions and other information specified on the drawing are limited to those required to be complied with during assembly. The *Bill of Material*, usually located at the lower right corner of the drawing, lists all the detail, component, and other piece parts that make up the assembly. Special assembly instructions and test requirements are also often included in this area of the drawing as well. As with detail drawings and component specifications, subassembly drawings also have their *next assembly* identified in the title block. This procedure (referencing of the *next assembly*) is followed up to the *final* or *top assembly drawing*, which depicts the deliverable end item (or product). Assembly drawings do not usually include detail part dimensions, but if the part is not complex it may be detailed on the assembly drawing and identified with a dash number. In this case the part number of the detail part will be identical to the assembly drawing number with the dash number added (e.g., 1234-3). A typical assembly drawing, sometimes referred to as a working assembly drawing, is shown in Fig. 1.4.

The installation drawing is another type of assembly drawing that is used to specify dimensions that must be maintained during installation of a deliverable end item. In this sense the term *installation* differs from *assembly* in that the end item installation is not accomplished by using all of the controls, processes, and personnel as would be true for an inplant assembly. An installation drawing often depicts the installation of an item in a facility or in an existing system located distant from the place where the item is manufactured. The installation drawing identifies all the assemblies and parts that make up the installation and often includes the installation procedure, when a fairly simple procedure is involved. Installation drawings are often included in parts catalogs and overhaul and repair procedure manuals.

The advent of computer-aided design (CAD) has had an enormous impact on the process of design and drawing preparation. Significant problems have often been encountered in trying to determine the actual configuration of a detail part or assembly because of changes made, but not incorporated on the drawing, due to time constraints. The vehicle used to make these changes is an Engineering Order (EO) or Change Order (CO), which is an $8\frac{1}{2} \times 11$ in. drawing on which the change to the part is detailed. Although these EOs are listed on the affected drawing and the symbols are noted on the drawing face identifying the area modified by the change, the nature of the change is shown only by the EO. Significant costs can be incurred after a project is completed in bringing the drawings up to

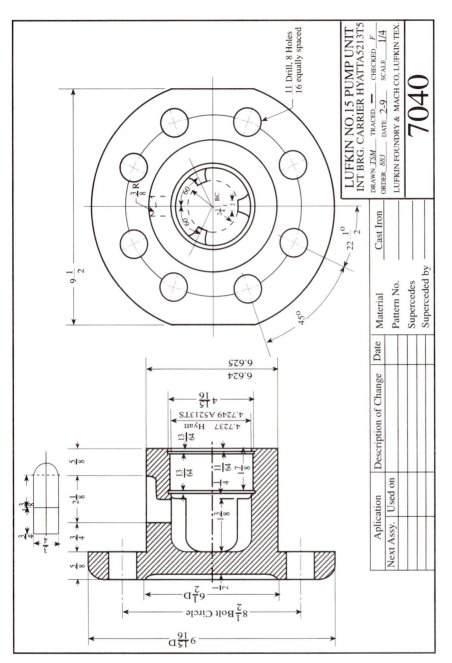

Figure 1.3 A detail drawing (from Carl L. Svenson, *Machine Drawing*, D. Van Nostrand, New York, p. 63, 1945). Courtesy Van Nostrand Reinhold, New York.

17

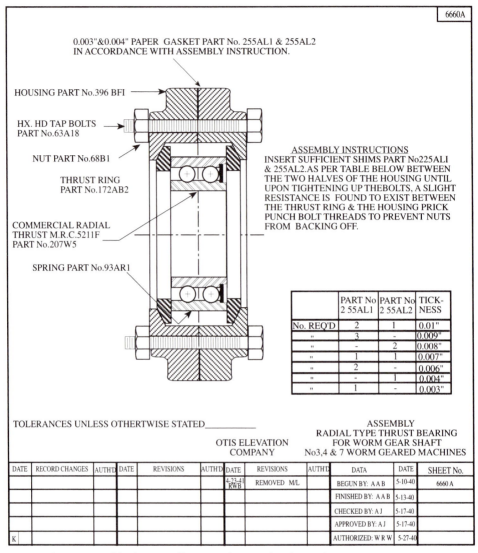

Figure 1.4 An assembly drawing (from Josef V. Lombardo *et al., Engineering Drawing*, p. 239, Barnes & Noble, Inc., New York, 1956). Reprinted by permission of Harper Collins Publishers.

the current, as-built configuration. Since design changes made using CAD can be instantaneously incorporated on the drawing(s) affected, they will always reflect the current configuration, which basically eliminates this problem. Drawing files are also much easier to manage with CAD, since drawing records are maintained on computer disk rather than on full-size vellums. The design itself can also be improved by using the capabilities of CAD for part evaluation, including function and stress/deflection analysis, and accessing the significant amount of material property and other information available through the use of data banks. The use of computer simulation in understanding and interpreting complex part or component interactions provides a capability that cannot be equaled, even through the use of testing, which is costly. The impact that computers have had on the integration of design and manufacturing may be even more significant. With the need to shorten product development time to remain competitive, an increasing number of firms are taking advantage of the capabilities of CAD/CAM (computer-aided manufacturing) to initiate production planning, tooling design, and manufacturing early during the detailed design phase of the project. For example, the John Deere Dubuque Works has implemented a method called *simultaneous engineering* to reduce product development time for farm equipment from 5 years to less than 30 months. This approach integrates design and manufacturing personnel on a single team to work on tasks concurrently, rather than serially. Other firms have used this approach to reduce the number of parts required to be manufactured for new products, thus, reducing fabrication time.[10] In spite of the many advantages of CAD systems, however, the student is cautioned that their use in no way diminishes the need to be able to prepare design drawings and to understand and use drawing systems. As indicated previously, the engineering drawing is the design engineer's vehicle of communication, and it is axiomatic that a good designer is a good communicator.

Development testing is an essential element in most detailed design processes. Development testing is used to help in determining design requirements, in evaluating new concepts, in obtaining information on the performance of project components and subassemblies, in validating computational models as well as the design itself, and in investigating interdisciplinary integration problems. The value of a well-planned and executed development test program can hardly be overstated, but it is important to remember that testing is costly and thus needs to be planned adequately and controlled diligently. The use of innovative cost control techniques should be considered, but care must be exercised to ensure that the probability of success and the potential savings justifies the decision to proceed. In one example of successful and innovative cost-control techniques, expenditures during detailed design of the Space Shuttle Orbiter were minimized by recycling major development test articles back into the production line after testing was completed. Although this is a somewhat unorthodox approach, it resulted in considerable savings for this project. It should be obvious that someone assigned to the project that has both technical and budgetary responsibility must have the final say as to the level of development testing accomplished. This is normally the Project Manager or his or her designee.

In most design processes of any significant magnitude, a design *freeze* is implemented at some point prior to completion. This is the point at which the design process is formalized and design changes are placed under strict and formal control, often by some sort of configuration control board. Configuration Control Boards (CCBs) normally include membership representing all the design disciplines, project management, the customer, safety, quality control, and other staff functions, as appropriate. The point in the overall design process at which the design is *frozen* is determined by customer requirements, by the need to control costs and configuration, by the need to inject greater discipline into the process, and by the need to forceably implement increased coordination among all the participants in the program. Although this is a necessary step in the overall process, it adds significant complication to the effort of the designer and restricts the ability to correct deficiencies in the design that become apparent after the *freeze* is implemented.

During the detailed design phase of the project the cost of the end item must continue to be authenticated. The importance of accurately knowing the cost of the

---

[10]J. Hartnett and R. Khol, "Component Design as an Engineering Discipline," *Machine Design*, November 1989.

product increases as the time to commit to production approaches. Although the expenditures may have been significant during the earlier phases of the project, tooling and production costs usually constitute the larger portion of a project's total cost. Fortunately, end-item definition is finalized during detailed design and thus cost estimating becomes increasingly accurate.

Qualification testing is the culmination of the design phase of the project. A satisfactory qualification test validates the design in the sense that it proves that the product meets the specification or contract requirements. Depending on the complexity of the product, qualification may be a simple test requiring only one test article or may consist of a series of tests with multiple test articles, with each test structured to verify a specific requirement. A rocket engine qualification test series structured in this manner might have a test series to validate the thrust of the engine at various propellant pressures and temperatures and another series to prove that the engine will perform for the required number of firing cycles. A separate qualification test series might be required for the propulsion system, which could include the rocket engine, propellant tanks, pressurization tanks, plumbing, and necessary instrumentation and control system.

With the completion of qualification testing the only remaining question regarding the design is, *can it be effectively produced using the materials, procedures and processes specified?* This is determined by the acceptance test of the first production article, the prototype. The qualification test validates the design and the prototype acceptance test validates that the design can be manufactured effectively by using the standard production department processes and procedures as well as those specified by the design.

1.2.9   Production Planning and Tooling Design

The decline in competitiveness of U.S. manufacturing beginning in the 1960s has been well documented. Various reasons have been suggested for this phenomenon, some of which are not connected to industry, itself. One occurrence that is related to this decline is academia's abandonment of courses having to do with manufacturing. With the emphasis on engineering science that began in the 1950s, universities across the nation added courses on the more theoretical subjects and, to keep from increasing the total semester hours required for graduation, dropped courses having to do with the more applied subjects. This resulted in elimination of most of the laboratory courses in machine shop, pattern making, welding, and foundry practice that introduced the student to manufacturing technology and tended to validate this discipline as an acceptable professional field for graduating engineering students. Lack of emphasis on manufacturing in education has undoubtedly contributed to the feeling that many engineering graduates seem to have that a career in this field is not challenging and holds little future. This is unfortunate for two reasons: first, as indicated, the nation desperately needs an infusion of new ideas and enthusiasm in manufacturing if it is to compete in the world market; second, 40 percent of the design engineering workforce is currently involved with production,[11] and this will probably increase with the growing use of CAD/CAM, which tends to force the integration of design and production.

Production planning is initiated by review of the design drawings to identify the machines and tooling required and to determine the machining operations to be

---

[11]R. Kohl, "New Research Shatters Stereotypes About Engineers," *Machine Design*, February 22, 1990.

used. Tooling is a term usually taken to mean all of the special equipment required to hold and position the part (jigs), guide the tool in performing the machining operation (fixtures), perform the actual metal removal and forming operations (tools), and check the part for dimensional accuracy (gages). Design engineers, often referred to as tool designers, perform many of these tasks by working with production planning and control and manufacturing personnel. The advantage of using the capabilities of CAD/CAM in integrating the detailed and tooling design functions to reduce the overall development time and to minimize cost is apparent. Once the information on the part design is in the computer memory, the logical next step is to use this information to program numerically controlled (NC) machines to accomplish the forming operations. With adequate data base support on materials properties and manufacturing processing information, the part design and/or the manufacturing process can be modified to optimize production. For example, a designer specifies an odd-sized hole in a part based on analysis. Because of the odd size, this hole requires two machining operations, drilling and boring, whereas a slightly smaller, nominally sized hole that would be perfectly satisfactory requires only one operation, drilling. If this manufacturing process information is available to the designer, the less expensive and time-consuming design approach can be selected.[10] This is known as producibility analysis, and several software programs and noncomputerized approaches are available to assist in performing this function.

Production planning is the task of selecting the machines to be used in the production of the part and in laying out the production line flow (preparing a flowchart). This is a function commonly performed by industrial engineers, who recognize the factors that influence plant layout and can develop a plan that integrates the part machining requirements and ancilliary support into a logical sequence with other ongoing production operations. This task was accomplished in the past by using templates on a large layout board but can now be performed by using CAD, a much more efficient and flexible approach.

Production control is the function of scheduling the work and controlling the inventory. It involves the routing of parts, components, and assemblies; establishing starting and finishing dates for each important component, assembly, and end item; and issuing the necessary instructions and other paperwork to ensure a smooth manufacturing operation. Production control is the heart of the production operation, providing all the supplies and parts to the various manufacturing workstations as well as the necessary instructions specifying what to do and how to do it. During the 1980s U.S. industry, under pressure to reduce costs and increase flexibility, began to adopt changes in the production control function. To speed up the preparation and issuance of manufacturing process documentation, some firms have turned to a computerized document generation and transmission system. These systems can be used to create, maintain, transmit, and store documentation governing essentially all of the various manufacturing operations required in a production line. Computer Integrated Manufacturing (CIM) is a term used to describe systems that integrate the entire manufacturing process. CIM not only provides for integration of CAD/CAM and all aspects of design engineering/manufacturing but includes functions such as materials management, cost accounting, shipping and receiving, and personnel training. An inventory management scheme identified as Just-In-Time (J-I-T) manufacturing has also been adopted by some companies. J-I-T has been referred to as this century's most important productivity enhancing management innovation.[12] By using this approach, a company attempts to minimize inventory

---

[12]R. Sobczak, "The ABC's of JIT," *Inbound Logistics*, November 1989.

and to receive supplies in small lots just as they are needed. With J-I-T, suppliers must be reliable. A glitch in a supplier's operation can shut down the production line in a short period of time, since there is no inventory buffer. With J-I-T suppliers operations are, in a sense, an extension of the production line. Suppliers can no longer be dealt with at arm's length as in the past when procurement was based on the lowest bid and satisfactory quality of the supplier's product was ascertained by receiving inspection. Suppliers must be qualified via an in-depth quality review before contract award, and some kind of continuous oversight must be maintained to ensure early detection of supply interruptions. The supplier's product quality must be based on a knowledge of their capabilities including facilities, equipment, and personnel, and confidence in the operation based on management goals and philosophy. For in-house fabricated parts that are required for subsequent production line assembly, personnel at the working level, and thus in a position to recognize potential problems, must feel an identity with, and responsibility for, the effort, since every person in the chain is critical in J-I-T.

Management information is an important aspect of the production control operation. Production control, in conjunction with production supervision and management, develops and maintains the schedule for the work and is thus in a good position to provide the appropriate schedule and other information necessary for management control. If the manufacturing process documentation is prepared by using computers with appropriate software the management information system (MIS) can use this capability to expedite the availability of information for management. Management can be provided with an interconnect to the computerized documentation system and can select information needed, as required. Management often will require special reports depicting specific items of interest. These can usually be prepared relatively easily by using the computerized system, but a good rule to follow is to minimize the number of reports generated and to try to use a report format that will satisfy multiple requirements. This is especially true when documentation is prepared without the use of computers.

Figure 1.5 depicts a typical manufacturing corporation organization chart, which shows the production control and planning functions and their relative positions in the organization.

1.2.10 Production

The design process is finalized with the completion of qualification testing and satisfactory acceptance testing of the prototype. Support of manufacturing by the design group will still continue however, since problems usually surface during production that need to be resolved by coordination between design and manufacturing. Product improvement efforts may also be authorized that involve redesign of some element(s) of the product. The acceptance test should be structured to disclose any discrepancies in the manufacturing process. The basic design has already been proven to be satisfactory by qualification testing; therefore, acceptance testing is used to provide a check on the manufacturing procedures and processes only. A quality control inspection plan must also be implemented during the production phase. This is not to be confused with overall product quality assurance, which involves segments of the operation in addition to production. The function involved here is the inspection of manufactured elements during the production process to ensure delivery of a functional product. The inspection may involve the measurement of certain dimensions, the use of a gage to check the dimensional

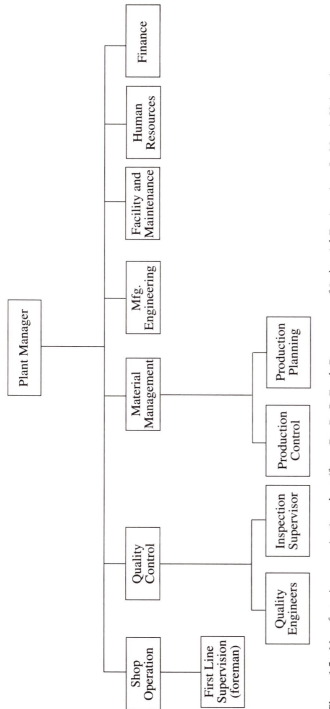

Figure 1.5 Manufacturing organization chart (from Dr. R. G. Ford, Department of Industrial Engineering, St. Mary's University, San Antonio, TX).

accuracy of some particular feature, the use of instruments to evaluate parameters, or a visual check of the workmanship. Quality control inspectors may be positioned at one location, as in the case of an assembly line or work center, or may be what are referred to as *roaming inspectors*, covering larger areas of the operation while watching for unsatisfactory parts as well as poor work practices. Inspection is normally concerned with part characteristics such as dimensions, composition, workmanship, finish, and function. The inspector tries to ensure that the fabricated part conforms to drawings and specifications, evidences quality workmanship, and functions properly. Proper function is normally ascertained via the acceptance test.

1.2.11 Design for Manufacture and Assembly

Design for manufacture and assembly is a concept that has grown out of the need for industry to compete on an international scale and to minimize the time period for developing new products and introducing them in the marketplace. Since as much as 70 percent to 80 percent of the final production cost can result from the design of the product, this approach is worthy of serious consideration. This concept emphasizes the need to consider the manufacturing and assembly processes during design and to incorporate those features that enhance the ability to fabricate and assemble the product. This can be accomplished by establishing interdisciplinary teams that design the product and develop the manufacturing process simultaneously, a so-called *simultaneous engineering* approach. Design for manufacture and assembly guidelines that have been developed over years of design and manufacturing experience are outlined below:[13]

1. ***Minimize the total number of parts.*** Reducing the number of parts that make up a product decreases the cost of manufacturing by eliminating fasteners, assembly instructions, and interfacing information. It normally contributes to reduced weight, reduced material requirements, and reduced complexity as well.

2. ***Use a modular design.*** Products made up of 4 to 8 modules with 4 to 12 parts per module can usually be automated most effectively. Maintain a generic product configuration through the assembly process insofar as possible and install the specialized modules during the latter stages. Elimination of the need for a separate housing or enclosure is also a good technique.

3. ***Minimize part variations.*** Standardized parts should be used whenever possible. This is especially true in regard to piece parts, such as nuts, bolts, screws, washers, fittings, and gaskets. Use of standardized parts eliminates the design, fabrication, and inspection of the specialized part as well as the need to carry an additional part number in the inventory system. Standard stock items are widely available and their reliability and quality are usually well established.

4. ***Use a multifunctional design.*** Design components to perform more than one function. For example, an electrical chassis can be designed to support circuit boards, act as an electrical ground, and function as a heat sink for

---

[13]J. Pearse and H. W. Stoll, "Design for Manufacturing," *Tool and Manufacturing Engineers Handbook*, SME, Dearborn, MI, 1989.

heat dissipation. An engine cylinder head can be designed to perform as a pressure vessel closure, to allow instrumentation and component mounting, to act as a heat exchanger, and to perform as a structural member. Features can be incorporated into a design to facilitate assembly, for example, self-alignment guides, detents, or nonsymmetric bolt and fastener patterns.

5. **Design parts for multiuse.** Many parts that perform very basic functions lend themselves to more than one use. A standardized parts program can be established, and new part designs that lend themselves to standardization can be included. The standard parts data base can then be consulted as new products are designed.

6. **Design to simplify fabrication.** The design of parts should make use of the lowest cost material that will satisfy the requirements and minimize waste and production processing time. Fabrication processes that avoid surface treatments (painting, plating, buffing, etc.) and secondary processes (grinding, reaming, polishing, etc.) should be used where possible. The design should be based on the use of simple fabrication processes.

7. **Use of fasteners.** Fasteners add significant cost to the manufacturing process. The cost of installing screws and other fasteners exceeds the cost of the fastener significantly. If fasteners must be used, the number and variation in size and type should be minimized. Fasteners and piece-parts that are small and hard to handle during fabrication should not be used. The use of *captured* washers and self-tapping screws will help in this regard.

8. **Minimize assembly directions.** To minimize assembly time, parts should be assembled from one direction. The best solution in this regard is to assemble parts in a top-down fashion along the Z axis, like making a sandwich.

9. **Maximize compliance.** To overcome problems associated with tolerance buildup, alignment during mating of parts, and part insertion during assembly, the design should include the use of generous tapers and chamfers, location points for manufacturing fixtures, guiding features, and generous radii.

10. **Minimize handling.** Positioning of parts during manufacture is costly. Thus, the part should be designed so that proper positioning is easy to attain. This can be achieved by using symmetrical designs or identifying the proper position by marking the part. The number of positions in which the part must be placed during fabrication should also be minimized.

11. **Eliminate or simplify adjustments.** Mechanical adjustments add to the cost of fabrication and cause assembly, test, and reliability problems. The need for these adjustments can often be negated by using stopping points, detents, notches, and spring-mounted components. If the designer understands why the adjustment has been recommended, a way of eliminating or reducing the need can often be found.

12. **Avoid flexible components.** Wiring and other flexible components are difficult to handle during assembly. The use of rigid or process-applied gaskets, plugs, or connectors to replace lead wires and circuit boards instead of electric wiring will help to minimize this problem.

1.3   OTHER IMPORTANT DESIGN CONSIDERATIONS

In addition to the design process structure (or morphology) that has been described in the preceding sections, there are several other considerations that are important when initiating a new design project. These are briefly discussed below and in later chapters of this textbook.

1.3.1   Product Use

A phrase often applied to sales activities is that *the customer is always right.* With the current emphasis on profitability, short investment payback periods, early product obsolescence, and intense competition, it seems that U.S. consumer products are developed under a philosophy that *the customer will buy the cheapest product available.* This is hopefully not as true for engineering design products where the customer is more sophisticated and will normally demand performance, reliability, serviceability, reasonable cost, safe operation, and good human engineering (ergonomics). Industries that are renewing their emphasis on quality as a result of competitive pressures, for example, the U.S. automobile industry, are returning to *the customer is always right* approach and are using a disciplined and structured method of determining customer expectations. These customer expectations are compared with the various product design characteristics to identify those that are not addressed and to develop an optimized product plan incorporating design modifications accommodating the dominant expectations.

Product liability in the United States is a design consideration that is growing in importance. Users of products are increasingly aware of their opportunities to sue manufacturers, and court sentiment often leans toward the consumer even when product misuse is involved. This situation has gotten so bad that some products are no longer manufactured in the United States, and the availability of some services is limited. Small airplane manufacturing in the United States has all but ceased, and women with high-risk pregnancies have a difficult time finding obstetricians willing to perform deliveries. The cost of this litigation in time and money is significant. One expert estimates the overall cost to be $300 billion a year on a national basis.[14] Officials in both large and small businesses spend an exorbitant amount of their time in defense against product and design liability claims instead of managing. The cost of this wealth-consuming activity to the nation is enormous in time and money as well as in reduced innovation, reduced economic and career opportunities, and inability to compete on an international basis.

1.3.2   Design Life

Many engineering design products are designed for a specific installation or assembly, and it is normally assumed that they will remain in this application for their useful life. For these instances the product specification will dictate the design life in terms of cycles or hours of operation, and it is a fairly straightforward task to design and test the product to meet these requirements. An aspect of design that has not received a lot of attention is the problem that arises when the product is removed from its original installation and is put to a *second use,* either before or after it has fulfilled its initial design life. The problem is that since this *second use* is not included in the specification it is not accounted for in the design and

---

[14]D. Ritter, "Litigation Pollution," *Engineering Times,* July 1990.

the result may be failure and personal injury leading to product liability litigation. The fact that the product was used in a way never intended by the original design may not have much influence on the court. This is an extremely difficult problem, and there is no real satisfactory solution. The courts seem to focus on whether the failure was *foreseeable* and not whether there was negligence or ignorance. About the best that the designer can do is to try to *foresee* both use and misuse and make provision in the design to provide for *credible* failures. Another technique that has seen some limited use is to permanently identify on the product its intended use and limited life, but there is no absolute assurance that the courts will consider this to be adequate warning. When tightly controlled operation of a product by a technically qualified user can be ensured, stringent life requirements can be enforced, for example, critical pressure vessels (as in spacecraft use) can be designed for a specified number of pressure cycles and be required to be recertified by retesting before additional use.

Some thought needs to be given to retirement of the product after completion of its design life. If the environment is to be treated as *surroundings* (in a thermodynamic context), the product must be capable of being refurbished and reused or the materials of construction must be recyclable or biodegradable. Some states have imposed a relatively high deposit on soft drink bottles to ensure that they are returned for subsequent reuse, and almost every community has a program for recycling aluminum soft drink cans. Plastic diaper manufacturers have developed a product that is supposedly biodegradable as a result of unfavorable publicity associated with diaper disposal in public landfills. The many automobile salvage yards around the country perform a service in recycling automobile parts. After the marketable components are removed from the vehicle chassis, the shell is sold for scrap, crushed, and subsequently used in new steel production. The automobile life cycle is a little like the old saying about pigs: "everything gets used but the squeal." Tire casings constitute a problem that has not been satisfactorily solved, however. This is the type of engineering challenge that manufacturers of products that do not biodegrade in a reasonable period of time and do not have a completely satisfactory recycling program need to face. Tire manufacturers should be concerned about the retirement of their product and should develop or encourage second uses or satisfactory recycling processes.

1.3.3   Human Factors Considerations

All designers should consider the operator of the device, component, or system under development and strive to produce *user-friendly* products. In this sense the term *operator* must include the person(s) that maintains and repairs the product as well as the person that uses it. A significant amount of anthropomorphic information is available that can be used to determine the size and location of manually actuated devices on the product as well as the optimum location of visual elements. Arm and leg actuation force data are available also. Certain general rules pertaining to things such as the meaning of the color of indicator and warning lights, direction of rotation of valves and electrical components, switch mounting, and the like, should also be followed. An individual's capacity for doing work should be considered in the design and, for repetitive tasks, a system layout should be selected that minimizes boredom but does not overload the individual's capacity to comprehend. Considerations pertaining to involuntary reaction as contrasted with intentional action must be accounted for in some designs as well.

Logic should be applied in all control systems layouts to maximize the efficiency of the operator. Switches, indicators, and the like, should be located in a logical sequence simulating the system being controlled. Indicators and devices requiring actuation should be located such that the operator can react appropriately, almost without thinking. A good systems control design philosophy may be to think of the operator as an extension of the system and to extend the systems design logic to include the operator's functions. The operator must also feel comfortable at his or her workstation and must not be subjected to undue fatigue. High maintenance items must be accessible (within reason). For example, it should not be necessary to remove half the components under the hood of an automobile to change the spark plugs.

Finally, the product should be aesthetically appealing. Many technical products are purchased primarily because of their customer appeal. The outward appearance of a product may be the feature that initially interests the potential buyer and opens the door to discussion of the technical features of the product.

1.4   ADVANCED TECHNOLOGY FOR THE DESIGN PROCESS

A fundamental change in the design process occurred in the United States beginning in the 1980s. This change was brought about as a result of several factors including market pressures in the United States and worldwide, the limited economic life of current consumer and high technology products, the need to shorten the development time for products, the need to improve product quality, and the need to improve communication between design engineering and manufacturing. The term that characterizes this change most effectively is improved communications. In many firms engineering design is thoroughly integrated with production planning, and the design is accomplished with design group personnel working with tool designers and manufacturing engineers. During this process the communication between design and manufacturing is greatly strengthened, eliminating a problem that plagued industry for years. But the changes affecting internal communications do not end here. The use of electronic data and documentation transmission grew dramatically during the 1980s, and it appears to be headed for even greater growth during the 1990s. Some forward looking firms have even constructed new facilities, at least in part, to enhance communications between employees. The 3-M Company recently established a new research and administration facility in Austin, Texas, which purposefully incorporated an architectural plan that strongly encouraged employees to get together informally to discuss ongoing projects. Using methods like this has allowed the 3-M Company to reduce their product development time from two years to a matter of weeks. Methods being adopted by U.S. industry to accelerate development time and to reduce product costs are briefly discussed in the following paragraphs.

1.4.1   Computer-Aided Design/Computer-Aided Manufacturing

CAD/CAM is one of the most significant new technologies affecting product design and manufacturing. Many companies are now performing design and production engineering concurrently rather than sequentially and are reducing product development time significantly. Traditionally, parts have been designed by a design engineer, possibly in conjunction with a detail draftsman. An engineering analyst would often then analyze the design for stress and deflection while another engineer was concerned with how to install the part in the next assembly. Drawings were then passed on to the manufacturing group where the tooling design, numerically controlled (NC) machine programming, and manufacturing plans were worked out. CAD/CAM has significantly changed all this. It is at least conceivable that all of these functions could now be accomplished by one individual working at a single computer terminal, and with a fully operational simultaneous engineering approach, several engineers could perform all of these functions at the same time by using networked computers.[10] The reduced costs of engineering when using CAD/CAM are obvious, but there are other advantages as well:[15]

1. The shortened product design cycle frees up engineering that can be assigned to other projects.
2. The quality of the product is improved.
3. The analytical and display capabilities offer the potential of reducing the number of test articles and mockups required.
4. Better communication between design and manufacturing results in improved designs and production methods.
5. Fewer engineering change orders are required.

To illustrate how CAD can be used to save time in the construction of engineering drawings refer to the sequence of steps shown in Fig. 1.6. The one-sixteenth

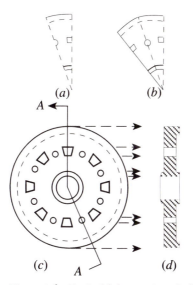

Figure 1.6 Typical labor-saving drafting procedures using CAD (from J. Ed Akin, *Computer Assisted Mechanical Design,* ©1990, p. 228. Reprinted by permission of Prentice Hall, Englewood Cliffs, NJ).

---

[15]J. Krouse, R. Mills, B. Beckert, and P. Dvorak, "CAD/CAM Planning: 1990," *Machine Design,* July 26, 1990.

segment of the component shown in (*a*) is drawn first. The one-eighth segment can then be created by adding a mirror copy to the inclined edge. A single command can then be used to make seven rotational copies of (*b*). By rotating about the center and incrementing by 45° after each copy is made, Fig.1.6*c* is obtained. Fig.1.6*d* is then added by using the capabilities of the system for providing horizontal lines at every line and arc that intersects section A-A in addition to the known thickness of the component segments.[16] By using computer software the component can be viewed at any angle desired, nongraphical data such as a bill of materials and notes can be added, information needed for manufacturing the part can be determined, and dimensioning can be added automatically. Possibly the greatest advantage in using CAD is the analytical capability available in various software packages that can be used to evaluate the design. Finite element analysis (FEA) has become the primary tool used in stress analysis and structural dynamics, and the ability to adapt it to use with CAD has contributed greatly to the proliferation of CAD systems in industry. FEA can be used in analyzing designs involving varying geometric shapes as well as nonhomogeneous materials. It also provides considerable flexibility in setting loading and support conditions. FEA is also used in heat conduction, electrical and magnetic fields, and fluid flow analysis.[15]

CAM is the term usually used to describe the transfer of information from the CAD drawing into a format usable by the numerically controlled (NC) machine(s) that forms the part. The numerical control program, which is a combination of standard machine tool code and machine specific instructions, links the part geometry and the machine tools used in manufacturing the part. NC programs can generally be identified by the manner in which they handle CAD data and the way in which they describe how a part is to be machined. For example, geometry can be defined for NC programming either during design, after it is sent to manufacturing, or through a translation program. Users must understand how engineers and programmers interrelate within the organization to select the method that best fits their operation. The ability to move directly from the CAD drawing to the actual product is a key element in reducing product development time. Software to convert design data into products is available but, for this to be effective, CAD programs must contain adequate information.[17] A high degree of interactivity is also required if a variety of software applications is to have widespread use. A fully interactive process will give designers instant feedback on essential elements such as strength, aesthetics, manufacturability, and cost, leading to development of optimum designs that can be quickly manufactured using automated NC programs.[18]

1.4.2 Solids or Surfaces

Although 2-D drafting continues to be the most widely used CAD/CAM application, the move by many manufacturing firms is to shift to solid (3-D) modeling. Solid modeling provides a complete geometric and mathematical description of part geometry, which is important if the model is to be used for design analysis, simulation, and generation of mass properties, or for developing NC data to machine the part. Thanks to advances in computers, software, and graphical displays, it is now possible to use 3-D representations in preference to the 2-D, $2\frac{1}{2}$-D, or wireframe models. The $2\frac{1}{2}$-D representations are 2-D renderings that include thickness data

---

[16]J. Akin, *Computer Assisted Mechanical Design*, Prentice Hall, Englewood Cliffs, NJ, 1990.
[17]N. Rouse, "NC Systems Close CAD/CAM Gap," *Machine Design*, December 11, 1986.
[18]"Tapping Advanced Technology," *Machine Design*, CAD/CAM 30, July 20, 1989.

for some regions of the 2-D part. Unlike surface representations, which use points, lines, and curves to define an object, solid modeling uses elements such as boxes, cones, cylinders, and manipulated 2-D shapes to generate models (see Fig. 1.7.) Solid models can be used for various purposes including creating realistic visual displays, analyzing the motion of components, including interference with other elements, and structural analysis. Possibly the greatest advantage that solid models offer over surface models is enhanced integration of design and manufacturing. Solid models simplify the NC programming task by eliminating the ambiguity that exists when surface models are used. With solid models there is only one surface, and the problem of defining the extent of the surface and surface interrelationships characteristic of surface models is eliminated. With surface representations the NC programmer must specify each section of the surface to be machined and must blend the surfaces together before a proper toolpath can be specified. With solid models the surface is more complete mathematically and does not need to be blended with other surfaces; thus, solid models are referred to as being more robust than surface models. Expert manufacturing programming systems are also being proposed that will automatically generate data and machine instructions to produce and validate parts. This approach is based on a method called volume decomposition, which involves breaking the part down into small delta and subdelta volumes that provide information about the features of the part, which are then used by the NC software to specify correct processes and toolpaths automatically.[19] A U.S. government standard, the Initial Graphics Exchange Standard (IGES), allows 2-D systems to transfer data between different vendors. This standard is being expanded to include 3-D data exchange.

There are several options for solid model generation. The most common techniques are known as boundary representation (BRep) and constructive solid geometry (CSG) representation. BRep can be thought of as a wireframe model with additional topological data that defines how the wireframe edges are blended into the connecting surfaces. CSG representation uses a Boolean data structure tree in which the leaves of the tree are elements of the object such as half spaces, blocks, or cylinders and the root represents the entire object. Figure 1.8 shows a Boolean CSG tree for an internal combustion engine piston.[15]

1.4.3   Computer-Integrated Manufacturing

It has been said that the importance of computer-integrated manufacturing (CIM) to the future of U.S. manufacturing cannot be overstated. CIM is a key ingredient in improving the productivity, efficiency, and profitability of the country's industrial base and in maintaining a competitive position in the world marketplace. CIM is based on the recognition that steps in the development of a manufactured product are interrelated and can be accomplished effectively and efficiently by using computers. This interrelationship comes about not only from the characteristics of the part being fabricated but also from the processes, specifications, instructions, and data that define and direct each step in the manufacturing process. By applying modern computer technology to control, organize, and distribute this information, all of the steps in the manufacturing process can be integrated into one coherent entity. This integration yields efficiencies not achieveable when using a more segmented manufacturing approach.[20]

---

[19]N. Rouse, "Designers Gain Insight into the Factory," *Machine Design*, June 8, 1989.

[20]T. J. Egan and A. Greene, "Computer-Integrated Manufacturing," *Tool and Manufacturing Engineering Handbook*, Society of Manufacturing Engineers, Dearborn, MI, 1989.

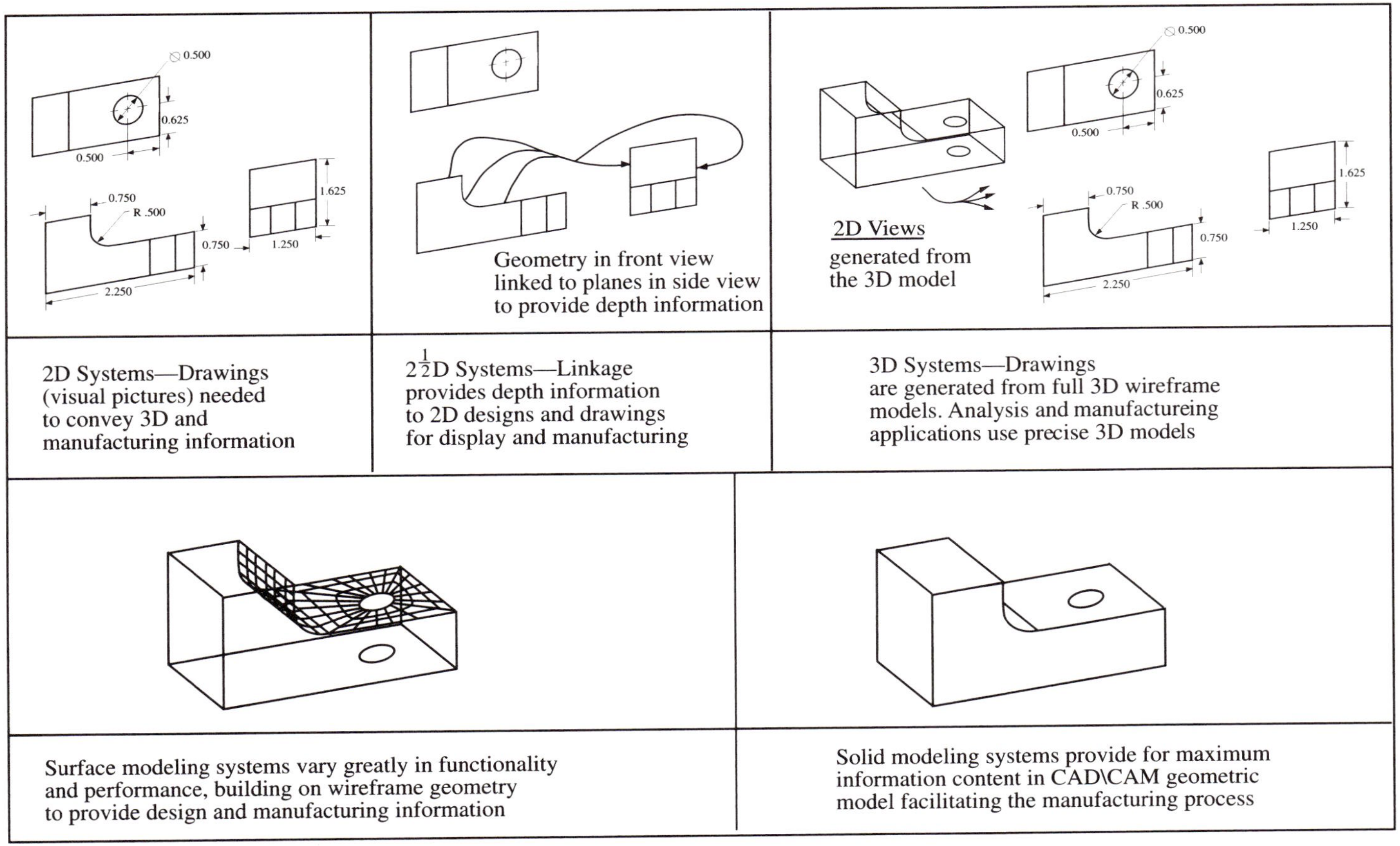
⌀ 0.500
0.625
0.500
0.750
R .500
0.750
2.250
1.625
1.250
Geometry in front view
linked to planes in side view
to provide depth information
2D Views
generated from
the 3D model
2D Systems—Drawings
(visual pictures) needed
to convey 3D and
manufacturing information
$2\frac{1}{2}$D Systems—Linkage
provides depth information
to 2D designs and drawings
for display and manufacturing
3D Systems—Drawings
are generated from full 3D wireframe
models. Analysis and manufactureing
applications use precise 3D models
Surface modeling systems vary greatly in functionality
and performance, building on wireframe geometry
to provide design and manufacturing information
Solid modeling systems provide for maximum
information content in CAD\CAM geometric
model facilitating the manufacturing process

Figure 1.7 Comparison of geometric modeling methods (from *Tool and Manufacturers Handbook,* Society of Manufacturing Engineers, pp. 3–5, 1989).

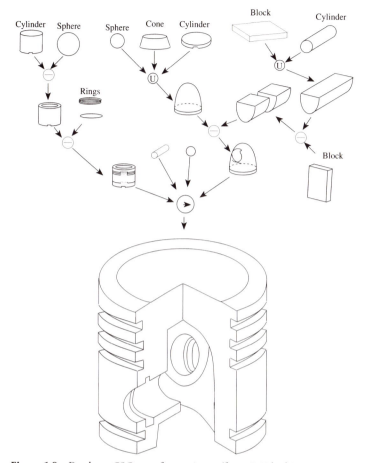

Figure 1.8   Boolean CSG tree for a piston (from J. Ed Akin, *Computer Assisted Mechanical Design,* ©1990, p. 109. Reprinted by permission of Prentice Hall, Englewood Cliffs, NJ).

Industries that have taken advantage of the potential benefits of CIM have generally been the larger companies under pressure to be competitive, productive, and responsive. Companies with the knowledge and resources to invest in new methodology that have targeted specific areas for improvement have been successful in implementing CIM. Although most CIM applications to date have only been partial, integrating only some functions within the organization or limiting the use of output data, experience has substantiated their cost effectiveness. A 1984 study by the Committee on the CAD/CAM Interface, formed by the Manufacturing Studies Board of the Commission on Engineering and Technical Systems of the National Research Council, examined the efforts of five companies in implementing CIM. The benefits realized by these companies during the integration process are indicated by Table 1.1. These data, which are representative of efforts over a 10- to 20-year time period, reflect the importance of CIM application to the future competitiveness of U.S. industry in the world marketplace.[19]

Table 1.1   The Benefits of CIM Implementation from *Tool and Manufacturing Engineering Handbook*, Society of Manufacturing Engineers, pp. 1–2, 1989.

Reduction in engineering design cost	15–30%
Reduction in overall lead time	30–60%
Increased production quality as measured by yield of acceptable production	2–5 times previous
Increased capability of engineers as measured by extent and depth analysis in same or less time than previously	2–35 times
Increased productivity of production operations) complete assemblies	40–70%
Increased productivity (operating time) of capital equipment	2–3 times
Reduction of work in process	30–60%
Reduction of personnel costs	5–20%

1.5   THE DESIGN PROCESS

The need for excellence in design has never been greater than it is at the close of the twentieth century. Worldwide competition in the production of consumer goods and high technology products and materials is intense, and this has spawned protectionist proposals and threats in various countries. The time period in history when industrialized countries could close their borders to imports and still sustain the quality of life that most of their citizens are accustomed to has long since passed. Like it or not, those of us living in the industrialized part of the world must compete in a worldwide economic market and must do it with increasing effectiveness. With this important fact in mind, consider the significance of Fig. 1.9. The design process, which has been described in detail in this chapter, must be supported by synthesis, analysis, testing, and proper material selection, just as the shaft in the figure is supported by the four-piece journal bearing shown. The modeling and optimization tasks, (Chapters 3 and 6 in this textbook) provide input into the synthesis, analysis, and testing elements of the design process. Chapter 4 provides information pertinent to the selection of materials. The design evolved during the design process has properties (teeth) that drive (determine) the availability, reliability, maintainability, dependability, manufacturability, and cost benefit ratio of the product. These characteristics all relate to the quality and value the product has to the ultimate user, the customer. Reliability is defined as the probability that a component will successfully perform for the specified life, given the required operating conditions and environment. High reliability will not ensure that the product will be operational (available) when needed, however. Thus, it is essential that the product be easily and inexpensively repaired when it fails. It is evident that the availability of a component, device, or system is a function of its reliability and maintainability. Dependability is another important characteristic that provides a measure of the component, device, or system condition, combining its reliability and maintainability. Manufacturability concerns the difficulty and cost of fabricating the product and the cost/benefit ratio provides a measure of the benefit of the product to the customer relative to its cost. Unless the product incorporates these characteristics effectively, it will not be competitive over any significant period of

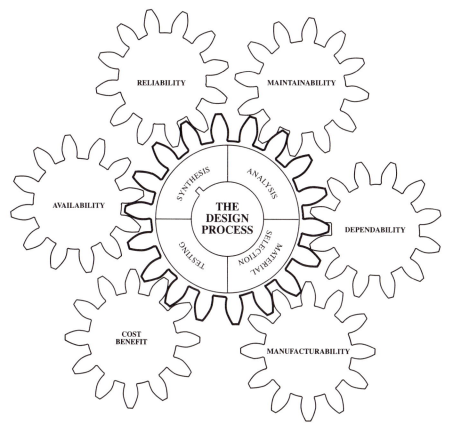

Figure 1.9 Important design parameters.

time in the world market. This textbook is dedicated to improving this effectiveness by helping engineering students to recognize the importance of these concepts and to understand how they are integrated into the design process.

BIBLIOGRAPHY

1. "Tool and Manufacturing Engineers Handbook," Society of Manufacturing Engineers, Dearborn, Michigan, 1989.
2. Akin, J., "Computer-Assisted Mechanical Design," Prentice Hall, Englewood Cliffs, NJ, 1990.
3. Dieter, G. E., "Engineering Design: A Materials and Processing Approach," McGraw-Hill Book Company, New York, 1983.
4. Cross, N., "Engineering Design Methods," John Wiley & Sons, New York, 1989.
5. Pugh, S., "Total Design," Addison-Wesley Publishing Company, Wokingham, England, 1991.
6. Corbett, J., Donner, M., Meleka, J., and Pym, C., "Design for Manufacture," Addison-Wealey Publishing Company, Wokingham, England, 1991.
7. Pahl, G. and Beitz, W., "Engineering Design, A Systematic Approach," The Design Council, London, England, 1988.

Chapter 2

Managing Design Projects

Everything that happens or that can happen in...
a business ought to be the direct result of what the
managers have specifically arranged. If the results are
good it will be because the judgement of the man-
agers has been good. The complete responsibility for
success or failure is with the management. **Henry Ford**

2.1 THE "BUCK" STOPS AT THE MANAGER'S DESK

During the early years of this century, when Henry Ford made the statement quoted above, it was undoubtedly *not* the generally agreed to philosophy of most of his peers or the managers that they employed. Many managers during this period insisted that the way to get people to perform work was by generating a feeling of fear. This was a time when production quotas were one of the principal methods of monitoring and providing incentive to workers, and when Frederick Taylor was developing the basic concepts associated with worker task time and motion studies. The productivity improvement studies during these early years usually concentrated on the worker and his environment and, if production was down, the workers got the blame, not the managers. These early productivity improvement efforts often led to unexpected results, such as the improved productivity recorded with increased work area lighting, which increased further when the lighting intensity was returned to the original level. It was subsequently concluded that the lighting level was only incidental, and the reason for the improved productivity was that the employees were encouraged by management interest and the feeling that they were considered to be an important part of the operation.

With the continuing loss of its manufacturing base to other countries, a condition that has developed during the latter half of this century, and with the inability to compete in many segments of the present global market, U.S. industry is turning

37

to the concepts of quality management propounded by W. Edwards Deming and others. Dr. Deming is best known for his work in Japan where he is credited with much of the success that that country's industry presently enjoys. Surprisingly, many of the Deming teachings are embodied in the statement made by Mr. Ford early in this century, that *management is responsible for the success or failure of industrial operations.*

Hopefully, it is not too late for industry in this country to turn itself around and get back to the basic principals of good management that will assure success. Make no mistake, however, even with the best managers applying the most effective techniques and good judgment, the task is formidable. There are many roadblocks to the revitalization of America's industry that do not exist to the same degree in other countries and are outside the control of industrial managers. It will take the dedicated commitment of a majority of America's people to make the political changes necessary to control or eliminate wealth-consuming policies and activities and to encourage wealth-producing endeavors that will ensure the reemergence of America as an industrial giant on a global scale. For this to occur there will have to be wide agreement that industry is the "engine" that makes America what it is (or was in the past), a point of view that seemingly does not enjoy wide acceptance by people in the 1990s.

The most important attribute of a manager is considered to be good judgment, especially in understanding and dealing with people. This is a capability that is more effectively gained by experience than through formal training and education; a capability that is demonstrated by the good, common sense application of accumulated experience rather than the use of any particular management technique or structured approach. Having said this, however, it must be recognized that the successful manager is the one who takes advantage of the many tools available to assist in getting the project started in the most effective way, in managing the ongoing effort, and in completing the project on schedule and within budget. This chapter describes some of these tools and methods.

2.2   MANAGEMENT TECHNIQUES

A few words about overall management style and technique are appropriate before describing the various tools available to managers. The approaches taken and the techniques practiced by managers have enormous potential for stimulating or depressing worker productivity. Managers must have the ability to understand people and must exercise acute awareness and perception in tailoring approaches to meet the various situations encountered. Since management accomplishes the goals of the organization through subordinates, its most important task is effective utilization of the workforce. Studies have shown that to supervise effectively, managers must exhibit a genuine interest in their employees and that this interest must be manifested by concern for the employee and his or her work.[1] In this study productive managers were characterized as those having a "unique balance of technical competence, people-oriented leadership skills, and sound administrative/business ability." When employees see that their work is recognized favorably and that they are properly rewarded, they will normally perform to the maximum of their ability.

[1]R & D Productivity, Study Report, Second Edition, Hughes Aircraft Company, Culver City California, 1978.

A second most important precept for managers is to eliminate fear on the part of their employees. If employees fear management and supervision, all sorts of unwanted and nonproductive action and inaction will result. Employees will be unwilling to express their ideas, especially if those ideas conflict with what they perceive to be management's position. In a design project, if management creates a working environment driven by fear, the company will be cheated out of the very thing the employee is paid for, ideas. In such a working climate, employees try to avoid contact with supervision and management, and if contact is unavoidable, they are reluctant to discuss their work and any problems that have been observed. This obviously inhibits communication and productivity and puts management in isolation from the work, a situation that has led to failure for more than one company. Establishing a working environment totally free of fear is not without cost, however. Because of the ready access to supervision and management by employees and the unintimidating environment achieved by elimination of fear, supervision and management will find a portion of their time committed to nonproductive and sometimes frustrating employee discussion. This is considered to be a small price to pay for having open communication with employees.

U.S. industry is undergoing a significant transformation in the 1990s as a result of growing global competition. The management technique known as Management by Objectives (MBO) that was initiated in the late 1970s is being replaced by the approach based on the teachings of W. Edwards Deming, Joseph Juran, Genichi Taguchi, and Philip Crosby, which emphasizes quality management methods. MBO was based on the assumption that employees needed a focus for their efforts and that rewards should be based on progress made toward achieving certain goals identified and agreed to by the employee and the supervisor. This management technique worked satisfactorily when employees had exclusive control over their actions and could obtain the necessary support from the organization to ensure that their assignments were not held up by other activities. However, over a period of time, problems with this management technique became apparent, which resulted in the move to methods that were considered to have contributed to the success of Japanese industry. The emphasis on individual objectives characteristic of MBO made it difficult to manage the overall effort when many individuals were involved whose activities were intertwined with other employees in their department and when other departments were involved.[2] Problems with MBO include lack of flexibility in adjusting to changed situations, the tendency for the employee to concentrate all of his or her energy toward achieving specified end results to the detriment of overall team performance, the similarity of this measurement technique to the production line productivity goals that reward quantity without regard for quality, the likelihood that goals are merely perfunctory, and the implication that the employee/supervisor exchange in setting goals substitutes for the continuing communication and oversight that the supervisor needs to exercise to ensure that the employee is productive and is made to feel a part of the company team in a participative style of management.

2.2.1 Total Quality Management

The management technique that embodies the product quality principles of the individuals mentioned above has been identified in many different ways, depending on the organization involved and the particular emphasis thought to be appro-

---

[2]J. N. Reddy, "Some Observations on Policy Management," ASI CEO Conference on Policy Management in American Industry, November 29, 1988.

priate. The term *Total Quality Management (TQM)* has been adopted by the U.S. Department of Defense (DOD) and will be used in this textbook to identify this management method which has been described as the "greatest management transition this country has ever witnessed."[3] Concepts in DOD's plan include:

1. Change in emphasis from defect correction to defect prevention.
2. Continuous improvement of the product rather than being satisfied with acceptable quality only.
3. Stabilization and improvement of the product rather than placing emphasis on intensive inspection.
4. Ensuring conformance with properly defined requirements and elimination of waivers.
5. Awarding contracts based on optimum life-cycle cost rather than the lowest procurement cost.
6. Emphasizing quality, cost, and schedule rather than just cost and schedule.

These goals can only be reached through a strong commitment on the part of management and what Dr. Deming calls *constancy of purpose.* The basis for this management technique is statistical process control, which provides a measure of the variations in products and processes and allows for their reduction and control, resulting in predictable and standardized production operations. Combined with management methods that allow the approach to be effective, this leads to improved product quality, reduced costs, new markets, and company longevity.[4] Some of the more important precepts associated with this management technique are listed here:[5]

1. Improved quality is a means to achieve higher profits and lower costs.
2. The process or system is the cause of 85 percent of the quality problems encountered by industry.
3. Only management can improve the system.
4. Employees will find their own defects and not pass them on.
5. Suppliers are critical members of the team and thus must be treated as an extension of the in-plant operation to ensure reliability and quality. Suppliers should not be selected on the basis of cost alone.
6. Every employee should have some understanding of statistical quality control.
7. Numerical quotas for the work force, numerical goals for management, and annual rating and merit systems must be eliminated.
8. A vigorous program of training and self-improvement for all employees must be instituted.

Among others, Ford Motor Company has implemented a quality management technique to improve product, process, and production planning as well as design.

[3] "DOD Eyes Total Quality Management," *Engineering Times*, September 1988.
[4] D. R. Katz, "Can Deming Work Another Miracle?," *Business Month*, October 1988.
[5] L. S. Aft, "Statistical Management and Problem Solving," ASME 90-Mgt-4, 1990.

This change was considered to be necessary to better satisfy consumer desires and demands. Thus, the system provides a method for translating these consumer *wants* into appropriate company requirements, which are then implemented in the design and manufacturing processes. The method is considered to be a planning tool to identify the critical items that need to be addressed by application of product improvement tools. Benefits include satisfied customers, shorter product development times, minimal product start-up problems, preservation of technical know-how and its relation to consumer demand, and greater customer emphasis within company engineering departments.[6] A key element in Ford's planning is the quality house, which is an orthogonal array or decision matrix that correlates specific customer desires with different actions that could be taken to satisfy these *wants*. These relationships are then weighted, compared with actions taken by the competition, and evaluated as to overall importance. Figure 2.1 shows the elements in a typical quality house. These orthogonal arrays are commonly used by industry in quality management to identify parameters that will improve the performance of components and subsystems and/or reduce their cost. The statistical approach used in these quality management processes to interpret data and make design and production decisions is *analysis of variance* (ANOVA). ANOVA is a statistically based decision tool that can detect differences in the performance of components and subsystems evaluated. This method is discussed in detail in Chapter 7.

EXERCISE 2.1

A What/How Orthogonal Array[6]

Organize students into groups and have each group prepare an orthogonal array as shown in Figure 2.2 on page 43 that lists customer desires for a cup of coffee (*What*) and identifies the corresponding substitute quality characteristics (*How*). Characterize the relationships as weak, medium, or strong. Think of as many customer *wants* as possible. Groups should be able to identify as many as 12 to 15. Decide on an appropriate weight for each customer desire and calculate the relationship weighting by multiplying the individual weights by the relationship rating (1, 3, or 9). The total of these scores for each What is then entered in the *gross weight* column. The *% total* column is completed by dividing individual gross weights by the total gross weight. Weights can be established by comparing each desire, in turn, against each of the others, counting the number of times each desire is selected, and multiplying the highest-ranked desire by the total number of desires (N), the second-ranked desire by $N-1$, and so on. Individual group results should be compared and discussed.

2.3   EFFECTIVE PROJECT MANAGEMENT

Projects are characterized by the fact that they have a clear-cut start and finish and a time frame in which the activity must be completed. Projects are established to accomplish a unique activity that did not exist before and will only be performed

---

[6]Ford Motor Company Seminar on Quality Function Deployment, Detroit, MI, 1988.

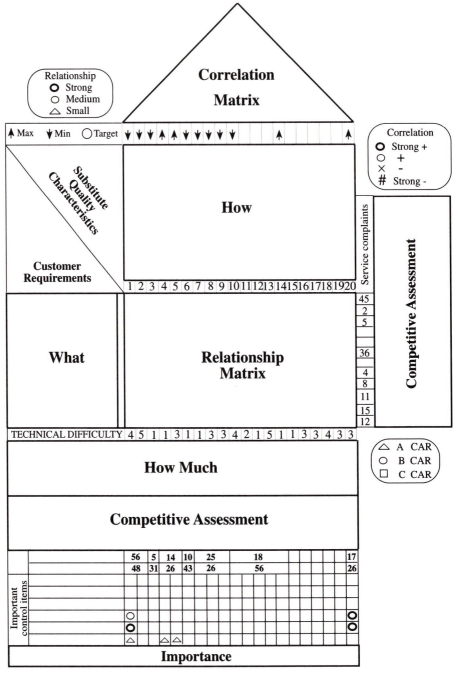

Figure 2.1 Quality house orthogonal array (from Ford Motor Co. seminar on quality function deployment).

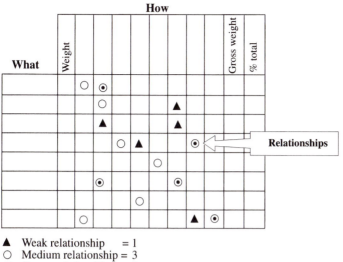

Figure 2.2 "What" versus "How" orthogonal array (from Ford Motor Co. seminar on quality function deployment).

one time. Personnel are normally assigned to projects from other ongoing activities and thus perform their responsibilities on an ad-hoc basis. Project resources are well defined and limited, requiring positive and effective management over the various phases and sequences of activities. The project manager must develop the project goals making sure that they are specific, well defined, and measurable. All project personnel must understand and agree with the goals and believe that they are realistic. The project manager must schedule the various project phases and activities with input and agreement from other team members. One of the project manager's most important tasks is to keep the project constantly moving toward successful completion and to ensure that project personnel are continuously mindful of the goals. In achieving this goal orientation philosophy the project manager must keep project team members focused on the relation between their objectives and the overall project goals. In this all of the tools available must be used including rewards, team member participation in decision making, and appropriate recognition for objectives achieved. The project manager must instill in team members an attitude of personal responsibility for proper job accomplishment and must take the lead in this by setting high technical and ethical standards personally. Team members will regard goals and objectives as worthy to the extent that they respect the person primarily responsible for setting them. If the project manager's conduct, either technically or ethically, does not engender respect and honor, the goals and objectives he or she has established will not garner the dedication and committment needed for a successful project.

The four primary processes involved in project management are as follows: [7]

1. **Planning:** Determining what needs to be done in what period of time and by what individual or team element.
2. **Directing:** Achieving objectives by implementing and carrying out approved plans through others.

[7]Project Management Short Course, TV Electric, Dallas, TX, 1990.

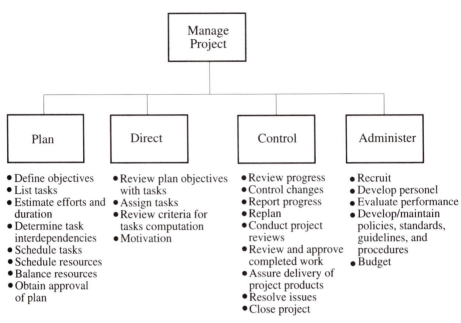

Figure 2.3 Project management processes (from Project Management Short Course, TU Electric).

3. **Controlling:** Evaluating what needs to be done, measuring progress, and taking corrective action to achieve objectives.

4. **Administrating:** Developing and implementing personnel policies and operational procedures necessary for the management of the project.

Figure 2.3 lists some of the various tasks within these processes.

EXERCISE 2.2

Project Planning (courtesy of TU Electric)

Note: Students will have to be organized into groups to complete this exercise.

The Situation

Your organization has just assigned you to a newly formed task team that is to take over a secret project presently being handled by Research and Development. Your team has been assigned responsibility and authority for designing a plan for managing the project and, after top management has reviewed and approved it, for implementing the project. Your team is made up of individuals with experience from a number of divisions because it is thought that a greater range of knowledge and skills is needed to develop the most effective plans. None of the team members have been told anything about the project other than it is expected to grow to sizeable proportions and require additional personnel.

The Task

Despite the lack of information regarding the project, your team must now design a preliminary plan for managing it. A list of 20 management activities (A through T) are listed in Table 2.1 in random order. Your task is to rank these activities according to the sequence you think should be followed in managing the project. This ranking will be reviewed by top management before you are given the go-ahead to begin work on the project.

Step 1. Go over the list of activities and without discussing it with anyone, rank the activities in the sequence you think should be followed in managing the project. Identify the first activity that should be accomplished as "1" and continue through the last activity, "20." Record your ranking under the column heading of "Individual Ranking."

Step 2. Now meet with your group and agree on a sequence of activities that should be followed. Record this ranking under the column heading of "Group Ranking."

Step 3. Enter the "Planning Experts' Ranking" shown in Appendix D.

Step 4. Enter the difference in columns 1 and 3 in this column.

Step 5. Enter the difference in columns 2 and 3 in this column.

Step 6. Total columns 4 and 5 and compare the individual ranking difference with the group ranking difference and discuss. The lower the score the better.

Table 2.1 Management Activities

	Step 1 Individual Ranking	Step 2 Group Ranking	Step 3 Planning Experts Ranking	Step 4 Difference Step 1 and 3	Step 5 Difference Step 2 and 3
A. Find quality people to fill positions.					
B. Measure progress to-ward and / or deviation from the project's goals.					
C. Identify and analyze the various tasks neces-sary to implement the project.					
D. Develop strategies (pri-orities, sequence, tim-ing of major steps).					
E. Develop possible alter-native courses of ac-tion.					

Table 2.1 Management Activities (*Continued*)

F. Establish appropriate policies for recognizing individual performance.				
G. Assign responsibilities / accountability / authority.				
H. Establish project objectives.				
I. Train and develop personnel for new responsibilities / authority.				
J. Gather and analyze the facts of the current project situation.				
K. Establish qualifications for positions.				
L. Take corrective action on project (recycle project plans).				
M. Coordinate on-going activities.				
N. Determine the allocation of resources (budget, facilities, etc.).				
0. Measure individual performance against objectives and standards.				
P. Identify the negative consequences of each course of action.				
Q. Develop individual performance objectives which are mutually agreeable to the individual and supervisor.				
R. Define scope of relationships, responsibilities and authority of new positions.				
S. Deside on a basic course of action.				
T. Determine measureable check points for the project and variations expected.				

(Adapted from Project Management Short Course, TU Electric.)

Activity	Jan	Feb	Mar	Apr	May	June	July	Aug	Sept
Contract negotiation	■								
Planning and WBS development	▬▬								
Component specification development		▬▬							
Detail design			▬	▭▭▭▭					
Development testing			▭▭▭						
Design review process						▭			
Qualification testing						▭			
Production planning						▭▭			
Production							▭▭▭▭▭		
First unit delivery									▭

Figure 2.4 Typical bar chart schedule.

2.4 PLANNING AND SCHEDULING THE EFFORT

2.4.1 The Gantt Chart

The simple bar chart (or Gantt chart) shown in Figure 2.4 is probably the most widely used project scheduling technique. In addition to its simplicity the advantages of this method include the following:

1. Direct correlation with time including the ability to reflect breaks in the effort for intermittant tasks.
2. Relatively straightforward task relationship for projects involving a limited number of tasks.
3. Straightforward integration of subtasks having separate scheduling charts.
4. Time schedule is flexible and can be expanded to show daily activities or compressed for longer term tasks.
5. Progress against the plan is easily reflected by filling in the bars.

The bar chart is used to reflect activities and displays them horizontally over some appropriate time scale. A similar chart used for displaying events is called a milestone chart (see Figure 2.5). Figures 2.4 and 2.5 actually show the same effort, only the emphasis is different. The bar chart shows the duration of the task by the length of the bar whereas the milestone chart normally only identifies the completion of the task. The method used depends on the purpose of the schedule and who is using it. For the person directly responsible for managing the effort, a bar chart might be more appropriate, whereas for personnel or organizations more remote from the task, the milestone chart may be more useful. In practice, the bar chart and milestone chart are normally combined to show task duration as well as significant milestones.[8]

[8]Q. W. Fleming, J. W. Bronn, and G. C. Humphreys, *Project and Production Scheduling*, Probus Publishing Co., Chicago, Illinois, 1987.

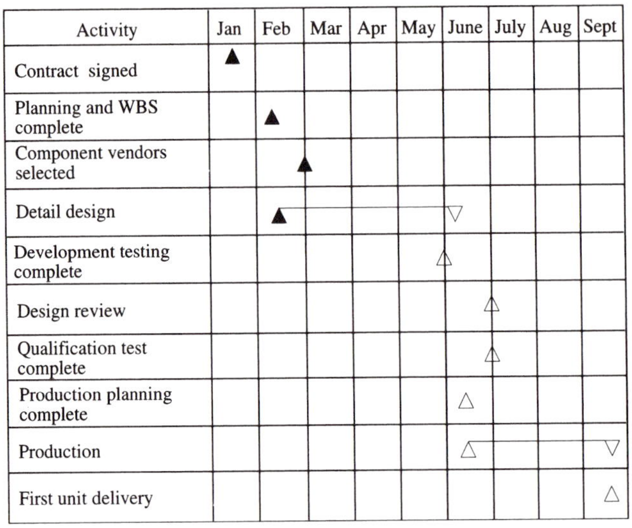

Activity	Jan	Feb	Mar	Apr	May	June	July	Aug	Sept
Contract signed	▲								
Planning and WBS complete		▲							
Component vendors selected			▲						
Detail design		▲					▽		
Development testing complete						△			
Design review							△		
Qualification test complete							△		
Production planning complete						△			
Production						△			▽
First unit delivery									△

Figure 2.5   Typical milestone schedule.

2.4.2   PERT Networks

Unfortunately bar charts have one significant shortcoming. For complex projects with interrelated tasks, bar charts do not reflect the dependence of one task on the completion of another. It was, at least partially, because of this inadequacy that the development of scheduling methods generally known as logic networks came about. One of the early logic network scheduling methods was identified as PERT (Program Evaluation and Review Technique). PERT was developed in 1957 to provide a method for scheduling the many activities associated with the U.S. Navy's Polaris missile program. Although PERT is given considerable credit for the success of this program, many of the early efforts to use this technique proved to be cumbersome and difficult. Any scheduling method used in a program should be a tool to assist in getting the job completed. If the effort required to make the tool useful approaches the benefit that the tool provides to the program, then utilization of the tool in that application is questionable. This may have been the problem with many of these early attempts to use PERT. PERT, as it was originally defined, is no longer used by industry; however, the term PERT has become synonymous with a wide variety of network scheduling techniques. Thus, when a company is said to be using PERT, it often means that they are using network scheduling techniques that employ critical path methodology. Usually, one of two acceptable network types is used: the arrow diagram method, or the precedence diagram method.[8]

2.4.3   The Critical Path Method

The critical path method (CPM), was developed in 1957 by E. I. duPont de Nemours & Co. working with computer specialists from Remington Rand. CPM was accepted by industry because of its activity-based orientation and the fact that it was easy to understand and adapt to graphic displays. The primary difference between PERT and CPM is the manner in which time estimates are made for activities. CPM relies on a single *most likely* time estimate for each activity whereas PERT uses statistical uncertainty compensated expected time estimates. In the late 1950s, CPM began to be referred to as the arrow diagram method (ADM). Arrow diagrams utilize an arrow (line with arrowheads to designate direction of flow) to represent activities in a project and nodes (circles or other geometric figures) to represent events that occur before and after each activity. An activity is a task that occurs over time and

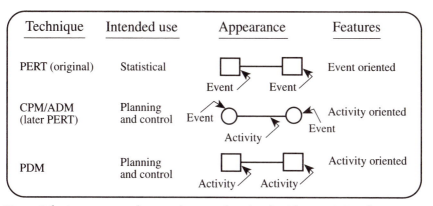

Figure 2.6   A summary of network types (reprinted with permission from Q. W. Fleming. et al., *Project and Production Scheduling,* Probus Publishing Co., Chicago IL, 1987, p. 79).

consumes resources such as labor, materials and equipment. An event occurs at a point in time and thus a milestone is an event. Both the bar (Gantt) chart and logic diagram show all the activities of a project, and the combination (bar/milestone) chart shows events as well. The principal differences are that the bar chart shows when the tasks will occur whereas the logic diagram does not; the logic diagram shows how the activities are related to each other and the bar chart does not.[8]

Another variation of the CPM network is the so-called precedence diagram method (PDM). In contrast to the CPM/ADM/later PERT, the PDM identifies the activity in the node, which is typically represented by a rectangular box. Lines tie the nodes together and thus depict the relationship between the activities. Figure 2.6 shows the basic difference between PERT (original), ADM, and PDM networks. PDM networks have the advantage of showing different types of relationships between activities. By connecting the lines between activities at different points on the node boxes, start-to-start, finish-to-finish, and other relationships can be displayed (see Figure 2.7).

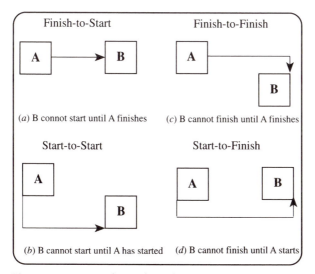

Figure 2.7   PDM relationships (reprinted with permission from Q. W. Fleming. et al., *Project and Production Scheduling,* Probus Publishing Co., Chicago, IL, 1987, pp. 113–115).

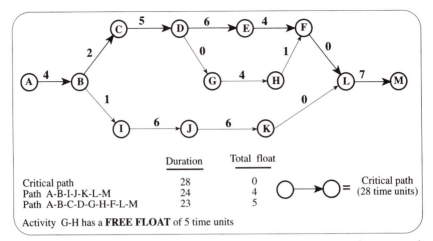

Figure 2.8 ADM schedule (reprinted with permission from Q. W. Fleming. et al., *Project and Production Scheduling,* Probus Publishing Co., Chicago, IL, 1987, p. 139).

Once the scheduling method has been selected, several steps must be taken to prepare the project schedule. All of the project (or subset thereof) activities and events must be identified and their relationships must be determined. Time spans for all the activities must be estimated and the critical path (the longest path in terms of time) must be located. This can best be accomplished by adding all the time durations in all of the possible paths in the network and identifying the one with the longest total time. Another method that is used to analyze the network for the critical path is to determine the earliest time and latest time that an activity can be accomplished without adversely impacting the total project schedule. Once the critical path has been defined, effort can be concentrated on completing the project in less time. Applying additional resources to critical path activities will allow early project completion, but increased effort on noncritical path tasks will not reduce the overall project schedule time.

Figure 2.8 depicts a typical ADM schedule with a critical path of 28 time units and float (slack time) for the two other paths of 4 and 5 time units. Activities G and H are shown to have a free float of 5 time units. Free float is the time that an activity can be delayed from its earliest start time until it begins to interfere with the earliest start time of the succeeding activity.

EXERCISE 2.3

PERT Network *(from Fundamentals of PERT, Penton/IPC Education Division, 1979)*

Figure 2.9 shows a PERT network schedule. A typical way in which these networks are analyzed is shown in Table 2.2. Starting at the final event, 9, the latest time T_L that event 9 can occur is determined to be 65 time units. The longest path from event 1 to event 9 is noted to be along path 1-2-4-5-9. The earliest time T_E that event 9 can be completed through predecessor event 8 is determined to be 43 time units along path 1-2-6-8-9. The float or slack time ($T_L - T_E$) along this path is determined to be 22 time units. The time required for activity 8-9, t_e, is noted to be 8 time units. For the path including event 7 and 9, T_E is noted to be 64 time units along path 1-2-4-5-7-9. Note that the longest schedule time path to event 7 is used to calculate T_E for the

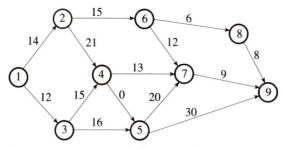

Figure 2.9   PERT network (from *Fundamentals of PERT,* Penton Education Division, Cleveland, OH, 1982).

path through activity 7-9. This is because activity 7-9 must wait for the last of the predecessor events to be completed. T_L for successor event 9 remains the same, 65 time units, and the slack time is 1 time unit. For successor event 8 the latest time that this event can occur without impacting the project completion date is the T_L for event 9 minus the time for activity 8-9, $65 - 8 = 57$ time units. The longest time to event 8 through predecessor event 6 is 35 time units and the slack for this path is 22 time units. When determining T_L for intermediate events, subtract the time associated with the longest path between the intermediate event and the final event from the T_L for the total network. For example, the T_L for event 6 is determined as follows: $65 - 9 - 12 = 44$. T_E for activity 2-6 is determined by identifying the longest path from the starting event 1 to event 6 through event 2. Thus, T_E for activity 2-6 is $14 + 15 = 29$ time units. Note that there is only one path from the start to event 6 through event 2 in this example.

Complete the information for the remaining paths in Table 2.2 and determine the critical path time. The calculations made in this exercise provide an easy way to analyze slack time. The slack time column indicates how important each activity is to timely project completion. If extra resources can be added to activities with zero slack, the project critical path can be shortened. The only activity that cannot be shortened by increasing resources is activity 4-5.

Table 2.2   PERT Activity Time Estimate

Activities		Estimates			
Successor Event	**Predecessor Event**	t_e	T_E	T_L	**Slack** $T_L - T_E$
9	8	8	43	65	22
9	7	9	64	65	1
9	5	30	65	65	0
8	6	6	35	57	22
7	6				
7	5				
7	4				
6	2	15	29	44	15
5	4				
5	3				
4	3				
4	2				
3	1				
2	1				

Adapted from *Fundamentals of PERT,* Penton Education Division, Cleveland, OH, 1982.

2.4.4 Summary

With the graphics capabilities of present-day computers, network planning, scheduling, and analysis are considerably more feasible for projects today than they were when they were first introduced in 1957. With the project management software available today, projects cannot only be planned and tracked, but dynamic adjustments can be made based on actual performance and status reports can be generated. Networks have proved their worth for the planning and management of one-time-only jobs like design projects, but they are of limited value for repetitive work such as production scheduling, where variations of the bar chart and other methods are more appropriate. For management presentations, network schedules should be converted to a more understandable form, such as the Gantt chart or milestone display. Most project management software packages include the capability for developing PERT, CPM, and bar chart schedules as well as for displaying data in various graphic formats to track task interdependence and timelines, resources, milestones, and cash flow.[9]

2.5 CLARIFYING THE PROJECT OBJECTIVES

When a manager is first assigned to a new project, one of the first tasks is to analyze and define the objectives so that the project team has a clear idea about what is to be accomplished and an understanding of the subobjectives embodied in the overall objective. The effort committed to this task pays off in several ways. It provides a focus for the project team as the activity level increases and new personnel not familiar with the project background are brought into the effort. It also provides a means for reaching a more comprehensive agreement and understanding with the client and/or upper management about the goals and objectives of the project and the purpose(s) of the end product. Finally, it provides a baseline from which to work when the project goals need to be changed over the duration of the project.

2.5.1 The Objectives Tree Method

The objectives tree method is a useful tool for defining the objectives of a project. It shows in diagrammatic format all of the identifiable objectives and their interrelationships organized in a hierarchical pattern of objectives and subobjectives. The procedural steps followed in developing objectives trees are as follows:[10]

1. Identify and list the design objectives from the design problem statement, from discussions with the client and/or upper management, and from discussions within the design team.

2. Organize the list into higher- and lower-level objectives and group them according to importance (into hierarchical levels).

3. Draw a diagrammatic tree of the objectives showing the hierarchical relationships and interconnections.

An example of an objectives tree for a regional transportation system is shown in Figure 2.10. The design team started with the client's vague definition for a

---

[9]K. Landis, "Critical Paths," *MacUser*, October 1989.
[10]N. Cross, "Engineering Design Methods," John Wiley & Sons, New York, 1989.

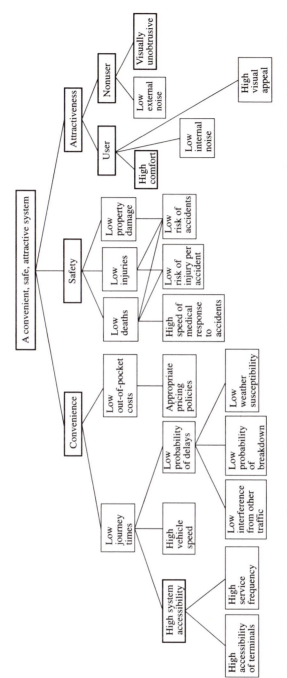

Figure 2.10 An objective tree for a regional transportation system (from N. Cross, *Engineering Design Methods*, John Wiley & Sons, New York, 1989, p. 52). Reprinted by permission of John Wiley & Sons, Ltd.

convenient, safe, attractive system and expanded the objectives. For example, *Convenience* was expanded into subobjectives of *Low journey times* and *Low out-of-pocket costs* for users. *Low out-of-pocket costs* can be satisfied by the subobjective of *Appropriate pricing policies.* The objective *Low journey times* is amplified by several subobjectives, as shown. The *Safety* objective was amplified to include *Low deaths, Low injuries,* and *Low property damage.* These subobjectives contribute more than one higher-level objective. The subobjective *Low risk of accidents* contributes to all three higher-level objectives. Although the effort expended in developing an objectives tree for a project uses resources that could be applied to getting on with the design task, the commitment is well justified. The example of the confusion that arose during a design program in regard to objectives and requirements described in Chapter 1 of this book could well have been eliminated by an in-depth discussion of the objectives and their meaning by the design team early in the project. Students often have difficulty identifying the essential objectives in laboratory experiments and design projects as well, which points to the fact that confusion between individuals on essential project elements is not uncommon and an appropriate amount of effort expended to eliminate this confusion is worthwhile.

2.5.2   Coordination with Client and Management

It is no less important that the client and/or upper management agree with the design team's interpretation of the objectives than it is for individual members of the design team to understand and agree on their meaning. Once the objectives have been established, work will begin on the design and resources will begin to be consumed at an accelerated rate. After this point in the design process any effort expended on misunderstood or invalid objectives is wasted. Therefore, it is essential that a mutual and agreed-to understanding of the objectives exist between the client, upper management, and the design team. If the project is being accomplished for a client, agreement should first be obtained in-house with upper management, followed by coordination with the client. An effective way in which to obtain the concurrence of management and, subsequently, the client, is to schedule meetings at which the objectives are identified and described and initial plans are presented as to how these objectives will be accomplished. This presentation should be made by the project manager and key design team members so that management and the client can get to know these key people and develop some confidence in their abilities.

Presentation of the project objectives and initial plans to the customer and/or management is a very significant milestone in the evolution of the project even though it comes at a very early time in the overall effort. It is absolutely essential that the project manager and the team members have a thorough and in-depth understanding of the objectives, and agreement on their meaning, before this presentation is scheduled. From this meeting upper management and the client must develop a feeling of confidence in the knowledge of the team members and in the fact that they can work as a team. Possibly even more important is for the presenters to have the knowledge and capability to be able to effectively counter suggestions that come up during the presentation that are not well thought out, as well as questions that lead into unprofitable discussions. This may be the first time that management and client have been forced to clarify their ideas about the project. The presenters must thus be able to keep the meeting properly focused so that valid observations can be thoroughly discussed and the others satisfactorily responded to, respectfully, but with minimum discussion.

2.6 DECISION MAKING

Design is a process involving constant decision making. Decisions are made on organization, objectives, requirements, materials, testing, systems, subsystems, components, suppliers, personnel assignment, work breakdown, manufacturing, and cost, to name a few. There are also many ways in which decisions can be made in the design organization. Some of these methods are very disciplined and use well-known and well-defined tools that are available to the decision maker. Decision matrices and decision trees fall into this category. When these methods are used, they are generally applied to the higher level and more significant project decisions, sometimes as a check on the normal decision making process. The principal decision-making approach used in design projects works as a function of the organizational structure itself, using formal and informal coordination and approval for decisions. In this type of decision making an individual is assigned to an area of design responsibility for which he/she has decision authority within guidelines and company policy, subject to supervisory approval. An example of this type of decision making follows:

> A designer on a defense-related project is assigned to design a test enclosure for testing liquid and pneumatic components used on a missile propellant-loading system. The design must be able to withstand a component failure within the enclosure, and appropriate pneumatic and liquid connections, component supports, instrumentation, and controls must be provided for. The enclosure must include the capability for visual access so that the operator can observe component operation within. The test enclosure must be portable and capable of operation by one technician. The designer reports to a first-level supervisor and analytical support is available from another group at the same organizational level. A small technical advisory group is also available made up of experienced designers.
>
> In this assignment the designer must first identify all the components in the missile system that will require testing in the enclosure. He must also determine the safety, personnel, and other procedures that apply to the design as well as the design philosophy. To do this the designer talks to the knowledgeable people within the division in which his branch is located. To assess the safety, quality control, and systems cleanliness requirements the designer must talk to organizations outside the division but within the missile department. To obtain information on the enclosure window the designer decides to talk to the aircraft people who are located in another department of the company. The designer then lays out an initial concept for the enclosure and asks the analytical group to do some calculations supporting the design. Detail design is subsequently initiated and material selection begins. During this period of time the designer meets with his immediate supervisor several times to review the task progress and to ask questions as they arise.
>
> In this assignment the designer is responsible for making the following decisions:
>
> 1. Determining who needs to be contacted for information.
> 2. Clarifying the objectives.
> 3. Establishing the design requirements based on the objectives.
> 4. Selecting the components and detailed parts.

5. Selecting the materials.
6. Determining whether the analytical support results are valid.
7. Deciding to perform the structural analysis himself.
8. Detail design decisions.
9. Determining what testing is required.
10. Assembly design decisions.
11. Procedural design decisions.
12. Determining when he needs to talk to his supervisor.
13. Determining when the design is completed and ready for review.

In this example, decision making is based on experience, judgment, testing, and analysis. If the designer does a good job, the design will be based on the experience of the organization. Judgment will be that of the designer and his supervision. The quality of analysis is a function of the performance of the analytical group and the judgment of the designer and supervisor.

The importance of being able to work with other people in accomplishing this task is very apparent. A designer must be able to obtain the enthusiastic support of other employees in providing information and advice. The designer must engender a desire to help on the part of the contacts he/she makes.

2.7   DECISION MATRICES

A decision matrix can vary from a simple chart consisting of rows and columns that allows the evaluation of alternatives relative to various design factors to a complex array such as the quality house (Figure 2.1) that allows consideration of the requirements; analysis of the requirements relative to design, manufacturing, and marketing characteristics; evaluation of solutions relative to competitor's and target values; technical difficulty associated with the solution; and important control items impacting or resulting from the various solutions. Design factors do not usually have the same degree of importance; therefore, some sort of weighting scale is normally assigned to account for this variation. The resulting grade for each alternative is, of course, a function of the weighting scheme adopted. For a situation such as that depicted by Table 2.3, in which various materials are evaluated for use in fabricating the test enclosure referred to in the previous example, the simple weighting scheme of Table 2.4 is applied to the various parameters of concern. In this case, stainless steel received the highest score and would therefore be a logical choice for this application. Aluminum 7075 would also be an acceptable material and, if the strength to density ratio is important enough, should be given serious consideration. Low-alloy steel also received a high grade in this evaluation, but the very poor rating in corrosion resistance would probably eliminate it from consideration. The important point to remember in this type of analysis is to use the evaluation to help make the proper decision. The final decision should not be made solely on the basis of a high decision matrix score but should always reflect the informed judgment of the designer, all factors considered.

The decision matrix can be an important tool in the overall design process, but it should be kept in mind that nothing takes the place of common sense and good judgment. For devices like the decision matrix to be viable, estimates must be made about the relative importance of the different evaluation factors. It is important to realize that these estimates have a certain built-in uncertainty that may result

Table 2.3 Decision Matrix for Test Enclosure Material Selection

Material	Strength Density Ratio	Weldability	Machinability	Corrosion Resistance	Availability	Cost	Score
Low alloy steel	122 / 3	4	2.5	0	4	4	17.5
Stainless steel	113 / 3	3	2.5	4	3	2.5	18.0
Aluminum 2011	60 / 2	2	3	3.5	3	2	15.5
Aluminum 7075	194 / 4	2.5	3	3.5	2.5	2	17.5
Titanium 6A14V	205 / 4	2	1	4	0	0	11.0

Table 2.4   Weighting Scheme for
Evaluating Design Alternatives

Weighting Category	Value
Far below average	0
Below average	1
Average	2
Above average	3
Far above average	4

in erroneous conclusions as to the best or most effective solution or alternative. It is always prudent to perform the evaluation of alternatives by using at least two independent methods. If two separate evaluation methods result in the same conclusion, and if the result seems to be logical using common sense and good judgment, proceeding with that approach is probably warranted. However, a good designer will maintain a questioning attitude, always seeking further confirmation that the decision was correct as the design process continues.

2.7.1   Decision Trees

The decision tree is another method that can be used to evaluate different alternatives. Decision trees are often used in evaluating business investment decisions by providing a structured analysis that takes into account the outcome of possible future decisions, and the effect of uncertainty, and allows the benefit of varying levels of present and future profit to be weighed against the concomitant commitments. Decision trees can also be used in the design process to evaluate different design alternatives. After a number of alternative design approaches have been identified, the designer selects the one that appears to offer the optimum solution. This approach is then investigated at a more detailed level, revealing other options that again are evaluated and the best one selected. This process continues to some logical end point that may be a certain critical period of time, (e.g., in the case of financial decisions) or to the lowest level of detail in a design decision. This top-down approach is more commonly used in design, but the process can also be started at the lowest level, building up to the overall concept. Athough the decision tree is based on selection of the best option at each branch or level of the tree, this does not ensure that the best possible overall concept will be obtained during the first iteration. Decisions made at any particular level may turn out to be less than optimal in the light of information disclosed at subsequent levels. The impact of later decisions on earlier decisions is important in the effective use of decision trees, and thus considerable iteration is necessary. It is important for the decision tree to be regarded as a dynamic device, one in which new (and better) information is integrated as it becomes available.[10]

Example 2.1

PLANT INVESTMENT DECISION USING A DECISION TREE.[11]

A chemical company is trying to decide whether to build a large or small plant to manufacture a new product with an expected life of 10 years. The decision must

[11] J. F. Magee, "Decision Trees for Decision Making," *Harvard Business Review,* July–August 1964.

be based on the size of the market for the product. Based on projections by the company, demand could be any of the following:

1. High during the first 2 years and if the product is found unsatisfactory by users, low thereafter.
2. High over the 10 year period.
3. Low over the 10 year period.

If the company builds a large plant and significant demand does not develop, they will be saddled with a large nonproductive facility. If a small plant is constructed, management has the option of expanding the plant after two years to accommodate high sustained demand. If the demand is low the company can maintain operations in the small plant profitably at the lower volume. The project engineer is in favor of proceeding with construction of the large plant. The chairman (principal stockholder) is concerned about the possibility of having unneeded plant capacity and favors construction of a smaller plant, recognizing that later expansion to accommodate increased demand will be more costly and result in a less efficient plant operation. The chairman also recognizes that the company must promptly fill the demand or competitors will develop equivalent products.

On the basis of data available, company management believes that the following prognosis is appropriate:

- Demand is expected to develop according to the following probabilities,

Initially high demand, low long-term demand	= 10%
Initially low demand, high long-term demand	= 0%
Initially high demand that is sustained	= 60%
Initially low demand that is sustained	= 30%

- The chance that initial demand will be high is 70% (60% + 10%). If initial demand is high, the company estimates that the likelihood that it will continue at a high level is 86% (60% ÷ 70%). A high initial level of sales will affect the probability of high sales in the future. If initial demand is low, the probability is 100% (30% ÷ 30%) that sales will be low in the future. The level of initial sales is thus an accurate indicator of the level of sales in future periods.

- Annual income is projected as follows:

1. Large plant, high demand yield = $1.0M annual cash flow.
2. Large plant, low demand yield = $100,000 annual cash flow.
3. Small plant, low demand yield = $400,000 annual cash flow.
4. Small plant, high demand yield = $450,000 annual cash flow initially, but only $300,000 per year in the long run because of competition.
5. Small plant expansion to meet high demand yield = $700,000 annual cash flow.

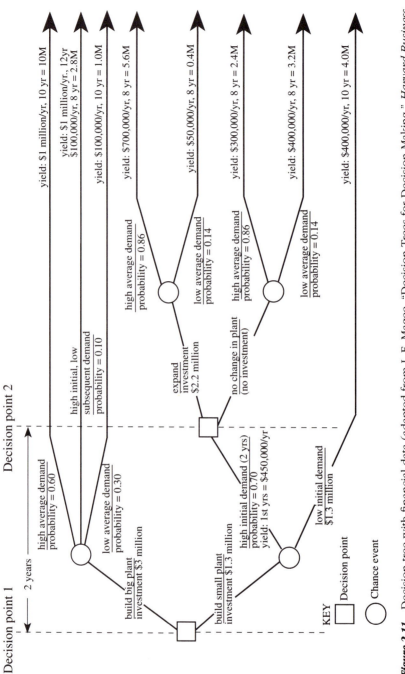

Figure 2.11 Decision tree with financial data (adapted from J. F. Magee, "Decision Trees for Decision Making," *Harvard Business Review*, July–August 1964).

60

Table 2.5   Decision 2 Analysis Using Maximum Expected Total Cash Flow as a Criterion

Choice	Chance Event	Probability (1)	Total-yield, 8 Years (thousands of dollars) (2)	Expected Value (thousands of dollars) (1) × (2)
Expansion	High average demand	0.86	$5,600	$4,816
	Low average demand	0.14	400	56
			Total	$4,872
			Less investment	2,200
			Net	$2,672
No expansion	High average demand	0.86	$2,400	$2,064
	Low average demand	0.14	3,200	448
			Total	$2,512
			Less investment	0
			Net	$2,512

From J. F. Magee, "Decision Trees for Decision Making," *Harvard Business Review*, July–August 1964.

6. Small plant expansion without sustained high demand yield = $50,000 annual cash flow.

• Construction estimates for the various options indicate that it will cost $3M to put the large plant in operation, $1.3M for the small plant and $2.2M to expand the small plant.

The decision tree based on this information is shown in Figure 2.11. Note that no information is included that the chemical company's executives did not know before developing the decision tree. However, the value of decision trees in *laying out* information in such a way that systematic analysis is enhanced provides an invaluable tool that improves decision making.

When management is in the process of deciding whether to build a large or small plant (decision 1), it does not have to concern itself with decision 2. However, it is important for management to be able to place a monetary value on decision 2 in order to compare the gain from taking the "build small plant" branch with the "build big plant" branch at decision 1. The analysis shown in Table 2.5 indicates that the total expected value of the expansion alternative is $160,000 greater than the no-expansion alternative ($2,672,000 − 2,512,000 = $160,000) over the eight-year life remaining after decision 2. Thus, management would choose to expand the plant if faced with decision 2 (considering financial return only).

Yields for the various decision branches are shown on the right-hand side of Figure 2.11. If these yields are adjusted for their corresponding probabilities less investment cost the following comparison can be made:

$$\text{Build big plant} = (\$10 \times 0.60) + (\$2.8 \times 0.10) + (\$1.0 \times 0.30) - \$3.0$$
$$= \$3,600,000$$

$$\text{Build small plant} = [(\$2.7 + 0.9) \times 0.70] + (\$4.0 \times 0.30) - \$1.3$$
$$= \$2,400,000$$

The choice that maximizes the expected total cash yield at decision 1 is to build the big plant initially. Note that this analysis does not account for the time value of money.

2.8   IMPROVING THE DESIGN PRODUCT

Many methods can be applied during the design process to analyze and synthesize the product to improve the design, lower or re-prioritize the allocation of costs, identify potential failure modes, and generally improve product quality. The use of TQM, or as identified in other references, QFD (Quality Function Deployment), has been referred to previously. This technique is more than a tool for improving the product however since it also implies a management approach that incorporates the concepts of W. Edwards Deming and others.

The use of TQM as a product improvement tool can be described by referring to Figure 2.12. Customer *wants* that are obtained through market research and competitor analysis are expanded into secondary and tertiary elements as shown. From this list a matrix of *product design requirements* is developed that will contribute to meeting the customer *wants*. The relationship between the customer *wants* and *product design requirements* are then ranked. This procedure identifies the customer *wants* that have not been addressed adequately and discloses areas where the product can be improved. Competitor products are also evaluated against the customer

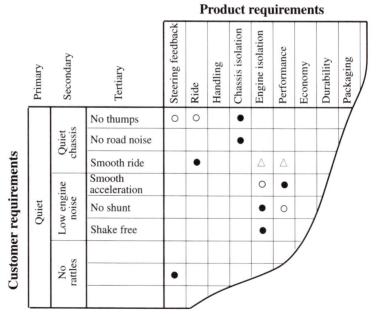

Figure 2.12   Customer *wants* versus *product design requirements* matrix (from S. Pugh, *Total Design,* Addison-Wesley, Wokingham, England, 1991, p. 202).

wants to obtain an assessment of the design product's standing against competition. Finally, competitive products are evaluated by in-house testing or other means and are given an appropriate score by which they are rated. From the completed matrix objective, target specification values for the various design parameters are selected, and a benchmark evaluation is completed ranking the in-house product against the competition. Using this target specification value, individual technicians involved in manufacturing and testing the product can understand the relationship between the specification requirement and the customer's *want*. In a similar manner, the relationship of the customer requirement to the product specification is traceable throughout the entire development process.[12]

The potential of the TQM method has been demonstrated by a major Japanese automotive manufacturer. Ten years after implementation of this method by its suppliers, start-up costs were reduced by approximately 60% and lead times for product development had been cut by one third. The number of product design changes were also cut dramatically during this same period.

2.8.1 Functional Cost Analysis

Functional cost analysis, or value engineering, is a technique that involves breaking the product down into its component parts and determining the cost effectiveness of these elements relative to the importance of the functions being provided. A matrix form similar to that shown in Figure 2.13 is used to perform this analysis. Use of this method requires a detailed level of knowledge about the product cost, and thus it can only be applied during the latter stages of the design process after the manufacturing plan has been generated. This technique allows the high cost areas of the design to be identified so that effort can be concentrated on cost reduction and/or enhanced functionability.[10] Figure 2.13a shows the cost analysis of an air valve and highlights the excessive cost of the *valve body* and *connect parts* function as well as the redundancy of some elements. Figure 2.13b shows the cost breakdown after redesign. Note elimination of costs associated with valve elements that were providing no useful function and reduction in the cost of the *connect parts* function.

2.8.2 Failure Mode and Effects Analysis

Failure Mode and Effects Analysis (FMEA) is a technique for identifying potential failure modes and their effect on components, subsystems, and systems, as well as the overall design product. FMEAs have been used in the aerospace industry for many years; they were enforced by NASA to a very disciplined degree during the early years of the space program. FMEAs can be applied at the component, subsystems, or systems level, depending on the product being designed and the purpose of the analysis. At the subsystems or systems level the FMEA is often accomplished by identifying the credible failure modes for the components and analyzing the effect of these failures on the operation and safety of the subsystems, systems, or end product. The systems design philosophy is then applied in conjunction with the FMEA to determine whether the design meets the failure criteria. Systems-level failure criteria can be specified in terms of continued operation or safe shutdown of the system in the event of component failure. This is often identified by the acronyms FS (fail safe) and FOFS (fail operational, fail safe).

---

[12] S. Pugh, *Total Design,* Addison-Wesley, Wokingham, England, 1991.

Parts	Functions	Stop air	Sense ram air	Sense servo air	Sense cabin air	Connect parts	Provide mounting	Resist corrosion	Provide support	Provide interchangeability	No function	$ total cost	%
Banjo assembly				0.2		0.4				0.47		1.07	5.5
Valve body		0.4	1.0			2.82	0.8	0.2	0.8		0.6	6.62	34.0
Spring											0.39	0.39	2.0
Diaphragm assembly		0.6	0.1	0.1	0.1	0.94		0.2	0.1			2.14	11.0
Cover				0.4		1.2	0.1	0.1	0.34	0.1		2.24	11.5
Lug											0.1	0.1	0.5
Nuts, bolts, and washers						2.14		0.1		0.1		2.34	12.0
Assembly						4.58						4.58	23.5
Total		1.0	1.1	0.7	0.1	12.08	0.9	0.6	1.24	0.67	1.09	19.48	100.0
% total		5.1	5.7	3.4	0.5	6.2	4.6	3.1	6.4	3.4	5.6		
High or low						H					H		

(a)

Parts	Functions	Stop air	Sense ram air	Sense servo air	Sense cabin air	Connect parts	Provide mounting	Resist corrosion	Provide support	Provide interchangeability	No function	$ total cost	%
Cover and connections		0.15	0.25	0.50	0.10	0.25	0.30		0.15	0.06		1.76	25.5
Body assembly		0.15	0.20	0.25	0.45	0.45	0.40		0.25	0.03		2.18	31.5
Diaphragm assembly		0.15	0.10	0.25	0.20	0.25	0.10		0.20	0.03		1.28	18.5
Valve assembly		0.05		0.05	0.05	0.15			0.31	0.05		0.66	9.5
Fasteners, nut bolts, etc.						1.04						1.04	15.0
Total		0.50	0.55	1.05	0.80	2.14	0.80		0.91	0.17		6.92	100.0
% total		7.2	7.9	15.1	11.6	30.9	11.6		13.2	2.5			
High or low						H							

(b)

Figure 2.13   (*a*) Function/cost analysis matrix for an air valve. (*b*) Function/cost analysis matrix for the redesigned air valve (from N. Cross, *Engineering Design Methods,* John Wiley & Sons, New York, 1989, p.132). Reprinted by permission of John Wiley & Sons, Ltd.

Failure of a component will cause shutdown of the system in a controlled (safe) manner with a FS philosophy. The system will remain operational after the first component failure and shutdown safely after the second with a FOFS philosophy. This type of analysis:

1. Provides an assessment of overall subsystems and systems safety.
2. Identifies critical components that may warrant upgrading or elimination.
3. Identifies areas of the design that should be provided with redundancy.
4. Provides information regarding the sensitivity of subsystems and systems physical location.

Component FMEAs are often performed to identify areas where redesign will provide the greatest benefit. For this type of analysis the component elements, functions, and the like are listed on the left-hand side of a matrix with the potential failure modes listed across the top. Figure 2.14*a* and *b* show typical FMEAs for an ordinary lead pencil and a wire paper clip, respectively. For Figure 2.14*a* the lead would warrant the greatest attention for possible improvements followed closely by the wood. In Figure 2.14*b* the function *holding the paper together* has the highest priority and should be emphasized in any product improvement program over such less important functions as *stress reliever*.[12]

Fault tree analysis can be used in conjunction with FMEA to provide an understanding of the interrelationships between failure modes. An example of fault tree analysis is shown in Figure 8.22, which depicts a fault tree network for a condensor system failure.

2.8.3 Taguchi Methods

The use of tolerances in the specification of parameters for component design and testing and in the stipulation of dimensions on drawings was first initiated about 1870, with the introduction of *go, no-go* tolerance limits. The need for the application of tolerances came about because manufacturers found that they could not make products alike in respect to a given quality, a phenomenon now referred to as variability. Furthermore, it was not necessary that all products be exactly alike and trying to make them so was too costly. The introduction of the *go, no-go* gage about this same time provided a simple device for the production technician to check his work quickly, resulting in lowered production costs.[13] Unfortunately, this led to an outlook that has returned to haunt us during the latter part of this century, *do not waste time trying to be exact.*

An approach that is preferred in today's environment is to designate target values for specified parameters to which processes are controlled. Deviations from these target values result in additional costs to both producer and consumer. Figure 2.15 shows two processes, both conforming to the tolerance requirements of a specification. From the parabolic cost plot, it is noted that process A incurs more loss than process B. This concept is referred to as process control, a technology attributable to Genichi Taguchi that is very much a part of the present industry emphasis on quality management.[14]

---

[13] W. A. Shewhart, "Statistical Method, From the Viewpoint of Quality Control," Department of Agriculture, 1939.

[14] G. Taguchi, "The Evaluation of Quality," *American Supplier Institute News*, September 1985.

Parts / Parts failure modes	Parts weight	Breaks	Falls apart	Smudges	Wears out	Gross weight	%
Lead	16	◉ 144		○ 48	◉ 144	336	42
Wood	16	◉ 144	◉ 144			288	36
Eraser	4	○ 12	○ 12	◉ 36	◉ 36	96	12
Eraser holder	9		◉ 81			81	10
						801	100

(a)

Functions / Parts failure modes	Function weight	Clip releases paper	Wire snaps	Snags on to other paper	Corrodes with time	Catches in clothing	Gross weight	%
Hold paper together	16	◉ 144		◉ 144	△ 16	○ 48	352	67
Reusability	9	○ 27	△ 9	○ 27	△ 9		72	14
Toothpick	4		○ 4			○ 12	16	3
Stress reliever (repetitive bending)	9		◉ 81				81	16
							521	100

◉ strong correlation = 9
○ medium correlation = 3
△ possible correlation = 1

(b)

Figure 2.14 FMEAs for (a) lead pencil and (b) wire paper clip (from S. Pugh, *Total Design*, Addison-Wesley, Wokinghm, England, 1991, p. 209).

The basic principle applied in this concept is the reduction of variability in the performance of the end product. To accomplish this the factors causing variability in the end product (or process) proposed must be well understood so that design sensitivity to these various causes can be minimized. The Taguchi methodology is based on the precept that the lowest cost to society represents the product with the highest quality. This higher quality is achieved by reducing variation in product characteristics. Thus, the difference between the conventional and Taguchi methods is that the conventional method proposes that higher quality costs more and the Taguchi method maintains that higher quality costs less. The *loss function* is a mathematical way to quantify the cost as a function of product variation. This loss

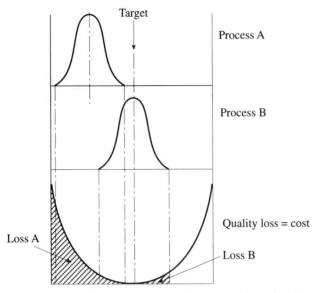

Figure 2.15 Conformance to specification and quality loss (from S. Pugh, *Total Design*, Addison-Wesley, Wokingham, England, 1991, p. 214).

function allows a determination to be made as to whether further reduction in the variation will continue to reduce costs. The Taguchi loss function, or *loss to society*, includes the costs of production as well as the costs incurred by the customer during use (repair, lost business, etc.). In this concept, minimizing the *loss to society* is the strategy that results in uniform products and reduces costs at both the production stage and at the point of consumption.[15]

The simplest loss function is known as the quadratic loss function which can be approximated by[16]

$$L(Y) = k(Y - m)^2 \tag{2.1}$$

where L is the loss associated with a particular performance characteristic Y, and m is the performance target value. The value of the loss parameter k can be determined by

$$k = \frac{A_o}{\Delta_o^2} \tag{2.2}$$

where A_o is the average loss to the customer when the performance characteristic is not within the limit, Δ_o (customer tolerance limit). If the product is rejected when it is not within the manufacturing tolerance limit, Δ, this results in a cost to the manufacturer. If the manufacturer's cost A is given, the manufacturing tolerance, Δ can be determined by

$$\Delta = \sqrt{\frac{A}{A_o}} \times \Delta_o \tag{2.3}$$

---

[15] P. J. Ross, *Taguchi Techniques for Quality Engineering,* McGraw-Hill, New York, 1988.
[16] G. Taguchi, *Introduction to Quality Engineering,* Asian Productivity Organization, Tokyo, 1986.

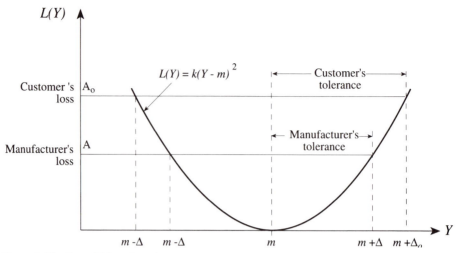

Figure 2.16 Taguchi loss function.

Figure 2.16 depicts the relationship between the loss L and the target value m. As can be seen from this figure, when the target value m is equal to the performance characteristic Y, the loss is zero. Loss occurs when $Y \neq m$. In this figure, the limits $m - \Delta_o$ and $m + \Delta_o$ designate the lower specification limit (LSL) and upper specification limit (USL) of customer tolerance, respectively. Similarly, $m \mp \Delta$ describes the LSL and USL of manufacturing.

▬ *Example 2.2* ▬▬▬▬▬▬▬▬▬▬▬▬▬▬▬▬▬▬▬▬▬▬▬▬▬▬▬▬▬▬▬▬▬▬

AUTOMOBILE HOOD CLOSING FORCE[15]

The hood of an automobile, when opened, is normally held open by a mechanism. The force required to close the hood from this position is of some significance to the owner of the vehicle. If the force is too great the owner will have difficulty closing the hood and will ask for the mechanism to be adjusted. If the force is too low the hood may be blown shut by a gust of wind and the owner will again ask for the mechanism to be adjusted. The engineering specifications and detail and assembly drawings call out a LSL and USL for the closing force. If the force is only slightly high or low the customer may be dissatisfied but may not ask for the mechanism to be adjusted. This may be more costly (in the long run) to the automobile dealer than the cost associated with adjusting the mechanism. A so-called goalpost view of this situation is shown graphically in Figure 2.17. The goalpost philosophy maintains that as long as the closing force is within the customer's zone of tolerance, no problem exists. If the closing force is less than the lower limit or greater than the upper limit, the hood mechanism would require adjustment under warranty at some cost, say, $50.

The closer the closing force is to the target value the more satisfied the customer is. If the force is a little high or low the customer feels some dissatisfaction (loss). The further the force is from the target value the greater the loss sensed by the customer. When the force reaches the customer's tolerance limits, the typical customer will demand correction of the problem. This scenario is depicted by points A and B on Figure 2.17. From the automobile manufacturer's point of view the difference in A and B is the cost of mechanism adjustment. From the customer's point of

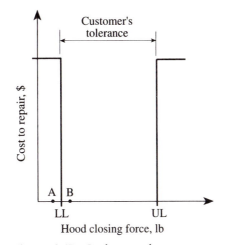

Figure 2.17 Goalpost tolerance syndrome (from P. J. Ross, *Taguchi Techniques for Quality Engineering*, McGraw-Hill, New York, 1988, p. 4).

view there is little difference in the force required to close the hood at point *A* versus point *B*. A better model for depicting the cost versus closing force is shown in Figure 2.18, the loss function curve. When the closing force is near the target value, there is little or no cost associated with hood closing. The further the force is from the target, the greater the cost is, until the customer's limit is reached. At this point the cost equals the mechanism adjustment cost. This approach quantifies the slight difference in cost associated with hood closing forces *A* and *B* and takes into account the customer's increasing dissatisfaction (loss) in going from *B* to *A*.

Let the customer loss A_o, when the force required to close the hood is outside of the customer's tolerance limit, be $60, the tolerance limit Δ_o, be 2 lb, and the

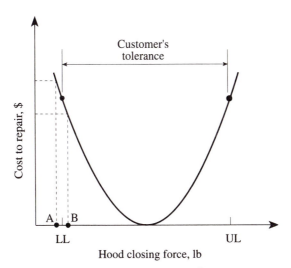

Figure 2.18 Taguchi Loss Function (from P. J. Ross, *Taguchi Techniques for Quality Engineering*, McGraw-Hill, New York, 1988, p. 4).

cost to the manufacturer for readjusting the mechanism be $50. The appropriate manufacturer's tolerance can be calculated as follows

$$\Delta = \sqrt{\frac{A}{A_o}} \times \Delta_o$$

$$= \sqrt{\frac{50}{60}} \times 2 = 1.83 \text{ lb}$$

Thus, the specification limits should be set to $\Delta = \mp 1.83$ lb, and quality control should reject the product when the force is outside of this range.

2.9 ENGINEER OR MANAGER?

Most of today's engineering graduates seem to have the goal of ultimately becoming managers. This has probably come about because most students believe that advancement up the ladder of success equates to holding a management position in their chosen field of endeavor. Although there is certainly some justification for this point of view, there are other considerations that should be factored into any decision to leave engineering to enter management.

All new graduates enter the work force as beginning engineers, but at some time early in their careers the opportunity will come to supervise or *to manage* an engineering task or small project. This first opportunity will probably be a very restricted form of management in that responsibility for the job will be assigned but not the responsibility for hiring, firing, or determining salaries. Thus, getting the job done well and on schedule will be largely a function of the new supervisor's ability to favorably influence employees assigned to the task. Regardless of the difficulties that these restrictions impose, this is an excellent opportunity to start evaluating management as a future profession. Additional *quasi-management* opportunities will open up as time goes on, and it is important to purposefully evaluate each experience. Periodic self-evaluation as to goals, interests, capabilities family considerations, and *quality of life* concerns will help to provide a focus for making a decision about entering management. Once the decision has been made to move into management, it is very difficult to return to engineering. This decision is like a one-way street: there are serious obstacles to going in the other direction.

Some of the considerations that should be factored into any decision about the future, whether to stay in engineering or to move into management and supervision, are outlined below:[17]

1. If money is of primary concern, consider the following:

$$\text{Top engineer salary} = \$60,000 \text{ per annum}$$
$$\text{At 40 to 50 hours/week} = \$30/\text{hr}$$
$$\text{Middle Manager salary} = \$80 - 90,000 \text{ per annum}$$
$$\text{At 60 hours/week} = \$30/\text{hr}$$
$$\text{Division-level manager salary} = \$120,000 \text{ per annum}$$
$$\text{At 70 to 80 hours/week} = \$30/\text{hr}$$

[17]E. A. Ohair, informal classroom notes, Texas Tech University, Lubbock, TX, 1989.

There is a considerable difference in the life-style of a top engineer as compared with a middle or upper manager, but little difference in hourly pay.

2. Managers get their work done through other people; they manage resources. Their concerns have to be associated with company image, business planning, and profit, not engineering. As a result, managers tend to lose their engineering skills over time.

3. Company executives will expect the manager to be loyal to the company and upper management first. This can, at times, place the manager in the middle as far as loyalty to the people working for him or her.

4. A decision involving personal integrity versus compromise will eventually come in most jobs, but probably sooner in management.

5. Engineering work can and should be documented so that proper credit is received. This will help to ensure that the engineer always has a job. This is not the case with managers.

6. There is an endless line of people who want to be managers, but there is a limited number of skilled engineers. Job mobility for engineers is much greater than for managers.

Whatever the decision is in regard to staying in engineering or becoming a manager, the most important consideration is to enjoy the work. One-third of a person's life is spent on the job. This is too great a commitment for anything that does not bring pleasure and enjoyment.

BIBLIOGRAPHY

1. DIETER, G. E., "Engineering Design: A Materials and Processing Approach," McGraw-Hill Book Company, New York, 1983.

2. CROSS, N., "Engineering Design Methods," John Wiley & Sons, New York, 1989.

3. PUGH, S., "Total Design," Addison-Wesley Publishing Company, Wokingham, England, 1991.

4. PAHL, G. and BEITZ, W., "Engineering Design, A Systematic Approach," The Design Council, London, England, 1988.

5. FLEMING, Q. W., BRONN, J. W. and HUMPHREYS, G. C., "Project and Production Scheduling," Probus Publishing Company, Chicago, IL 1987.

6. ROSS, P. J., "Taguchi Techniques for Quality Control," McGraw-Hill Book Company, New York, 1988.

7. TAGUCHI, G., "Introduction to Quality Engineering," Asian Productivity Organization, Tokyo, 1986.

Chapter 3

Modeling and Simulation

Through modeling engineers expand and enrich their vision, exercise their sensibilities, formulate unique and personal interpretations, process and optimize a large number of alternative solutions, originate innovations, and enhance their understanding of physical problems—in short, they develop the qualities which we marvel at in children.

3.1 MODELING IN ENGINEERING

One of the fundamental activities in which engineers are involved is model building. The process of model building is a way to present knowledge and to explore alternative solutions. A scale model of a physical system can be used to predict accurately the performance of the prototype. Model studies have proved useful in design and development for many years, especially when experimental testing of a full-size prototype is either impossible or prohibitively expensive. The advantages of model testing in experimental design are as follows:

1. When the problem is too complex for an analytical solution, an empirical solution can be developed.

2. Analytical techniques can be substantiated by correlating the predicted model behavior with the actual behavior of the model.

3. Prototypes with nonattainable characteristics can be studied, such as those with:
   (a) Large structures
   (b) Molecular structures
   (c) An environment that cannot be simulated

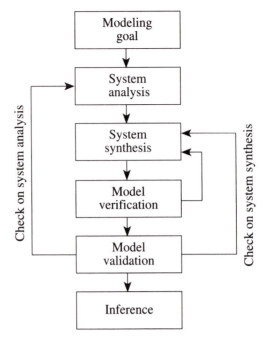

Figure 3.1 Modeling process (from G. Arthur Mihram, "The Modeling Process," ©197-*IEEE Trans. Sys. Man. Cybern.,* vol. SMC-2, no. 5, pp. 621–629, 1972).

(d) High-speed reactions

(e) Dangerous situations

The selection of a proper model and its implementation requires a priori knowledge about the system to modeled. Figure 3.1 shows a general modeling process for model development. As shown in the figure, the system goal must be defined and studied in the early stage of the modeling effort. At this preliminary model development stage, the system analyst must determine the need for the model, what kinds of analyses to perform, and an appropriate measure of performance. The second stage is the system analysis required for model development. During this stage, salient components, interactions, relationships, and dynamic behavior mechanisms of a system are isolated. System synthesis, which is the next stage in the modeling process, deals with structuring and implementing the various models in accordance with the findings from the system analysis stage. During this stage, the cost effectiveness and method of implementation of the various models are determined. Selection of the optimum model is then made by comparing model accuracies, implementation approaches, maintainability, and projected costs of the experimentation in subsequent stages of the modeling process. Once the model has been selected, verification of the model response is required. As shown in Figure 3.1, the verification stage of a modeling process serves as a check on the system synthesis stage. Before the model can be considered acceptable, similar checks on the system analysis stage are necessary. This is done during the validation stage by comparing responses from the model with corresponding responses recorded from the actual (modeled) system. Experimentation should be conducted with both the model and with the actual system. If the actual system is not available for experimentation, validation tests will not be dependable; thus, the analyst takes a risk in drawing conclusions regarding the actual system. In the inference stage of a modeling process, experiments are conducted solely with the verified and validated model.

3.2 HEURISTIC MODELING

This phrase is used to describe what might be called *common-sense* or *minimum cost* physical modeling. The word heuristic means to discover or to learn and, by implication, it embodies the concept of learning by comprehending the total problem. It is used here to emphasize the value of simple, inexpensive models in helping to grasp the relative size and interrelationship of individual elements in a design and in resolving problems associated with interfaces and interferrences.

The product of design is normally a material entity that did not exist previously. Most inexperienced and many experienced designers have difficulty envisioning the interrelationships of the various elements in a design, and when the total project involves elements from different disciplines, the potential problems are compounded. Fortunately, with some of the analytical features incorporated in computer-aided design programs, this difficulty can be minimized. However, many design organizations do not have this capability. Furthermore, many interface and interrelationship problems are not recognized until after the design is completed and construction begins. An example of what is meant by heuristic modeling is given in the following description:

> The Atlas F missile emplacement program was a large effort in involving many different contractors with contracts managed by both the U.S. Air Force (USAF) and the U.S. Army Corps of Engineers (COE). The purpose of this effort was to install the Atlas missile and all supporting equipment in a vertical underground concrete silo. The silo was first poured and then the structure surrounding the missile was constructed and supporting equipment was installed. Finally, the missile was lowered through open doors at the ground surface and interconnections to the ground support equipment were made. A significant number of propellant and pressurization lines provided interconnection to ground support equipment within the silo and supply vehicles at the surface. These lines were located in an appendage (propellant systems shaft) to the silo that entered the silo wall 30 to 40 feet below the ground surface. Propellant lines were delivered to the construction site already fabricated, cleaned for propellant service and sealed for protection from contamination. Shortly after the installation of these lines was initiated the COE, the agency responsible for overseeing the construction contractor's effort, contacted the USAF, who had the overall responsibility for completing the effort, about a problem in installing one of the propellant lines. One of the prefabricated piping sections could not be maneuvered into the propellant systems shaft. The design of this piping section was complicated by several bends at various angles and the construction contractor maintained that the overall configuration was such that it could not be passed into the propellant systems shaft. The contractor that manufactured the piping sections had been contacted and the cost of remanufacturing the piping sections for the 72 missile sites was going to cost

over $300,000. Thus, the incentive to find some other solution was high. The USAF had a small contingent of consulting engineers on their staff and the problem was passed on to two of them. After considerable discussion and analysis of the drawings, the two engineers decided that the piping section configuration was too complicated for analysis from the drawings and they decided to make a small model of the silo, propellant shaft, and piping section to get a better understanding of the interference problem. A crude model of the silo/propellant shaft was constructed out of cardboard and a pipe cleaner was used to construct a model of the piping section. As had been indicated by the construction contractor, there was definitely a problem working the piping section into the propellant shaft. However, using the model one method for manipulating the piping section was devised that indicated that it could be passed into the shaft, barely. The next question was, how closely did the model actually reflect the dimensional configuration of the piping section and silo? To verify this, a demonstration was arranged at one of the missile sites. When the two engineers arrived for the demonstration, the construction contractor already had the piping section rigged, hanging from a crane, and was demonstrating how the piping section could not be passed into the propellant shaft. The USAF Colonel in charge of the operation was glum, whereas the COE Colonel had an *I told you so* expression on his face. Upon inspection by the two engineers it was noted that the piping section was rigged in a different manner than that required to duplicate the method devised using the model. The piping section was rerigged accordingly, and another attempt was made to pass the piping section into the shaft. To almost everyone's surprise (including the two engineers) the piping section passed into the shaft with essentially no room to spare. The USAF Colonel was so delighted that he almost danced a jig. The COE Colonel was now glum.

Although the use of this modeling technique is not likely to result in any scenerio as dramatic (or accompanied by the theatrics) characteristic of this example, nevertheless there are several lessons that can be learned by using heuristic modeling:

1. Physical modeling of the object in question provides a grasp of the problem that cannot be achieved by any other technique, even using sophisticated computer analytical tools.

2. Crude models can be developed from basic materials (wood, fiberglass, sheet metal, paper, cardboard, glue, rubber bands, paper clips, and even pipe cleaners) that are available almost everywhere. The cost of these models is minimal.

3. A good bit of caution needs to be applied to any conclusions reached on the basis of using such models, but for gross indications relative to appropriate overall proportion, one structural member being too large relative to another, interferences, interfaces, and installation difficulties, as in the example, techniques of this kind can prove to be invaluable.

4. It is almost always helpful to develop a crude model first before spending time and effort on more sophisticated models. Development of the crude model will provide insight as to what should be included in the sophisticated model.

3.3   MATHEMATICAL MODELING

As the name implies, mathematical modeling is a process of writing mathematical expressions describing the behavior of a physical system under consideration. The physical system to be modeled may range from a spacecraft in orbit under the mutual gravitation of earth and moon, concentrations of reactants in a chemical reactor, air flow around an object in a wind tunnel or behavior of structures, and machine elements when acted on by external forces. Mathematical modeling begins by assuming that the system obeys certain constitutive laws—these are the basic laws of physics. In the case of the orbiting spacecraft, the laws governing the dynamics are Newton's laws; the total external forces are equated to the inertial forces and the equations of motion are written in terms of acceleration of the spacecraft. Other laws such as the conservation of mass, energy momentum, and the like, may also be needed. For instance, to describe the reactions in a chemical reactor, one starting point might be to write the equations of the reactions, and supplement them with the equations for mass and energy conservation; the second law of thermodynamics determines the conditions under which the reactions are possible. In addition to the contitutive laws, simplifying assumptions can be made. These assumptions make the problem under consideration mathematically tractable. Natural processes are very complex; however, interest is normally confined to a particular behavior observed, for example, in an experiment. It is therefore important that the simplifying assumptions are such that, on the one hand, they make the problem mathematically simple while, on the other, they include all the physical behavior to be investigated. From the example of the orbiting spacecraft, it might be assumed that the motion takes place in an elliptical orbit, and the problem could be formulated in such a way as to investigate small variations in the original elliptical motion. For the chemical reactor, a similar simplifying assumption could be postulated by stating that the concentrations vary about the original chemical equilibrium point.

Once a reliable mathematical model has been formulated, the next step is to obtain the solution. For a large class of problems, numerical solutions obtained by computer are acceptable. There is a variety of numerical schemes available to deal with all types of problems. To obtain the solution for the orbiting spacecraft, a step-by-step time integration algorithm, such as the Runge–Kutta scheme, can be used to obtain the trajectory of the spacecraft. For large structural problems, finite element and finite difference formulations are often used. For very large and complex problems, the computer is the only reliable tool available.

In recent years, another tool called symbolic manipulation has become available. Compared to numerical simulation, which manipulates numbers to obtain solutions, symbolic manipulators use symbols, and all of the mathematical operations are performed on symbols. The advantage of using symbolic manipulators is that they can be incorporated in the modeling process itself to derive the equations. In fact, it is desirable to use symbolic maupulators to derive the equations because this eliminates a great deal of tedious algebra and the accompanying algebraic mistakes.

As an example, suppose that the natural frequencies and mode shapes of a cantilever beam shown in Figure 3.2 are to be taken from any elementary book on vibration. The equation of motion for the beam calculated is

$$\frac{\partial^4 y}{\partial x^4} - \beta^4 y = 0 \tag{3.1}$$

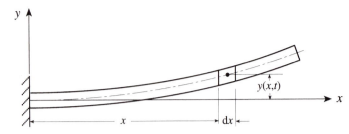

Figure 3.2   Cantilever beam model.

where

$$\beta^4 = \frac{\rho A \omega^2}{EI}$$

ρA = mass of the beam per unit length

ω = natural frequency

In addition to the differential equation the boundary conditions are needed. For the cantilever beam, the deflection and the slope at $x = 0$ are zero, that is:

$$y(x = 0) = y'(x = 0) = 0 \qquad (3.2)$$

where prime denotes the derivative with respect to x. At the free end, the shear force and the bending moment are zero; that is,

$$y'''(x = l) = y''(x = l) = 0 \qquad (3.3)$$

Equation 3.1, with the boundary conditions from Eq. 3.2 and Eq. 3.3, completely model the vibration of the cantilever beam. At this state, the symbolic manipulation software packages such as MAPLE, MACSYMA, or MATHEMATICA can be used. For this problem, the software package MAPLE was used. The program was written to mimic the solution procedure described below.

The general solution of the fourth-order beam equation is given as

$$y(x) = A \cosh \beta x + B \cos \beta x + C \sinh \beta x + D \sin \beta x \qquad (3.4)$$

By applying the boundary conditions at $x = 0$, we get

$$B = -A \quad D = -C \qquad (3.5)$$

Substituting in Eq. (3.4),

$$y(x) = A(\cosh \beta x - \cos \beta x) + C(\sinh \beta x - \sin \beta x) \qquad (3.6)$$

To apply the bending moment and the shear force conditions at $x = l$, differentiate Eq. 3.6 twice and three times with respect to x and substitute $x = l$ to obtain

$$y''(x = l) = Ac_{11} + Cc_{12} = 0$$
$$y'''(x = l) = Ac_{21} + Cc_{22} = 0 \qquad (3.7)$$

or in matrix form

$$\begin{bmatrix} c_{11} & c_{12} \\ c_{12} & c_{22} \end{bmatrix} \begin{bmatrix} A \\ C \end{bmatrix} = \begin{bmatrix} 0 \\ 0 \end{bmatrix}$$

(3.8)

where
$c_{11} = \cosh \beta l + \cos \beta l$
$c_{12} = \sinh \beta l + \sin \beta l$
$c_{21} = (\sinh \beta l - \sin \beta l)\beta$
$c_{22} = (\cosh \beta l + \cos \beta l)\beta$

the determinant of the 2×2 matrix is the characteristic polynomial

$$p(\lambda l) = \cosh \beta l \cos \beta l + 1$$

(3.9)

The roots of the characteristic polynomial are the eigenvalues β from which the natural frequency ω of the beam can be obtained. To find the roots of $p(\beta l)$ requires considerable numerical effort; however, with MAPLE, the roots can easily be evaluated. Equation 3.9 has infinite roots, however, the few lower roots only are of interest. The following steps are used to find the roots.

1. Plot the polynomial as shown in Figure 3.3,
2. Find the approximate location of the intersection with the β axis,
3. To find the roots accurately, solve the polynomial in the neighborhood of the approximate root locations by using MAPLE.

Once the roots are known, they can be substituted in Eq. 3.6 to obtain the corresponding mode shape of the beam as shown in Figure 3.4. The complete symbolic program along with the output is given in Appendix D.

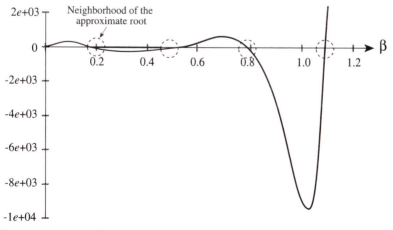

Figure 3.3   Roots of the characteristic polynomial.

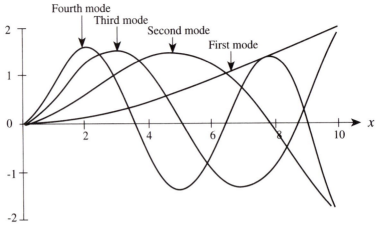

Figure 3.4 Mode shape of the beam.

3.4 DIMENSIONAL ANALYSIS

When the functional relation between parameters or the governing equations of the system are unknown, dimensional analysis is used to obtain a valid scaling law. The Buckingham π theorem can be used to develop a dimensionless parameter, which indicates the state of a given physical system. To obtain a correct relationship the correct physical characteristics must be identified. The following are considered physical characteristics:

1. Characteristic length
   - Length
   - Cross-sectional area
   - Volume
   - Moment of inertia

2. Characteristic motion
   - Velocity
   - Acceleration
   - Angular velocity
   - Moment of inertia

3. Characteristic material properties
   - Young's modulus
   - Density
   - Viscosity
   - Strength

4. Characteristic phenomenon involved
   - Surface tension coefficient
   - Heat transfer coefficient
   - Temperature difference

To determine the dimensionless ratio, π terms, the following steps can be used. Any dimensionless physical characteristic, such as the width/length ratio, can be selected as a π term.

1. Select basic dimensions such as mass (M), length (L), time (T), and the like, for use as repeating variables from the physical characteristics. As a rule of thumb, select the repeating variable so that one is a characteristic length (length, cross-sectional area, volume, moment of inertia, etc.), one is a characteristic motion (velocity, acceleration, angular velocity, etc.), one is a characteristic metarial (Young's modulus, density, viscosity, tensile strength, etc.), and one is a characteristic of the phenomenon involved (surface tension coefficient, heat transfer coefficient, mass transfer coefficient, etc.).

2. Use the repeating variables together with one of the remaining physical characteristics for each π term. Equate the exponents of the basic units using unknowns for the repeating variables, unity for the remaining characteristics, and zero for the dimensionless π term. Solve for the unknowns and form the π terms.

The number of independent dimensionless parameters required to obtain π terms can be determined by

$$S = n - b \qquad\qquad (3.10)$$

where S is the number of π terms, n is the total number of physical characteristics, and b is the number of basic units involved. The following example illustrates how to calculate the necessary π terms.

—— *Example 3.1* ———————————————————————————————

As Figure 3.5 shows, assume that a sphere of diameter d falls through a fluid in time t with initial velocity v. Considering the resistance of the fluid through which the sphere falls,

(a) Determine the number of π tems necessary.

(b) Form the π terms.

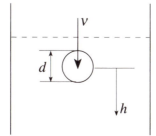

Figure 3.5   Sphere falling in a fluid media.

Table 3.1   Physical Characteristics

Physical Characteristic	Symbol	Basic Dimension
Distance	b	L
Time	f	T
Velocity	v	LT^{-1}
Gravity	g	LT^{-2}
Fluid density	ρ	ML^{-3}
Fluid viscosity	μ	$ML^{-1}T^{-1}$
Sphere mass	m	M
Sphere diameter	d	L

Solution

(a)   The first step is to select the important physical characteristics involved with the problem. These characteristics and the corresponding basic units are shown in Table 3.1.

From Table 3.1, the number of selected physical characteristics is $n = 8$, and the repeating basic dimensions are M, L, and time T; therefore, $b = 3$. Hence, the number of necessary π terms S is

$$S = n - b$$
$$= 8 - 3 = 5$$

(b)   The physical characteristics can be written in a functional equation as

$$F(b, t, v, g, \rho, \mu, m, d) = 0 \tag{3.11}$$

and the functional equation can be written in dimensional form as

$$F(L, T, LT^{-1}, LT^{-2}, ML^{-3}, ML^{-1}T^{-1}, M, L) = 0 \tag{3.12}$$

The repeating variables are $d(L)$, $v(LT^{-1})$, and $\mu(ML^{-1}T^{-1})$. As mentioned previously, repeating variables are selected, one from each physical characteristic. Following step 3, the π terms can be formed as shown in Table 3.2. Referring to Table 3.2, the dimensionless form for the π_4 term can be written in the form of an equation as

$$\pi_4 = \mu^{x_4} d^{y_4} v^{z_4} \rho \tag{3.13}$$

Table 3.2   Forming π Terms

π Terms	Repeating Variables			Remaining Physical Characteristics
$\pi_1 =$	μ^{x_1}	d^{y_1}	v^{z_1}	b
$\pi_2 =$	μ^{x_2}	d^{y_2}	v^{z_2}	t
$\pi_3 =$	μ^{x_3}	d^{y_3}	v^{z_3}	g
$\pi_4 =$	μ^{x_4}	d^{y_4}	v^{z_4}	ρ
$\pi_5 =$	μ^{x_5}	d^{y_5}	v^{z_5}	m

Following step 3, Eq. 3.13 can be written in basic units as

$$M^0 L^0 T^0 = (ML^{-1}T^{-1})^{x_4}(L)^{y_4}(LT^{-1})^{z_4}(ML^{-3})^1 \qquad (3.14)$$

Because the π_4 term has no unit, the left-hand side of the above equation should be dimensionless; thus, the exponents of M, L, and T are assumed to be zero. Equating exponents for M, L, and T yields

$$\text{for } M, 0 = x_4 + 1$$
$$\text{for } L, 0 = -x_4 + y_4 + z_4 - 3$$
$$\text{for } T, 0 = -x_4 - z_4$$

Solution of the above equations yields $x_4 = -1$, $y_4 = 1$, and $z_4 = 1$. Hence, from Eq. 3.13, the π_4 term is written as

$$\pi_4 = \mu^{-1} d^1 v^1 \rho^1 = \frac{\rho v d}{\mu} \qquad (3.15)$$

The π_4 term is known as Reynold's number. Following the same procedure, other π terms can be determined.

The Reynold's number found in the above example is the ratio of inertia forces to viscous forces. The critical value of Reynold's number can be used to distinguish between laminar and turbulant flow. Some of the other dimensionless parameters that frequently occur in the fluid flow studies include:

1. The Mach number M is used as a parameter to characterize the compressibility effects in fluid flow and is defined by

$$M = \frac{v}{c} \qquad (3.16)$$

2. The Froude number F_R is used for flows with free surface effects. This is a key parameter in the design of ship and hydraulic structures and is defined by

$$F_R = \frac{v^2}{lg} \qquad (3.17)$$

3. The Weber number W_E is another important dimensionless parameter that is used in studies involved with gas–liquid or liquid–liquid interfaces.

$$W_E = \frac{\rho l v^2}{\sigma} \qquad (3.18)$$

In Eqs. 3.16, 3.17, and 3.18, the physical characteristics v, c, l, g, ρ, and σ represent the velocity, speed of sound, length, acceleration of gravity, mass density, and surface tension, respectively. These equations (Eqs. 3.16, 3.17, and 3.18) can be obtained by the same procedure that is used to find the Reynold's number equation (Eq. 3.15).

3.5 SIMILARITY LAWS IN MODEL TESTING

Dimensional analysis is an important tool often used to increase the accuracy and the efficiency of experimental design. To predict prototype behavior from measurements on the model, there must be similitude between the model and the prototype. The laws that quantify the scale model behavior are called the laws of similitude. Similitude is used to establish a set of scaling factors to obtain the relationship between the model and the prototype. To obtain the correct relationship, the correct or important parameters that affect the experimental design must be identified. To select the important parameters requires some experience. If a wrong parameter is selected or an important one is left out, no subsequent analysis of the basic units can correct the error. The most common types of similarity are geometric, kinematic, and dynamic.

3.5.1 Geometric Similarity

The model and prototype are assumed to be geometrically similar if, and only if, all pairs of points on the model and prototype have the same ratio of distances in all three coordinates. Hence, ratios of model lengths to the corresponding prototype lengths must be the same. For example, consider the two cantilever beams (a) and (b) depicted in Figure 3.6. In order for the cantilever beams be geometrically similar, the following equality must be satisfied:

$$\lambda = \frac{l_p}{l_m} = \frac{w_p}{w_m} = \frac{b_p}{b_m} \tag{3.19}$$

where λ is called the scale factor and the subscript m refers to the model and p to the prototype.

3.5.2 Kinematic Similarity

The motions of the model and prototype are kinematically similar if the ratio of the corresponding velocities v/u in the flow fields are constant. For example, kinematic similarity of a prototype and a model of a wind turbine (Figure 3.7) requires the following velocity relation:

$$\lambda = \frac{v_p}{u_p} = \frac{v_m}{u_m} \tag{3.20}$$

where $v = r\omega$, r is the radius and ω is the angular velocity.

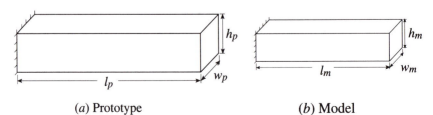

(a) Prototype   (b) Model

Figure 3.6   Cantilever beams.

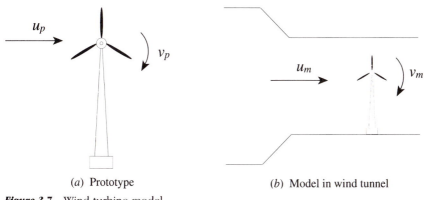

(a) Prototype   (b) Model in wind tunnel

Figure 3.7   Wind turbine model.

3.5.3   Dynamic Similarity

Dynamic similarity requires geometric similarity and ensures kinematic similarity between the model and the prototype. Dynamic similarity exists if prototype and model force and pressure coefficients are identical. In the wind turbine example for dynamic similarity to exist, the Reynold's numbers of the prototype and model must be equal, that is,

$$R_{e(m)} = R_{e(p)} \tag{3.21}$$

3.6   WIND AND WATER TUNNELS

Aerodynamics plays a vital role in many engineering fields such as aerospace, architectural, automotive, and marine. Aerodynamic testing can be conducted in either a wind tunnel or a water tunnel, depending on the facility available and the information required. These tunnels are used to examine the streamlines and to determine the behavior patterns of aerodynamic forces acting on test objects. These facilities allow designers to model new body shapes without going to the expense of constructing full-scale units for the test. Scale models are usually necessary due to the constraint of tunnel size. For example, full-scale testing of a Boeing 747 would be impractical. In the aerospace industry, the use of scale models has resulted in the saving of countless lives and enormous expense.

A water tunnel uses the same principle as a wind tunnel but uses water as the fluid instead of air. Any closed testing system or tunnel has the following four major components, regardless of the fluid being employed:

1. A contoured duct to control and direct fluid flow.
2. A drive system to move the fluid through the duct.
3. A model of the object to be tested.
4. Instrumentation to measure forces exerted on the model.

Because the kinematic viscosity of water is 16 times greater than that of air, model studies at high Reynolds numbers should be conducted in water tunnels rather than wind tunnels. Linear tow tanks are another means of performing aerodynamic

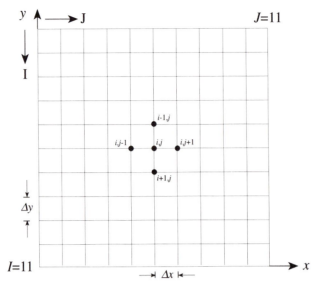

Figure 3.8   Arangement of grid points.

testing in water. The major disadvantage of this type of testing is that the model can only be subjected to intermittent, limited-duration testing. Another common water testing method uses a thin layer of water flowing over a flat table on which a model is placed. The disadvantage of water table testing is that it can only be used for two-dimensional analysis and cannot be used when accurate Reynolds number testing is required. A water tunnel overcomes these problems by providing a three-dimensional flow field.

3.7   NUMERICAL MODELING

The solution of partial differential equations (PDE) that govern a structure's response may not always be possible; thus, numerical techniques must be utilized. In general, there are two alternative methods of approximating the solution of partial differential equations: (1) the finite difference method, and (2) the finite element method.

3.7.1   The Finite Difference Method

The familiar finite difference model of a problem gives a pointwise approximation to the exact solution of a PDE. To find the approximate solution of a PDE using the finite difference technique, a network of *grid points* is established through the domain of interest defined by the independent variables. As shown in Figure 3.8, x and y are the two independent variables, and Δx and Δy are the respective grid spacings. Subscripts i and j are used to denote space points having coordinates $i\,\Delta y, j\,\Delta x$.

   Let the exact solution to the PDE be $U(x, y)$, and its approximation determined at each grid point be ϕ_{ij}. Each partial derivative of the original PDE is replaced by an appropriate finite difference expression involving Δx, Δy, and ϕ_{ij}, which leads to a set of algebraic equations in ϕ_{ij}. By solving this system of equations, the values of ϕ_{ij} can be determined at each grid point (i, j). To obtain a close approxima- tion, sufficiently small grid spacings should be selected. Note that, improvement in accuracy by using small grid spacing is achieved at a price, because it increases the computational time. Moreover, a very small grid size may increase the round-off error, thus, leading to a less accurate solution.

There are basically three finite difference expressions used in the finite difference method—the forward, backward, and central difference expressions. The central difference expressions that approximate the first and second order derivatives of the original PDE are formulated as

$$\phi_1' = \frac{\partial \phi}{\partial x} = \frac{\phi_{i+1} - \phi_{i-1}}{2\,\Delta x} \tag{3.22}$$

$$\phi_1'' = \frac{\partial^2 \phi}{\partial x^2} = \frac{\phi_{i+1} - 2\phi_i + \phi_{i-1}}{2\,\Delta x^2} \tag{3.23}$$

To illustrate this method, consider the problem of determining the temperature distribution in a thin homogeneous rectangular plate with prescribed boundary conditions (B.C.) along the edges of the plate as shown in Figure 3.9. The top B.C. is held at 100°C, and the temperature of the other three B.C.'s are changing linearly as shown in the figure. The problem can be mathematically formulated as

$$\frac{\partial^2 \phi}{\partial x^2} + \frac{\partial^2 \phi}{\partial y^2} = \frac{Cp}{k}\frac{\partial \phi}{\partial t} \tag{3.24}$$

It is convenient to start by nondimensionalizing the independent variables x and y as

$$X = \frac{x}{L} \quad \text{for} \quad 0 \le X \le 1 \tag{3.25}$$

$$Y = \frac{y}{B} \quad \text{for} \quad 0 \le Y \le 1 \tag{3.26}$$

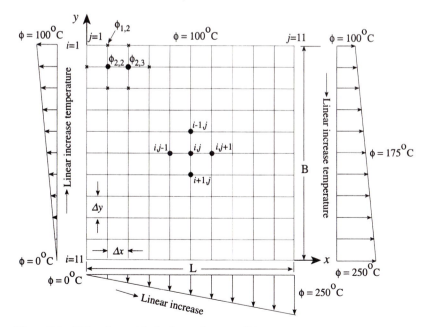

Figure 3.9 Thin plate with the boundary conditions.

Equation 3.24 is then rearranged as

$$\frac{\partial^2 \phi}{\partial X^2} + \frac{L^2}{B^2}\frac{\partial^2 \phi}{\partial Y^2} = \frac{L^2 C\rho}{k}\frac{\partial \phi}{\partial t} \tag{3.27}$$

Nondimensionalizing time t

$$t = \frac{k}{L^2 C\rho}T \tag{3.28}$$

yields

$$\frac{\partial^2 \phi}{\partial X^2} + \frac{L^2}{B}\frac{\partial^2 \phi}{\partial Y^2} = \frac{\partial \Phi}{\partial T} \tag{3.29}$$

where

$$\frac{\partial \Phi}{\partial T} = \frac{L^2 C\rho}{k}\frac{\partial \phi}{\partial t} \tag{3.30}$$

L/B is called the *aspect ratio* and must be given for a specific problem. For the steady-state solution, the right-hand side (time-dependent term) drops out. The problem can then be described by the following differential equation

$$\frac{\partial^2 \phi}{\partial X^2} + \left(\frac{L}{B}\right)^2\frac{\partial^2 \phi}{\partial Y^2} = 0 \quad \text{for } 0 \leq X \leq 1 \text{ and } 0 \leq Y \leq 1 \tag{3.31}$$

Using Eqs. 3.22 and 3.23, the derivatives of Eq. 3.31 are now replaced by the central difference expressions

$$\frac{\partial^2 \phi}{\partial X^2} = \frac{\phi_{i,j+1} - 2\phi_{i,j} + \phi_{i,j-1}}{\Delta X^2} \tag{3.32}$$

$$\frac{\partial^2 \phi}{\partial Y^2} = \frac{\phi_{i+1,j} - 2\phi_{i,j} + \phi_{i-1,j}}{\Delta Y^2} \tag{3.33}$$

Substituting in Eq. 3.31 yields

$$\frac{\phi_{i,j+1} - 2\phi_{i,j} + \phi_{i,j-1}}{\Delta X^2} + \left(\frac{L}{B}\right)^2\frac{\phi_{i+1,j} - 2\phi_{i,j} + \phi_{i-1,j}}{\Delta Y^2} = 0 \tag{3.34}$$

or

$$\phi_{i,j+1} - 2\phi_{i,j} + \phi_{i,j-1} + \left(\frac{L}{B}\right)^2\left(\frac{\Delta X}{\Delta Y}\right)^2\phi_{i+1,j} - 2\phi_{i,j} + \phi_{i-1,j} = 0 \tag{3.35}$$

Let

$$R = \left(\frac{L}{B}\right)^2\left(\frac{\Delta X}{\Delta Y}\right)^2 \tag{3.36}$$

and solve for $\phi_{i,j}$

$$\phi_{i,j} = \frac{\phi_{i,j+1} + \phi_{i,j-1} + R\phi_{i-1,j} + R\phi_{i+1,j}}{2(1 + R)} \tag{3.37}$$

For simplicity, assume that $\Delta X = \Delta Y$ and $L/B = 1$. Then Eq. 3.37 reduces to

$$\phi_{i,j} = \frac{\phi_{i,j+1} + \phi_{i,j-1} + \phi_{i-1,j} + \phi_{i+1,j}}{4} \qquad (3.38)$$

Equation 3.38 indicates that the temperature at each grid point is the average of temperatures at surrounding grid points; hence, no interior grid point can have a temperature greater than the hottest boundary temperature. The first step in determining the grid point temperature by using Eq. 3.38 is to initialize all of the interior grid point temperatures to some value and start iteration. For this example, assume that all the interior grid point temperatures are equal to zero. Then, the temperature at $\phi_{2,2}$ is

$$\phi_{2,2} = \frac{\phi_{1,2} + \phi_{2,3} + \phi_{3,2} + \phi_{2,1}}{4}$$

$$\phi_{2,2} = \frac{100 + 0 + 0 + 90}{4} = 47.5°C$$

Knowing the temperature at $\phi_{2,2}$, the temperature at $\phi_{2,3}$ can be calculated

$$\phi_{2,3} = \frac{\phi_{1,3} + \phi_{2,4} + \phi_{3,3} + \phi_{2,2}}{4}$$

$$\phi_{2,3} = \frac{100 + 0 + 0 + 47.5}{4} = 36.875°C$$

After all the values of grid points are calculated, use these values as the initial value and start the second iteration. This iteration process is repeated until the largest change in the $\phi_{i,j}$ component is less than the assumed convergence criterion ε. It is obvious that the number of iterations required for convergence depends on how good the initial estimate was and on the assumed convergence criterion ε. In most instances, setting all of the interior grid points equal to the mean of all of the boundary temperatures is usually a good initial estimate for starting the iteration.

3.7.2   The Finite Element Method

A finite element model of a problem gives a piecewise approximation to the governing equations of a structure response. The basic premise of the finite element method is that the domain can be analytically modeled or approximated by replacing it with an assemblage of discrete elements. Because these elements can be put together in several ways, they can be used to represent extremely complex shapes. Although both the finite difference method and the finite element methods can give accurate results if used properly, one advantage of the latter is the relative ease with which the boundary conditions of the problem are handled. Many physical problems have irregularly shaped boundaries. Boundaries of this type are relatively difficult to handle using the finite difference method.

The first step in the finite element method is to divide the region into a finite number of elements and to label the elements and the node numbers, as shown in Figure 3.10. The points at which the elements are connected are called the grid points. They are identified by a grid point number to define their location in space. The accuracy of the approximation to a problem solution depends on the characteristics and number of elements that are used in the domain. The most

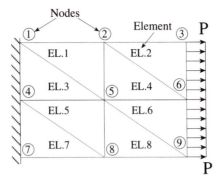

Figure 3.10 Subdivision of a plate into triangular elements.

common finite elements used in modeling are shown in Figure 3.11a and b, one-dimensional elements with two and three nodes, respectively. They have a single degree of freedom at each grid point and are used for one-dimensional problems involving two force (truss) members. There are two general element families used in modeling the two-dimensional domain—triangular and quadrilateral elements. Figure 3.11c and d shows linear elements that have straight sides. Higher-order two-dimensional elements (quadratic or cubic) have either straight or curved sides or both (see Figure 3.11e and f). Two-dimensional elements are used to model membranes, plates, and shells. Because of the greater accuracy, quadrilateral elements are preferred over triangular elements. Figure 3.11 g, h, and i depicts three-dimensional elements, also called *solid elements,* which are used to model thick plates, thick shells, and three-dimensional solid media, respectively.

3.8 MONTE CARLO SIMULATION

Monte Carlo simulation belongs to the family of Monte Carlo methods. Monte Carlo methods constitute a branch of experimental mathematics that is concerned with experiments on random numbers. In essence, this is an elegant approach to approximating physical and mathematical problem solutions by simulation of random quantities.[1]

During the mid-1940s, von Neumann and his colleagues introduced the term "Monte Carlo" as a code name for their classified research on neutron diffusion problems.[2] The name Monte Carlo was probably chosen because of the method's random number basis, which can supposedly be duplicated by roulette wheels. Such roulette wheels are often found in gambling houses, of which, in the mid-1940s, the most famous were in the Principality of Monte Carlo. In some literature, Monte Carlo methods are also referred to as methods of statistical trials, since they involve a scheme of producing random events where each trial is independent of the rest.

There are various forms of Monte Carlo methods that are used depending on the problem to be solved. The simplest form is that used for the direct simulation of

[1] J. M. Hammersley and D. C. Handscomb, *Monte Carlo Methods,* Wiley, New York, 1964.
[2] P. D. Spanos and L. D. Lutes, "A Primer of Random Vibration Techniques in Structural Engineering," *Shock Vib. Digest,* vol. 18, pp. 3–10, 1986.

One-dimensional elements

Two-dimensional elements

Three-dimensional elements

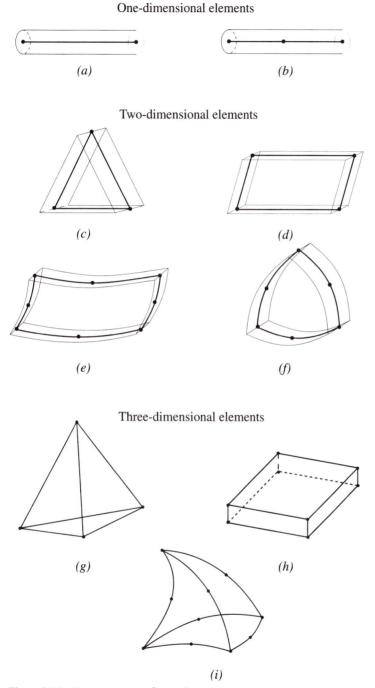

Figure 3.11   Some common finite elements.

a probabilistic problem. In the most general form, Monte Carlo simulation can be characterized by three steps:[3]

1.   Application of random numbers to simulate representative samples of random variables with prescribed distributions.

[3] A. Kareem, "Wind Effects on Structures: A Probabilistic Viewpoint," *Probabilistic Eng. Mechanics,* vol. 20, pp. 166–200, 1987.

2. Solution of the problem on the basis of large realizations.

3. Statistical analysis of results (i.e., derivation of means, mean squares, and variances, etc.).

Depending on the complexity of the problem to be solved, these steps are used accordingly.

Because Monte Carlo simulation involves statistical trials, its accuracy is proportional to the number of trials. Increasing the number of trials or observations is not always an economic course of action, since at some point the computer-time cost becomes prohibitive. For example, if n is the number of trials, it has been shown that the computer cost increases in proportion to n while the statistical uncertainly decreases in proportion to $1/\sqrt{n}$.[4] Due to the nature of Monte Carlo simulation, it is most effective in solving problems in which the required accuracy of the sought statistics are within 5 to 10 percent of the actual values.[5] In some instances, variance-reducing techniques can be used to reduce the uncertainty of the solutions. In essence, these techniques improve the accuracy of the results without increasing the number of tests (trials) and without sacrificing reliability.

Monte Carlo simulation can be used to simulate most processes influenced by random factors. Some of the problems that can be solved by this method include the following:

1. Solution of linear algebraic equations.

2. Engineering design problems.

3. Structural dynamics problems.

4. Operational research problems (e.g., queueing systems).

5. Control of floodwaters and construction of dams.

6. Ecological competition among species.

7. Chemical kinetics.

8. Diffraction of waves on random surfaces.

Consider a simple example of the application of the Monte Carlo method.[5] Assume that a unit square circuit board has been designed. The available area P on the circuit board is to be estimated (Figure 3.12). Choose $n = 40$ points at random within the square. Note that the selected coordinates of the points are random and uniformly distributed over the entire unit square. The random coordinates can be generated by using a computer as shown in Table 3.3. In order to assign the numbers 0 to 9 to the coordinates of the points, the unit square is divided into 10 parts along both the x and y axes. Only the first digit of the respective two-digit random numbers is used for the location of the points within the unit square. In

[4] L. J. Branstetter, G. Jeong, J. T. P. Yao, Y. K. Wen, and Y. K. Lin, "Mathematical Modelling of Structural Behaviour during Earthquakes," *Probabilistic Eng. Mechanics,* vol. 3, pp. 130–145, 1988.
[5] I. M. Sobol, *The Monte Carlo Method,* The University of Chicago Press, Chicago, 1974.

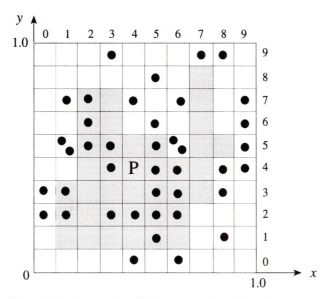

Figure 3.12 Shape of available area on circuit board.

the table, let columns 1, 3, 5, and 7 and columns 2, 4, 6, and 8 be the x and y coordinates of the points, respectively. In Figure 3.12, the number of points enclosed in the shape are $n' = 17$. From geometrical considerations, the area P can be approximated by the ratio $n'/n = 17/40 = 0.43$. The latter value is close to the true area of P of 0.38. Because this experiment involves trials, the accuracy will be increased by increasing the number of times the experiment is performed and also by increasing the number n of random points. Hence, if the experiment is repeated a number of times, the estimate areas can be shown to closely cluster around the true area. As a word of caution, the Monte Carlo method is certainly not the most efficient way to estimate an area of a plane surface, but as the space dimensions of the body increase, so does the efficiency of this method. Therefore, this method is well suited to the solution of multidimensional problems.

Table 3.3 Generated Random Numbers

1	2	3	4	5	6	7	8
x	y	x	y	x	y	x	y
03	27	43	73	56	16	96	47
97	74	24	67	62	42	81	14
16	72	62	27	66	56	50	26
12	56	85	99	36	96	96	68
55	59	56	35	64	38	54	82
16	22	77	94	39	49	54	43
84	42	17	53	31	57	24	55
63	01	63	78	49	26	95	55
33	21	12	34	29	78	64	56
57	60	87	32	44	09	07	37

3.9   DISCRETE EVENT SIMULATION

Discrete event simulation, or simply simulation, is a form of descriptive modeling. Simulation experiments explore the behavior of the intended system. A discrete event simulation model is basically a computer program that can be used to simulate the behavior of the system under study. An engineer developing a simulation model has at his or her disposal various special purpose simulation languages such as Simscript, Simula, or the General Purpose Simulation System (GPSS). Alternatively, a simulation program can be developed by using general purpose programming languages such as Fortran, Pascal, or C. A simulation model can be used for describing new systems, predicting the behavior of systems, demonstrating a system's behavior and training. The advantages of simulation modeling are:[6]

1. Ease in performing controlled experiments.
2. Time compression in the sense that a simulation experiment takes a small fraction of time compared with actual system operation time.
3. Sensitivity analysis for observation of the behavior limits of the system.
4. Experimentation without disturbing the real system.
5. Use of the simulation model as an effective training tool.

Consider the single-server oil change station shown in Figure 3.13 as an example to simulate. In this model, cars arrive for service according to some distribution such as Poisson, and they are serviced according to some service time distribution. If the service station is busy, the arriving car waits in the queue for service. If the service station is idle, the car moves into the station, the oil is changed, and the vehicle departs. In both of these cases, systems statistics are collected. The flow of the simulation model of this simple oil change service station is shown in Figure 3.14.

The above simulation example can be used to support a variety of studies such as:

1. Evaluating different tasks to achieve balanced workloads and improve system efficiency.
2. Performing sensitivity analyses to determine which parameter has the most impact on the system efficiency.
3. Determining which parameter combination will provide the most feasible operation of the system.
4. Analyzing the performance impact of different degrees of automation.

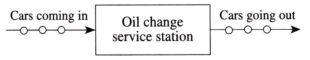

Figure 3.13   Oil change service station.

---

[6] W. T. Graybeal and U. W. Pooch, *Simulation: Principles and Methods*, Little, Brown, Boston, 1980.

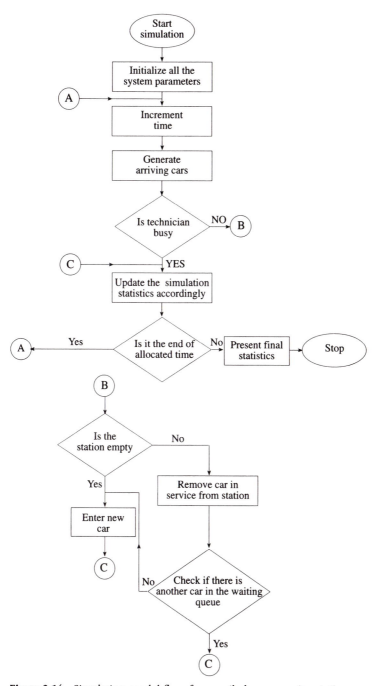

Figure 3.14 Simulation model flow for an oil change service station.

3.10 KNOWLEDGE-BASED SYSTEMS IN THE DESIGN PROCESS

The idea of capturing the knowledge of an expert in an automated medium and then using this knowledge when needed, has been one of the most productive areas in the application of artificial intelligence. The class of technical activities representing the knowledge of an expert in an electronic medium for later use are collectively called expert systems technology. The systems produced with the aid of this technology are collectively called expert systems, knowledge-based systems, or decision systems.

Because of their flexibility in handling incomplete and inconsistent information as human experts would, the use of knowledge-based systems is gaining importance in the engineering design process.[7,8,9] The following capabilities characteristics of expert systems typically separate them from conventional systems:

1. Solving problems normally solved by experts in the field.
2. Producing solutions for problems that are not suitable for traditional techniques.
3. Functioning with incomplete and inconsistent data.
4. Producing quick prototypes.
5. Providing explanations of decisions reached.

Production of an expert system with the above capabilities requires that knowledge be gathered (knowledge acquisition), knowledge needs to be represented in an electronic medium (knowledge representation), knowledge needs to be used to derive inferences (knowledge manipulation), and interfacing with a user of the systems needs to be continuous throughout the process (user interface).[10]

These necessities for knowledge acquisition, knowledge representation, knowledge manipulation, and user interface also dictate the overall components of expert systems. As shown in Figure 3.15, the components of a typical expert system are user interface, knowledge base (rule base), interpreter, and working memory. The user interface is the obvious component to be able to achieve input to the system and to present the output to the user. Knowledge base is the location holding the collective set of rules capturing the expertise obtained from experts. Interpreter is a program that uses logical inferences in arriving at conclusions dictated by logic. Working memory performs logical inference operations on selected rules from the knowledge base.

3.10.1 An Expert System for Preliminary Design of Structures

Designing a structure, such as the tension leg platform (TLP) shown in Figure 3.16, is an engineering process that requires creativity, experience, and knowledge of

[7] M. M. Tanik and E. C. Chan, *Fundamentals of Computing for Software Engineers,* Van Nostrand Reinhold, New York, 1991.

[8] J. Flemming, E. Elghadamsi, and M. M. Tanik, "A Knowledge-Based Approach to Preliminary Design of Structures," *J. Energy Resources Technol.* vol. 112, pp. 213–219, 1990.

[9] M. M. Tanik and A. Ertas (eds.), *Expert Systems and Applications,* Proceedings of ASME, New York, 1991.

[10] "Expert Systems—Guest Editors," introduction by M. M. Tanik and R. T. Yeh in *IEEE Software,* pp. 15–16, 1988.

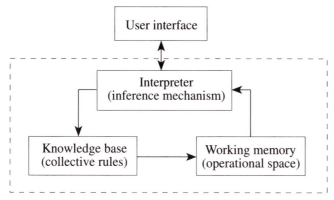

Figure 3.15 Components of a typical expert system.

engineering principles.[11] As in any engineering activity, there is usually more than one solution that is considered appropriate. Typically, the speed and success of a structural designer in selecting a suitable structural scheme depends on his or her experience.

The structural design process follows the standard stages of design, including preliminary design, detailed design, and drawing production. Currently, computers are extensively used in most of the phases involved in the production of structures. In this example, the use of computers with expert systems technology in the preliminary design stage will be demonstrated. The analysis, detailed design, and drawing production stages present problems suitable for conventional solutions described elsewhere in the book.

The preliminary design stage has two main goals. The first is the conceptual design in which the framing scheme is selected. The second is preliminary member sizing in which the sizes of the various components of the selected design are determined. Typically, the engineer may have to perform an analysis to select preliminary member sizes. Experienced engineers are normally able to make reasonable estimates of the required sizes. An expert system, as shown in Figure 3.15, can be used by novice engineers in the preliminary design phase. In this example,

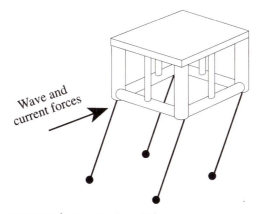

Figure 3.16 Tension leg platform.

[11]J. Flemming, E. Eighadamsi, and M. M. Tanik, "A Knowledge-Based Approach to Preliminary Design of Structures," *J. Energy Resources Technol.* vol. 112, pp. 213–219, 1990, permission from ASME.

the prototype of an expert system used for selecting initial structural member sizes is discussed.

For the design of an expert system for structural member selection, three key issues need to be decided: (1) representation of experience, (2) finding matching structures, and (3) selection of member sizes.

There are various methods representing experience. One of the most widely used techniques is the rule-based approach. In this approach, attributes related to the problem are stored in the knowledge-base. In case of the platform problem, examples of attributes are as follows:

1. Geographical location of platform.
2. Purpose of platform.
3. Water depth.
4. Soil type.
5. Wave and current loading descriptions.
6. Wind-loading descriptions.
7. Seismic zone.
8. Design specification.
9. Structural framing cost.
10. Foundation cost.

To be able to utilize previous experience, designs of previously built structures that have matching features to the intended new design need to be found. Table 3.4 shows a set of criteria for the TLP example that can be used to find matching structures. For example, water depth is a required criterion. If the new TLP is to be designed for a water depth of 1000 feet, structures designed for depths considerably less than 1000 feet (300 feet) or considerably more than 1000 feet (3000 feet) should not be considered. The expert system should not find (through the matching process) these overconservative or unreliable structures for the knowledge base for the new structure design.

Table 3.4 **Example Criteria**

Criteria	Comments
Geographical location of platform	Desirable criteria
Purpose of platform	Essential criteria
Water depth	Essential criteria
	The tolerance for a match will vary within the water depth
Soil type	Desired criteria
Wind speed	Essential criteria
Wave loads	Essential criteria
Current loads	Essential criteria
Seismic zone	Essential criteria
Design specification	Desirable criteria
Structural framing cost	Desirable criteria
Foundation cost	Desirable criteria

The last phase of the preliminary design process is to select the initial member sizes. Member attributes such as type, length, location, and special loading conditions can be used to identify matching members. Once the members are identified, they should be ranked in a manner similar to the ranking of the matching structures. To determine the required section properties for desired members, the rankings of the similar members, their sizes, and the ranking of the structure are used.

The design of the expert system is completed at this point. The process involves the selection of an expert system implementation language (tool, shell) and the encoding of the above developed rules using the selected implementation language and computer.

3.10.2   Selection of Expert System Tools

To be able to actually implement the expert system in question, a suitable implementation tool must be selected. This is comparable to selecting a programming language and a compiler for conventional programming purposes. Selection of implementation tools is by itself a complicated process.[12] Some of the features of tools which seem promising for designers are

1.   Capability to query the user during the inference process.
2.   Existence of an explanation support mechanism.
3.   Graphical display support system.
4.   Ease of use.
5.   Capability in detecting incomplete and inconsistent data.

3.10.3   Determining the Suitability of the Candidate Problem

The first step in selecting an expert system for a particular problem is to determine whether the problem is likely to be suitable for solution by an expert system method. This determination is based on *essential* and *desirable* features of the problem. Each of these features is assigned a weight indicating its relative importance in assessing the candidate problem.[13] These weights are not problem dependent, instead, they are related to the capabilities of expert system technology. The *essential* and *desirable* features and associated weights used to evaluate the candidate problem and associated weights are listed in Tables 3.5 and 3.6, respectively.

3.10.4   Knowledge Representation Schemes

Separating domain-specific knowledge from the program is the fundamental idea of expert systems. The term *knowledge base* refers to the collection of domain knowledge, and the term *inference engine* refers to the program reasoning through that knowledge. The specification of knowledge representation comprises both of these components.

The inference engine works on the knowledge base and generates solutions to the problems of the user. A knowledge representation supported by a powerful inference scheme allows the user to solve a problem by simply storing the facts

---

[12]S. Mills and M. M. Tanik, "Selection of Expert System Tools for Engineering Design Applications," *Proc. ASME Expert Systems and Applications,* PD-vol. 35, pp. 41–45, 1990.
[13]S. Mills, C. Ertas, Y. Hurmuzlu, and M. M. Tanik, "Selection of Expert System Development Tools for Engineering Applications," *J. Energy Resources Technol.,* vol. 114, no. 1, pp. 38–46, 1992.

Table 3.5   Essential Features of Candidate Problem

Description	Weight (1–10)
The problem is not natural language dependent	10
The problem is not knowledge intensive	7
Solution of the problem requires use of heuristics	8
Test cases for the problem are available	10
The problem is decomposable into subproblems	7
"Common sense" is not needed to solve the problem	10
An optimal solution is not required	8
Problem solution will take place in the future	10
A domain expert exists and is available and cooperative	10
Transfer of expertise for the problem domain is difficult	7
Solution should not depend upon physical senses	10

in the knowledge base. There are several techniques being used for knowledge representation. In this section, *production systems,* which is a commonly used scheme will be discussed.[14]

The terms *production-rule systems* and *rule-based systems* are used to refer to the knowledge representation scheme. In this scheme, knowledge is expressed as a set of rules. A production system consists of three parts:[15]

1. ***Rule base.*** The rule base is composed of a set of production rules. A production rule is an expression in the form

   IF<condition-part>THEN<action-part>

   The <condition-part> states the conditions that must be satisfied for the production to be applicable and <action-part> defines the actions to be executed.

2. ***Context.*** The context is the short-term memory buffer that contains the initial inputs by the user and the deductions that have been made by the inference engine. The actions of the production rules that are invoked can expand the context with the addition of current deductions.

Table 3.6   Desirable Features of Candidate Problem

Description	Weight (1–10)
The candidate problem should have previously been identified	4
Solutions should be explainable and require interaction	5
Similar applications have been successfully implemented	8
The problem should be solved at many different locations	5
The environment in which the system will operate is hostile	3
Solution of the problem requires subjective judgment	4
The expert will not be available in the future	3

[14]E. Demirors et al., "Rule Based Expert Systems: A PC + Experience," Technical Report 91-cse-19, Department of Computer Science and Engineering, Southern Methodist University, May 1991.

[15]Avron Barr/Edward Feigenbaum, *The Handbook of Artificial Intelligence,* © 1989, by Addison-Wesley Publishing Company, Inc. Reprinted with permission of the publisher.

Figure 3.17 Basic cycle of the rule interpreter.

3. **Interpreter.** The interpreter decides the order of rules to be invoked. The production systems operate according to the cycles of the interpreter. A basic cycle of the interpreter is given in Figure 3.17.

In the following example "On–CL X" refers to the fact that the symbol X is currently in the context list CL and "Put–On–CL X" is used to refer to X being added to the current context list.

PRODUCTION

Rule 1 ⇒ IF On–CL green
　　　　　THEN Put–On–CL produce

Rule 2 ⇒ IF On–CL packed in small container
　　　　　THEN Put–On–CL delicacy

Rule 3 ⇒ IF On–CL refrigerated OR On–CL produce
　　　　　THEN Put–On–CL perishable

Rule 4 ⇒ IF On–CL weighs 15 lb AND On–CL inexpensive AND NOT
　　　　　　　On–CL perishable
　　　　　THEN Put–On–CL staple

Rule 5 ⇒ IF On–CL weighs 15 lb each AND On–CL produce
　　　　　THEN Put–On–CL watermelon

In the above example, the following steps are performed to differentiate the staple food from produce:

1. If more than one production is applicable, then deactivate any production whose action adds duplicate deductions to the CL.
2. If more than one production rule is applicable, then execute the action with the lowest (arbitrarily chosen)-numbered application production.

The above production system can be used to identify food items. Let the existing knowledge about the unknown food be "it is green" and "weighs 15 lb." Therefore, at the beginning the context list is

$$CL = (green, weighs\ 15\ lb)$$

1. Cycle 1 ⇒ since rule 1 is applicable, the action part of rule 1 is executed. This adds "produce" to the context representing a new fact about the unknown food item. Therefore, the updated context list is

$$CL = (produce, green, weighs\ 15\ lb)$$

2. Cycle 2 $\Rightarrow$ rule 1, rule 2, and rule 3 are applicable. Since rule 1 adds "produce," which is duplication, rule 1 is eliminated from the execution. Hence, rule 3 is executed. Therefore, the undated context list is

$$CL = \text{(perishable, produce, green, weighs 15 lb)}$$

3. Cycle 3 $\Rightarrow$ rule 1, rule 2, and rule 3 are applicable. Rule 5 is executed. Therefore, the updated context list is

$$CL = \text{(watermelon, perishable, produce, green, weighs 15 lb)}$$

4. Cycle 4 $\Rightarrow$ No redundant productions to activate.

BIBLIOGRAPHY

1. DOEBELIN, E. O., *System Modeling and Response*. Wiley, New York, 1980.
2. FOX, R. W. and McDONALD. A. T., *Introduction to Fluid Mechanics*. Wiley, New York, 1973.
3. GIORDANO, R. F. and WEIR M., *A First Course in Mathematical Modeling*, Brooks/Cole, Monterey, CA, 1985.
4. GORDON, G., *System Simulation*. Englewood Cliffs, N.J., Prentice-Hall, 1969.
5. MARTIN, F. F., *Computer Modeling and Simulation*. Wiley, New York, 1968.
6. WELLSTEAD, P. E., *Introduction to Physical System Modelling*. Academic Press, New York, 1979.

Chapter 4

Design Analyses for Material Selection

Engineering has never been easy. The speed of introduction of new materials, tools, and techniques are increasing. We are approaching a human processing bottleneck for effective use of these inventions in better engineering of cost-effective, timely, useful and reliable artifacts.

4.1 MATERIAL SELECTION

Selecting proper materials and understanding the fabrication processes associated with design are two of the most important responsibilities of a design engineer in the evolution from concept to completed product. A primary design requirement in the selection of a material for a specific application is that the material be capable of meeting the design service life requirement at the least cost. For a defined service life, the designer normally narrows available choices to a few candidate materials. The final selection of a particular material is usually made from past experience with similar materials. Selecting materials based on past experience is still popular because the designer feels confident in using a tried and proven material. However, rapidly changing advances in technology are demanding better performance from engineering materials. Hence, selecting the proper engineering material for a given application requires a broad knowledge of the state of the art in materials development.

Material selection can occur at varying times during the design process. Normally, materials are selected during detail design, but for some projects, candidate materials may be identified during the concept and preliminary design stages of the effort. Material changes can sometimes be required during the early production phases as a result of unanticipated machining and assembly problems. However, the need for material changes this late in the process is decreasing as a result of the integration of design and production with simultaneous engineering and other similar concepts. Infrequently, material changes are required after the product has been placed in operation. In such situations, a product recall is often required that may result in negative publicity for the company. Material selection decisions usually have significant ramifications and, therefore, must be made by knowledgeable people with adequate data and material test resources.

During the selection of the optimum solution to a design problem, an attempt should be made to determine if the concept is unduly limited by the materials selected. Based on the design requirements, certain classes of materials may be eliminated from consideration, and other classes may be selected as suitable candidates for the final choice because of their characteristics and properties. Conversely, in some instances, it may be advisable to modify the design concept if it is limited by the available materials.

At the detail design stage, candidate materials identified earlier in the design process are subjected to an in-depth analysis. At this stage, analysis and determination of the material requirements should be accomplished, selection and evaluation of candidate materials should be completed, and the materials selection decision should be finalized.

4.1.1   Analysis of Material Requirements

During this step, the maximum and minimum requirements necessary for proper performance of the material must be fully defined. It is also essential that selection criteria be established for materials for the given application.

Material Design Requirements   Material properties and characteristics cover a wide range of parameters and play an important role in meeting the design requirements. Material selection properties and characteristics normally considered in design are summarized in Table 4.1 and are discussed (in part) in this chapter.

In general, material strength has three distinct elements:

1. ***Static strength.*** The ability to withstand a constant load at ambient temperature.

2. ***Fatigue strength.*** The ability to withstand a time-dependent fluctuating load.

3. ***Creep strength.*** The ability to withstand a load at elevated temperatures over a sufficiently long time.

Although high strength is desired for many design applications, it should be remembered that machining and processing of high strength material are costly.

Stiffness is related to the ability of a material to store the energy of deformation. The stiffness of a material is important when a part must remain relatively rigid under load or when the part must exhibit flexibility to sustain sudden loads without fracture. Generally, lower strength materials can be used when stiffness is the primary design requirement. Materials having high stiffness/density ratio are often used when light weight is one of the design requirements.

To illustrate material selection criteria for stiffness when light weight is one of the design requirements, consider the case of a solid cylinder in bending, as shown in Figure 4.1. The basic stiffness is given by

$$\delta = \frac{PL^3}{48EI} = \frac{PL^3}{48E\left(\frac{\pi D^4}{64}\right)} \tag{4.1}$$

Table 4.1   Material Selection Properties and Characteristics

Static characteristics	**5. Manufacturing**
a. Strength	a. Producibility
Ultimate strength	b. Availability
Yield strength	c. Processing characteristics
Shear strength	*Machinability*
b. Ductility	*Weldability*
c. Young's modulus	*Moldability*
d. Poisson's ratio	*Heat treatability*
e. Hardness	*Formability and forgeability*
2. Fatigue characteristics	*Hardenability*
a. Corrosion fatigue	d. Minimum handling thickness
b. High load	e. Joining techniques
c. Low load/extended life	f. Quality assurance
d. Constant amplitude load	**6. Hostile environments**
e. Spectrum load	a. Moisture
f. Fatigue strength	b. Temperature
3. Fracture characteristics	c. Acidity/Alkalinity
a. Fracture toughness	d. Salt solution
b. Flaw growth	e. Ammonia
c. Crack instability	f. Hydrogen attack
4. Thermal properties	g. Nuclear hardness
a. Coefficient of linear thermal	**7. Damping characteristics**
expansion	**8. Anisotropy**
b. Melting and boiling points	**9. Others**
c. Heat transfer coefficient	a. Electrical properties
d. Specific heat	b. Magnetic properties
e. Thermal conductivity	c. Chemical properties
f. Thermal Shock resistance	d. Corrosion properties
g. Heat of sublimation	

where P, L, D, E, I, and δ are the applied load, length of the distance between supports, diameter of the solid cylinder, modulus of elasticity, area moment of inertia, and max deflection, respectively. Then

$$D = \left(\frac{4PL^3}{3E\pi\delta} \right)^{1/4} \qquad (4.2)$$

The weight W of the solid cylinder is

$$W = V\rho = L\frac{\pi D^2}{4}\rho \qquad (4.3)$$

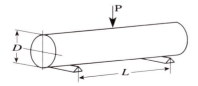

Figure 4.1   Solid cylinder in bending.

where ρ and V are density and volume, respectively. Combining Eq. 4.2 with Eq. 4.3, we have

$$W = \frac{\pi}{2(3\pi)^{1/2}} L^{5/2} \left[\frac{P}{\delta}\right]^{1/2} \frac{\rho}{E^{1/2}} \tag{4.4}$$

For a given stiffness P/δ, the weight of the solid beam is minimized when $\rho/E^{1/2}$ is minimized. Hence, $\rho/E^{1/2}$ is the material selection criterion.

Consider the same case for material selection for strength. The stress equation for bending is given by

$$\sigma = \frac{Mc}{I} \tag{4.5}$$

where, for the maximum fiber stress, $c = D/2$, $I = \pi D^4/64$, and maximum bending $M = PL/2$. Then

$$\sigma = \frac{16PL}{\pi D^3} \tag{4.6}$$

Therefore,

$$D = \left[\frac{16PL}{\pi\sigma}\right]^{1/3} \tag{4.7}$$

Using Eq. 4.3, the weight of the solid cylinder can be written as

$$W = (4\pi)^{1/3}(L^5 P^2)^{1/3}\left(\frac{\rho}{\sigma^{2/3}}\right) \tag{4.8}$$

For a minimum weight at constant L and P, the above result shows that $\rho/\sigma^{2/3}$ should be minimized. Therefore, $\rho/\sigma^{2/3}$ is the material selection criterion. Similar derivations for different structures and constraints can be found in reference [1].

Material Selection Factors   Several factors need to be considered in selecting the suitable material for a given application. The most important factors are:

1. Availability
2. Manufacturability
3. Repairability
4. Reliability
5. Service environment
6. Compatibility
7. Cost

Factors to be considered in determining the availability of materials are as follows:

1. Thicknesses
2. Widths

3. Minimum quantities

4. Number of sources

Failure to factor these considerations into the overall design process will result in serious scheduling difficulties.

Manufacturability should not be overlooked by the designer. The designer is responsible for making certain that the design is producible with the selected material. As shown in Figure 4.2,[1] material selection may significantly affect the manufacturing cost. Since high-performance alloys have a low machinability rating, the relative cost for machining alloy steel is high in comparison with that of carbon steels. This is due to alloy steel's high strength, high resistance to shear load, and rapid work hardening. Therefore, careful attention is necessary when making the selection of materials in the early stages of design. Once the product is manufactured and is in service, the material of construction must be easily and economically repaired. Often, part repair must be done within a short period of time and procedures must be as simple as possible. Selected material candidates should require a minimum of special processes, tools, and skills for repair. Therefore, repair techniques must be considered during the material selection process.

Other factors that should be considered during the selection process include past experience service history, mechanical and physical material properties as affected by mill and manufacturing processes, quality control requirements, and correlation

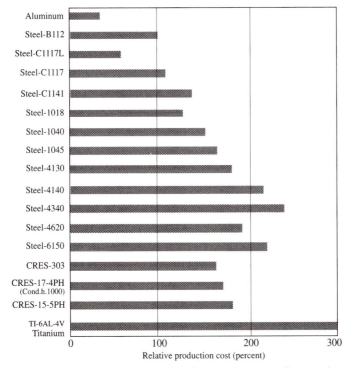

Figure 4.2   Relative production cost for various types of materials.

[1] H. E. Trucks, "Design for Economical Production," Society of Manufacturing Engineers Publications Development Department, Dearborn, MI, 1987, p.2.

of laboratory test data, such as fracture toughness, impact, and stress corrosion, to the service requirements. Prior service history is a very important criterion for material selection. Although some new materials offer significant improvement in physical and chemical properties, designers must be cautious in using them until their performance is proven. Designers should recognize the risk of failure when a new unproven material is selected for an application. Selected materials must perform under the specified environmental conditions; thus, the designer should review the material properties and the predicted performance in the environment in which the product will function. The compatibility of two or more materials used in an application is another important factor in selection. For example, two materials performing in a high-temperature environment may fail if their coefficients of thermal expansion are significantly different.

Perhaps the most important consideration of all in the selection process is the cost of the material for the proposed application. Cost is a most important selection factor in reducing the number of possible material candidates to a manageable level. The total cost of a material used for a specific application must be compared with the total cost of alternate materials in finalizing the choice. Cost and weight can be incorporated into property parameters (strength or stiffness) to facilitate comparisons.

Consider again the case of a solid cylinder in bending to illustrate a cost-performance index. From Eq. 4.6, the stress due to maximum bending is

$$\sigma = \frac{16PL}{\pi D^3} \tag{4.9}$$

To facilitate comparison, consider two solid cylinders A and B with different material properties and cross sectional areas. For the same load P and length L, the load carrying ability of two solid cylinders can be written as

$$\frac{\pi D_A^3 \sigma_A}{16L} = \frac{\pi D_B^3 \sigma_B}{16L} \tag{4.10}$$

or

$$\frac{D_B}{D_A} = (\frac{\sigma_A}{\sigma_B})^{1/3} \tag{4.11}$$

Using Eq. 4.3 yields

$$\frac{W_B}{W_A} = \frac{D_B^2 \rho_B}{D_A^2 \rho_A} \tag{4.12}$$

or

$$\frac{W_B}{W_A} = \left(\frac{\sigma_A}{\sigma_B}\right)^{2/3} \frac{\rho_B}{\rho_A} \tag{4.13}$$

Suppose that the unit costs of materials A and B are C_A \$/lb and C_B \$/lb, respectively. The total cost C_T of the solid cylinders to withstand the same load P is given by

$$C_{TA} = W_A \times C_A \quad \text{and} \quad C_{TB} = W_B \times C_B \tag{4.14}$$

Table 4.2   Formulas for Minimum-Cost Criterion

Components	Strength Based	Stiffness Based
Solid cylinder in tension or compression	$C\frac{\rho}{\sigma}$	$C\frac{\rho}{E}$
Solid cylinder in bending	$C\frac{\rho}{\sigma^{2/3}}$	$C\frac{\rho}{E^{1/2}}$
Solid cylinder in torsion	$C\frac{\rho}{\tau^{2/3}}$	$C\frac{\rho}{G^{1/2}}$
Solid rectangular beam in bending	$C\frac{\rho}{\sigma^{1/2}}$	$C\frac{\rho}{E^{1/3}}$
Thin-walled pressure vessels under internal pressure	$C\frac{\rho}{\sigma}$	$C\frac{\rho}{E}$
Thin wall shaft in torsion	$C\frac{\rho}{\tau}$	$C\frac{\rho}{G}$

From Eq. 4.13,

$$\frac{C_{TB}}{C_{TA}} = \left(\frac{\sigma_A}{\sigma_B}\right)^{2/3} \frac{\rho_B}{\rho_A} \frac{C_B}{C_A} \tag{4.15}$$

From Eqs. 4.8 and 4.15, we conclude that $C(\rho/\sigma^{2/3})$ is the cost per unit of strength of a solid cylinder loaded in bending. This is often referred to as a *minimum-cost criterion*. From Eq. 4.4, it can be seen that the minimum-cost criterion based on stiffness is $C(\rho/E^{1/2})$. Formulas for *minimum-cost criterion* for different loading conditions are given in Table 4.2. Formulas from Table 4.2 can be rearranged to give criteria for maximum stiffness-weight and strength-weight ratios, as shown in Table 4.3.

4.1.2   Selection and Evaluation of Candidate Material

After the material analysis step is completed, the designer then narrows the candidate list to a few materials. The elimination process is then initiated for the final material selection. The rating system shown in Table 4.4 can be used for this purpose. As shown in the table, two design requirements and three selection factors are adopted using this approach. Since all the design requirements and selection factors are not of equal importance, weighting factors $a, b, \ldots$, are used to find the overall rating.[2] The overall rating, G_i, can be calculated by

$$G_i = \frac{aR_1 + bR_2 + cR_3 + \ldots}{a + b + c + \ldots} \tag{4.16}$$

Table 4.3   Formulas for Optimum Strength and Stiffness

Components	Strength	Stiffness
Solid cylinder in tension or compression	$\frac{\sigma}{\rho}$	$\frac{E}{\rho}$
Solid cylinder in bending	$\frac{\sigma^{2/3}}{\rho}$	$\frac{E^{1/2}}{\rho}$
Solid cylinder in torsion	$\frac{\tau^{2/3}}{\rho}$	$\frac{G^{1/2}}{\rho}$
Solid rectangular beam in bending	$\frac{\sigma^{1/2}}{\rho}$	$\frac{E^{1/3}}{\rho}$
Thin-walled pressure vessels under internal pressure	$\frac{\sigma}{\rho}$	$\frac{E}{\rho}$
Thin wall shaft in torsion	$\frac{\tau}{\rho}$	$\frac{G}{\rho}$

[2] F. A. Crane and J. A. Charles, *Selection and Use of Engineering Materials*, Butterworth & Co., Boston, 1984, p. 213.

Table 4.4   Final Evaluation of the Candidate Materials

Material	Design Requirements				Material Selection Factors						Overall Rating
Candidates	**1**		**2**		**3**		**4**		**5**		$aR_1 + bR_2 + \dots$
	A_1	R_1	A_2	R_2	A_3	R_3	A_4	R_4	A_5	R_5	$a + b + \dots$
Material$_1$	—	—	—	—	—	—	—	—	—	—	G_1
Material$_2$	—	—	—	—	—	—	—	—	—	—	G_2
Material$_3$	—	—	—	—	—	—	—	—	—	—	G_3
Material$_4$	—	—	—	—	—	—	—	—	—	—	G_4

A: Absolute value.

G: Overall rating.

R: Relative value respect to largest value.

(Adapted from F. A. Crane and J. A. Charles).

A high value of G_i identifies the best material. However, a low design requirement or selection factor value can be an indication of the most suitable material to meet a specific requirement such as cost. In this instance, assume that R_5 is the relative value of cost, then Eq. 4.16 is modified to

$$G_i = \frac{aR_1 + bR_2 + cR_3 + dR_4 + e(1 - R_5)}{a + b + c + d + e} \tag{4.17}$$

4.1.3   Decision

The previously described screening processes will point to the best material. The decision will be based on selecting the one material that meets the requirements with the best balance of properties.

—— *Example 4.1* ——

Table 4.5 shows the materials that can be used for manufacturing the landing gear cylinder shown in Figure 4.3. Considering strength, rigidity, and cost, determine the appropriate material choice from the given alternates.

Solution

To determine the overall rating, for simplicity, assume that the weighting factors for the design requirements and the selection factor are unity. Also, assume that the landing gear cylinder is a thin-walled pressure vessel under internal pressure. Consequently, Table 4.6 can be constructed by the use of the information provided in Tables 4.2 to 4.5.

Table 4.5   Mechanical Properties of Material Candidates

Material	Yield Strength S_y (MPa)	Density ρ (tons/m^3)	Elastic Modulus E (GPa)	Cost C ($/ton)
High strength steel	1700	7.7	200	650
Aluminum alloy	400	2.7	71	700
Titanium	1100	4.5	120	5000

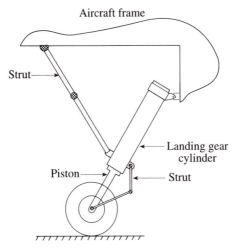

Figure 4.3   Aircraft landing gear mechanism.

The results in Table 4.6 substantiate that the stiffness/weight ratio will not affect the material selection decision, since the values for all choices are close (within 4%). Clearly, aluminum alloy can be eliminated from the selection process because of its low strength/weight ratio. The question is which material to select from the remaining two. Titanium has a high strength/weight ratio and low weight compared with high strength steel. However, because of the cost, overall rating of the steel is considerably higher than the titanium. Although the cost/weight ratio of the titanium is relatively high, the desire to reduce the weight of the aircraft may make it the favorable choice over the high strength steel.

Now consider the following typical cost analysis of a landing gear cylinder.[3]

- Net weight:

Steel	259 lb
Titanium	188 lb
Saving in weight	71 lb

Table 4.6   Design Requirements and Selection Factors for Overall Rating

Material	S_y/ρ		E/ρ		$C \times \frac{\rho}{S_y}$		Overall Rating $G = \frac{aR_1 + bR_2 + c(1 - R_3)}{a + b + c}$
	A_1	R_1	A_2	R_2	A_3	R_3	
High strength steel	221	0.91	26	0.96	2.9	0.14	0.91
Aluminum alloy	148	0.61	26	0.96	4.7	0.23	0.78
Titanium	244	1	27	1	20.5	1	0.67

[3] E. D. Verink, Jr., *Methods of Material Selection*, Gordon and Breach, New York, 1968.

- Relative cost (material plus manufacturing):

Steel	$4,710
Titanium	$9,900
Total increase in cost	$5,190

Hence, the increase in cost per pound of weight saved is

$$\text{Unit increase in cost} = \frac{5190}{71} = \$73 \text{ per pound}$$

From the above results, if the increase in cost can be justified, the choice would be titanium because of its high strength and light weight. Cost may be a less important factor for a military aircraft than for a commercial aircraft; hence, the choice of titanium might be appropriate for one but not the other. However, as discussed previously, before a final choice is made, the reliability, availability, manufacturability, and maintainability must be studied in detail.

4.2 DESIGN ANALYSIS FOR FATIGUE RESISTANCE

Fatigue resistance of a material (long life) is one of the most important design requirements for the selection of the materials. Materials are often found to have failed under the action of cyclic loading. Analysis often shows that the maximum stresses are below the ultimate strength of the material and frequently below the yield strength. Normally, the fatigue strength of materials is documented by S–N diagrams, obtained from constant amplitude fatigue tests. Fatigue life under variable amplitude loading is estimated by Miner's rule.[4] This theory is stated as follows:

$$\sum_{1}^{i} \left(\frac{n_i}{N_i} \right) = C^*$$

(4.18)

where n is the number of cycles of stress σ applied to the specimen and N is the life corresponding to σ. The value of the constant C^*, determined experimentally, is found to be in the range $0.7 \le C^* \le 2.2$. Many authorities suggest the use of $C^* = 1.0$. This gives a gross estimate of fatigue life over the load spectrum. Miner's rule is simple to use in evaluating the cumulative damage of structural components at stress levels exceeding the endurance limit. According to Miner's rule, if a specimen is subjected to a constant amplitude alternating load and fails after N_f cycles, each cycle expended $1/N_f$ fraction of the specimen's life. Suppose that a machine component is subjected to n_1 reversed stress cycles at a stress range level σ_1; n_2 reversed stress cycles at a stress range level σ_2, and so on. Assume that the number of cycles to failure at stress level σ_1 to be N_1 cycles, σ_2 to be N_2 cycles, and so on (Figure 4.4). Then, by Miner's rule, the machine component would be predicted to fail when

$$\frac{n_1}{N_1} + \frac{n_2}{N_2} + \ldots + \frac{n_i}{N_i} = 1$$

(4.19)

---

[4] M. A. Miner, "Cumulative Damage in Fatigue," *Transactions, ASME, Journal of Applied Mechanics,* vol. 12, pp. A159–A164, 1945.

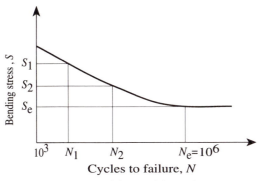

Figure 4.4 Method of fatigue life prediction.

Fatigue life prediction can be modified to include the effect of mean stresses (Figure 4.5). This is accomplished by making use of the modified Goodman diagram. The equivalent stress amplitude, σ_{eq}, is considered equal to the combination of stress amplitude, σ_a, and mean stress, σ_m, which results in the same fatigue life. The equivalent stress amplitude is given by

$$\sigma_{eq} = \sigma_a \left(\frac{S_{ut}}{S_{ut} - \sigma_m} \right) = \sigma_a F_{cr} \tag{4.20}$$

where
S_{ut} = ultimate tensile strength of the material
F_{cr} = $\frac{S_{ut}}{S_{ut} - \sigma_m}$ is called the correction factor

—— *Example 4.2* ————————————————————————

As shown in Figure 4.6, a riser is one of the major components of a system used for drilling offshore wells from floating vessels. Essentially, a riser is a long tensioned beam continuously loaded by waves, currents, and vessel motion. In deep-water drilling, in the upper and lower portions of the riser string, upper and lower ball joints are used to provide flexibility. It is obvious that the drill pipe undergoes a relatively sharp bend owing to deflection of the ball joints caused by vessel

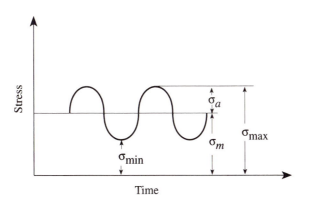

Figure 4.5 Sinusoidal fluctuating stress-time history.

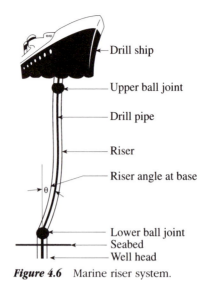

Figure 4.6 Marine riser system.

movement and hydrodynamic loads on the riser string. Bending stresses associated with the bend at the ball joints can cause fatigue damage to the drill pipe inside the riser.[5] By using classical fatigue theory, find the cumulative fatigue damage inflicted on the drill pipe during one pass through. One pass is defined as vertical movement of the drill pipe through 144 in. while being rotated. Assume that the drill pipe is subjected to a constant axial tension load of 100,000 lbf to prevent failure by buckling.

Given Design Data

Modulus of elasticity of drill-pipe material, E = 30 × 106 psi
Ultimate strength of drill-pipe material, S_u = 100 kpsi
Riser upper ball-joint angle, θ = 1°
Outside diameter of drill pipe, D_o = 5.00 in.
Inside diameter of drill pipe, D_i = 4.276 in.
Penetration speed of drill pipe, V = 120 in./h
Rotational speed of drill pipe, n = 100 rpm
Tension load, T = 100 kips (1 kip = 1000 lb)

Solution

An element of the deformed drill pipe is acted on by the forces shown in Figure 4.7, for which the bending moment equation is[5]

$$M(x) = \frac{T\theta}{k}(\cosh kx - \sinh kx) \tag{4.21}$$

where θ is the upper ball-joint angle and

$$k = \sqrt{\frac{T}{EI}} \tag{4.22}$$

[5] Adapted from A. Ertas et al., "The Effect of Tool Joint Stiffness on Drill Pipe Fatigue in Riser Ball Joints," *Transactions, ASME, Journal of Engineering for Industry*, vol. 111, no. 4, pp. 369–374, 1989.

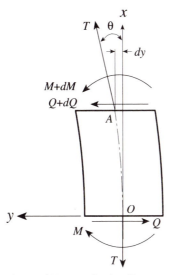

Figure 4.7 Free-body diagram of a drill pipe.

While the drilling operation continues, each element of the drill pipe is subjected to alternate bending stresses (compression and tension) due to rotation. From Eq. 4.21, this bending stress is maximum at the ball joint and decays exponentially with distance from the ball joint, both below and above. Consequently, when the drill pipe makes one pass through, the upper ball joint, as each joint of the drill pipe passes through the ball joint, suffers cumulative fatigue damage from stress cycles of widely varying amplitudes.

Computation of Constants

The moment of inertia of the drill pipe is given by

$$I = \frac{\pi(D_o^4 - D_i^4)}{64} = \frac{\pi(5^4 - 4.276^4)}{64} = 14.27 \text{in.}^4$$

For $T = 100,000$ lb and $E = 30 \times 10^6$ psi,

$$k = \sqrt{\frac{T}{E \times I}} = \sqrt{\frac{100,000}{(30 \times 10^6)(14.27)}} = 0.01528 \text{ in.}^{-1}$$

The cross-sectional area of the drill pipe is

$$A = \frac{\pi(D_o^2 - D_i^2)}{4} = \frac{\pi(5^2 - 4.276^2)}{4} = 5.275 \text{ in.}^2$$

The mean stress σ_m, due to tension load of 100 kips, is

$$\sigma_m = \frac{T}{A} = \frac{100}{5.275} = 18.96 \text{ kpsi}$$

The correction factor F_{cr}, which takes into account the mean stress, is

$$F_{cr} = \frac{S_{ut}}{S_{ut} - \sigma_m} = \frac{100}{100 - 18.96} = 1.234$$

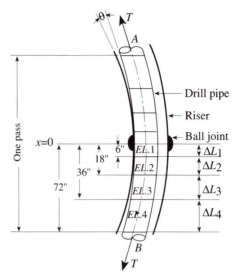

Figure 4.8 Shematic of a drill pipe in a ball joint.

The following steps are performed to determine the cumulative fatigue damage inflicted on the drill pipe during one pass through the upper ball joint:

Step 1. Divide a joint of the drill pipe into a number of finite elements. As shown in Figure 4.8, smaller elements are taken close to the higher stress region. As we will learn later, the fatigue damage calculation will show that the bending moment beyond a distance of 72 in. of the ball joint has no significant contribution to the life of a drill pipe.

Step 2. Calculate the bending moment using Eq. 4.21. For $\theta = 1°(0.01744$ rad) the bending moment M at $x = 0$ is

$$M(x = 0) = (100,000)\left(1° \times \frac{\pi}{180}\right)\left(\frac{1}{0.01528}\right)[\cosh(0.01528)(0)$$
$$- \sinh(0.01528)(0)]$$
$$M(x = 0) = 114.20 \text{ kips} - \text{in.}$$

Similarly, the calculated bending moment at $x = 6$ in. is $M(x = 6) = 104.22$ kips-in. The average bending moment at the first element, M_{ave} (El.1), can now be calculated as

$$M_{ave}(\text{EL.1}) = \frac{114.20 + 104.22}{2} = 109.21 \text{ kips} - \text{in.}$$

Step 3. Determine the average alternating (reversed) bending stress $\sigma_{av}(x)$, from the following relation

$$\sigma_{av}(\text{EL.1}) = \frac{M_{ave}(\text{EL.1}) \times D_o}{2 \times I} = \frac{109.21 \times 5}{2 \times 14.27} = 19.13 \text{ kpsi}$$

Step 4. Determine the equivalent stress, σ_{eq}, by the use of the equation representing the modified Goodman line to account for the mean stress due to nominal tension.

$$\sigma_{eq}(\text{El.1}) = F_{cr}\sigma_{av}(\text{EL.1}) = 1.234 \times 19.13 = 23.61 \text{ kpsi}$$

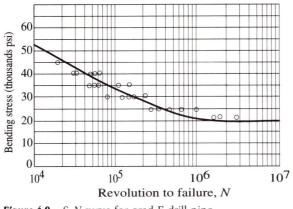

Figure 4.9 S–N curve for grad E drill pipe.

This is done because a tensile load superimposed on a drill pipe experiencing alternating bending stress reduces its endurance limit. Hence, the fatigue effect of bending becomes more severe because of this tensile load.

Step 5. Evaluate the number of stress reversals n_1 for the first element from the following relations.

$$n_1 = \frac{\text{rpm} \times 60 \times \Delta L_1}{V} = \frac{100 \times 60 \times (6 \text{ in.} - 0 \text{ in.})}{120} = 300 \text{ cycles}$$

Step 6. Find the number of cycles N_1 for failure at the stress level of 23.61 kpsi for the first element from the S–N curve shown in Figure 4.9.

$$N_1 = 500,000 \text{ cycles}$$

Step 7. Apply Miner's rule to evaluate the cumulative fatigue damage for the first element.

$$\frac{n_1}{N_1} \times 100 = \frac{300}{500,000} \times 100 = 0.06\%$$

This procedure is repeated for the remaining elements, and the results are summarized in Table 4.7. Note that Figure 4.9 clearly shows that the life is infinite at stresses below 20 kpsi. Consequently, at these stresses the damage is zero.

Table 4.7 Fatigue Damage Calculation

EL. No.	x (in.)	$M(x)$ (kips-in.)	M_{ave} (kips-in.)	σ_{av} (kpsi)	σ_{eq} (kpsi)	n (cycle)	N (cycle)	$(n/N) \times 100$ %
1	$x = 0$	114.20	109.22	19.13	23.61	300	5×10^5	0.06
	$x = 6$	104.22						
2	$x = 6$	104.22	95.49	16.73	20.64	600	inf.	0.0
	$x = 18$	86.76						
3	$x = 18$	86.76	76.33	13.37	16.50	900	inf.	0.0
	$x = 36$	65.90						
4	$x = 36$	65.90	51.95	9.10	11.23	1800	inf.	0.0
	$x = 72$	38.00						

The total damage to the drill pipe during its passage through the location beginning 72 in. above and ending 72 in. below (one pass) the upper joint can be calculated as follows:

$$\text{Total damage} = 2 \times \left(\frac{n_1}{N_1} + \frac{n_2}{N_2} + \frac{n_3}{N_3} + \frac{n_4}{N_4}\right) \times 100$$

$$= 2 \times \left(0.06 + 0 + 0 + 0\right) = 0.12\%$$

4.3 DESIGN ANALYSIS USING FRACTURE MECHANICS

Fracture mechanics is used to investigate the failure of brittle materials: to study material behavior and design against brittle failure and fatigue. Catastrophic failure can occur in engineering structures because of brittle fracture of ferrous materials and aluminium alloys as well as other materials. These failures arise as a consequence of unstable crack propagation from a preexisting defect owing to material processing or fabrication. Even when the designer considers the likelihood of catastrophic failure as a result of material defects, it is almost impossible to eliminate brittle failure totally because of the complex interrelation between the material, the design, and fabrication. Failures by fracture usually occur under loadings below the yield strength of the material. Three possible material separation modes for crack extension under external load are shown in Figure 4.10. In engineering practice, mode II, in-plane shear, and mode III, out-of-plane shear, have limited application. Hence, only mode I, tensile opening mode, is discussed in the following section.

4.3.1 Mode I Stress State in a Crack

The proper employment of linear elastic fracture mechanics (LEFM) is used to provide preliminary design guidelines and material selection in the following discussion. Consider a plate, as shown in Figure 4.11, subjected to a tensile stress σ

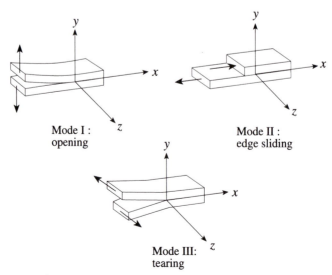

Figure 4.10   Fracture modes.

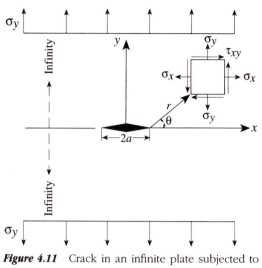

Figure 4.11 Crack in an infinite plate subjected to a tensile stress.

at infinity with a crack of length $2a$. An element $dx\,dy$ of the plate at a distance r from the crack tip and at an angle θ with respect to the crack plane experiences the following stress field[6]

$$\sigma_x = \frac{K_I}{\sqrt{2\pi r}} \cos\frac{\theta}{2}\left[1 - \sin\frac{\theta}{2}\sin\frac{3\theta}{2}\right] \tag{4.23}$$

$$\sigma_y = \frac{K_I}{\sqrt{2\pi r}} \cos\frac{\theta}{2}\left[1 + \sin\frac{\theta}{2}\sin\frac{3\theta}{2}\right] \tag{4.24}$$

$$\tau_{xy} = \frac{K_I}{\sqrt{2\pi r}} \sin\frac{\theta}{2}\cos\frac{\theta}{2}\cos\frac{3\theta}{2} \tag{4.25}$$

$$\sigma_z = \nu(\sigma_x + \sigma_y), \quad \tau_{xy} = \tau_{yx} = 0 \quad \text{for plane strain} \tag{4.26}$$

$$\sigma_z = 0 \qquad\qquad\qquad\qquad\qquad \text{for plane stress} \tag{4.27}$$

where K_I is called the *stress intensity factor*. It relates the load, crack size, and structural geometry to stresses near a crack tip, and can be expressed as follows:

$$K_I = \sigma\sqrt{\pi a} \tag{4.28}$$

For a ductile material, when the normal stress σ equals S_y, the material becomes unstable and plastic deformation occurs. A similar analogy can be applied to brittle material such that when the stress-intensity factor K_I reaches the critical stress-intensity factor K_{IC}, significant crack growth occurs. Therefore, the designer must keep the K_I value below K_{IC} in the same manner that the normal stress σ is below the yield strength S_y. Once the value of K_{IC} for a specific material of a particular thickness is known, the designer can determine the crack size that can be tolerated in structural members for a given stress level. Conversely, the design stress level can be determined for an existing crack size in the structural member.

---

[6] G. R. Irwin, "Analysis of Stresses and Strains Near the End of a Crack Traversing a Plate," *Transactions, ASME, Journal of Applied Mechanics*, vol. 24, 1957.

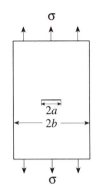

Figure 4.12 Finite-width plate with a through-thickness crack.

4.3.2 Stress–Intensity Factor Equations

For the simplest case in which loading is in the elastic range (linear elastic fracture mechanics), the stress intensity factor K_I, given in Eq. 4.28 for *plane strain* can be modified for different geometries as follows:[7]

Central Through-Thickness Crack The stress–intensity factor for a plate with finite width $2b$, which contains a central, through-thickness crack of length $2a$, subjected to uniform uniaxial tension (Figure 4.12), is given by

$$K_I = \sigma \sqrt{\pi a} f(a/b) \tag{4.29}$$

where $f(a/b)$ is called the correction factor given as

$$f(a/b) = 1 - 0.1\left(\frac{2a}{2b}\right) + \left(\frac{2a}{2b}\right)^2 \tag{4.30}$$

Equation 4.30 is accurate to within 1% over the range of $2a/2b$ between 0 and 0.6.

Double-Edge Through-Thickness Crack The stress-intensity factor for a double-edge notched plate subjected to uniform uniaxial tension, as shown in Figure 4.13, is also given by Eq. 4.29. The correction factor $f(a/b)$ is

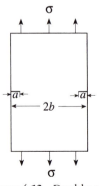

Figure 4.13 Double-edge through-thickness crack.

[7] William F. Brown Jr. and John E. Srowley, "Plane Strain Crack Toughness Testing of High Strength Metallic Materials," *ASTM Special Technical Publication,* No. 410, 1967, pp. 11–12.

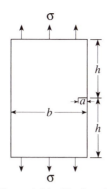

Figure 4.14 Single-edge through-thickness crack.

$$f(a/b) = 1.117 + 0.203\left(\frac{2a}{2b}\right) - 1.196\left(\frac{2a}{2b}\right)^2 + 1.930\left(\frac{2a}{2b}\right)^3 \qquad (4.31)$$

Eq. 4.31 is accurate to within 1% over the range of $2a/2b$ between 0 and 0.7.

Single-Edge Through-Thickness Crack   Equation 4.29 can also be used to find the stress–intensity factor for a single-edge notched plate subjected to uniform uniaxial tension as shown in Figure 4.14. The correction factor $f(a/b)$ is

$$f(a/b) = 1.223 - 0.231\left(\frac{a}{2b}\right) + 10.550\left(\frac{a}{2b}\right)^2 - 21.710\left(\frac{a}{2b}\right)^3 + 30.382\left(\frac{a}{2b}\right)^4 \quad (4.32)$$

Eq. 4.32 is accurate for all values of $a/2b$ up to 0.6 to within 0.4%.

—— *Example 4.3* ——————————————————————————

As shown in Figure 4.15, a wide and long steel plate with small thickness is subjected to a tensile stress. If the steel has a critical stress–intensity factor $K_{IC} = 28$ MPa $\sqrt{m}$, calculate the tensile stress for failure. The steel plate has a through-thickness crack with $a = 45$ mm.

Solution

Since the plate is long and wide, assume a through crack at the center of the plate. Hence, using the infinite plate formula the stress–intensity equation $K_I = \sigma\sqrt{\pi a}$

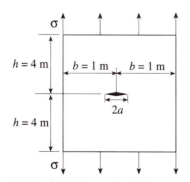

Figure 4.15 Through thickness-crack in an infinite plate.

can be modified to calculate the critical stress–intensity factor, K_{IC},

$$K_{IC} = \sigma_{max} \sqrt{\pi a} \tag{4.33}$$

where σ_{max} is the critical failure stress; hence,

$$\sigma_{max} = \frac{K_{IC}}{\sqrt{\pi a}} = \frac{28 \sqrt{10^3 \text{ mm}}}{\sqrt{\pi \times 45 \text{ mm}}} = 74.49 \text{ MPa}$$

Elliptical Crack in an Infinite Plate   An analysis of an embedded elliptical or semielliptical flaw in structural members is of great importance for many practical engineering applications. Flaws easily develop in structural components, such as pressure vessels and pipelines, from small defects and material imperfections. The widely used approximation[8] for the stress–intensity factor K_I at any point along the perimeters of elliptical or circular flaws that are embedded in an infinite body subjected to uniform uniaxial tension (Figure 4.16) is expressed by

$$K_I = \frac{\sigma \sqrt{\pi a}}{\Phi} \left(\sin^2 \theta + \frac{a^2}{c^2} \cos^2 \theta \right)^{1/4} \tag{4.34}$$

where Φ is the elliptical integral given by

$$\Phi = \int_0^{\pi/2} \sqrt{1 - k^2 \sin^2 \phi} \, d\phi \quad \text{and} \quad k^2 = 1 - \frac{a^2}{c^2} \tag{4.35}$$

The elliptical integral Φ can be expressed in an expansion series as

$$\Phi = \frac{\pi}{2} \left[1 - \frac{1}{4} \left(\frac{c^2 - a^2}{c^2} \right) - \frac{3}{64} \left(\frac{c^2 - a^2}{c^2} \right)^2 - \cdots \right] \tag{4.36}$$

Neglecting higher-order terms, Φ can be approximated as

$$\Phi = \frac{3\pi}{8} + \frac{\pi}{8} \frac{a^2}{c^2} \tag{4.37}$$

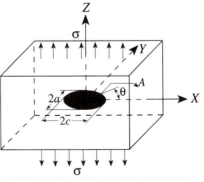

Figure 4.16   Embedded crack in an infinite body.

[8] G. R. Irwin, "The Crack Extension Force for a Part Through Crack in a Plate," *Transactions, ASME, Journal of Applied Mechanics,* vol. 29, no. 4, pp. 651–654, 1962.

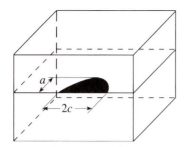

Figure 4.17   Semi-elliptical surface flaw.

Including the back crack free-surface correction factor and the plastic zone correction factor, Eq. 4.34 can be modified for the semielliptical surface flaw shown in Figure 4.17.

$$K_I = 1.1 \frac{\sigma}{\Phi} \sqrt{\pi(a + r_p^*)} \left(\sin^2 \theta + \frac{a^2}{c^2} \cos^2 \theta \right)^{1/4} \tag{4.38}$$

where the plastic zone correction r_p^* is given by

$$r_p^* = \frac{K_I^2}{4\pi \sqrt{2} S_y^2} \tag{4.39}$$

Equation 4.38 can be further modified for the maximum value of the stress–intensity factor at the minor axis ($\theta = \pi/2$) as[8]

$$K_I = 1.1\sigma \sqrt{\pi \frac{a}{Q}} \tag{4.40}$$

where Q is the flow-shape parameter given as

$$\sqrt{Q} = \sqrt{\Phi^2 - 0.212(\frac{\sigma}{S_y})^2} \tag{4.41}$$

Finally, a magnification factor M_k for deep flaws can be used to find the maximum stress intensity for a surface flaw

$$K_I = 1.1 M_k \sigma \sqrt{\pi \frac{a}{Q}} \tag{4.42}$$

To obtain the magnification factor M_k, Figure 4.18 can be used.[9] For catastrophic failure, the initial crack depth a must reach the critical crack depth a_{cr}. This critical crack depth can be evaluated by using Eq. 4.42 yielding

$$a_{cr} = \left(\frac{K_{IC}}{1.1 M_k \sigma_{max}} \right)^2 \frac{Q}{\pi} \tag{4.43}$$

[9] A. S. Kobayashi, M. Zii, and L. R. Hall, "Approximate Stress Intensity Factor for an Embedded Elliptical Crack Near to Parallel Free Surface, *International Journal of Fracture Mechanics,* vol. 1, pp. 81–95, 1965.

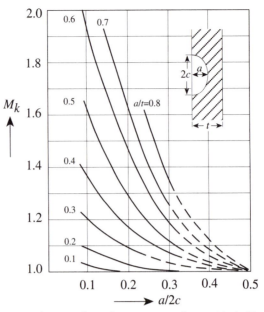

Figure 4.18   Kobayashi correction factor M_k (with the permission of Kluwer Academic publisher).

4.3.3   Leak before Break

The *leak-before-break* criterion was first proposed by Irwin et al.[10] This concept is used to estimate the material fracture toughness K_{IC}, required for a surface flaw to grow through the thickness t, thus allowing the pressure vessel to fail from leakage rather than total fracture. As shown in Figure 4.19, the first mode of failure (leakage) assumes that a flaw twice the wall thickness in length should be stable at a stress equal to the design stress.[11] That is, the critical crack size at the nominal design stress level of a material should be greater than the wall thickness.

— *Example 4.4* —

Compressed natural gas (CNG) is an alternative to petroleum-based fuels for the transportation sector. Because of its very low energy density, CNG must be compressed in a cylinder to a pressure of 16–25 MPa to store it on board a vehicle. If the inside diameter and the length of a CNG pressure vessel are 0.3 m and 1.3 m, respectively, using the principles of fracture mechanics, select the most suitable

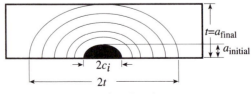

Figure 4.19   Leak-before-break criteria.

[10]   G. R. Irwin, *Materials for Missiles and Spacecraft*, McGraw-Hill, New York, 1963.

[11]   S. T. Rolfe and J. M. Barsom, *Fracture and Fatigue Control in Structures; Application of Fracture Mechanics*, Prentice-Hall, Englewood Cliffs, NJ, 1977.

Table 4.8   Materials for Pressure Vessel Design

Material	S_y (MPa)	K_{IC} (MPa $\sqrt{m}$)	ρ tons/m³	E (GPa)	Cost $/ton
Steel 1	1700	83	7.7	210	600
Steel 2	1525	58	7.7	210	450
Steel 3	750	120	7.7	210	400
Al. alloy 1	420	24	2.7	71	850
Al. alloy 2	420	40	2.7	71	900
Titanium alloy	1075	51	4.5	120	5000

material from Table 4.8. Assume an initial surface flaw of depth 2 mm and an $a/2c$ ratio of 0.25 that remains constant. Perform the analysis based on a constant mean pressure of 24 MPa.

Solution

The mode of failure due to a flaw on the surface of a pressure vessel is that of tensile separation. The wall of the vessel can be modeled as a plate of infinite length and finite thickness with a semielliptical surface crack propagating through the thickness of the plate (Figure 4.20).

Analysis of Steel 1   To find the design stress σ_d that the pressure vessel can withstand, assume an allowable design stress of

$$\sigma_d = 0.5 S_y$$

The stress–intensity factor from Eq. 4.42 can be modified to a critical stress–intensity factor by using the maximum design stress that the CNG vessel can withstand, as

$$K_{IC} = 1.1 M_k \sigma_{max} \sqrt{\pi \frac{a}{Q}} \tag{4.44}$$

Let $\sigma_{max} = \sigma_d$ and rearrange Eq. 4.44,

$$\sigma_d = \frac{\sqrt{Q} K_{IC}}{1.1 M_k \sqrt{\pi a}}$$

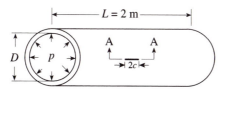

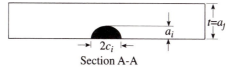

Figure 4.20   Semielliptical flaw in pressure vessel wall.

where

$$\sqrt{Q} = \sqrt{\Phi^2 - 0.212(\sigma_d/S_y)^2}$$

and

$$\Phi = \frac{3\pi}{8} + \frac{\pi}{8}\left(\frac{a}{c}\right)^2 = \frac{3\pi}{8} + \frac{\pi}{8}(0.5)^2 = 1.2763$$

Then

$$\sqrt{Q} = \sqrt{(1.2763)^2 - 0.212(0.5)^2} = 1.2554$$

Since the design stress, σ_d, is a function of assumed values of M_k and ratio of σ_d/S_y, an iterative procedure must be used. Assuming an error criterion, $\epsilon_{er} \leq 2\%$, the following steps are used to determine the converged values of design stress. For the first iteration, assume that $M_k = 1.0$; hence,

$$\sigma_d = \frac{(1.2554)(83)}{(1.1)(1.0)\sqrt{\pi 0.002}} = 1195 \text{ MPa}$$

From the thin-walled pressure vessel theory

$$\sigma_d = \frac{pD}{2t}$$

rearranging yields

$$t = \frac{pD}{2\sigma_d} = \frac{(24)(0.3)}{2(1195)} = 0.00301 \text{ m}$$

which yields

$$\frac{a}{t} = \frac{0.002}{0.00301} = 0.664$$

Referring to Figure 4.18, $M_k = 1.36$, and the new ratio is

$$\frac{\sigma_d}{S_y} = \frac{1195}{1700} = 0.703$$

The next step is to recalculate $\sqrt{Q}$ by using the new value of σ_d/S_y

$$\sqrt{Q} = \sqrt{(1.2763)^2 - 0.212(0.703)^2} = 1.234$$

Thus

$$\sigma_d = \frac{(1.234)(83)}{(1.1)(1.36)\sqrt{\pi 0.002}} = 864.3$$

Now, the new calculated values can be compared with the old values to find the error:

$$\% \quad \text{Error} \quad \sigma_d = \frac{|1195 - 864.3|}{1195.2} \times 100 = 27.7\%$$

$$\% \quad \text{Error} \quad M_k = \frac{|1.0 - 1.36|}{1.0} \times 100 = 36\%$$

For the next iteration, bisect the range of σ_d/S_y as well as the range of M_k for the initial guesses:

$$\frac{\sigma_d}{S_y} = \frac{0.5 + 0.703}{2} = 0.601$$

$$M_k = \frac{1.0 + 1.36}{2} = 1.18$$

The same procedure is used for the next iterations. After three iterations, convergence is achieved, and the results for steel 1 are

$$\sigma_d = 959 \text{ MPa}$$
$$\sigma_d/S_y = 0.56$$
$$M_k = 1.240$$
$$t = 0.00375 \text{ m}$$

Then, the cost of constructing the pressure vessel out of Steel 1, neglecting the end caps, is

$$V = \frac{\pi}{4}\left(D_o^2 - D_i^2\right) \times L$$

where

$$D_o = D_i + 2t = 0.3 + 2(0.00375) = 0.3075 \text{ m}$$

thus

$$V = \frac{\pi}{4}\left(0.3075^2 - 0.3^2\right) \times 1.3 = 0.00465 \text{ m}^3$$

and the mass of the pressure vessel is

$$m = \rho V = (7.7)(0.00465) = 0.0358 \text{ tons}$$

Consequently, the cost is

$$\text{Cost} = (600)(0.0358) = \$21.48$$

The same calculations are repeated for the other materials given in Table 4.8, and the results are presented in Table 4.9. From the results it is evident that an increase in strength allows less thickness; consequently, less material is used. An

Table 4.9 Calculations of Design Parameters

Material	S_y	K_{IC}	σ_d	S_y/σ_d	M_k	Q	t (mm)	m (tons)	Total Cost ($)
Steel 1	1700	83	959	1.8	1.24	1.56	3.75	0.0358	21.48
Steel 2	1525	58	753	2.0	1.11	1.59	4.78	0.0459	20.64
Steel 3	750	120	1057	0.71	1.37	1.11	3.41	0.0352	13.01
Alluminum alloy 1	420	24	329	1.3	1.02	1.49	10.94	0.0375	31.91
Alluminum alloy 2	420	40	495	0.85	1.05	1.21	7.27	0.0246	22.17
Titanium alloy	1075	51	673	1.6	1.08	1.55	5.35	0.0300	150.10

increase in toughness yields higher design stress, which provides satisfactory service life. However, caution is essential when high toughness is required for a design. For example, the high toughness of steel 3 yields a higher design stress, which gives a design factor of safety less than unity. The same argument can be made for aluminum alloy 2. Hence, these two materials are eliminated as candidates. Among the remaining materials, titanium gives the lowest weight. The weight and bulk of onboard storage are often cited as drawbacks to CNG use. Although titanium seems to be the solution, a pound of weight saved does not justify the high cost. This eliminates titanium as a choice.

Now consider the weighting factors (in order of importance) for the final rating of the remaining three materials as shown in Table 4.10. It must be pointed out that the weight is one of the most important factors affecting the vehicle's performance. Hence, a high weighting factor (4) is assigned to the weight. The second highest weighting (3) is applied to the design–safety factor. Toughness-to-stress ratio is assumed to be less important in the overall rating. Hence, a weighting factor of 1.0 is used for this ratio. From Table 4.10, the results of the overall evaluation show that high strength steel 1 appears to be the best material to use. Although the higher safety factor of steel 2 is attractive, the increased weight is significant for the vehicle performance. Steel 1 offers good all-around characteristics with moderate cost, a reasonable safety factor, and low weight. Hence, for this application steel 1 is chosen.

4.3.4 Fatigue Crack Propagation

Classical analytical tools like *S–N* curves and Goodman diagrams are primarily based on fatigue life predictions from statistical failure data. In recent years, fracture mechanics has been employed in modeling more accurate analytical tools for

Table 4.10 Overall Rating

Material	S_y/ρ		E/ρ		$\frac{K_{IC}}{S_y}\sqrt{mm}$		S_y/σ_d		$C(\rho/S_y)$		m (kg)		Overall Rating $G = \frac{2R_1+2R_2+R_3+3R_4+2(1-R_5)+4(1-R_6)}{14}$
	A_1	R_1	A_2	R_2	A_3	R_3	A_4	R_4	A_5	R_5	A_6	R_6	
Steel 1	221	1.00	27.3	1.00	1.54	0.85	1.77	0.87	2.72	0.50	35.8	0.78	0.67
Steel 2	198	0.90	27.3	1.00	1.20	0.66	2.03	1.00	2.27	0.42	45.9	1.00	0.62
Aluminum alloy 1	156	0.71	26.3	0.96	1.81	1.00	1.28	0.63	5.46	1.00	37.5	0.82	0.50

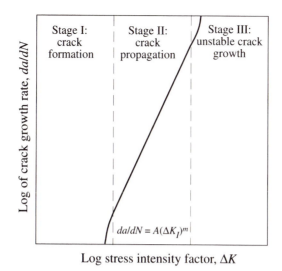

Figure 4.21 Fatigue crack propagation stages.

prediction of fatigue life. These models describe the three stages of fatigue crack propagation (Figure 4.21) as follows:

1. Stage I: crack formation (nonpropagating fatigue cracks).
2. Stage II: crack propagation.
3. Stage III: unstable crack growth and fracture.

Stage II represents the fatigue–crack propagation behavior, which permits the use of the power–law relationship to calculate fatigue life. By using the power–law relationship, the fatigue–crack growth rate da/dN of a material in terms of the range of applied stress–intensity factor ΔK can be written as

$$\frac{da}{dN} = A(\Delta K_I)^m \qquad (4.45)$$

where
  a = crack depth
  N = number of cycles
  ΔK_I = stress–intensity factor occurring at the crack tip

A and m are the constants for a particular material, environment, and loading condition.

Procedure to Analyze Crack Growth

1. Assume an initial flaw size a_i.
2. Calculate the critical crack size a_{cr}, that would cause catastrophic failure.
3. Assume an increment of crack growth Δa.

4. Determine ΔK_I, using the proper equation for K_I, with $\Delta\sigma$ and a_{av}, where

$$a_{av} = a_i + \frac{\Delta a}{2} \qquad (4.46)$$

5. Determine the number of cycles at a given stress level by using Eq. 4.45. Direct integration should continue until a reaches a_{cr}. If a_{cr} is larger than the thickness of the material, integration should stop when a is equal to the material thickness.

▬ Example 4.5

Using the theory of fracture mechanics, find the cumulative fatigue damage inflicted on the drill pipe of Example 4.2 during one pass through the riser upper ball joint.[12]

Given Design Data

Modulus of elasticity of drill-pipe material, E = 30×10^6 psi

Critical stress–intensity factor, K_{IC} = 56 kpsi $\sqrt{\text{in.}}$

Yields strength of drill-pipe material, σ_{ys} = 90 kpsi

Crack growth (remains constant), $\frac{a}{c}$ = 0.5

Penetration speed of drill pipe, V = 120 in./h

Outside diameter of drill pipe, D_o = 5.00 in.

Inside diameter of drill pipe, D_i = 4.276 in.

Initial crack size, a_i = 0.0312 in.

Rotational speed of drill pipe, n = 100 rpm

Tension load, T = 100 kips

Solution

The drill pipe is assumed to have a semielliptical (part-through) surface crack, as shown in Figure 4.22. This type of crack is typical of the starting-crack shape customarily found. As shown, the drill pipe is assumed to be cut along a longitudinal plane and opened into a plate. On the basis of quality inspection, the estimated initial flaw size a_i present in the drill pipe is assumed to be 0.0312 in. The critical stress–intensity factor K_{IC} is assumed to be 56 kpsi $\sqrt{\text{in.}}$

The sinusoidal fluctuating stress-time history for a drill pipe is shown in Figure 4.23. From this figure, the stress range of the drill pipe is observed to be

$$\sigma_r = \Delta\sigma = \sigma_{max} - \sigma_{min}$$

where

$$\sigma_{max} = \sigma_m + \sigma_{ave}$$

[12] A. Ertas et al., "A Comparison of Fracture Mechanics and S–N Curve Approaches in Designing Drill Pipe," *ASME, Offshore and Arctic Operations Symposium Proceedings*, PD-vol. 26, pp. 45–50, 1989.

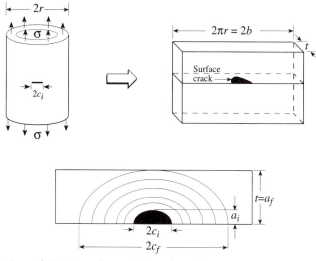

Figure 4.22 Semielliptical (part-through) surface crack in drill pipe.

and

$$\Delta\sigma = \sigma_r = 2\sigma_{ave}$$

Table 4.11 is obtained by using σ_m = 18.96 kpsi (see Example 4.2) and the values of σ_{av} from Table 4.7. The following steps are performed to determine the cumulative fatigue damage.

Step 1.   Divide a joint of the drill pipe into a number of finite elements as shown in Figure 4.8.

Step 2.   Calculate the average stress amplitude σ_{av} for each of these finite elements (see Table 4.11).

Step 3.   Calculate the critical flaw size a_{cr} that would cause failure by brittle fracture for the first element as follows:

As discussed above, for catastrophic failure, the initial crack depth a_i must reach the critical crack depth calculated using

$$a_{cr} = \left(\frac{K_{IC}}{1.1 M_k \sigma_{max}}\right)^2 \frac{Q}{\pi}$$

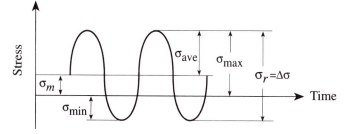

Figure 4.23   Stress-time history of a drill pipe.

Table 4.11 Drill-Pipe Stress Calculations

EL No.	σ_{av} (kpsi)	$\Delta\sigma$ (kpsi)	σ_{max} (kpsi)
El 1	19.13	38.26	38.09
El 2	16.73	33.46	35.69
El 3	13.37	26.74	32.33
El 4	9.10	18.20	28.06

where σ_{max} = 38.09 kpsi, for element 1. Using Eq. 4.37,

$$\Phi = \frac{3\pi}{8} + \frac{\pi\,a^2}{8\,c^2} = \frac{3\pi}{8} + \frac{\pi}{8}(0.5)^2 = 1.276$$

For simplicity, M_k is assumed to be a unity, and

$$\sqrt{Q} = \sqrt{\Phi^2 - 0.212\left(\frac{\sigma_{max}}{S_y}\right)^2}$$

then

$$Q = (1.276)^2 - 0.212\left(\frac{38.09}{90}\right)^2 = 1.59$$

Hence,

$$a_{cr} = \left(\frac{56}{1.1 \times 38.09}\right)^2 \frac{1.59}{\pi} = 0.905 \text{ in.}$$

Since the critical crack size is larger than the wall thickness, the drill pipe will fail by leaking and not by catastrophic propagation to failure.

Step 4. Assume an increment of crack growth Δa and determine the number of failure cycles for the crack to grow by this increment at the given stress level for each finite element by using Eq. 4.45 with $A = 0.614 \times 10^{-10}$ and $m = 3.16$.[13]

$$\frac{da}{dN} = 0.614 \times 10^{-10}(\Delta K_I)^{3.16} \tag{4.47}$$

where ΔK_I can be obtained from Eq. 4.42 as

$$\Delta K_I = 1.1\Delta\sigma\sqrt{\pi\frac{a_{av}}{Q}} \tag{4.48}$$

To determine the incremental crack growth, Eq. 4.47 can be modified to

$$\frac{\Delta a}{\Delta N} = 0.506 \times 10^{-9}\left[\Delta\sigma\sqrt{\frac{a_{av}}{Q}}\right]^{3.16} \tag{4.49}$$

---

[13] E. Kral et al., "Fracture Mechanics Estimates," *Oil and Gas Journal Technology*, pp. 50–55, 1984.

where ΔN is the number of cycles required to propagate the incremental crack growth Δa, and

$$a_{av} = a_i + \frac{\Delta a}{2}$$

Hence, the number of cycles ΔN, required to propagate the crack from initial depth a_i to depth $a_i + \Delta a$, can be written as

$$\Delta N = \frac{1.976 \times 10^9 \Delta a}{\left(\Delta \sigma \sqrt{\frac{a_i + 0.5 \Delta a}{Q}}\right)^{3.16}}$$

Assume $\Delta a = 0.01$ in.; the first iteration gives

$$\Delta N = \frac{1.976 \times 10^9 (0.01)}{\left(38.26 \sqrt{\frac{0.0312 + 0.5 \times 0.01}{1.59}}\right)^{3.16}} \approx 77,600 \text{ cycles}$$

For the second iteration, initial crack size will be $a_i = 0.0312 + 0.01 = 0.0412$ in. Hence, the number of cycles for the next $\Delta a = 0.01$ increment is

$$\Delta N = \frac{1.976 \times 10^9 (0.01)}{\left(38.26 \sqrt{\frac{0.0412 + 0.5 \times 0.01}{1.59}}\right)^{3.16}} \approx 52,800 \text{ cycles}$$

Since the critical crack size was found to be larger than the wall thickness of the drill pipe, iteration for incremental crack growth should stop when it reaches a final crack size a_f, which is equal to the wall thickness of 0.362 in.

$$N(\text{EL.1}) = \sum \Delta N = 77,600 + 52,800 + \cdots = 400,281 \text{ cycles}$$

The damage for the first element is

$$\frac{n_1}{N_1} = \frac{300}{400,281} \times 100 = 0.075\%$$

The same procedure is repeated for the remaining elements, and the summary of the results is given in Table 4.12. Note that, since the stress is changing, Q and a_{cr} should be recalculated for each element.

The total damage to the drill pipe during one pass through the upper joint is

$$\text{Total damage} = 2 \times \left(\frac{n_1}{N_1} + \frac{n_2}{N_2} + \frac{n_3}{N_3} + \frac{n_4}{N_4}\right) \times 100$$

$$= 2 \times (0.075 + 0.098 + 0 + 0) = 0.346\%$$

Table 4.12 Fatigue Damage Calculations

EL No.	σ_{av} (kpsi)	$\Delta \sigma$ (kpsi)	σ_{max} (kpsi)	n (cycles)	N (cycles)	n/N %
EL 1	19.13	38.26	38.09	300	400,281	0.075
EL 2	16.73	33.46	35.69	600	611,994	0.098
EL 3	13.37	26.74	32.33	900	1.24×10^6	0
EL 4	9.10	18.2	28.06	1800	4.18×10^6	0

The results of Examples 4.2 and 4.5 show that the fracture mechanics approach provides conservative results in determining the fatigue damage of drill pipe at low-tension load and small riser angles θ.

4.3.5 Design with Materials

Since material selection is an integral part of design, the designer must be aware of each class of material and its advantages, disadvantages, and design limitations. As can be seen from Table 4.13,[14] composites offer impressive properties and can be used in many design applications. However, cost and problems related to the manufacturing of composite structures are primary barriers to large-scale use.

Most materials are subjected to environmental effects such as sunshine, temperature, rainfall, and wind; metals show the least resistance against corrosion. Although the toughness of polymers is often low, they find a wide range of applications because of their superior corrosion resistance and low coefficient of friction. Polymers can be formed and manufactured in almost any shape, and the components can be designed to snap together, making assembly fast and cheap.

In the past decade, research has been under way on the development of high-performance engineering ceramics. These materials have an important role as tool and die materials as well as in engine components such as turbochargers and valves. Ceramics have the highest hardness of all the solid materials. For

Table 4.13   Design-Limiting Properties of Materiels

Material	Good	Poor
Metals High E, K_{IC}	Stiff ($E \approx 100$ GPa) Ductile ($c_f \approx 20\%$) → formable	Yield (pure, $S_y \approx 1$ MPa)→ alloy Hardness ($H \approx 3S_y$) → alloy
Low S_y	Tough ($K_{IC} > 50$ MPa m$^{1/2}$) High MP ($T_m \approx 1000°$C) T-shock (ΔT_i 5000°C)	Fatigue strength ($S_e = \frac{1}{2}S_y$) Corrosion resistance→ coatings
Ceramic High E, S_y	Stiff ($E \approx 200$ GPa) Very high yield, hardness ($S_y > 3$ GPa)	Very low toughness ($K_{IC} \approx 2$ MPa m$^{1/2}$) T-shock ($\Delta T_i \approx 200°$C)
low K_{IC}	High MP ($T_m \approx 200°$C) Corrosion resistant Moderate density	Formability→ powder methods
Polymers Adequate S_y, K_{IC} Low E	Ductile and formable Corrosion resistant Low Density	Low stiffness ($E \approx 2$ GPa) Yield ($S_y = 2$-100 MPa) Toughness often low (1 MPa m$^{1/2}$)
Composites High E, S_y, K_{IC} but cost	Stiff ($E > 50$ GPa) Strong ($S_y \approx 200$ MPa) Tough ($K_{IC} > 20$ MPa $\sqrt{m}$) Fatigue resistant Corrosion resistant Low density	Formability Cost Creep (polymer matrices)

(From M. F. Ashby and D. R. H. Jones.)

[14] Reprinted with permission from M. F. Ashby and D. R. H. Jones, "Engineering Materials 2: An Introduction to Microstructures, Processing, and Design," *International Series on Materials Science and Technology*, Vol. 39, Pergamon Press, Elmsford, NY, 1986, p. 268.

example, corundum (Al_2O_3), silicon carbide (SiC), and diamond (C) are used to cut, grind, and polish a wide variety of materials. Table 4.13 shows that metals have very low hardness compared with ceramics. However, the ceramics have very low toughness because of their brittleness. Moreover, ceramics always contain small surface flaws. Hence, the design strength of ceramic materials is determined by their fracture toughness and by the size of preexisting cracks. If the longest flaw size $2a$ is known, referring to Eq. 4.28, the fracture toughness K_{IC} of a ceramic can be determined by

$$K_{IC} = \sigma_{\max} \sqrt{\pi a} \tag{4.50}$$

where $\sigma_{\max}$ is the tensile strength of the ceramic. As we can see from this equation, the strength of a ceramic can be improved by decreasing the crack length through careful quality control or by increasing K_{IC} by making the ceramic into a composite.

4.4   DESIGN ANALYSIS FOR COMPOSITE MATERIALS

Material composites are not a new concept. Nature is abundant with examples of composites. In fact, practically everything in the world is made of composite materials,[15] from bone to common metals. Originally, the man-made composites started as laboratory curiosities. But during the 1940s, with the advent of the glass-reinforced plastics, the engineering application potential became evident.[16] It took two decades before composite material science became a distinct discipline. Since the 1960s, the demand for materials with high strength-to-weight ratio has steadily increased. This demand has contributed to advances in the development of composites, such as polymers and lightweight metal matrix composites.

A composite can be simply viewed as a combination of material components, usually a reinforcing agent and a binder. The regions formed by the materials involved should be large enough to be regarded as a continuum. The process of component bonding in composites is usually done to maximize the favorable properties of the components while mitigating the effects of some of their less desirable characteristics. The properties to be optimized may be physical, chemical, or mechanical. Many authors have proposed classifications of composite materials based mostly on the ideas and concepts that are discussed in the paragraphs that follow. Hull[17] proposes a simple and broad classification of composite materials as shown in Table 4.14.

The elements involved in composite fabrication are fibers or reinforcing elements, matrices, and interface bonding. The fibers can be made from glass, boron, carbon, organic material, ceramic, or metal. The matrix materials are normally polymers, metals, or ceramics. The interfacial bonding can be either mechanical or chemical, but generally chemical bonding is most widely used. There are many techniques available for composite fabrication.[16] Fabrication techniques include hand and automated tape lay-up, resin injection, vacuum bag and autoclave molding,

[15] K. K. Chawla, *Composite Materials*, Springer-Verlag, New York, 1987.
[16] S. Luce, *Introduction to Composite Technology*, Society of Manufacturing Engineers, Dearborn, MI, 1988.
[17] D. Hull, *An Introduction to Composite Materials*, Cambridge University Press, Cambridge, England, 1981, p.1.

Table 4.14   Broad Classification of Composite Materials

Materials	Examples
Natural composite materials	Wood Bone Bamboo Muscle and other tisue
Microcomposite materials	Metallic alloys; e.g., steels Toughened thermoplastics; e.g., impact polystyrene, ABS Sheet-molding compounds Reinforced thermoplastics
Macrocomposites (engineering products)	Galvanized steel Reinforced concrete beams Helicopter blades Skin

From D. Hull, *An Introduction to Composite Materials*, Cambridge University Press, Cambridge, England, 1981, p. 1.

pultrusion, and filament winding. The technique used depends on the kind and quality of composite to be manufactured. For example, for matrix composites, diffusion bonding is usually used.

Composites have found wide use in practically all engineering disciplines.[18] The applications have mainly been driven by the fact that composites can be tailor-made per specification for an optimum design. This means that the properties can be easily modified to suit the design. From a design point of view, composites also exhibit superior material properties as compared with conventional monolithic materials. Composites have less weight, less thermal expansion, greater stiffness, greater strength, and higher fatigue resistance. In aerospace, composites are used for helicopter rotor blades, rocket nozzles, and reentry shields. In automotive engineering, applications range from doors to body moldings. In chemical engineering, composites are used for containers, pipe-work, pressure vessels, and the like. In civil engineering, composites have found important use as glass-reinforced plastics and form work for concrete. In electrical engineering, the applications range from high strength insulators to printed circuit boards. Bioengineering uses carbon fiber composites as prosthetic devices, and other composites have been used to manufacture heart valves.

The solid mechanics of composites is more involved than that of conventional metals (which, for simplicity, are often assumed to be isotropic). Generally, there are two types of information that determine the properties of a composite material: the internal phase geometry and the physical characteristics of the phases, that is, their stress–strain relations. Hashin[19] provides a detailed survey of the analysis of the properties of composite materials. The properties presented in this survey include static strength, fatigue failure, elasticity, thermal expansion, moisture swelling, viscoelasticity, and conductivity. Among the composite structures, the laminates have

[18]B. Harris, *Engineering Composite Materials*, The Institute of Metals, Brookfield, 1986.
[19]Z. Hashin, "Analysis of Composite Materials—A Survey," *Transactions, ASME Journal of Applied Mechanics*, vol. 50, pp. 481–505, 1983.

been most thoroughly investigated. The methods used to analyze the properties of laminates are based on the well-developed laminate theory.[15,17,20,21]

4.4.1 Stress–Strain Relations in Composite Materials

In the elementary analysis of composite materials, the stress–strain convention is chosen such that the orthogonal axis corresponds to three mutually perpendicular planes of material symmetry (Figure 4.24). This axis is also refered to as the principal material axis. For the following analysis, the fiber direction is assumed to be parallel to axis 1, and the direction transverse to it is axis 2. In general, in linear elasticity, the stress–strain relations are expressed in the form

$$
\begin{Bmatrix} \sigma_1 \\ \sigma_2 \\ \sigma_3 \\ \tau_{23} \\ \tau_{31} \\ \tau_{12} \end{Bmatrix} = \begin{bmatrix} C_{11} & C_{12} & C_{13} & C_{14} & C_{15} & C_{16} \\ C_{21} & C_{22} & C_{23} & C_{24} & C_{25} & C_{26} \\ C_{31} & C_{32} & C_{33} & C_{34} & C_{35} & C_{36} \\ C_{41} & C_{42} & C_{43} & C_{44} & C_{45} & C_{46} \\ C_{51} & C_{52} & C_{53} & C_{54} & C_{55} & C_{56} \\ C_{61} & C_{62} & C_{63} & C_{64} & C_{65} & C_{66} \end{bmatrix} \begin{Bmatrix} \varepsilon_1 \\ \varepsilon_2 \\ \varepsilon_3 \\ \gamma_{23} \\ \gamma_{31} \\ \gamma_{12} \end{Bmatrix}
\tag{4.51}
$$

where the coefficients C_{ij} are called *elastic constants* or the *stiffness matrix* of the material, ε is the strain, γ is the shear strain, σ is the normal stress, and τ is the shear stress. As can be seen from the symmetric matrix [C], there are 15 off-diagonal terms plus 6 diagonal terms, resulting in a total of 21 independent elastic constants for most anisotropic materials. For isotropic materials only 2 elastic constants C_{11} and C_{12} are independent and the elastic properties are the same in all directions. The stiffness matrix is reduced to

$$
[C] = \begin{bmatrix} C_{11} & C_{12} & C_{12} & 0 & 0 & 0 \\ C_{12} & C_{11} & C_{12} & 0 & 0 & 0 \\ C_{12} & C_{12} & C_{11} & 0 & 0 & 0 \\ 0 & 0 & 0 & \frac{(C_{11}-C_{12})}{2} & 0 & 0 \\ 0 & 0 & 0 & 0 & \frac{(C_{11}-C_{12})}{2} & 0 \\ 0 & 0 & 0 & 0 & 0 & \frac{(C_{11}-C_{12})}{2} \end{bmatrix}
\tag{4.52}
$$

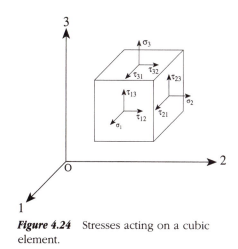

Figure 4.24 Stresses acting on a cubic element.

[20] L. A. Carlsson and R. B. Pipes, *Experimental Characterization of Advanced Composite Materials*, Prentice-Hall, Englewood Cliffs, NJ, 1987.

[21] L. N, Phillips (ed.), *Design of Composites in Design with Advanced Composites Materials*, Springer-Verlag, New York, 1989.

Using Eqs. 4.51 and 4.52, the stress–strain relationship for isotropic materials can be formulated as

$$\{\varepsilon\} = [S]\{\sigma\} \tag{4.53}$$

where the square matrix $[S]$, called matrix of elastic compliances, is equal to the inverse of the $[C]$ matrix, and given as

$$[S] = \begin{bmatrix} S_{11} & S_{12} & S_{12} & 0 & 0 & 0 \\ S_{12} & S_{11} & S_{12} & 0 & 0 & 0 \\ S_{12} & S_{12} & S_{11} & 0 & 0 & 0 \\ 0 & 0 & 0 & 2(S_{11} - S_{12}) & 0 & 0 \\ 0 & 0 & 0 & 0 & 2(S_{11} - S_{12}) & 0 \\ 0 & 0 & 0 & 0 & 0 & 2(S_{11} - S_{12}) \end{bmatrix} \tag{4.54}$$

For isotropic materials composed of a number of sufficiently thin laminae so that the through-thickness stresses are zero, Eq. 4.54 can be reduced to

$$\begin{Bmatrix} \varepsilon_1 \\ \varepsilon_2 \\ \gamma_{12} \end{Bmatrix} = \begin{bmatrix} S_{11} & S_{12} & 0 \\ S_{12} & S_{22} & 0 \\ 0 & 0 & S_{66} \end{bmatrix} \begin{Bmatrix} \sigma_1 \\ \sigma_2 \\ \tau_{12} \end{Bmatrix} \tag{4.55}$$

The relationships between compliances S and elastic constants E, G, and v are given as follows:

$$S_{11} = \frac{1}{E}$$
$$S_{22} = S_{11}$$
$$S_{12} = -\frac{v}{E}$$
$$S_{66} = 2(S_{11} - S_{12}) = \frac{1}{G} \tag{4.56}$$

where the shear modulus of elasticity G can be defined in terms of Young's modulus E and Poisson's ratio v as

$$G = \frac{E}{2(1 + v)} \tag{4.57}$$

For thin material, it is customary to replace the stiffness coefficients C by the symbol Q. Then Eq. 4.51, for plane-stress isotropic composite materials, can be reduced to

$$\begin{Bmatrix} \sigma_1 \\ \sigma_2 \\ \tau_{12} \end{Bmatrix} = \begin{bmatrix} Q_{11} & Q_{12} & 0 \\ Q_{12} & Q_{22} & 0 \\ 0 & 0 & Q_{66} \end{bmatrix} \begin{Bmatrix} \varepsilon_1 \\ \varepsilon_2 \\ \gamma_{12} \end{Bmatrix} \tag{4.58}$$

where the reduced stiffnesses Q_{ij} are given as

$$Q_{11} = \frac{E}{1 - v^2}$$
$$Q_{22} = Q_{11}$$

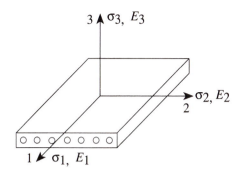

Figure 4.25 Principal material directions for an orthotropic material.

$$Q_{12} = \frac{\nu E}{1 - \nu^2}$$

$$Q_{66} = \frac{1}{2}(Q_{11} - Q_{12}) = G \qquad (4.59)$$

Equations 4.55 and 4.58 describe the stress–strain relationships of the lamina assuming plane-stress conditions for isotropic materials such as thin aluminum plates stacked on top of each other. However, laminae containing fiber–matrix materials are not isotropic. They behave as orthotropic materials having three mutually perpendicular planes of material symmetry. As shown in Figure 4.25, material properties in the three axes are different. However, for a thin lamina, it can be shown that the elastic properties are independent of direction in the plane normal to axis 1 (plane stress). Hence, elastic compliances S_{ij} in Eq. 4.55 for such an orthotropic thin lamina are defined by independent elastic constants as

$$S_{11} = \frac{1}{E_1}$$

$$S_{22} = \frac{1}{E_2}$$

$$S_{66} = \frac{1}{G_{12}}$$

$$S_{12} = -\frac{\nu_{12}}{E_1} = -\frac{\nu_{21}}{E_2} \qquad (4.60)$$

Similarly, reduced stiffnesses Q_{ij} in Eq. 4.58 are defined as

$$Q_{11} = \frac{E_1}{1 - \nu_{21}\nu_{12}}$$

$$Q_{22} = \frac{E_2}{1 - \nu_{21}\nu_{12}}$$

$$Q_{12} = \frac{\nu_{12}E_2}{1 - \nu_{21}\nu_{12}} = \frac{\nu_{21}E_1}{1 - \nu_{21}\nu_{12}}$$

$$Q_{66} = G_{12} \qquad (4.61)$$

In the above equation, Poisson's ratio ν_{12} refers to the strain produced along axis 2 when the line of action of the load application is in axis 1 direction. It should be realized that Poission's ratio for orthotropic materials can be greater than the maximum of 0.5 that is inherent in isotropic materials.

—— *Example 4.6* ——————————————————————

Consider a lamina with material properties $E_1 = 20 \times 10^6$ psi, $E_2 = 1.5 \times 10^6$ psi, $G_{12} = 1.0 \times 10^6$ psi, $\nu_{12} = 0.2400$, $\nu_{21} = 0.018$, $\theta = 0°$, and let the strains be $\varepsilon_1 = 3.313 \times 10^{-4}$, $\varepsilon_2 = 0.91 \times 10^{-4}$, and $\gamma_{12} = 0.00$. Calculate the stresses in the lamina.

Solution

As indicated in Eq. 4.58 the reduced stiffness matrix [Q] is

$$[Q] = \begin{bmatrix} Q_{11} & Q_{12} & 0 \\ Q_{12} & Q_{22} & 0 \\ 0 & 0 & Q_{66} \end{bmatrix}$$

Using Eq. 4.61 and substituting the respective values in the above matrix yields

$$[Q] = \begin{bmatrix} 20.09 & 0.36 & 0 \\ 0.36 & 1.51 & 0 \\ 0 & 0 & 1.00 \end{bmatrix} 10^6 \text{ psi}$$

Using Eq. 4.58, the stress in the lamina can be calculated as

$$\left\{ \begin{array}{c} \sigma_1 \\ \sigma_2 \\ \tau_{12} \end{array} \right\} = \begin{bmatrix} 20.09 & 0.36 & 0.00 \\ 0.36 & 1.51 & 0.00 \\ 0.00 & 0.00 & 1.00 \end{bmatrix} \left\{ \begin{array}{c} 3.313 \\ 0.91 \\ 0.0 \end{array} \right\} 10^2 \text{ psi}$$

Solution of the above matrix yields $\sigma_1 = 66.89 \times 10^2$ psi, $\sigma_2 = 2.57 \times 10^2$ psi, $\tau_{12} = 0.0$ psi.

————————————————————————————————

Equation 4.55 for orthotropic materials shows that there is no coupling between tensile and shear strains. However, this is not true when the principal material axes of a lamina are oriented at an angle θ with respect to the (x, y) reference axes (Figure 4.26). In this case, both the stresses and strains need to be transformed to the reference axes. For example, the matrix relationship of the stresses and strains in the principal-material axes $(1, 2)$ and the reference axes (x, y) can be formulated as

$$\left\{ \begin{array}{c} \sigma_1 \\ \sigma_2 \\ \tau_{12} \end{array} \right\} = [T] \left\{ \begin{array}{c} \sigma_x \\ \sigma_y \\ \tau_{xy} \end{array} \right\} \tag{4.62}$$

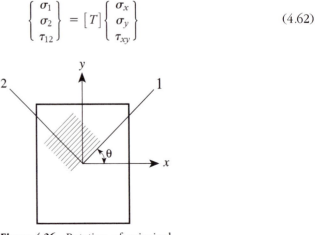

Figure 4.26 Rotation of principal material axes (1,2) from arbitrary (x, y) axes.

or

$$\left\{ \begin{array}{c} \sigma_x \\ \sigma_y \\ \tau_{xy} \end{array} \right\} = [T]^{-1} \left\{ \begin{array}{c} \sigma_1 \\ \sigma_2 \\ \tau_{12} \end{array} \right\}$$ (4.63)

and

$$\left\{ \begin{array}{c} \varepsilon_1 \\ \varepsilon_2 \\ \frac{\gamma_{12}}{2} \end{array} \right\} = [T] \left\{ \begin{array}{c} \varepsilon_x \\ \varepsilon_y \\ \frac{\gamma_{xy}}{2} \end{array} \right\}$$ (4.64)

or

$$\left\{ \begin{array}{c} \varepsilon_x \\ \varepsilon_y \\ \frac{\gamma_{xy}}{2} \end{array} \right\} = [T]^{-1} \left\{ \begin{array}{c} \varepsilon_1 \\ \varepsilon_2 \\ \frac{\gamma_{12}}{2} \end{array} \right\}$$ (4.65)

where the transformation matrices are

$$[T] = \begin{bmatrix} m^2 & n^2 & 2mn \\ n^2 & m^2 & -2mn \\ -mn & mn & (m^2 - n^2) \end{bmatrix}$$ (4.66)

and

$$[T]^{-1} = \begin{bmatrix} m^2 & n^2 & -2mn \\ n^2 & m^2 & 2mn \\ mn & -mn & (m^2 - n^2) \end{bmatrix}$$ (4.67)

where $m = \cos\theta$ and $n = \sin\theta$. The stress–strain relation becomes

$$\left\{ \begin{array}{c} \sigma_x \\ \sigma_y \\ \tau_{xy} \end{array} \right\} = \begin{bmatrix} \overline{Q}_{11} & \overline{Q}_{12} & \overline{Q}_{16} \\ \overline{Q}_{12} & \overline{Q}_{22} & \overline{Q}_{26} \\ \overline{Q}_{16} & \overline{Q}_{26} & \overline{Q}_{66} \end{bmatrix} \left\{ \begin{array}{c} \varepsilon_x \\ \varepsilon_y \\ \gamma_{xy} \end{array} \right\}$$ (4.68)

The matrix $[\overline{Q}_{ij}]$ is called the transformed reduced-stiffness matrix and the stiffnesses $\overline{Q}_{ij}$ are given by

$$\overline{Q}_{11} = m^4 Q_{11} + 2m^2 n^2 (Q_{12} + 2Q_{66}) + n^4 Q_{22}$$
$$\overline{Q}_{12} = m^2 n^2 (Q_{11} + Q_{22} - 4Q_{66}) + (m^4 + n^4) Q_{12}$$
$$\overline{Q}_{22} = n^4 Q_{11} + 2m^2 n^2 (Q_{12} + 2Q_{66}) + m^4 Q_{22}$$
$$\overline{Q}_{16} = m^3 n(Q_{11} - Q_{12}) + mn^3 (Q_{12} - Q_{22}) - 2mn(m^2 - n^2) Q_{66}$$
$$\overline{Q}_{26} = mn^3 (Q_{11} - Q_{12}) + m^3 n(Q_{12} - Q_{22}) + 2mn(m^2 - n^2) Q_{66}$$
$$\overline{Q}_{66} = m^2 n^2 (Q_{11} + Q_{22} - 2Q_{12} - 2Q_{66}) + (m^4 + n^4) Q_{66}$$ (4.69)

where $m = \cos\theta$ and $n = \sin\theta$

Similarly, the compliance relation becomes

$$\left\{ \begin{array}{c} \varepsilon_x \\ \varepsilon_y \\ \gamma_{xy} \end{array} \right\} = \begin{bmatrix} \overline{S}_{11} & \overline{S}_{12} & \overline{S}_{16} \\ \overline{S}_{12} & \overline{S}_{22} & \overline{S}_{26} \\ \overline{S}_{16} & \overline{S}_{26} & \overline{S}_{66} \end{bmatrix} \left\{ \begin{array}{c} \sigma_x \\ \sigma_y \\ \tau_{xy} \end{array} \right\}$$ (4.70)

The transformed compliances $\bar{S}_{ij}$ have the following values ($m = \cos\theta$, $n = \sin\theta$):

$$
\begin{aligned}
\bar{S}_{11} &= m^4 S_{11} + m^2 n^2 (2S_{12} + S_{66}) + n^4 S_{22} \\
\bar{S}_{12} &= m^2 n^2 (S_{11} + S_{22} - S_{66}) + (m^4 + n^4) S_{12} \\
\bar{S}_{22} &= n^4 S_{11} + m^2 n^2 (2S_{12} + S_{66}) + m^4 S_{22} \\
\bar{S}_{16} &= 2m^3 n (S_{11} - S_{12}) + 2mn^3 (S_{12} - S_{22}) - mn(m^2 - n^2) S_{66} \\
\bar{S}_{26} &= 2mn^3 (S_{11} - S_{12}) + 2m^3 n (S_{12} - S_{22}) + mn(m^2 - n^2) S_{66} \\
\bar{S}_{66} &= 4m^2 n^2 (S_{11} - S_{12}) - 4m^2 n^2 (S_{12} - S_{22}) + (m^2 - n^2)^2 S_{66}
\end{aligned}
\tag{4.71}
$$

4.4.2 Failure Criteria of Composite Laminates

The analysis of failure in composite materials is more involved than that for isotropic materials. This is mainly because the anisotropic property of composite materials leads to different stiffnesses in varying loading directions. Also, the failure mode is greatly dependent on the composite materials. The difficulty of failure analysis of composite materials is further compounded by the fact that the ultimate strength behavior of composite materials is often different in tension and compression.

Most of the commonly used failure theories are set forth to predict the strength of a unidirectional lamina based on the strength in the fiber direction, the strength in the transverse fiber direction, and the strength under in-plane shear. The strength in the fiber direction is either tensile, σ_1^t, or compressive, σ_1^c; the strength in the transverse fiber direction is either tensile, σ_2^t, or compressive, σ_2^c; and the in-plane shear is in plane 1 in direction 2, τ_{12}^* (Figure 4.26). This implies that in developing the failure criteria for composite materials, the stress conditions must be considered in relation to the normal and shearing components relative to their principal axes. For simplicity, the fiber-reinforced lamina will be treated as a homogeneous, orthotropic material. Generally, three main failure criteria are used for composite materials:[15,17,20,21]

(a) Maximum Stress Criterion   This criterion assumes that failure will occur when any one of the stress components is equal to or greater than its corresponding critical value. Thus, a lamina fails if

$$
\begin{aligned}
\sigma_1 &\geq \sigma_1^t & \sigma_1 &\leq \sigma_1^c \\
\sigma_2 &\geq \sigma_2^t & \sigma_2 &\leq \sigma_2^c \\
\tau_{12} &\geq \tau_{12}^* & \tau_{12} &\leq \tau_{12}^*
\end{aligned}
\tag{4.72}
$$

where σ_1^t and σ_1^c are the ultimate tensile and compressive strengths, respectively, in direction 1; σ_2^t and σ_2^c are the ultimate tensile and compressive strengths, respectively, in direction 2; and τ_{12}^* is the ultimate planar-shear strength in plane 1 in direction 2.

(b) Maximum-Strain Criterion   The maximum-strain criterion assumes that failure occurs when any one of the strain components is equal to or greater than the corresponding ultimate value. In this case the modes of failure are

$$
\begin{aligned}
\varepsilon_1 &\geq \varepsilon_1^t & \varepsilon_1 &\leq \varepsilon_1^c \\
\varepsilon_2 &\geq \varepsilon_2^t & \varepsilon_2 &\leq \varepsilon_2^c \\
\gamma_{12} &\geq \gamma_{12}^* & \gamma_{12} &\leq \gamma_{12}^*
\end{aligned}
\tag{4.73}
$$

where ε_1^t and ε_1^c are the ultimate tensile and compressive strains, respectively, in direction 1; ε_2^t and ε_2^c are the ultimate tensile and compressive strains, respectively, in direction 2; and γ_{12}^* is the ultimate planar-shear strain on plane 1 in direction 2.

(c) Tsai–Hill Criterion   This criterion is derived from the von Mises failure criterion used for homogeneous isotropic materials. The Tsai–Hill criterion states that failure of an orthotropic lamina will occur under a general stress state when

$$\frac{\sigma_1^2}{(\sigma_1^t)^2} - \frac{\sigma_1\sigma_2}{(\sigma_1^t)^2} + \frac{\sigma_2^2}{(\sigma_2^t)^2} + \frac{\tau_{12}^2}{(\tau_{12}^*)^2} \le 1 \qquad (4.74)$$

where σ_1^t and σ_2^t are the respective lamina strengths in tension; and τ_{12}^* is the laminate planar shear strength. If the stresses are compressive, then the corresponding compressive failure strength (σ_1^c and σ_2^c) is to be used in Eq. 4.74. The Tsai–Hill failure criterion has been shown to be more realistic than the previous two criteria.

▬▬ *Example 4.7* ▬▬

Consider an orthotropic laminate under simple uniaxial tension σ_o as shown in Figure 4.27 with properties $E_1 = 25 \times 10^6$ psi, $E_2 = 1.0 \times 10^6$ psi, $G_{12} = 0.5 \times 10^6$ psi, $\nu_{12} = 0.2500$, and $\theta = +45°$, and let the stresses in the lamina be $\sigma_1^t = 200$ kpsi, $\sigma_2^t = 10$ kpsi, $\tau_{12}^* = 20$ kpsi. Using the Tsai–Hill criterion, find the least σ_o for which the laminate failure will occur.

Solution

The stresses in the principal directions are given by the following transformation.

$$\begin{Bmatrix} \sigma_1 \\ \sigma_2 \\ \tau_{12} \end{Bmatrix} = \begin{bmatrix} m^2 & n^2 & 2mn \\ n^2 & m^2 & -2mn \\ -mn & mn & (m^2 - n^2) \end{bmatrix} \begin{Bmatrix} \sigma_x \\ \sigma_y \\ \tau_{xy} \end{Bmatrix}$$

where $m = \cos\theta$ and $n = \sin\theta$. For the given plane stress problem

$$\begin{Bmatrix} \sigma_1 \\ \sigma_2 \\ \tau_{12} \end{Bmatrix} = \begin{bmatrix} m^2 & n^2 & 2mn \\ n^2 & m^2 & -2mn \\ -mn & mn & (m^2 - n^2) \end{bmatrix} \begin{Bmatrix} 0.0 \\ \sigma_o \\ 0.0 \end{Bmatrix}$$

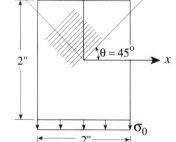

Figure 4.27 Material axes of a lamina oriented at an angle θ.

For $\theta = 45°$, $m = n$. It then follows that

$$
\left\{
\begin{array}{c}
\sigma_1 \\
\sigma_2 \\
\tau_{12}
\end{array}
\right\}
=
\left\{
\begin{array}{c}
1 \\
1 \\
1
\end{array}
\right\}
n^2 \sigma_o
$$

Using the Tsai–Hill criterion

$$
\frac{(n^2\sigma_o)^2}{(200)^2} - \frac{(n^2\sigma_o)^2}{(200)^2} + \frac{(n^2\sigma_o)^2}{(10)^2} + \frac{(n^2\sigma_o)^2}{(20)^2} = 1
$$

and rearranging, we find that

$$
(n^2\sigma_o)^2 \left[\frac{1}{10^2} + \frac{1}{20^2} \right] = 1
$$

Hence,

$$
\sigma_o^2 = \frac{80}{n^4}
$$

$$
\sigma_o = 4\sqrt{20} \text{ kpsi}
$$

4.5 RESIDUAL (INTERNAL) STRESS CONSIDERATIONS

Residual stresses, also referred to as internal or locked-in stresses, are the self-equilibrating stresses that exist in a body in the absence of external forces or constraints. The self-equilibrating nature of residual stresses means that the resultant force and the resultant moment produced by these stresses must be zero.

Residual stresses can be divided into two main categories:

1. Macro residual stresses.
2. Micro residual stresses.

Macro residual stresses refer to a system of internal stresses that varies continuously through the volume of the body and acts over large regions. Micro residual stresses are limited to regions as small as a unit cell and may extend as far as several grains.

4.5.1 Sources of Residual Stress

Residual stresses arise from various steps involved in the processing and maufacturing of materials. For example, during most metal forming operations, such as rolling and extrusion, the material undergoes nonuniform plastic deformation resulting in a pattern of residual stresses throughout the material's cross section. To illustrate this point, consider a metal sheet being rolled as shown in Figure 4.28. During the rolling process, plastic flow occurs only near the surfaces that are in direct contact with the rollers. The ensuing nonuniform deformation results in elongation of the surface fibers while the fibers near the center of the sheet remain unchanged. However, for the sheet to remain as a continuous body, the center fibers tend to restrain

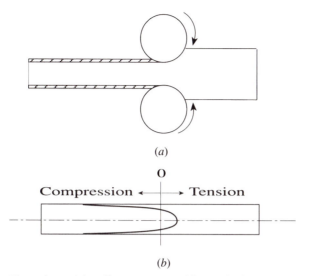

(a)

(b)

Figure 4.28 (a) Rolling operation. (b) Residual-stress pattern in rolled metal sheet.

the surface fibers from stretching, whereas the surface fibers tend to elongate the central fibers. This results in a pattern of residual stresses throughout the sheet with a compressive value at the surface and a tensile value at the center of the sheet (Figure 4.28b).

Similarly, during extrusion processes (Figure 4.29), because of the high friction values between the material and the extrusion chamber/dies, the fibers on the outer surface of the billet travel at a slower rate than the fibers near the center of the billet. The nonuniform deformation throughout the body (Figure 4.29c) results in tensile stresses at the outer surface and compressive stresses near the center of the extruded product.

Residual stresses are sometimes introduced during the cooling of hot products owing to the temperature differences between the surface and the center of metal. Evolution of residual stresses during the cooling of a block of metal is illustrated in Figure 4.30. Since the metal fibers on the outer surface of the block cool at a faster rate than the fibers at the center, there will be more contraction on the outer surface than at the center of the block. This nonuniform contraction induces compressive strain in the center fibers while the outer fibers will be subjected to a tensile strain (Figure 4.30). However, since the hot center fibers have a lower yield stress than the cooler outer fibers, they cannot support the imposed compressive strains and yield in compression. The compressive plastic deformation of the center fibers results in the shrinkage of the center of the block in order to relieve some of the stress (Figure 4.30c). Upon cooling of the entire block, the total shrinkage of the center fibers will be greater than that of the outer fibers because of both cooling and compressive plastic deformation. This nonuniform deformation throughout the block produces compressive stresses at the outer surface and tensile stresses at the inner surface (Figure 4.30d). More extensive coverage of the formation of residual stresses can be found in references [17–20].

4.5.2 Effect of Residual Stresses

When a mechanical component is subjected to external forces, the material behaves in a manner that is governed by the total stress acting on the material. Since the total

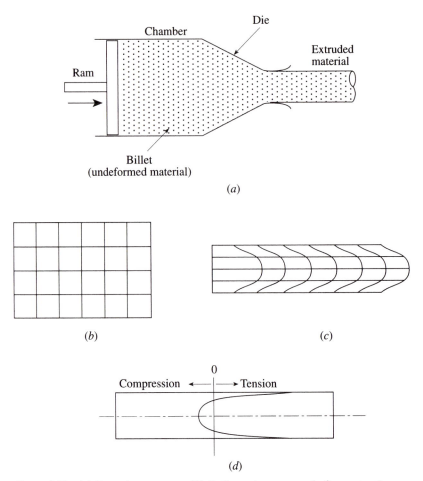

Figure 4.29 (a) Extrusion process. (b) Deformation pattern before extrusion. (c) Material grid pattern after extrusion. (d) Residual-stress pattern in extruded metal.

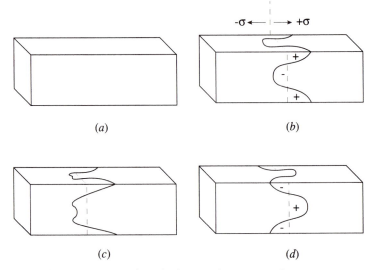

Figure 4.30 Evolution of residual stress during quenching process.

stress experienced by a material involves the superposition (or summation) of the external stresses and the residual (internal) stresses, knowledge of residual stresses plays an important role in designing and/or predicting the behavior of materials under various loading conditions.

___ *Example 4.8* _____

As an example, let us consider the case of a block of A96061-T6 aluminum that has been hot forged and quenched. Let us assume that because of the quenching process, a residual-stress pattern exists in the block as is shown in Figure 4.31. If the block is subjected to a tensile force of 50,000 lb, determine the factor of safety guarding against yielding.

Solution

The yield strength of Al A96061-T6 is $S_y = 40$ kpsi. The stress due to the force is

$$\sigma = \frac{F}{A} = \frac{50,000}{2 \times 1} = 25,000 \text{ psi}$$

The factor of safety without considering the residual stresses is

$$n = \frac{S_y}{\sigma} = \frac{40}{25} = 1.6$$

However, since the center of the block contains tensile residual stresses, the total stress at the center of the block is

$$\sigma_{\text{center}} = 25,000 + 18,000 = 43,000 \text{ psi}$$

which exceeds the yield stress of the material. Therefore,

$$n = \frac{40}{43} = 0.93$$

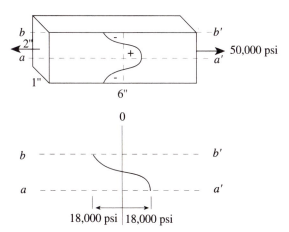

Figure 4.31 Hot-forged and quenched aluminum black.

The above example illustrates the importance of accounting for residual stresses during the design process. Without considering the residual stresses, the analysis shows that the block will be capable of handling the applied load. The addition of tensile residual stresses to the total stress shows that the block will not be able to withstand the operating load.

Other effects of residual stresses include warping of components during machining and stress–corrosion cracking. When machining components contain residual stresses, the remaining residual stresses are enhanced. Since residual stresses are in self-equilibrium, removing part of the material disturbs the equilibrium of internal forces and moments. The component material responds to this by adjusting its shape (distorting) in order to establish a new equilibrium condition. Stress–corrosion cracking is a type of failure where the effect of a corrosive medium is accentuated in the presence of stresses. Residual stresses can be particularly detrimental in conjunction with corrosive environments (such as ammonia compounds with brass and chlorides with austenitic stainless steel).

Not all types of residual stresses are detrimental. In fact, many types of residual stresses are beneficial and serve as a strengthening mechanism for the material. For example, autofrettage is a process routinely used in the pressure vessel industry to induce compressive residual stresses at the bore of pressure vessels. This is done by straining the vessel beyond its elastic limit at the bore. When the vessel is unloaded, the inner surface (bore) will contain compressive residual stresses. Upon application of internal pressure, the resulting total stress (tensile) at the bore will be lower because of the presence of compressive residual stresses.

4.5.3   Measurement of Residual Stresses

To assess the effect of residual stresses on the intended service capability of the component, one must be able to quantify the magnitude and distribution of residual stresses. Experimental methods for measurement of residual stresses are divided into two main categories:

1. Nondestructive methods.
2. Destructive methods.

Examples of nondestructive techniques are

1. Acoustoelastic.
2. X-ray.

All destructive methods of residual stress measurement take advantage of the principle that when part of a residual stressed material is removed, the remaining material distorts in an attempt to reach a new equilibrium condition. This change in shape (strain) can be measured via strain gauges. By utilizing appropriate equations, the measured strains are converted to residual stresses in the removed material. By removing small consecutive layers of material, the magnitude and distribution of residual stresses throughout the entire cross section can be determined. Examples of destructive methods are as follows:

1. Hole-drilling methods.
2. Sach's boring-out method.

3. Sectioning method.

4. Deflection method

Comprehensive description of nondestructive and destructive methods of residual-stress measurement techniques can be found in references [19–20] or in handbooks on experimental stress analysis.

4.6 MATERIAL STANDARDS AND SPECIFICATIONS

Material standards and specifications are used to identify the physical and chemical characteristics and requirements for materials used in design. Specifications are normally used in the procurement process to describe accurately the essential requirements for materials, including the procedures that are used to determine that the requirements have been met. A standard is a document used by general agreement to determine whether or not something is as it should be. Standards are used to control the variability of materials and to provide uniformity in quality and performance over a wide range of applications. Standards describe the general requirements for a material in terms of its constituents and properties. Many material standards fall into the category called *consensus standards*. A consensus standard is a standard developed by a body representing all of the parties interested in using and complying with the standard. Standards are implemented in procurement through specifications. Thus, procurement specifications commonly refer to standards to identify material or other requirements.

4.6.1 Organizations Involved in Standards and Specifications Preparation

There are several hundred organizations in the United States involved in the preparation of voluntary consensus standards. They include branches of the government, professional and technical societies, trade associations, public service and consumer groups, and testing and inspection bodies.[22] Two of these, the American National Standards Institute (ANSI) and the American Society for Testing and Materials (ASTM), are involved almost exclusively with the preparation of voluntary consensus standards. The Standards Development Section of the National Bureau of Standards (NBS) within the U.S. Department of Commerce performs a similar function as do major parts of other organizations including the Codes and Standards Division of the American Society for Mechanical Engineers (ASME).

ANSI standards are used widely throughout industry. ANSI is the coordinator of the U.S. national standards system and assists participants in the program in reaching agreements on standards, needs, and priorities, arranges for standards development efforts, and assures that fair and effective procedures are used in standards development. All standards developed by this organization have an alphanumeric code that begins with the prefix ANSI.[23]

ASTM is concerned with the development of standards having to do with the characteristics and performance of materials, products, systems, and services. It is the world's largest source of voluntary consensus standards. All of its standards are identified with the prefix ASTM followed by an alphanumeric code. ASME and ASTM have cooperated for many years in the preparation of material specifications

[22] "Materials and Process Specifications and Standards," National Academy of Sciences, 1977.
[23] "Guide to Materials Engineering Data and Information," American Society for Metals, 1986.

for ferrous and nonferrous materials. In 1969, the American Welding Society (AWS) began publishing specifications for welding rods, electrodes, and filler metals. ASME now works with the AWS on these specifications as well. All identical specifications are identified by ASME/ASTM symbols or ASME/AWS symbols. When changes are required to make the specification acceptable for ASME Code usage only, the ASME symbol is used. The ASME Code, which includes a large number of material specifications, has been adopted into law by 45 states and all of the Canadian provinces.[24]

The use of the ASME Codes relative to weld joint efficiency is demonstrated by the following example.[25]

___ *Example 4.9*

As shown in Figure 4.32, a horizontal vessel 60 ft long is fabricated from carbon steel SA 442 using 6 rings 10 ft long. The vessel is supported by 120° saddles located 2 ft 6 in. from each head attachment seam. The heads are ellipsoidal and attached using type no. 2 butt joints. The shell courses have type no. 1 longitudinal joints, which are spot radiographed in accordance with UW-52. The circumferential welds joining the courses are type no. 2 with no radiography. Given the following design parameters, determine the required shell thickness.

Design Parameters

Assumed joint efficiency (circumferential seams), E	= 0.85
Assumed joint efficiency (long seams), E	= 0.65
Maximum allowable stress, S (see Table UCS-23 in ASME Boiler and Pressure Vessel Code)	= 13,800 psi
Design pressure, P including static head	= 60 psi
Reaction at each saddle, Q	= 175,000 lb
Saddle to tangent line, A	= 30 in.
Vessel diameter, OD	= 120 in.
Design temperature, T	= 100° F
Assumed shell thickness, t	= 0.3125 in.
Shell length, L	= 720 in.
Weight of vessel, W	= 30,000 lb
Weight of contents, W_c	= 320,000 lb
Total weight, W_T	= 350,000 lb
Head depth, H	= 30 in.

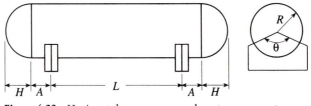

Figure 4.32   Horizontal pressure vessel on two supports.

[24]"ASME Boiler and Pressure Vessel Code," 1988.

[25]Section VIII, Division 1, Appendix L, "Examples Illustrating the Application of Code Formulas and Rules," pp. 714–716.

Solution

The following three cases must be investigated:

Case 1. *Circumferential tensile stress due to internal pressure.* Using the equation in UG-27(c)(1) p. 23, Pressure Vessels code, Section VIII-1, radius R is given by

$$R = \frac{OD}{2} - t = \frac{120}{2} - 0.3125 = 59.6875 \text{ in.}$$

$$t = \frac{PR}{SE - 0.6P}$$

$$= \frac{60(59.6875)}{13,800(0.85) - 0.6(60)} = 0.306 \text{ in.} \tag{4.75}$$

Case 2. *Longitudinal tensile stress due to bending and internal pressure.* The following equation combines the longitudinal tensile stress due to pressure with the longitudinal tensile stress due to bending at the midpoint between the saddles.

$$t = \frac{PR}{2SE + 0.4P} \mp \frac{QL}{4\pi R^2 SE} \times \left[\frac{1 + \frac{2(R^2 - H^2)}{L^2}}{1 + \frac{4H}{3L}} - \frac{4A}{L} \right] \tag{4.76}$$

$$= \frac{60(59.6875)}{2(13,800)(0.65) + (0.4)(60)} + \frac{175,000(720)}{4\pi(59.6875)^2(13,800)(0.65)}$$

$$\times \left[\frac{1 + \frac{2(59.6875^2 - 30^2)}{720^2}}{1 + \frac{4(30)}{3(720)}} - \frac{4(30)}{720} \right] = 0.199 + 0.248 = 0.447 \text{ in.}$$

This is greater than the actual thickness, so we must either thicken the shell or increase the efficiency of the welded joint by changing the weld type or the amount of radiography.

> *Action.* Spot radiograph to increase the efficiency of the welded joint from 0.65 to 0.80. Recalculating t by using $E = 0.80$ from Eq. 4.78 yields a thickness of 0.364 in. Since the calculated thickness is still greater than the actual thickness, increase the value of E further by changing the circumferential seam to a type 1 fully radiographed weld. From Table UW-12, p. 104 of Rules of Construction of Pressure Vessels, Section VIII-1, choose $E=1.0$. Recalculation of t with the new value of $E = 1.0$ yields a thickness of 0.291 in. Since this value is less than the actual thickness, it is safe.

> *Conclusion.* The circumferential joint at the center of the vessel must be type no. 1 fully radiographed. This is at the point of maximum positive moment. The maximum negative moment is at the supports, but there is no joint there. Other circumferential joints must be investigated by using the moment at the joint in calculating the combined stresses. It should be noted that many other areas of stress due to saddle loadings exist and should be investigated (see Appendix G of Section VIII-1).

Case 3. *Longitudinal compressive stress due to bending.* First determine the allowable compressive stress (see UG-23(b), p. 20 of Section VIII-1): Using the assumed value of t and R, calculate the value of the constant A using the following formula:

$$A = \frac{0.125}{R_o/t} \tag{4.77}$$

$$= \frac{0.125}{60/0.3125} = 0.000651$$

This value of A is now used to find the allowable compressive stress for a given material and design temperature from Fig. 5-UCS-28.2 (Appendix 5, p. 520 of Section VIII). For A = 0.000651 and design temperature of 100°F, we obtain the allowable stress, B = 9446 psi.

The general equation for thickness is the same as for longitudinal tensile stress except that the pressure portion drops out, and the most severe condition occurs when there is no pressure in the vessel. Then the thickness calculation for this case for the allowable stress of 9446 psi is

$$t = \frac{QL}{4\pi R^2 SE} \times \left[\frac{1 + \frac{2(R^2 - H^2)}{L^2}}{1 + \frac{4H}{3L}} - \frac{4A}{L} \right] \tag{4.78}$$

$$= \frac{175,000(720)}{4\pi(59.6875)^2(9446)(1.0)} \times \left[\frac{1 + \frac{2(59.6875^2 - 30^2)}{720^2}}{1 + \frac{4(30)}{3(720)}} - \frac{4(30)}{720} \right] = 0.236 \text{ in.}$$

Required design thickness, t = 0.306 in. governed by circumferential tensile stress.

4.7   CORROSION CONSIDERATIONS

There is no argument among scientists today that corrosion is a major factor in the design, cost, and implementation of engineering applications. Corrosion costs U.S. industries and the American public an estimated $170 billion per year.[26] Also, according to estimates from the International Association of Drilling Contractors, about 75 percent to 85 percent of all drill pipe losses in the field are caused by some type of corrosion.[27] What is in dispute, however, is the way in which the various forms of corrosion are categorized. Their overlapping similarities make this a difficult task. A general approach must be utilized keeping in mind that each type of corrosion is not necessarily limited to its assigned major category.

The different types of corrosion are grouped according to the ASM Handbook on Corrosion as indicated in Table 4.15.[26] Among the many types of corrosion,

Table 4.15   Types of Corrosion

General Corrosion	Localized Corrosion	Metallurgically Influenced Corrosion	Mechanically Assisted Degradation Corrosion	Environmentally Induced Cracking Corrosion
Atmospheric	Filform	Intergranular	Erosion	SCC
Galvanic	Crevice	Dealloying	Fretting	Hydrogen
General	Pitting		Cavitation	damage
biological	Localized		Fatigue	Liquid metal
Molten salt	biological			embrittlement
High				Solid metal-
temperature				induced
				embrittlement

[26] *Metals Handbook*, Ninth Edition, Vol. 13, ASM International, 1987, p. 79.
[27] M. A. Al-Marhoun and S. S. Rahman, "Treatment of drilling Fluid to Combat Drill Pipe Corrosion," *Journal of Science and Engineering Corrosion*, National Association of Corrosion Engineers, Houston, TX, 1990, p. 778.

some are encountered frequently in industry. The following is a brief explanation of these prevalent types.

4.7.1 Atmospheric Corrosion

Atmospheric corrosion can be defined as the deterioration of a material when exposed to air and its impurities. As one might expect, atmospheric corrosion is very widespread and accounts for more of the cost due to failures than any other form. Some important factors that influence the rate of atmospheric corrosion are temperature, climate, and relative humidity.

4.7.2 Galvanic Corrosion

All metals have the ability to act as either an anode or a cathode. The potential difference between two dissimilar metals causes galvanic corrosion. Current flows from the anode to the cathode, so the less-resistant metal (anode) is corroded relative to the cathodic metal. There are two ways to minimize galvanic corrosion. One way is to use metals that are close together in the galvanic series. Another way is to insulate the dissimilar metals from each other.

4.7.3 Crevice Corrosion

This form of corrosion is highly localized and occurs within gaps or spaces between metals. Crevice corrosion may occur at cracks, seams, or other metallurgical defects. This type of corrosion is usually associated with small volumes of stagnant liquid that become trapped, allowing corrosion to occur. Eliminating possible crevices as well as keeping parts clean helps to prevent crevice corrosion.

4.7.4 Pitting Corrosion

Pitting corrosion is extremely localized and forms cavities or pits. These cavities are small compared with the overall size of the surface but can cause equipment failure after only minimal weight loss of the part. The depth of pitting (pitting factor) is defined as the ratio of the deepest metal penetration to the average metal penetration. Pitting corrosion can be minimized by specifying a clean and smooth metal surface.

4.7.5 Intergranular Corrosion

This type of corrosion occurs at the grain boundaries of metals. The limited area of grain boundary material acts as an anode, and the larger grain area acts as a cathode. The flow of energy from the anode to the cathode results in penetrating, corrosive attack. Intergranular corrosion is common to austenitic steels that have been heated to their sensitizing range of 950°F to 1450°F. Reducing the occurrence of this type of corrosion is possible by annealing the steel after it has been sensitized.

4.7.6 Erosion Corrosion

Material degradation can be caused by relative movement between a corrosive fluid and a surface. This type of attack is called erosion corrosion. Erosion destroys the protective surface films of the material, making it easier for chemical attack to occur. Two types of erosion are cavitation and fretting. Cavitation arises from the formation and collapse of vapor bubbles near the metal surface, resulting in local deformation.

Fretting occurs between two surfaces under cyclic load and results in surface pits or cracks. Reducing fluid velocity as well as eliminating instances where direct surface contact may occur are ways to control erosion corrosion.

4.7.7   Stress–Corrosion Cracking

Stress-corrosion cracking (SCC) is a term used to describe a type of accelerated corrosion caused by internal residual stresses or externally applied stresses coupled with a corrosive reaction. Only certain combinations of metals and corrosive environments will cause SCC to occur. Presently, more than 80 different combinations of alloys and corrosive environments are known to cause stress–corrosion cracking. To prevent the occurrence of SCC, one should select alloys that are less susceptible to cracking in service environment and that will maintain low stress levels.

BIBLIOGRAPHY

1. *Metals Handbook*, Vol. 1, American Society for Metals, Metals Park, OH, 1961.

2. DIETER, G. E., *Engineering Design: A Materials and Processing Approach*, McGraw-Hill, New York, 1983.

3. MAGAD, E. L. and AMOS, J. M., *Total Materials Management*, Van Nostrand Reinhold, New York, 1989.

4. CRANE, F. A. a. and CHARLES, J. A., *Selection and Use of Engineering Material*, Butterworth & Co,. London, 1984.

5. TRUCKS, H. E., *Designing for Economical Production*, Society of Manufacturing Engineers, Dearborn, MI, 1987.

6. VERINK, E. D., *Methods of Materials Selection*, Gordon and Breach, New York, 1966.

7. KERN, R. F. and SUESS, M. E., *Steel Selection, a Guide for Improving Performance and Profits*, Wiley, New York, 1979.

8.   ASHBY, M. F. and JONES, D. R. H., *Engineering Materials 2, an Introduction to Microstructures, Processing and Design*, Pergamon Press, 1986.

9. PATTON, W. J., *Mateials in Industry*, Prentic-Hall, Englewood Cliff, NJ, 1986.

10. PSARAS, P. A. and LANGFORD, H. D., *Advancing Materials Research*, National Academy Press, Washington D.C. 1987.

11. ROLFE, S.T. and BARSOM, J. M. *Fracture and Fatigue Control in Structures, Applications of Fracture Mechanics*, Prentic-Hall, Englewood Cliffs, NJ, 1977.

12. TAIT, R. B. and GARRETT, G. G., *Fracture and Fracture Mechanics* pergamon Press, Elmaford, NY, 1984.

13. BROEK, D., *Elementary Engineering Fracture Mechanics*, Martinus Nijhoff Publishers, The Hague, 1982.

14. WEI, R. P. and STEPHENS, R. I., "Fatigue Crack Growth under Spectrum Loads," American Society for Testing and Materials, Philadelphia, PA, 1975.

15. GOEL, V. S., "Analyzing Failures, the Problems and Solutions," American Society of Metals, Metals Park, OH, 1986.

16. GOEL, V. S., "Fatigue Life Analysis and Prediction", American Society of Metals, Metals Park, Oh, 1986.

17. ALMEN, J. O. and BLACK, P. H., *Residual Stresses and Fatigue in Metals*, McGraw-Hill, New York, 1963.

18. HORGER, O. J., "Residual Stresses," *Handbook of Experimental Stress Analysis* (M. Hetenyi, Ed.) John Wiley & Sons, New York, 1950.

19. BARRET, C. S., "A Critical Review of Various Methods of Residual Stress Measuement," *Proceedings of Society for Experimental Stress Analysis*, Vol. II, no. 1, pp. 147–156, 1945.

20. "Methods of Residual Stress Measurement," SAE report J936, Society for Automotive Engineers Inc., New York, 1965.

Chapter 5

Engineering Economics

It would be well if engineering were less generally thought of, and even defined, as the art of constructing. In a certain important sense it is rather the art of not constructing; or, to define it rudely but not inaptly, it is the art of doing that well with one dollar which any bungler can do with two after a fashion. **A. M. Wellington**

5.1 PROJECT/PRODUCT COST AND THE ENGINEER

During the last half of this century significant changes have occurred in the United States and the rest of the world in regard to product cost and the meaning of competition. Through the end of World War II and into the 1950s U.S. industry seemed to operate with a philosophy wherein new, modified and unique products were developed within a particular product line with pricing based on development, manufacturing, and marketing costs, plus a reasonable profit. Competitive product costs were considered, of course, but this comparison had more to do with product cost/utility ratio than just lower cost. Industry seemed to take pride in the *product* of their efforts and little of the *copycat* mentality that seems so prevalent today was present. Patents for unique product ideas usually meant monetary returns for the inventor and long-term profit for the company with the manufacturing rights. Today, many companies that have innovative ideas for products or processes puposefully do not make patent application because they do not want to disclose their idea publicly. They know that as soon as the innovation is made public it will be copied, manufactured in a cheaper labor location, and sold at lower cost than the original. Unfortunately, the copycat seldom has any real commitment to the product, and after the initial market demand has been met will often go on to copy some other product or, possibly worse, close the manufacturing plant down and declare bankruptcy for protection against creditors. Fortunately, this is not a fair representation of all of the consumer product industry nor of industry involved in developing highly technical products. There are many reputable manufacturers who are concerned with producing a quality product and standing behind the product after the sale. Unfortunately, these reputable manufacturers operate in the same economy as the less reputable ones and are thus driven to compromises and cost-cutting practices they might normally shun. Who gains from all this pencil sharpening? The buyer gains in the sense that he or she is paying less (initially) for the product, but with lower cost (in general) comes lower quality, reliability, and product support, and the ultimate cost to the consumer is, more often than not, equal to or greater than the higher-quality product.

Another significant influence on the cost of products and the potential market came about as a result of the United States becoming part of a global economy. When U.S. industry was more or less isolated from industry worldwide, producers competed on a so-called *level playing field* in that labor costs were pretty well controlled by management, labor unions, and legislation, and government subsidies and support were applied somewhat equitably. In competing in a global economy, no such even-handedness prevails and many U. S. industries have found themselves in difficulty as a result. To maintain their competitiveness, many companies have located manufacturing facilities in low-cost labor regions of the world to take advantage of lower manufacturing costs. There are several disadvantages that should be considered in taking such action, however, including capricious foreign governments, loss of effective control, availability of management capable of directing the operation while maintaining loyality to the company, availability of an adequately trained work force, product quality, transportation, availability of supplies, and bad publicity in the United States.

Product cost and facility investment decisions in today's environment are certainly more complex and frought with more uncertainty than those of 40 to 50 years ago. Since much of the data input to these decisions comes from engineers it is especially important that graduates have some grasp of the factors affecting such decisions. For years almost all engineering curricula in the United States included a requirement for three semester hours (equivalent) of economics, usually at the sophomore level. Although the courses offered to satisfy this requirement (more often than not) presented a broad view of economics, with no intentional emphasis on engineering considerations, they did provide engineering students with a general understanding of economic principles and awakened a vague awareness that the engineering profession was governed by these principles just as was the rest of the business community. With increasing emphasis on other subjects that were considered too important to be left out of the curriculum, and with the desire to be competitive as to the required number of hours necessary for graduation, many universities subsequently eliminated this course and often substituted an engineering economics course as an elective. In recognition of the increasing importance of economics and cost in engineering, most universities have more recently made an engineering economics course a requirement.

Although it can be argued that few newly graduated engineers will have the opportunity to apply economic analysis and decision making during the early years of their employment, it is nevertheless essential that they have some understanding of the significance of cost to the success or failure of the projects to which they are assigned. To this end, some of the many and varied cost and economic analyses to which beginning, as well as journeyman, engineers are exposed are outlined in this chapter. Typical design situations and job assignments in which these analyses are necessary are listed below:

1. Obtaining an accurate cost for a significant element in the design of a system to which the new engineer is assigned. This is a simple task but it requires a good understanding of the supplier's operation so that the validity and long-term reliability of the estimate can be ascertained. For significant elements it is especially important that confidence in the integrity of the supplier's management be ensured. Establishing a reasonable delivery schedule is also important in providing an environment in which the supplier can effectively comply with the contract requirements.

2. Estimating the fabrication costs for relatively simple design elements and components. This is a straightforward task of gathering information on the cost of materials and supplies used in the design (usually from outside sources) and in determining the cost of the various production tasks (usually in-house). Manufacturing firms will have a breakdown of the various machining, assembly, and inspection costs that can be utilized in compiling this estimate.

3. Estimating the cost of modifications to previously designed components and systems accomplished by the company's work force either in-house, in the field, or by an outside contractor in-plant or at the field location. This task is considerably more complex, even when the modification effort is accomplished by the in-house work force, since it usually involves making material and component takeoffs from drawings prepared by other designers and working with unfamiliar detail design and assembly drawings. This type of estimating is further complicated by the fact that in-house labor task cost data is not usable for tasks accomplished in another contractor plant or in the field. Estimating is also complicated by general unfamiliarity with the field situation: what tools and equipment, labor skills, supervision, materials, supplies, and other resources are available and at what cost?

    Estimates for contract changes can have significant ramifications, resulting in bad feelings between the customer and the contractor as well as other parties involved in the contract arrangement. The difficulties usually arise from assessments as to who is at fault. The customer may feel that the change was required due to inadequate initial design, whereas the contractor will usually attribute the change to a modification of the customer's requirements. It is especially important in this environment to have a thorough and valid estimate of the cost and a willingness to negotiate a modification cost and schedule that is mutually agreeable to both parties.

4. Evaluating the estimated costs on proposals and bids received from contractors. Engineers usually do a comparative evaluation, comparing the various estimates and determining which constitutes the best offer. If it is a response to a request for bid (RFB), cost is the principal evaluation criterion and the analysis consists of evaluating the cost elements and how they combine to make up the total cost. It is important to ensure that the bid complies with the RFB and, if so, that the contract is awarded to the lowest qualified bidder. For a request for proposal (RFP) the analysis is more complicated. In this case both the cost and the proposed solutions of the various proposers must be evaluated. An RFP competition usually involves additional negotiation with the two to three top proposers to establish comparability for cost evaluation and to bring the proposals to a

state of acceptability to the contracting agency. In-house cost estimates are often made to establish a basis for evaluation.

5. Preparing proposals, either as a result of an RFP or through effort on an unsolicited proposal. These situations require more than a valid estimate of the cost of the effort or product. Questions such as: How important is the project to the company? Is this effort likely to lead to further contracts? Will this effort place the firm in a unique competitive position? Does this work integrate well with the company's long-term marketing and technology goals? Will this effort have a serious impact on staffing and other company resources?–all need to be answered. Whether the company has the financial resources and cash flow to handle the contract is all-important and must be addressed. Large and well respected companies have been forced out of business for failure to ensure adequate cash flow, even when a large backlog of work existed. During the mid-1960s, Douglas Aircraft Company, Santa Monica, CA, a firm with a significant work backlog, was forced to sell out to become part of Mc-Donnell Aircraft Company, St. Louis, MO, to secure adequate cash flow to meet its aircraft delivery commitments.

6. Internal company requests or assignments may also precipitate proposal effort involving engineers. Because these situations involve internal company operations, the data developed are primarily technical, although departmental goals and capabilities have to be considered. The costs associated with these efforts are more often than not reflected in types, manhours, and quantities of materials, and other special requirements.

7. Managing outside contracts and studies. Feasibility studies constitute a special situation in which cost is usually one of the primary factors on which a decision is made to proceed with a project. In the role as one of the team helping to manage a feasibility study, the engineer is required to exercise experience and knowledge in directing the design philosophy and approach adopted, in part, to ensure that cost limitations and/or goals are met.

8. Participating in (or making) decisions as to whether the firm should make certain capital investments. Investment decisions not only involve the time value of money but the long-term future of the company as well. This usually involves a decision as to where the company wants to be a number of years in the future and what other competing investment options are available for the funds. This is the case even though the funds are not readily available and will have to be raised by borrowing or other means. Investments always involve projections of what may happen in the future and are thus frought with uncertainty. Shorter investment time periods mean generally, safer and more predictable outcomes. Interest rates vary and the longer the investment period the greater the uncertainty. Until the early 1970s it was common for capital investment periods in the United States to be 10 to 15 years or longer. As interest rates rose from 6 to 8 percent at that time to 18 to 20 percent in the early 1980s, firms reduced the period over which they were willing to make investments. In today's environment, firms require recovery of invested funds in a period of two to three years. This, of course, significantly reduces the number and variety of investment options open to most firms, and makes the engineer's task considerably more difficult.

5.2   COST ANALYSIS AND CONTROL

In today's world of rapidly advancing technology and intensifying competition, both worldwide and at the local, state, and national levels, it is essential that industrial organizations have viable and responsive cost analysis and control systems. Effective analysis of product or project costs and the ability to implement cost control measures are like two sides of a coin, one is of no use without the other. The term *management* includes management of the product or project cost, and this requires a knowledge and understanding of the cost elements and their sensitivity to various control parameters. Cost analysis forms the basis for cost control, and without accurate and timely cost data, effective cost control is impossible. Management must have the authority to implement cost control procedures and policies, as well. The most accurate and timely cost data are useless unless coupled with an effective cost control mechanism.

In the early stages of a project, cost control is usually accomplished at the work package level. As indicated in Chapter 1, work packages are commonly identified based on the organizational breakdown, and cost control is accomplished through the appropriate organizational entity when this is true. This may result in some difficulty in controlling cost at the project level; however, since the project manager does not have direct management control over the organizational entity responsible for the work package. An additional problem that occurs at this stage of the project is that cost estimates made up to this time are projections based largely on experience with similar projects. Budget allocations to the various organizational entities for direct labor based on these cost estimates, thus, have no relation to the actual work accomplished since they are made before this level of effort is initiated. In addition, the contribution of indirect labor to project output is very difficult to measure. For example, the costs of labor associated with management, supervision, personnel, procurement, labor relations, and the like, can be identified accurately from the payroll of the various departments, but relating the output of this effort to completion of the project defies measurement. Thus, data essential to effective cost control are often lacking at this stage of the program. Because of the lack of definition at this early stage, it is very difficult to determine what percentage of the work has been completed and whether the effort is behind or ahead of schedule. Fortunately, the rate at which costs are incurred early in the program is usually slow, but increasing, and is primarily associated with the cost of labor charged against a particular work package.

As the project evolves, cost data, and the ability to use these data for analysis, continues to improve and to be better defined. Unfortunately (in a cost control sense), the activity and number of personnel involved in the project also increases and, even though better cost data are available, cost control will be even more difficult than at the earlier stages of the project. For large projects, cost control is closely integrated with project planning and the management tools described in Chapter 2. When a planning network is utilized to manage the project, all the cost-incurring activities should be included in the network, and changes in the network must provide for adjustments in the project cost. Planning is an activity that looks to the future and tries to predict the most effective way to accomplish the required activities. Control considers what has been accomplished relative to the plan, including the costs that have been incurred, and implements any changes required to complete the effort within cost and schedule.[1] Regardless of the size of

---

[1]"Project Cost Control Using Networks," The Operational Research Society and The Institute of Cost and Works Accountants, London, 1969.

the project, some degree of planning and control must be implemented. For small projects, or projects of relatively short life, sophisticated planning and cost control procedures are usually not required (or desired). Nevertheless, basic cost tracking and control principles should be understood and applied, as appropriate.

5.2.1   Cost Categories

Several cost categories are available to serve as a basis for economic and cost analysis. Classification into these categories provides an understanding of the effect of the various types of costs on the project and helps to ensure that all project costs are accounted for.[2]

First Cost (Investment Cost)   First cost is the cost of initiating an activity or project. It is usually limited to one-time costs only: those that occur only one time for any given undertaking. First cost includes elements such as the procurement cost of equipment, shipping and installation costs, and any required training costs. For an item that is not off the shelf, it includes design and development costs and construction or production costs as well as shipping, installation, and training costs. First cost is important in that it is a measure of whether a project or activity can be undertaken. Projects that appear to be profitable over a period of time may entail such a high first cost and concomitant level of investment that they are beyond the financial capabilities of the organization.

Operations and Maintenance Costs   Operations and maintenence (O&M) costs are the costs incurred in operating and maintaining entire plants, systems, subsystems, items of equipment, or individual components. This cost category includes labor, fuel, and power costs, materials and supplies, spare parts, repairs, insurance, taxes, and an allocated portion of indirect costs called overhead or burden. O&M costs occur over the life of the item being operated and maintained and usually increase with time. A very common evaluation method for determining when an item should be replaced uses a plot of maintenance and capital costs associated with replacement versus the economic life of the asset. Figure 5.1 shows a typical plot of this type. O&M costs generally increase with the age of the equipment, whereas the cost of investment decreases. The ideal time to replace the asset using this type of analysis is at the minimum life-cycle cost.

Fixed and Variable Costs   Fixed costs are costs that remain relatively constant over the operational life of the facility, system, subsystem, equipment, or component. Fixed costs are independent of the production volume or output and include elements such as depreciation, taxes, insurance, interest on invested capital, sales, and some administrative expenses. Production volume can change rapidly affecting labor costs, materials and supplies, utilities, and the like. Fixed costs vary at a much slower rate and thus must be established based on a production volume and mode of operation that are considered appropriate by management. Fixed costs arise from predictions about where the company wants to be in the future. Equipment purchased now may allow reduction of labor costs in the future or may provide for product improvement or diversification. Research is conducted that is not related to ongoing production in the hope that some future payoff may justify the expenditure. Fixed cost investments are made in anticipation that profit will be increased in the future by reduced variable costs or by increased sales and income.

---

[2]W. J. Fabrycky, and B. S. Blanchard, *Life-Cycle Cost and Economic Analysis,* PrenticeHall, Englewood Cliffs, NJ, 1991

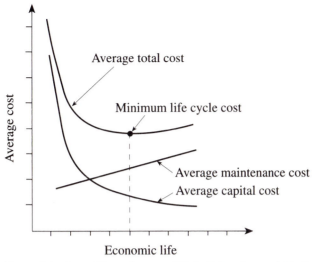

Figure 5.1   Asset life-cycle cost (W. J. Fabrycky, and B. S. Blanchard, *Life-Cycle Cost and Economic Analysis*, 1991, p. 250. Adapted by permission of Prentice Hall, Englewood Cliffs, NJ).

Variable costs are a function of the level of production or activity. As production levels or activity increase, certain (variable) costs will increase in response. Direct labor, materials, utilities, and other costs that can be allocated on a per production unit basis increase with increased production levels. Materials costs would normally vary directly with the number of units produced, but increased purchases due to the increased production may result in lower costs due to discounts, reduced transportation costs, and the like. Labor costs vary based on how much the production line efficiency can be increased from greater volume, use of overtime, and requirements for additional staffing. Figure 5.2 shows the typical relationship between fixed, variable, and total costs.

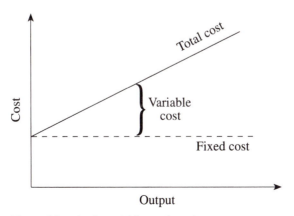

Figure 5.2   Fixed, variable, and total costs versus output (W. J. Fabrycky, and B. S. Blanchard, *Life-Cycle Cost and Economic Analysis*, 1991, p. 24. Adapted by permission of Prentice Hall, Englewood Cliffs, NJ).

Incremental or Marginal Cost   Incremental or marginal cost both refer to increases in cost. Marginal cost is a term used to describe the situation in which the cost of an incremental level of output is just covered by the income generated. Incremental costs are normally applied to unit costs and are referred to as incremental cost per pound, incremental cost per unit of production, and so forth. Incremental costs are applied in manufacturing operations to project the added cost of increased production. In this case it is necessary to estimate the cost per unit for increasing the output by some increment. Figure 5.3 shows the relationship between total (fixed and variable) cost and output for a hypothetical manufacturing operation. The cost of increasing production from Q units to $Q + \Delta Q$ units is shown to be ΔC dollars.

Direct and Indirect Costs   The costs incurred in all business and industrial operations are assigned to two principal categories—direct and indirect costs. Direct costs include material, labor, and any subcontracts that provide elements used in the product. The quantity and type of direct materials used in manufacturing processes can be determined from the engineering drawing bill of materials. The cost of this material is obtained from suppliers. Direct labor for manufacturing processes usually includes only the hourly workers involved and excludes salaried personnel, even though they may be supervising and managing the production process. Thus, the indirect labor cost includes supervision and management as well as other support operations labor necessary to produce the product. Indirect material costs include materials used in the overall production operation but not used in the product itself. Other indirect expenses include the cost of utilities, building rentals, depreciation, taxes, insurance, maintenance, and the like.

Recurring and Nonrecurring Costs   Recurring costs are those that continue to occur over the life of the project and include categories such as manufacturing costs, engineering support required during production or construction, customer support over the life of the product, and ongoing project management. Nonrecurring costs are costs that occur only one time during the project and include cost categories such as product applied research, design and development, testing other than acceptance testing, new construction, and manufacturing tools and equipment.

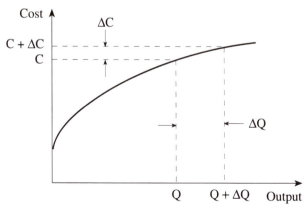

Figure 5.3   Incremental costs from increased output (W. J. Fabrycky and B. S. Blanchard, *Live-Cycle Cost and Economic Analysis,* 1991, p. 25. Adapted by permission of Prentice Hall, Englewood Cliffs, NJ).

When evaluating the costs associated with product and project changes, schedule modifications, and increased or decreased production levels, the evaluation of recurring and nonrecurring costs is helpful in providing insight as to the actual costs and their source.

When evaluating the life-cycle cost of the product, all recurring and nonrecurring costs are included. During design and development costs incurred are primarily nonrecurring. As the product moves into the production and utilization phases, recurring costs are incurred. Life-cycle cost analysis considers all of these costs, in addition to the investment costs.

Sunk (Past) Cost   Sunk or past costs are project expenditures that have occurred in the past and cannot be recovered by any future action. Since these sunk costs are not relevant to cost analyses concerned with future events, they must be ignored when project decisions are made based on the projection of events.

5.2.2   Cost Estimating

The ability to estimate costs accurately is essential if a firm is to stay in business. Unfortunately, cost estimating is not an exact science and in the best of circumstances will only provide an approximation of the cost that will actually be incurred. It is thus essential that the various methods and tools available for this task be understood and be applied so that the degree of approximation will be minimized. Various methods for estimating costs are described in the following discussion.[2]

The Engineering Estimate   An engineering estimate is made up of the individual costs of many elements of the design at a low level of detail. The estimate includes costs for every element of the design and fabrication process including the cost of production tools, jigs, and fixtures. Components, lengths of piping and tubing, the number and size of welds, the quantities and types of materials, surface treatments, test requirements, and the like, are identified from the engineering drawings and are multiplied by the element unit cost that may come from vendors, or from production and in-house cost standards. Element costs are determined for each of these elements at the lowest reasonable level of detail. Some elements of the overall cost, such as inspection, production control, and so forth, are estimated by taking a percentage of direct labor, which invariably results in uncertainty in the final estimate. The overall estimate is made by adding up the total of all these individual element costs.

There are two basic drawbacks to the use of engineering estimates—the time required to prepare the estimate and the difficulty in *seeing the forest and not just the trees.* When a cost estimate is made up of many individual cost elements, it is difficult to keep an overall perspective and not get lost in all the detail. When this cost estimating technique is used it is often augmented by the use of other (more gross) techniques to ensure that the overall estimate is valid. Engineering estimates are often used in construction projects, especially when field changes are being negotiated.

Estimating by Use of Analogy   Gross estimates are often made by using cost data from similar programs. When little detailed information is available on which to base the cost, using data from similar programs may be the only way to develop a reasonable estimate. This is the type of estimating that companies use when initiating new product lines or projects. Cost data from the analogous program are adjusted to correspond to changes in performance, numbers of major elements, differences in size, and so forth, for the new program. An example of this type of estimating is documented in the experience of the aircraft companies during the missile development era of the mid-1950s. Since no aircraft company had any experience in the development and production of missiles the extensive cost data that were available from aircraft programs were adapted for use in estimating the cost of missile development and production.

This type of estimating is also used at lower levels of detail such as basing the direct labor hours on the number of hours required to fabricate a similar part or on the cost of direct material.

Statistical Estimating Methods   Statistical cost estimating is based on determining the relationship between cost and the factor(s) on which it depends. Factors, such as power rating, flowrate, mass, volume, and production quantities are used to make statistical cost estimates. For example, rough estimates of the fabrication costs for certain materials are estimated based on the mass of the direct material used. Thus, if a product made of steel weighs 1000 lb, its cost of fabrication might be estimated as $1000 or $1.00/lb.

This estimating technique is often used for long-range planning. When little information is available, as at the beginning of a project, it will normally be necessary to include a cost for contingencies in the estimate to provide for unforeseen events that result in changes to the design. At this point in the effort, the cost estimate is based on the total cost of the end-product with little or no cost detail below the system or end product level. Cost estimates can be made by using statistical data and can be validated by using cost data from analogous programs or products. As the project evolves and the end product becomes better defined, estimates can be made with less contingency cost, since there is less uncertainty in the overall design and corresponding cost. At this stage, estimates can be based on the costs of the elements included in the overall system by making an engineering estimate. As in the earlier program phase, the engineering estimate can be validated by making an estimate using cost data from analogous programs. Figure 5.4 shows how these estimating techniques can be used relative to the various phases of a typical engineering design project.

5.2.3   Life-Cycle Costing

Life-cycle costing refers to the cost estimating approach that encompasses the costs of all phases in the life of the project or product, including research and development costs, production costs, operation and support costs, and retirement and disposal costs.[2] In the early stages of a design effort firm cost targets should be established and used to manage the various project phases. One technique that has been used successfully in this regard is the *design-to-cost* approach wherein cost is established as a project constraint similar to performance or schedule. As with other design and development program management approaches, design-to-cost requires the assignment of cost targets to significant elements of the program. The primary difference is in the aggressive manner in which these elements are held to these cost targets. Trade-off studies are used extensively in an attempt to find other ways in which effort can be accomplished with reduced cost. System performance requirements are constantly reviewed to identify areas of potential cost reduction. Test requirements are scrutinized to ensure that only essential development testing is

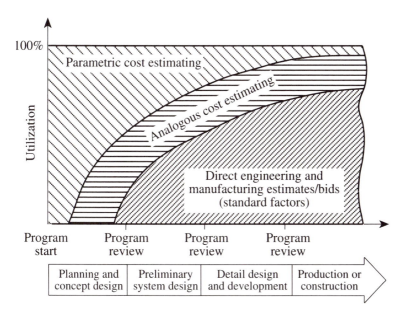

Figure 5.4 Use of estimating methods versus project phase (W. J. Fabrycky and B. S. Blanchard, *Life-Cycle Cost and Economic Analysis,* 1991, p. 148. Adapted by permission of Prentice-Hall, Englewood Cliffs, NJ).

accomplished. This often places the program manager at odds with the engineering staff since some testing thought to be essential by engineering may be eliminated by the program manager to reduce costs. When this occurs, the program manager must be especially sensitive to engineering input and try to gain a consensus decision based on the criticality of cost relative to the need for information derived from the test. Development of the NASA Space Shuttle is a notable example of a design-to-cost effort wherein the balance between cost and development testing was managed effectively. Unfortunately, this same effectiveness did not extend to the operations phase of this program. The result was a serious and widely publicized accident that caused a significant launch schedule delay and program cost increase. When the cost of the program is paramount, the use of this management/costing technique is essential, but excellent judgment must be applied in making decisions regarding test and performance trade-offs, especially when personnel safety is involved.

One method of managing cost in a major program is to use a cost breakdown structure. A cost breakdown structure ties the project activities to the available resources by subdividing the total cost into logical categories such as the functional areas and major tasks as shown by Figure 5.5. For a program to be managed effectively, the categories shown in Figure 5.5 must be expanded to include greater detail and to a lower (subsystems and component) level, such as that provided by the work breakdown structure shown in Figure 1.2.

The level of definition used in the cost breakdown structure is tailored to the manner in which cost is managed. It must provide the necessary cost data to the organizational level charged with managing each element of the effort. In accomplishing this, the cost breakdown structure should have the following characteristics:[2]

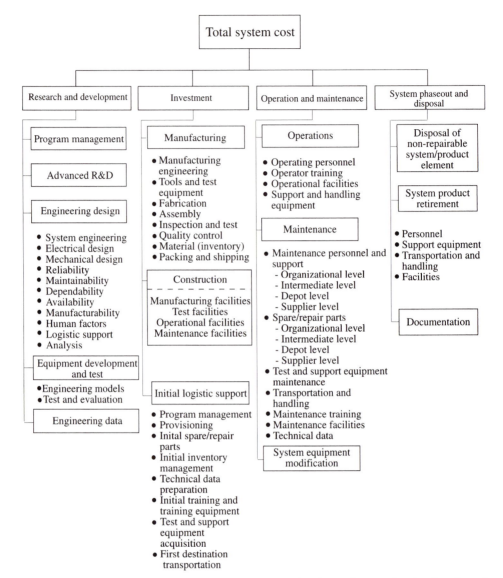

Figure 5.5   Typical cost breakdown structure (W. J. Fabrycky and B. S. Blanchard, *Life-Cycle Cost and Economic Analysis*, 1991, pp. 29, 333. Adapted by permission of Prentice-Hall, Englewood Cliffs, NJ).

1. All costs associated with the project must be included.

2. Cost categories must be well defined and clearly understood by all project personnel.

3. The cost breakdown structure must be broken down to the level necessary for management to identify areas of high cost and cause-and-effect relationships.

4. The cost breakdown structure must include a cost coding scheme that allows the collection of cost data for analysis of specific areas of interest, such as distribution costs as a function of manufacturing.

5. The cost coding scheme must allow for separation of producer, supplier, and consumer costs expeditiously.

6. The cost breakdown structure must be compatible (through the use of the cost coding scheme and other techniques) with planning documentation, the work breakdown structure, the organizational structure, scheduling techniques, and management information systems used.

Neglecting the time value of money, the life-cycle cost of an asset can be determined by using Eq. 5.1

$$\text{ALCC} = \frac{P}{n} + O + (n-1)\frac{M}{2} \tag{5.1}$$

where
ALCC = average annual life-cycle cost

O = constant annual operating cost (equal to first year operating cost including a portion of maintenance)

M = annual increase in maintenence costs

n = life of the asset in years

P = first cost of asset

The minimum cost life can be found by differentiating 5.1 and setting it equal to zero.

$$\frac{d\text{ALCC}}{dn} = -\frac{P}{n^2} + \frac{M}{2} = 0 \tag{5.2}$$

Hence,

$$n^* = \sqrt{\frac{2P}{M}} \tag{5.3}$$

Substituting Eq. (5.3) into Eq. (5.1) yields the minimum life-cycle cost as

$$\text{ALCC}^* = \sqrt{2PM} + O - \frac{M}{2} \tag{5.4}$$

As an example of the use of Eq. (5.3), consider the purchase of a new automobile with a first cost of $12,150 and a first year operating and maintenance cost of $1000, with maintenance costs increasing by $300/year. The minimum-cost life from Eq. 5.3 is,

$$n^* = \sqrt{\frac{2(12,150)}{300}} = 9 \text{ years}$$

The minimum annual life-cycle cost for this period of time using Eq. 5.4 is

$$\text{ALCC}^* = \sqrt{2(12,150)(300)} + 1000 - 150$$
$$= \$3550$$

Table 5.1   Average Annual Cost of an Asset

End of Year	Maintenance Cost at End of Year	Cumulative Maintenance Cost	Average Maintenance Cost	Average Capital Cost	Average Life-Cycle Cost
A	B	$C = \sum B$	$D = C/A$	$E = \$12,150/A$	$F = D + E$
1	$1,000	$ 1,000	$1,000	$12,150	$13,150
2	1,300	2,300	1,150	6,075	7,200
3	1,600	3,900	1,300	4,050	5,350
4	1,900	5,800	1,450	3,038	4,488
5	2,200	8,000	1,600	2,430	4,050
6	2,500	10,500	1,750	2,025	3,775
7	2,800	13,300	1,900	1,736	3,636
8	3,100	16,400	2,050	1,519	3,569
9	3,400	19,800	2,200	1,350	3,550
10	3,700	23,500	2,350	1,215	3,565

From W.J. Fabrycky and B. S. Blanchard, *Life-Cycle Cost and Economic Analysis*, 1991, p. 249. (Adapted by permission of Prentice Hall, Englewood Cliffs, NJ.)

This example can also be solved by tabulating the cost components as shown in Table 5.1. As indicated, the minimum average life-cycle cost occurs at nine years.

Unfortunately, salvage values and annual maintenance cost increases do not usually change in linear fashion. For these situations the tabular approach *must* be used.

5.3   IMPORTANT ECONOMIC CONCEPTS

When the consequences of economic decisions are immediate, or occur over a short period of time, the change in the value of money with time does not have to be accounted for. The positive and negative aspects of the decision can be evaluated and the decision can be made and implemented. Most business economic decisions involve the element of time, however, and it is thus necessary to consider factors such as interest, the time value of money, and depreciation of equipment.

5.3.1   Interest

Interest can be thought of as the rental charge levied by financial institutions on the use of money. Like other rental charges, the level or rate of interest is determined by supply and demand. If individuals as a whole increase the portion of their earnings put into savings the supply of money will increase and rates will generally go down. If savings are reduced, the amount of money available for financial institutions to provide for loans will decrease and rates will normally rise. Governments usually believe that it is in the best interest of their citizens to influence the money market, however. Thus, if the U.S. government believes that the economy is sliding into recession, it may, through the Federal Reserve Board, reduce the interest rate charged to member banks (referred to as the *prime rate*) and thus strongly influence the rate to remain low to stimulate the economy. In this sense the money market is unlike other free market commodities that operate unencumbered by government influence.

The interest rate is the ratio (expressed as a percentage) of the rental charge on the borrowed money to the total amount of money borrowed over a period of time, usually one year. For example, if $100 is charged for the use of $1000 for one year the interest rate is 10 percent per annum. For simple interest, the amount of interest paid on repayment of a loan is proportional to the length of time that the principal sum has been borrowed. Simple interest can be calculated using

$$I = Pni \qquad (5.5)$$

where
  I = interest earned,$
  P = amount of money loaned (principal),$
  n = interest time period, usually years
  i = interest rate.

For this example, the interest earned over a period of one year is

$$I = \$1000(1)(0.10) = \$100$$

At the end of the year, both the principal ($1000) and interest ($100) would be due.

A simple interest loan can be negotiated for any period of time. The principal and interest become due only at the end of the loan period, which can be for less than one year or for multiple years. For loans of less than one year the interest would be calculated using Eq. 5.5 with n equal to the loan period in days divided by the number of days in a year.

When a loan is made that covers several interest periods, interest is determined at the end of each period. This interest could be paid at the end of each period or could be accumulated until the loan is due for repayment. If the interest is accumulated and the borrower is charged interest on the total amount owed, including principal and interest, the interest is said to be compounded. Tables 5.2 and 5.3 show a comparison between a four-year loan of $1000 at an interest rate of 10 percent under simple and compound interest. In Table 5.3 the interest is paid at the end of each period and thus payment of interest on the previously earned interest is avoided.[3]

5.3.2 Equivalence and the Time Value of Money

To make a selection between two alternative pieces of equipment, it is necessary for the capabilities to be defined in identical units so that a meaningful comparison can

Table 5.2 Calculation of Interest When Interest Is Paid Annually

Year	Amount Owed at Beginning of Year	Interest To Be Paid at End of Year	Amount Owed at End of Year	Amount to Be Paid by Borrower at End of Year
1	$1,000	$100	$1,100	$ 100
2	1,000	100	1,100	100
3	1,000	100	1,100	100
4	1,000	100	1,100	1,100

From G. J. Thuesen and W. J. Fabricky, *Engineering Economy*, Prentice Hall, 1989, p.36. (Adapted with permission of Prentice Hall, Englewood Cliffs, NJ.)

[3]G. J. Thuesen and W. J. Fabrycky, *Engineering Economy*, Prentice Hall, Englewood Cliffs, NJ, 1989

Table 5.3   Calculation of Interest When Interest Is Allowed to Compound

Year	Amount Owed at Beginning of Year (A), $	Interest To Be Paid at End of Year (B), $	Amount Owed at End of Year (A + B), $	Amount to Be Paid by Borrower at End of Year
1	1,000	$1,000(0.1) = 100$	$1,000(1.1) = 1,100$	0.00
2	1,100	$1,100(0.1) = 110$	$1,000(1.1)^2 = 1,210$	0.00
3	1,210	$1,210(0.1) = 121$	$1,000(1.1)^3 = 1,331$	0.00
4	1,331	$1,331(0.1) = 133.10$	$1,000(1.1)^4 = 1,464$	1,464

From G. J. Thuesen and W. J. Fabricky, *Engineering Economy*, Prentice Hall, 1989, p.36. (Adapted by permission of Prentice Hall, Englewood Cliffs, NJ.)

be made. Thus, if a natural gas compressor is to be selected from two alternatives such as,

1. capacity = 35 cfm at 3000 psi with a suction pressure of 50 psi or
2. capacity = 1000 scfm, maximum discharge pressure of 4000 psi, suction pressure of 5 psi

it would be necessary to convert the volumetric flowrates to some equivalent basis to make a valid comparison. Economic decisions are no different in this regard. To make a decision between economic alternatives involving several options, it is necessary for the comparison to be made on an *equivalent* or *apples-to-apples* basis. For financial decisions this involves the monetary amounts, the time periods over which the sums are received or paid out, and the interest rate charged or earned. Table 5.4 shows two repayment schemes with an equivalent value of $10,000 now, with interest at a rate of 10 percent. One plan will be preferred over the other depending on whether a larger annual payment is desired over the five year period or whether smaller annual payments are desired with one large payment at the end of the five-year period.

5.3.3   Cash Flow

A cash flow diagram provides a graphical description of an alternative's receipts and disbursements over a selected period of time. Receipts are reflected by upward-pointing arrows (an increase in cash) and disbursements are shown by downward-pointing arrows (a decrease in cash). Figure 5.6 shows cash flow diagrams for both

Table 5.4   Alternatives for Repayment of $10,000 in Five Years at 10% Interest

Year	Plan 1	Plan 2
1	−$ 3,000	−$ 1,000
2	− 2,800	− 1,000
3	− 2,600	− 1,000
4	− 2,400	− 1,000
5	− 2,200	− 11,000
Total	−$13,000	−$15,000

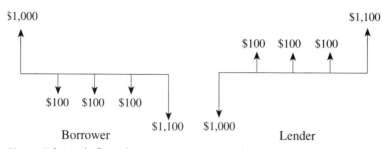

Figure 5.6 Cash flow diagrams. From G. J. Theusen and W. J. Fabrycky, *Engineering Economy,* 1989, p.38. (Adapted with permission of Prentice Hall, Englewood Cliffs, NJ).

the borrower and the lender for a loan of $1000. The borrower receives $1000 initially at time $t = 0$ and pays $100 in interest at the end of every period over the life of the loan agreement. The $1000 principal is also repaid at the end of the loan period with the last interest payment. The lender experiences a negative cash flow initially as shown by the downward-pointing arrow and receives the $100 interest payments at the end of each period in addition to the $1000 principal at the end of the loan agreement.

If both negative and positive cash flows are experienced during an interest period, the net value is shown and the length and direction of the arrow for that interest period are determined by the algebraic sum of the positive and negative cash flows. Disbursements made to initiate an investment are shown at the beginning of the investment period. Receipts and disbursements that occur during the life of the investment are considered to have transpired at the end of the period in which they occur. Table 5.5 summarizes the three groups of interest formulas for the various types of payments and interest compounding methods normally encountered.

5.3.4 Present Worth Analysis

Present worth is the amount of money that represents the difference between the present equivalent receipts and the present equivalent disbursements for an investment at a given interest rate. The present worth amount at an interest rate i over a period of n years is found by using

$$PW(i) = \sum_{t=0}^{n} F_t(1 + i)^{-t} \qquad (5.6)$$

where
$PW(i)$ = present worth at interest rate i, and
$\quad F_t$ = future amount at time t, years.

The present worth amount includes the time value of money through the value of i selected. It also provides an equivalent value of the net cash flow at any point in time from $t = 0$ to $t = n$. Finally, a single unique value for the present worth is associated with each interest rate used no matter what the cash flow pattern may be. Present worth is thus a valuable tool in determining the viability of alternatives in the decision-making process. Figure 5.7 shows the range of present-worth values versus

Table 5.5 Summary of Interest for Various Payment in Interest Compounding Method

	Factor	Find	Given	Discrete Payments, Discrete Compounding	Discrete Payments, Continuous Compounding	Continuous Payments, Continuous Compounding
Single Payment	Compound amount	F	P	$F = P(1+i)^n = P^{(F/P,i,n)}$	$F = Pe^{rn} = P^{(F/P,r,n)}$	$F = Pe^{rn} = P^{(F/P,r,n)}$
	Present worth	P	F	$P = F/(1+i)^n = F^{(P/F,i,n)}$	$P = F/e^{rn} = F^{(P/F,r,n)}$	$P = F/e^{rn} = F^{(P/F,r,n)}$
Equal-Payment Series	Compound amount	F	A	$F = A\left[\dfrac{(1+i)^n - 1}{i}\right]$ $= A^{(F/A,i,n)}$	$F = A\left[\dfrac{e^{rn} - 1}{e^r - 1}\right]$ $= A^{(F/A,r,n)}$	$F = \bar{A}\left[\dfrac{e^{rn} - 1}{r}\right]$ $= \bar{A}^{(F/\bar{A},r,n)}$
	Sinking fund	A	F	$A = F\left[\dfrac{i}{(1+i)^n - 1}\right]$ $= F^{(A/F,i,n)}$	$A = F\left[\dfrac{e^r - 1}{e^{rn} - 1}\right]$ $= F^{(A/F,r,n)}$	$\bar{A} = F\left[\dfrac{r}{e^{rn} - 1}\right]$ $= F^{(\bar{A}/F,r,n)}$
	Present worth	P	A	$P = A\left[\dfrac{(1+i)^n - 1}{i(1+i)^n}\right]$ $= A^{(P/A,i,n)}$	$P = A\left[\dfrac{1 - e^{rn}}{e^r - 1}\right]$ $= A^{(P/A,r,n)}$	$P = \bar{A}\left[\dfrac{e^{rn} - 1}{re^{rn}}\right]$ $= \bar{A}^{(P/\bar{A},r,n)}$
	Capital recovery	A	P	$A = P\left[\dfrac{i(1+i)^n}{(1+i)^n - 1}\right]$ $= P^{(A/P,i,n)}$	$A = P\left[\dfrac{e^r - 1}{1 - e^{rn}}\right]$ $= P^{(A/P,r,n)}$	$\bar{A} = P\left[\dfrac{re^{rn}}{e^{rn} - 1}\right]$ $= P^{(\bar{A}/P,r,n)}$
Gradient Series	Uniform gradient series	A	G	$A = G\left[\dfrac{1}{i}\dfrac{n}{(1+i)^n - 1}\right]$ $= G^{(A/G,i,n)}$	$A = G\left[\dfrac{1}{e^r - 1} - \dfrac{n}{e^{rn} - 1}\right]$ $= G^{(A/G,r,n)}$	
	Geometric gradient	P	F_1	$P = \dfrac{F_1}{1+g}\left[\dfrac{(1+g')^n - 1}{g'(1+g')^n}\right]$ $= F_1^{(P/A,g',n)}$		

From G. J. Thuesen and W. J. Fabrycky, *Engineering Economy*, 1989, p. 68. (Adapted by permission of Prentice Hall, Englewood Cliffs, NJ.)

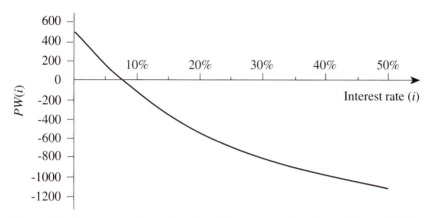

Figure 5.7 Present worth as a function of interest rate for the cash flow of Table 5.6 From G. J. Theusen and W. J. Fabrycky, *Engineering Economy*, 1989, p. 163. (Adapted by permission of Prentice Hall, Englewood Cliffs, NJ).

interest rates for the cash flow depicted by Table 5.6.[3] Note that the equivalent receipts are greater than the disbursements for interest rates through 7.5 percent. Stated in another way, the investment provides $23 in equivalent profit if the time value of money is 7.5 percent. At higher interest rates the equivalent receipts are less than the disbursements, which results in a negative present worth. The financial arrangement may still be worth pursuing, however, depending on the outlook for future interest rate changes and other lender considerations.

5.3.5 Internal Rate-of-Return Analysis

Rate-of-return analysis is one of the most widely used methods of evaluating investment opportunities. It provides a measure of the economic viability of an investment alternative that can be compared to other opportunities for which the rates of return are known. Rate of return is defined as the interest rate that causes the equivalent receipts of a cash flow to be equal to the equivalent disbursements of that cash flow. Stated another way, this method of analysis yields the interest

Table 5.6 Present Worth from a Cash Flow for a Range of Interest Rates

$$PW(i) = \sum_{t=1}^{n} F_t (1 + i)^{-t}$$

$$= -2,000 + 500[\sum_{t=1}^{n} (1 + i^*)^{-t}]$$

End of Year	Cash Flow ($)	i (%)	
0	2000	0	$500
1	500	5	165
2	500	7.5	23
3	500	15	−32
4	500	30	732
5	500	50	−1132
		∞	−2000

G. J. Thuesen and W. J. Fabrycky, *Engineering Economy*, 1989, p. 163. Adapted by permission of Prentice-Hall, Englewood Cliffs, NJ.

rate that reduces the present, annual, or future worth of a series of receipts and disbursements to zero. Thus, the internal rate of return for an investment alternative is the interest rate i^* that satisfies Eq. 5.6 when $PW(i) = 0$ as shown below:

$$0 = PW(i^*) = \sum_{t=0}^{n} F_t(1 + i^*)^{-t}$$

where the alternative has a life of n years.

To calculate the rate of return for the cash flow shown in Figure 5.8, the i^* that causes the present equivalent worth to equal zero is determined. This requires the use of a trial-and-error solution. Assuming that $i = 30$ percent gives

$$\$5,000 + \$15,000(0.7692) \neq \$10,000(0.5917) + \$10,000(0.4552)$$
$$+ \$10,000(0.3501) + \$15,000(0.2693)$$
$$\$16,538 \neq \$18,010$$

For $i = 35$ percent,

$$\$5,000 + \$15,000(0.7407) \neq \$10,000(0.5487) + \$10,000(0.4064)$$
$$+ \$10,000(0.3011) + \$15,000(0.2230)$$
$$\$16,111 \neq \$15,907$$

Interpolating gives

$$i = 30\% + 5 \left[\frac{(18,010 - 16,538) - 0}{(18,010 - 16,538) - (15,907 - 16,111)} \right]$$

$$= 34.39\%$$

Figure 5.9 depicts the results of these calculations and indicates that the investment of $5000 and $15,000 at the beginning of years 1 and 2, respectively, will yield a rate of return of 34.39 percent over the five-year period.

5.3.6 Payback Period

When the required rate of return on an investment is known, the length of time needed for the investment to pay for itself can be determined. Investments that tend to pay for themselves in short time periods are more desirable than long-period

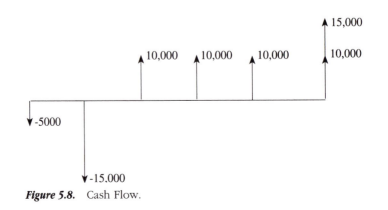

Figure 5.8.   Cash Flow.

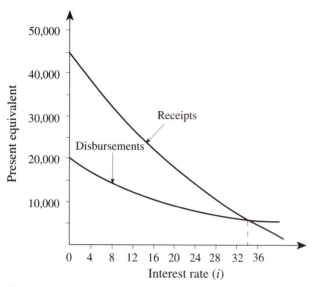

Figure 5.9   Present equivalent versus interest rate graph for rate of return analysis. From W. J. Fabrycky and B. S. Blanchard, *Life-Cycle Cost and Economic Analysis,* 1991, p.59. (Adapted by permission of Prentice Hall, Englewood Cliffs, NJ).

investments since the return of capital investment is quicker and there is less uncertainty with the shorter time period. Payback periods are often quoted without considering the time value of money ($i = 0$ percent), as well. This approach may be warranted in situations where uncertainty as to the appropriate interest rate is high, but one should be aware of the limitations of this method of analysis. For this case, Eq. 5.7

$$\sum_{t=0}^{n} F_t \geq 0 \tag{5.7}$$

can be used to calculate the minimum number of periods, or years, required to recover the initial investment. Table 5.7 shows the cash flow for three investment

Table 5.7   Three Alternatives with a Three-Year Payback Period

End of Year	A	B	C
0	−$1000	−$2000	−$900
1	500	600	−300
2	300	500	600
3	200	900	600
4	200	1000	0
5	200	2000	0
6	2000	2000	0
Present worth, $i = 0$	$PW(0)_A = \$600$	$PW(0)_B = \$5,000$	$PW(0)_C = \$0$
Payback period	3 years	3 years	3 years

(G. J. Thuesen and W. J. Fabrycky *Engineering Economy,* Prentice Hall, p. 177. 1989. Adapted by permission of Prentice Hall, Englewood Cliffs, NJ).

alternatives for which the payback period is three years.[3] Although the payback period is identical, the present worth of these investments varies considerably. The use of payback analysis with zero interest has serious drawbacks.

To include consideration of the time value of money, a method known as the discounted payback period can be used. This method allows the determination of the time required for an investment's present equivalent receipts to equal or exceed the present equivalent disbursements. Using F_t and t as defined previously, the discounted payback period is the smallest value of n^* that satisfies the expression

$$\sum_{t=0}^{n^*} F_t(1 + i)^{-t} \geq 0 \qquad (5.8)$$

Using an interest rate of 12 percent for Alternative A in Table 5.7,

$$-\$1000 + \$500(0.8929) + \$300(0.7972)+$$
$$\$200(0.7118) + \$200(0.6356) + \$200(0.5674) \geq 0$$
$$\$68 \geq 0$$

Thus, the shortest time period that will satisfy the inequality is an n^* of five years. This compares with a period of three years when the time value of money is not accounted for. At an interest rate of 12 percent, the discounted payback period for investment B in Table 5.7 can be determined to be four years. At this interest rate, alternative C never recovers its investment. The importance of using the discounted payback period method is thus apparent whenever the interest rate can be estimated, even approximately.

5.3.7   Depreciation and Taxes

Depreciation can be defined as the reduction in the value of an asset with the passage of time. Thus, an income producing asset can be depreciated over its expected life to reflect reduced ability to perform its intended service. The primary causes of physical depreciation are deterioration because of the action of the elements and wear and tear from use. Functional depreciation is the reduction in value of an asset as a result of changed demand for the services the asset performs. This can come about as a result of technology improvement that causes the asset to become obsolete, as a result of a change in the type of work required, or as a result of the need to increase capacity beyond the capability of the asset.

If an asset is used to produce income, depreciation can be taken into account for tax purposes. Assets can be composed of either tangible or intangible property. Intangible property includes such things as designs, licenses, patents, and copyrights. A deduction for depreciation of such assets can be taken whether or not income was actually derived from the asset. Thus, tax law in the United States is structured to encourage investment. Unfortunately, changes in the allowable depreciation schedules in 1987 seriously reduced the flexibility in use of the various depreciation options and seriously reduced the advantage of investment in some types of assets. At present the only options allowable by the federal government under the current Modified Accelerated Cost Recovery System (MACRS) are to use the specified depreciation schedule provided by the Internal Revenue Service (IRS) or the straight line method. For example, Tables 5.8 and 5.9 depict the taxes paid on an asset under these two alternatives. The first cost of the asset is $2000, and it is projected

Table 5.8   Straight Line Method of Depreciation for an Asset

Income Taxes for Straight-Line Method of Depreciation and 10-Years Life

Year End	First Cost	Income Before Depreciation and Income Tax	Annual Book Depreciation	Income Less Depreciation (Taxable Income) $C - D$	Income Tax Rate	Income Tax $E \times F$
A	B	C	D	E	F	G
0						
1	$2000	$500	$100	$400	0.32	$128
2		500	200	$300	0.32	96
3		500	200	$300	0.32	96
4		500	200	$300	0.32	96
5		500	200	$300	0.32	96
6		500	200	$300	0.32	96
7		500	200	$300	0.32	96
8		500	200	$300	0.32	96
9		500	200	$300	0.32	96
10		500	200	$300	0.32	96
11		500	100	$400	0.32	128
		$5500	$2000			$1120

$$\text{Present worth of income taxes} = \sum_{t=1}^{11}\left[\text{Col.}\,G \times \left(P/F,12,i\right)\right] = \$598$$

G. J. Thuesen and W. J. Fabrycky, *Engineering Economy*, 1989, p. 420. Adapted by permission of Prentice-Hall, Englewood Cliffs, NJ.

Table 5.9   MACRS Prescribed Percentage Depreciation for an Asset

Income Taxes for MACRS Method of Depreciation and 10-Years Life

Year End	First Cost	Income Before Depreciation and Income Tax	Annual Book Depreciation	Income Less Depreciation (Taxable Income) $C - D$	Income Tax Rate	Income Tax $E \times F$
A	B	C	D	E	F	G
0						
1	$2000	$500	$200	$300	0.32	$96.00
2		500	360	$140	0.32	44.80
3		500	288	$212	0.32	67.84
4		500	230	$270	0.32	86.40
5		500	184	$316	0.32	101.12
6		500	146	$354	0.32	113.28
7		500	132	$368	0.32	117.76
8		500	132	$368	0.32	117.76
9		500	132	$368	0.32	117.76
10		500	132	$368	0.32	117.76
11		500	64	$436	0.32	139.52
		$5500	$2000			$1120

$$\text{Present worth of income taxes} = \sum_{t=1}^{11}\left[\text{Col.}\,G \times \left(P/F,12,i\right)\right] = \$564$$

G. J. Thuesen and W. J. Fabrycky, *Engineering Economy*, 1989, p.420. Adapted by permission of Prentice-Hall, Englewood Cliffs, NJ.

to have zero salvage value. The property class as specified by the IRS requires that the asset be depreciated over a life of 10 years. The asset is expected to bring in an income of $500 per year before depreciation and income taxes, and a tax rate of 32 percent is projected over the life of the machine. The taxpayer normally earns a rate of 12 percent return on investments and wants to compare the MACRS specified depreciation schedule and the straight-line approach.[3]

For the two cases shown (Tables 5.8 and 5.9), the total income tax paid over the life of the asset is the same. However, the present worth of the taxes at the beginning of year 1 differs by $34 in favor of the prescribed percentage method. Note the higher depreciation under the MACRS method during the early years, which results in a faster write-off and an overall improved approach over the straight-line method for calculating taxes for this asset. Thus, the MACRS method is an accelerated depreciation method that will result in postponing the payment of taxes.

5.3.8   Inflation and Deflation

Inflation and deflation are terms used to describe the change in price levels of goods and services with time. Inflation has been a much more common occurrence in world economies during the past half century but deflation was a major malady during the 1920s and 1930s. The cost of goods and services is driven by various factors. When the supply of goods and services increases without a corresponding increase in demand, prices go down. In general, the opposite is true when supply decreases, but in recent years government policies also have had a significant and overriding inflationary effect. Government policies that tend to reduce the value of the dollar, such as price supports, deficit financing, and increasing the money supply, all result in inflationary pressure on the economy. The embargo placed on the supply of oil by the oil-producing nations in the early 1970s is a good example of how reduced supply can affect the price of goods. The price of oil experienced a tenfold increase in price in a matter of a few months as a result of this action, which was strictly arbitrary and not a result of any systemic cost increases.

To provide a way to account for these price changes over time, a price index is used. A price index is the ratio of the price of a commodity at a point in time relative to the price at some earlier time. The federal government prepares various composite price indexes including the Consumer Price Index (CPI), the Producer Price Index (PPI), and the Implicit Price Index for the Gross National Product (IPIGNP), which measure historical price level changes in the economy. Figure 5.10 depicts the Consumer Price Index for the United States for the years 1967 to 1988.[3]

The price index is usually expressed as a number corresponding to the ratio of prices for the years of interest multiplied by 100. For example, if the price of a loaf of bread was $0.25 in 1967 and in 1991 the price was $1.19, the price index would be ($1.19/$0.25)(100) = 476, which indicates that the 1991 price is 476 percent times the price in 1967.

The effect of inflation (or deflation) on the rate of return for an investment depends on how the future returns respond. If the return is in constant dollars that are not increased to reflect inflation, the effect will be to reduce the before-tax rate of return. If the amount of the return increases to keep up with inflation, the before-tax rate of return will not be affected by inflation. This will not be the case when an after-tax analysis is made. Even when the future returns increase

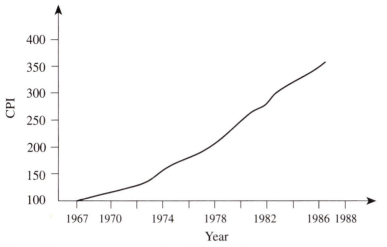

Figure 5.10   Consumer Price Index (1967 = 100).

to reflect inflation, the allowable depreciation schedule will not change, and the result will be increased taxable income. This obviously reduces the after-tax return on the investment and points out the importance of accounting for inflation when evaluating capital expenditure proposals.[4]

5.4   SELECTING AN APPROPRIATE RATE OF RETURN

Economic alternatives available to a firm must be compared with other available investment opportunities so that the maximum return is ensured. To allow a meaningful comparison, a minimum attractive rate of return (MARR) must be established that should be equal to the greatest of (1) the cost of borrowing money, (2) the cost of capital, or (3) the rate of return that the firm can earn from other investments. It is obvious that money should not be borrowed at 10 percent to fund a project with only an 8 percent return on investment. Also, the return on investment of a project should not be less than the cost of raising funds by capital restructuring. The ratio between capital and debt is significant to a firm's ability to borrow money and therefore must be actively managed so that banks and other financial entities will make favorable evaluations of the firm's solvency. Finally, business organizations usually have several options or alternatives for investment. One alternative is to not invest in any project but to place surplus funds in financial instruments totally unrelated to the firm's primary business and withdraw them when more attractive investment opportunities present themselves. This is known by economists as the *do-nothing alternative*. When money is invested in this manner it is not considered to be idle. Rather, the assumption is made that the funds yield a rate of return equal to the MARR and would be withdrawn and invested in a project only if the opportunity offered a higher rate of return.

Investments are always concerned with future events that cannot be predicted accurately, therefore probability, risk, and uncertainty should always be considered.

[4]D. G., Newnan, *Engineering Economic Analysis,* Engineering Press, San Jose CA., 1988.

If reasonable probabilities can be established for the future outcomes of an investment opportunity, then expected values for each outcome can be determined and the investment can be evaluated accordingly. This is known as risk analysis, where there are two or more possible outcomes and the probability of each outcome is known. The expected value for an investment can be determined using[4]

$$\text{Expected value} = \text{outcome } A \times P(A) + \text{outcome } B \times P(B) + \dots \qquad (5.9)$$

where outcome A or B = value of outcome A or B, and $P(A or B)$ = probability of A or B.

—— *Example 5.1* ——

A friend wants to bet $10 on which football team will win the Southwest Conference in 1992. The probabilities of the likely contenders winning are considered to be as shown. As in horse racing, the outcome represents the $10 bet plus the amount won. What is the expected value of a bet on Texas A&M to win the conference?

Solution

Team	Probability of Winning	Outcome of Bet If You Win
Texas Tech	0.25	25.00
Texas	0.20	30.00
Texas A&M	0.40	15.00
Houston	0.15	40.00

$$\text{Expected value} = \text{outcome if A\&M wins} \times P(\text{A\&M winning})$$
$$+ \text{outcome if A\&M loses} \times P(\text{A\&M losing})$$
$$= \$15.00(0.40) + \$0(0.60) = \$6.00$$

The expected value of the $10 bet on Texas A&M is thus equal to $6.00.

The lowest rate at which firms can borrow money is known as the prime interest rate. The prime interest rate is the rate that banks and other lending institutions charge their best customers for the use of money. This rate varies over time and is widely reported in the news media. It is primarily a function of the interest rate that the Federal Reserve charges member banks and is one of the principal ways in which the federal government manages and controls the economy of the nation.

The cost of capital for a corporation is conventionally higher than the cost of borrowed money. The cost of capital must include consideration of the market value of the common and other stock of the firm, which varies widely. The return on total capital earned by the corporation has an impact on the market value of the stock as well as the cost of borrowed money. *Fortune* magazine reports that the after tax rate of return on investment for individual firms averages 8 percent. The after tax rate of return on common stock and retained earnings as reported by

Business Week magazine averages 14 percent.[4] Firms that are struggling for survival cannot invest funds in any but the most critical projects with high rates of return and for short periods of time, often with payback periods of one year or less. For a project with a life of several years a very high rate of return would be required. Struggling firms also cannot usually borrow money at the prime rate because of their weak financial condition. These conditions result in the need for such firms to establish a high MARR, which limits the number of investment opportunities and usually results in riskier investments. The more stable companies, which make up the bulk of all industrial enterprises, take a longer-range view of their capital investments. With a greater supply of funds to work with they can invest in projects that the struggling firm cannot consider. With a solid financial foundation these firms can also qualify for lower interest rates when borrowing money, which allows the firm to operate with a lower MARR. For projects requiring only a small investment, payback periods are usually short, normally one to two years. Larger investments are analyzed by evaluating the rate of return. For projects with average risk, the aftertax MARR for these firms typically varies from 12 percent to 15 percent. For projects with greater risk, the MARR will be considerably higher, depending on the degree of risk. These values for the MARR are obviously opportunity costs rather than the cost of borrowed money or the cost of capital, which indicates that these firms have adequate high rate of return opportunities and have not been required to consider projects with lower rates of return nearer to the cost of borrowed money or the cost of capital. Although this may be interpreted to mean that good projects are going unfunded, it reflects the fact that most firms are reluctant to invest in projects expected to earn only slightly more than the cost of borrowing money or the cost of capital because of the inherent risk and uncertainty about the future.

5.5 EVALUATION OF ECONOMIC ALTERNATIVES

To evaluate engineering proposals, decision criteria must be established so that the proposal that best satisfies the desired objectives can be selected. The option of rejecting all proposals under consideration, or *doing nothing*, should also be retained as a possibility. The decision criterion adopted will drive the selection but will not ensure that company objectives will be realized unless they are thoroughly understood. Decision criteria can be based on the following:

1. The economical equivalence of a present, annual, or future investment cash flow.
2. Optimization of selection decision variables that affect the investment cost, periodic costs, or project life.
3. Rate of return on incremental investment.

In all of these approaches, a decision must be made as to whether to consider risk and uncertainty or to make the simplifying assumption that the future is known with certainty. Like other decision-making processes, the decision matrix is a valuable tool to assist in evaluating the various alternatives. By using a decision matrix (see Chapter 2), the interaction between a number of alternatives can be evaluated relative to a wide variety of future outcomes over which the decision maker has no control.

Table 5.10   Differences Between Mutually Exclusive Alternatives

End of	Alternatives		Differences
Year	A_1	A_2	$(A_2 - A_1)$
0	$-2000	$-2500	$-500
1	750	500	-250
2	750	1200	450
3	750	1200	450
4	750	1200	450

G. J. Thuesen and W. J. Fabrycky, *Engineering Economy*, 1989, p.295. Adapted by permission of Prentice Hall, Englewood Cliffs, NJ.

5.5.1   Evaluation by Economic Equivalence

Evaluation by economic equivalence is represented by the function[2]

$$PE, AE, \text{ or } FE = f(F_t, i, n) \tag{5.10}$$

The present equivalent, annual equivalent, and future equivalent amounts provide adequate bases for evaluation of a single alternative or for the comparison of mutually exclusive alternatives. Although few proposals are totally independent of each other, they can usually be arranged into mutually exclusive alternatives for evaluation purposes. Mutually exclusive proposals can be compared by evaluating their differences. This is demonstrated by refering to Table 5.10.[3]

To compare the two alternatives, A_1 and A_2, the cash flow difference is determined. The cash flows representing Alternative A_1 and A_2 are shown in Figure 5.11

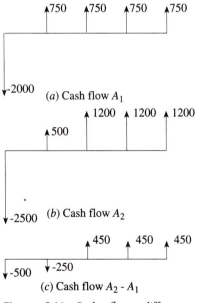

(a) Cash flow A_1

(b) Cash flow A_2

(c) Cash flow $A_2 - A_1$

Figure 5.11 Cash flow differences between two alternatives (G. J. Thuesen and W. J. Fabrycky, *Engineering Economy*, 1989, p.206. Adapted by permission of Prentice-Hall, Englewood Cliffs, NJ).

a and b. The cash flow shown in Figure 5.11 (c) represents the difference between A_1 and A_2. The decision to undertake alternative A_2 in lieu of A_1 requires an additional investment of $500 now and $250 one year hence. The receipts from this extra investment amount to $450 at the end of years 2, 3, and 4. The question that must be answered is whether the extra receipts justify the extra investment.

5.5.2 The Economic Optimization Function

The use of this method allows evaluation of both economic equivalence and economic optimization, which is appropriate when investment cost, periodic costs, and/or project life are functions of the decision variables. An economic optimization function is a mathematical model that links an evaluation measure E with controllable decision variables X and uncontrollable system parameters Y. It thus allows for evaluating decision variables in the presence of system parameters by using a mathematical test that results in an optimized value for E. This relationship is expressed as

$$E = f(X, Y) \tag{5.11}$$

An example of this method of evaluation is the determination of an optimal procurement quantity for inventory. The evaluation measure in this case is cost, and the objective is to select a procurement quantity based on demand, procurement cost and warehouse operational cost so that the total cost is minimized. Procurement quantity is directly under the control of the decision maker but demand, procurement cost, and warehousing costs are not. By applying the optimization function method, the decision maker can determine the procurement quantity that trades off the other cost elements and results in a minimum total cost.

This approach can also be extended to operational and design decisions involving alternatives. In this application, decision-dependent system parameters Y_d must be isolated from decision-independent parameters Y_i. Equation 5.12 above can then be written as

$$E = f(X, Y_d, Y_i) \tag{5.12}$$

An application of this version of the decision evaluation function is the establishment of a procurement and inventory system for an item that is available from several sources. The procurement level and quantity would be the decision variables in this case. For each separate source there would be a set of source-dependent parameters including the unit cost, procurement cost per procurement, the replenishment rate, and the procurement lead time. Uncontrollable system parameters would include the demand rate, warehousing cost per unit per period, and the shortage penalty cost. The objective is to minimize the total system cost by proper selection of procurement level, quantity, and source.[2]

5.5.3 Evaluation Using the Rate of Return on Incremental Investment

In evaluating alternatives using this method, the rate of return on the increment or difference cash flows between alternatives is compared with the MARR. Alternatives are usually arranged in order of increasing first year cost and are analyzed in pairs beginning with the first two alternatives. The most favorable alternative from this initial analysis is then compared with the next alternative, and the process is continued until all alternatives have been evaluated. The analysis is accomplished as follows:[4]

1. Compute the rate of return (ROR) for each alternative. Eliminate any alternatives with a rate of return less than the MARR.

2. Arrange the remaining alternatives in increasing order of first year cost.

3. Make a two-alternative analysis of the first two alternatives by using

$$
\begin{bmatrix} \text{Higher-cost} \\ \text{Alternative Y} \end{bmatrix} = \begin{bmatrix} \text{Lower-cost} \\ \text{Alternative X} \end{bmatrix} + \begin{bmatrix} \text{Difference} \\ \text{(Y - X)} \end{bmatrix} \quad (5.13)
$$

Compute the incremental rate of return (ΔROR) on the increment of investment ($Y - X$) and apply the following test

- If ΔROR $\geq$ MARR, retain the higher cost alternative Y.
- If ΔROR $<$ MARR, retain the lower cost alternative X.
- Reject the other alternative used in the analysis.

4. Using the preferred alternative from step 3, and the next alternative in the list from step 2, proceed with another two-alternative analysis.

5. Continue this process until all alternatives have been evaluated and the best alternative has been identified.

—— *Example 5.2* ——

The five mutually exclusive alternatives listed below have 10 year useful lives. Which alternative should be selected if the MARR is 8 percent?

- Alternative 1–This alternative requires an initial expenditure of $5000 and has an annual benefit (receipts less disbursements) of $885.

- Alternative 2–This alternative requires an initial expenditure of $3000 and has an annual benefit of $488.

- Alternative 3–This alternative requires an initial expenditure of $8000 and has an annual benefit of $1416.

- Alternative 4–This alternative requires an initial expenditure of $1000 and has an annual benefit of $149.

- Alternative 5–This alternative requires an initial expenditure of $10000 and has an annual benefit of $1424.

Solution

Step 1— Calculate the rate of return for each alternative and eliminate any that have a ROR $<$ MARR. To calculate the ROR for each alternative, use the following equation and solve for the interest rate i using a trial-and-error method,

$$
\frac{P}{A} = \left[\frac{(1 + i)^n - 1}{i(1 + i)^n} \right] = \frac{\$5,000}{\$885} = 5.65
$$

or

$$
5.65i(1 + i)^{10} = (1 + i)^{10} - 1
$$

Assume that $i = 12$ percent:

$$5.65(0.12)(1 + 0.12)^{10} = (1 + 0.12)^{10} - 1$$
$$2.1058 = 2.1058$$

therefore,

$$ROR = 12\%$$

Using the same procedure the *ROR* for the other alternatives can be determined.

Step 2— Arrange the remaining alternatives in increasing order of first year cost. Since the ROR for Alternative 5 (7 percent) is less than the MARR it can be eliminated from further consideration.

	4	2	1	3
Cost	$1000	$3000	$5000	$6000
Uniform annual benefits	$ 149	$ 488	$ 885	$1018
Rate of return	8%	10%	12%	12%

Step 3— Make a two alternative analysis beginning with the first two alternatives.

	Increment 2-4	Increment 1-2	Increment 3-1
Δ Cost	$2000	$2000	$1000
Δ Annual benefit	$ 339	$ 397	$ 133
Δ Rate of return	11%	14.9%	5.6%

From analysis of increment 2-4, a ΔROR of 11 percent is computed; thus alternative 2 is preferred over alternative 4. When increment 1-2 is evaluated, a ΔROR of 14.9 percent is realized; thus alternative 1 is preferred over alternative 2. Since the 3-1 increment has a rate of return less than the MARR, alternative 3 can be discarded, which leaves alternative 1 as the best alternative and the one that would be selected for investment. Note that if adequate resources were available, there were no other mitigating circumstances and these were the only investment options, alternatives 2, 3, and 4 would be undertaken as well.

5.5.4   Cost Model for Decision Making

For many proposed new designs, engineers must prepare a cost model, and they may be expected to demonstrate a *proof of concept*. A typical application of the use of cost models is the use of this technique for evaluating energy savings projects. Energy savings cost models require that certain conditions be estimated to determine the validity of the project. These estimates concern the identity of targeted end users of the technology, degree of awareness by the public, likelihood of adoption, appropriate timing, and technical feasibility. Equation 5.15 can be used to determine the expected energy savings for a proposed new design.[5]

[5]"A Guide for Estimating Energy Savings, Energy Research in Applications Program," Texas Higher Education Coordinating Board, 1990.

$$\begin{aligned}
\text{Energy savings} \ = \ & \text{(targeted energy consumption)} \\
& \times \text{(savings factor)} \\
& \times \text{(feasibility factor)} \\
& \times \text{(penetration factor)} \\
& \times \text{(adoption factor)} \quad\quad\quad\quad\quad\quad (5.14)
\end{aligned}$$

Savings Factor   The savings factor, SF, represents a percentage of the energy that will be saved by the installation of a proposed system or design. The savings factor can be calculated using the following equation:

$$SF = 1 - \left(\frac{E_{new}}{E_{old}} \right) \quad\quad\quad\quad (5.15)$$

where

E_{new} = Energy consumption for the proposed new design

E_{old} = Energy consumption for the old design

For example, the savings factor may be used to represent an expected reduction in fuel costs due to using an alternative fuel such as methanol to run an automobile's engine. The savings factor would correspond to the expected reduction in fuel consumption brought about by using the alternative fuel.

Feasibility Factor   The feasibility factor deals with the probability that the desired technical goals for a project can be met within an allotted time period. As the time for demonstration and proof of a concept draws near, the numerical value for the feasibility factor should increase. A timetable for completion can be set up on a yearly basis so that progress can be tracked. The feasibility factor has a range of 0.01 to 1 with 1 corresponding to project completion.

By carefully examining progress reports, reviewers may be able to make an educated guess as to the probability of success. An indicator of whether the project is on schedule is the time lag between any projected and achieved accomplishments. The time lag provides a clue as to the likelihood that the project goals will be met within the specified time frame.

Penetration Factor   The penetration factor represents the portion of the targeted population that is being reached annually by a program. The penetration factor PF, can be calculated by using the following equation:

$$PF = \frac{R_{pop}}{T_{pop}} \quad\quad\quad\quad (5.16)$$

where

R_{pop} = reached population

T_{pop} = targeted population

In general, the penetration factor varies from year to year depending on type of product and market position. The penetration factor may reflect the future success of marketing and advertisement in making the product desirable after proof of concept. For example, the targeted population may be the total number of trucks on the road. The manufacturers of trucks that use compressed natural gas (CNG) could advertise, hoping to convince owners to change to trucks that run on CNG. For this case, the penetration factor would be those reached by the marketing efforts for a year divided by the number of trucks on the road for that same year.

Adoption Factor   The adoption factor is the fraction of the penetrated population who will use the design to save energy. The adoption factor AF can be determined as the number of adoptors, A_{pop}, divided by the reached population, R_{pop}.

$$AF = \frac{A_{pop}}{R_{pop}} \tag{5.17}$$

Targeted Energy Consumption   This term refers to the amount of energy that a targeted population will consume in a year. Using this type of analysis requires that a population be identified and that an estimate be made of how energy consumption will vary for future years. A good way to estimate yearly energy consumption is in terms of millions (10^6) or billions (10^9) of British Thermal Units (BTUs). Btus can be readily determined for all forms of energy including electricity, natural gas, or barrels of oil consumed.

The targeted population of energy users may consist of either a part of a sector of the economy or an entire sector. The more accurately the target population is defined the better the projection of cost savings. The following example will demonstrate how to develop a model and to make a decision by the use of the above methodology.[6]

—— **Example 5.3** ——

Peanuts, after having been dug up, have a high moisture content, which must be reduced for marketing and safe storage. Conventional drying is usually ac-complished by increasing the dry-bulb temperature of the drying air by using gas heaters as shown in Figure 5.12. An alternative drying concept is to dehumidify the air by using a liquid desiccant solution (lithium chloride mixed with water). Desic-cants are materials that have a high affinity for water vapor due to the vapor pressure

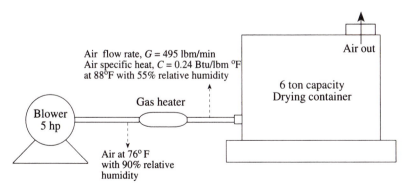

Figure 5.12   Conventional peanut drying system.

[6]A. Ertas, "Economical Models for Hybrid Cooling and Drying Systems by Using Desiccant Tech-nology," Report ERAP-2- #309, submitted to the Texas Higher Education Coordinating Board, Dec. 2, 1991.

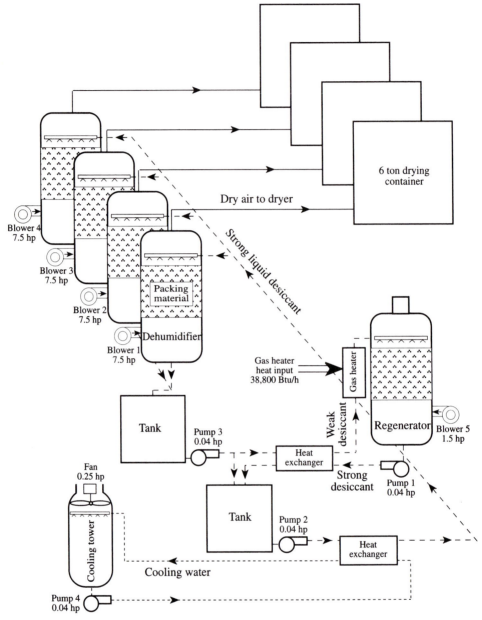

Figure 5.13   Proposed new peanut drying system.

difference between the desiccant and the moisture in the air. The lower the vapor pressure of the desiccant the better the dehumidification process is. A proposed peanut drying system using a liquid desiccant is shown in Figure 5.13. This system operates by using forced air obtained from the dehumidifier. When the humid air and liquid desiccant are brought into contact in the dehumidifier, moisture in the air will move from the humid air to the liquid desiccant. As shown in Figure 5.13, a regeneration tower is needed to restore the diluted liquid desiccant to the desired concentration. Regeneration is achieved by heating the diluted liquid desiccant in a heater and removing the moisture by passing air over the desiccant in the regeneration tower. Since the liquid desiccant is heated, its vapor pressure is higher than that of the moisture in the air; hence, the moisture in the desiccant will move from the desiccant to the air. The hot liquid desiccant must then be cooled down to the dehumidifier operating temperature (usually 90°F) in the cooling tower. Figure 5.13 shows the energy consumption of all the components for this system.

An economical model for the system is to be developed and the feasibility of the system is to be compared with that of the conventional system. The initial investment for a 6-ton conventional system including blower and gas heater is $2070 and the initial investment for a liquid desiccant system is $14,709. The cost of the drying containers and the maintenance costs for both systems are assumed to be the same.

Energy and Cost Analysis of the Conventional System   The time required to dry 6 tons of peanuts by circulating $G=495$ lbm/min air through the container is estimated to be 38 hrs. The blower work W_{bl} during this time is

$$W_{bl} = (5 \text{ hp})(0.746 \text{ kW}/\text{hp})(38 \text{ h}) = 141.74 \text{ kWh}/\text{container}$$

or

$$= 483,759 \text{ Btu}/\text{container}$$

The heat required to for the temperature increase, $\Delta T = 88 - 76 = 12°F$, at the gas heater is

$$Q_{gas} = G \times C \times \Delta T$$
$$= (495)(0.24)(12) = 1425.6 \text{ Btu}/\text{min}$$

or

$$= 85,536 \text{ Btu}/\text{h}$$

or

$$= (38)(85,536) = 3,250,368 \text{ Btu}/\text{container}$$

The overall energy requirement, E_{conv}, of the conventional system for a complete drying process is

$$E_{conv} = W_{bl} + Q_{gas}$$
$$= 483,759 + 3,250,368$$
$$= 3,734,127 \text{ Btu}/\text{container}$$

The cost of natural gas per 10^6 Btu is assumed to be $3.84, and the cost of electricity per kWh is $0.04. The drying operation cost DC_{conv} for each container can be calculated as

$$DC_{conv} = (3.25)(3.84) + (141.74)(0.04)$$
$$= \$18.15/\text{container}$$

In a drying season, which is three months, each drying unit can dry 58 containers, hence, the annual operating cost, AC_{conv}, of the conventional system is

$$AC_{conv} = (18.15)(58) = \$1052.7/\text{container}$$

Table 5.11 Power Requirement of the New System

Component	Each (hp)	Number	Total (hp)
Pumps	0.125	4	0.5
Blower (5)	1.5	1	1.5
Blower (1-4)	7.5	4	30
Fan	0.25	1	0.25
Total			32.25

Energy and Cost Analysis of the New System   Table 5.11 shows the total power requirement of the new system, Assuming all the pumps and fans work at maximum power. The total energy W_T required for the overall drying process is

$$W_T = (32.25)(0.746)(38) = 914.2 \text{ kWh/run}$$

or

$$= 3,120,164.6 \text{ Btu/run}$$

From Figure 5.13, the total heat energy input to the gas heater to obtain the proper regeneration temperature is

$$Q_T = (38,800)(38) = 1,474,400 \text{ Btu/run}$$

The overall energy requirement E_{new}, of the new system for a complete drying process is

$$W_T + Q_T = 3,120,164.6 + 1,474,400$$
$$= 4,594,564.6 \text{ Btu/run}$$

The drying operation cost, DC_{new} for the new system that has four containers is

$$DC_{new} = (914.2)(0.04) + (1.474)(3.84)$$
$$= \$42.23/\text{run}$$

Assuming that the new drying system can also dry 58 times in a drying season, the annual operational cost of this system is

$$AC_{new} = (42.23)(58) = \$2,449.30/\text{yr}$$

Payback Calculation   Initial investment for the conventional system is (4)($2070) = $8280. Note that since the new system has four containers, the cost of the conventional system is multiplied by four. The initial cost investment difference P between the two systems is

$$P = 14,709 - 8280 = \$6429$$

The annual saving F_t is the difference in the cost of operations between the two systems

$$F_t = 4(AC_{conv}) - AC_{new}$$
$$= 4(1052.7) - 2449.3 = \$1761.5$$

Using an annual interest rate $i = 11$ percent for large businesses and substituting the value of F_t in the following equation,

$$\sum_{t=1}^{n} F_t(1 + i)^{-t} \geq P \tag{5.18}$$

the payback period n is found to be approximately five years. Therefore, the proposed liquid desiccant drying system employing packed tower regeneration and a natural gas liquid heater is a feasible option to the conventional natural gas system.

Targeted Energy   A total of 1.872 million tons of peanuts are harvested annually in the United States of which about 0.187 million tons are grown in Texas. From the energy calculation for the conventional system the targeted energy E_{targ} in Texas is

$$E_{targ} = \left(\frac{3,734,127 \text{ Btu/container}}{6 \text{ tons/container}}\right)(0.187 \times 10^6 \text{tons})$$

$$= 0.11638 \times 10^{12} \text{ Btus}$$

Saving Factor   From Eq. 5.15, the saving factor, SF is

$$SF = 1 - \left(\frac{E_{new}}{E_{old}}\right)$$

$$= 1 - \left(\frac{4,594,564.6}{(3,734,127)(4)}\right)$$

$$= 0.69$$

Thus, the new system promises up to 69 percent energy savings.

Feasibility Factor   Suppose that a contract to develop this new drying concept is awarded as a project in January 1992. The concept is to be proved by January 1994. If the project stays on schedule, then feasibility factor values of 0.35, 0.6, and 0.8 are considered reasonable for 1992, 1993, and 1994, respectively. The feasibility factor for the following years is assumed to be 1.

Penetration Factor   Assume that the number of peanut drying facilities in Texas is T_{pop}. After the proof of concept, 5 percent of the peanut facilities are assumed to be reached by the advertisment. The penetration factor for 1994 is

$$PF_{94} = \frac{R_{pop}}{T_{pop}}$$

$$= \frac{(0.05)(T_{pop})}{T_{pop}} = 0.05$$

During the following years, assume that the reached peanut facilities will grow linearly. Hence, for the year, n, the penetration factor will be

$$PF_n = 0.05 \times (n - 1993)$$

Using this equation, the penetration factors presented in Table 5.12 can be determined.

Table 5.12   Penetration Factors for a Drying System

Year	1992	1993	1994	1995	1996	1997	1998	1999	2000	2001
PF	0.0	0.0	0.05	0.10	0.15	0.20	0.25	0.30	0.35	0.40

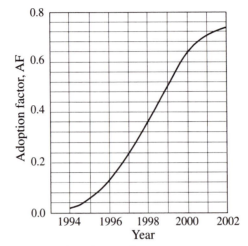

Figure 5.14   Adoption factor versus year.

Adoption Factor   A typical adoption factor versus year curve is shown in Figure 5.14. Based on Figure 5.14, adoption factors for 10 subsequent years are given in Table 5.13.

As can be seen from this table, marketing and the acceptance of the new drying system starts from the proof of concept date. By using Eq. 5.14, the energy savings ES for 1994 (the proof of concept year) can be determined:

$$ES_{94} = (0.11638 \times 10^{12})(0.69)(0.8)(0.05)(0.02)$$
$$= 0.0642 \times 10^9 \text{ Btus}$$

Similarly, the energy savings for subsequent years are calculated and presented in Table 5.14. The summation of all the energy savings will give the total Btus saved. Note that savings for 1992 and 1993 are equal to zero. Introducing a new drying concept employing a liquid desiccant will save 65.43×10^9 Btus in 10 years.

Table 5.13   Adoption Factors for a Drying System

Year	1992	1993	1994	1995	1996	1997	1998	1999	2000	2001
AF	0.0	0.0	0.02	0.06	0.14	0.24	0.34	0.50	0.64	0.70

Table 5.14   Energy Savings in 10^9 Btus

Year	1994	1995	1996	1997	1998	1999	2000	2001
AS	0.0642	0.4818	1.6863	3.8545	6.8257	12.0453	17.9877	22.4846

BIBLIOGRAPHY

1. NEWNAN, D. G., *Engineering Economic Analysis.* Engineering Press, San Jose, CA, 1988.

2. RIGGS, J. L, *Engineering Economics,* McGraw-Hill, New York, 1977.

3. WHITE, J. A., AGEE, M. H., CASE, K. E., *Principles of Engineering Economic Analysis.* John Wiley & Sons, Inc., New York, 1984.

4. FABRYCKY, W. J. and BLANCHARD, B.S., *Life-Cycle Cost and Economic Analysis.* Prentice Hall, Englewood Cliffs, NJ 1991.

5. THUESEN, G.J. and FABRYCKY, W.J. *Engineering Economy,* Prentice Hall, Englewood Cliffs, NJ, 1989.

Chapter 6

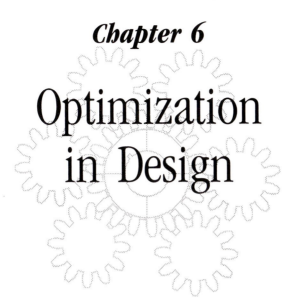

Optimization in Design

The designer must consider the optimum solution to a design problem so that a better, more efficient, less expensive solution that is in harmony with the laws of man and nature can be achieved.

6.1 INTRODUCTION

Problems associated with optimum design have been the subject of considerable attention for a number of years. During the last several decades the field of optimum design has made remarkable progress in systems design, control of dynamics, and engineering analysis. With recent advances in the field of computer technology, many modern optimization methods have been developed, and designing complex systems with the optimum configuration has become possible within a reasonable computation time.

There are two kinds of *effects* inherently associated with any mechanical element or system:

1. Undesirable effects, such as high cost, excessive weight, large deflections, and vibrations.
2. Desirable effects, such as long useful life, efficient energy output, good power transmission capability, and high cooling capacity.

Optimum design can be defined as the best possible design from the standpoint of the most significant effects, that is, minimizing the most significant undesirable effects and/or maximizing the most significant desirable effects.

Application of optimization to a design problem requires formulation of an *objective function* such as weight, cost, or shape, and the expression of design *constraints* as equalities or inequalities. The objective function, U, in terms of the *independent variables* (design variables) from which an optimum solution is sought can be written as

$$U = U(x_1, x_2, \cdots x_n) \tag{6.1}$$

where x is the design variable and n is the number of design variables.

In an optimum design solution there are certain restrictions to be satisfied by the design variables for an acceptable design. These restrictions or limitations are called constraints. One seeks the optimum solution for a given objective function that fulfils the following equality and inequality constraints

$$h_i(x) = h_i(x_1, x_2, \cdots, x_n) = 0 \qquad i = 1, 2, \cdots, m \tag{6.2}$$

$$g_j(x) = g_j(x_1, x_2, \cdots, x_n) \geq 0 \qquad j = 1, 2, \cdots, p \tag{6.3}$$

where $h_i(x)$ and $g_j(x)$ are the equality and inequality constraints, respectively. The design variable can be a material property, structural dimension (thickness, width, length), or geometric data (coordinates of node points of a structure).

—— Example 6.1

Show the equation for the objective function for the minimum volume of a simply supported beam subjected to a concentrated load of 2F as shown in Figure 6.1.

Solution

Since a minimum-volume design is sought, the objective function is the volume of the beam, which can be defined in terms of the four design variables, x_1, x_2, x_3, and x_4 as

$$U(x) = 2\left[\frac{\pi(x_2^2 - x_1^2)}{4}\right]x_4 + \frac{\pi(x_3^2 - x_1^2)}{4}(2L - 2x_4)$$

Similarly, the objective functions can be written for factors like those for static deflection and cost.

If the design requires the beam to resist a maximum load of 16,000 N, the equality constraint can be written thus:

$$h_1 = 2F = 16,000N$$

If a permissible stress σ_p of 400 MPa is used, the beam will be safe against bending

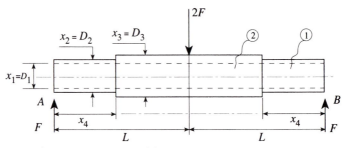

Figure 6.1 Simply supported beam.

failure. The inequality constraint is

$$g_1 = \sigma_p = 400 \text{ MPa} \geq \sigma_b$$

where σ_b is the bending stress. The appropriate optimization method can now be applied to the above objective function (with the given constraints) to define the volume region that includes the optimum-design point.

Optimum design problems may involve the following:

1. Only equality constraints.
2. Only inequality constraints.
3. Both equality and inequality constraints.
4. No constraint.

For case (4), which is called an unconstrained-optimization problem, there are no restrictions imposed on the design variables.
In general there are two types of optimization problems:

1. *Single-criterion optimization problems*—in which the designer's goal is to minimize or maximize one objective function. In this situation the optimum point is simply the minimum or maximum.
2. *Multicriterion optimization problems*—in which the designer's goal is to minimize or maximize more than one objective function simultaneously. In this situation, all the objective functions should be considered to find the optimum solution.

6.2   MATHEMATICAL MODELS AND OPTIMIZATION METHODS

The mathematical methods for optimum design can be divided into two categories:

1. *Analytical methods.* This method includes differentiation, variational methods, and the use of Lagrange multipliers.
2. *Numerical methods.* This method includes linear (simplex method) and nonlinear programming methods like these given in Figure 6.2[1]

In this chapter, some of the common analytical and numerical optimization methods used in design are discussed. Additional information on these methods can be found in references 1–3.

6.2.1   The Differential Calculus Method

The differential calculus method uses the first and second derivatives to find the maximum and minimum values of a given differentiable function $U(x)$. To find a

[1]J. Farkas, *Optimum Design of Metal Structures,* Halsted Press, John Wiley & Sons Inc., New York, 1984.

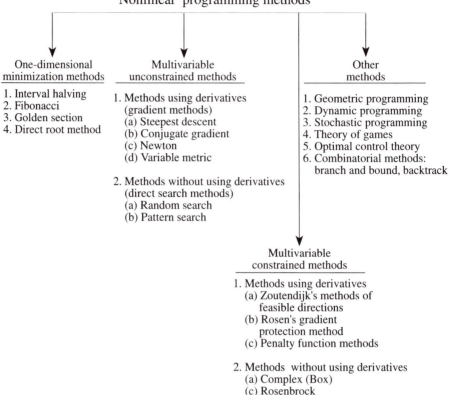

Figure 6.2   Selected nonlinear programming methods.

maximum or minimum using this method, take the first derivative of the objective function $U(x)$ and set it equal to zero; then, solve for the independent variable x.

$$\frac{dU}{dx} = 0 \tag{6.4}$$

Equation 6.4 gives the value of the critical points (optimum points). If the second derivative evaluated at the critical point is less than zero, then the point is a maximum. The point is a minimum if the second derivative is greater than zero.

$$\frac{d^2U}{dx^2} < 0 \qquad \text{implies a local maximum.} \tag{6.5}$$

$$\frac{d^2U}{dx^2} > 0 \qquad \text{implies a local minimum.} \tag{6.6}$$

As shown in Figure 6.3, if the slope and the second derivative at a point are zero, as at point x_2, this is an inflection point. The function must be continuous to have a maximum or minimum. Thus, point x_6 in Figure 6.3 is not a maximum.

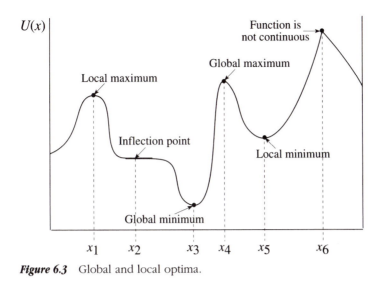

Figure 6.3 Global and local optima.

Example 6.2

United Parcel Service (UPS) charges by weight to ship a parcel unless the sum of the dimensions as shown in Figure 6.4 is greater than 96 in.; otherwise, there is an extra charge per unit volume. Determine the dimensions that will maximize the volume without exceeding the 96 in. requirement.

Solution

The objective function is the volume; hence,

$$U = x^2 H$$

The equality constraint is as follows:

$$h(x) = H + 4x - 96 = 0$$

By eliminating the free variable H from the above equations,

$$U = -4x^3 + 96x^2$$

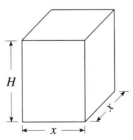

Figure 6.4 UPS required parcel.

Consider the first derivative:

$$\frac{dU}{dx} = -12x^2 + 192x = 0$$

The solution is, $x_1 = 0$ and $x_2 = 16$. To ensure that $x_2 = 16$ is a maximum, take the second derivative

$$\frac{d^2U}{dx^2} = -24x + 192$$

and evaluate at $x_2 = 16$

$$\frac{d^2U}{dx^2}\Big|_{x_2=16} = -192$$

Indicating a maximum for U at $x_2 = 16$, where $dU/dx = 0$, use $x_2 = 16$ to solve for H from the equality constraint:

$$H = 96 - 4x$$
$$= 96 - 4(16) = 32 \text{ in.}$$

Then maximum volume is

$$V = x^2H = (16)^2(32) = 8192\text{in.}^3$$

6.2.2 The Lagrange Multiplier Method

The Lagrange multiplier method of optimization is named for its developer, Joseph Louis Lagrange (1736–1814), a French mathematician and astronomer. This method is important when dealing with nonlinear optimization problems. It uses a function called the Lagrange expression, LE, which consists of the objective function, $U(x, y, z)$, and constraint functions $h_i(x, y, z)$, multiplied by Lagrange multipliers, λ_i.

$$\text{LE} = U(x, y, z) + \lambda_1 h_1(x, y, z) + , \cdots, + \lambda_i h_i(x, y, z) \tag{6.7}$$

The additional unknown, λ_i, is introduced into the Lagrange expression so that in determining the optimum values of x, y, and z, the problem can be treated as though it was unconstrained. The conditions that must be satisfied for the optimum points are as follows:

$$\frac{\partial \text{LE}}{\partial x} = \frac{\partial \text{LE}}{\partial y} = \frac{\partial \text{LE}}{\partial z} = 0 \tag{6.8}$$

$$\frac{\partial \text{LE}}{\partial \lambda_1} = \frac{\partial \text{LE}}{\lambda_2} = \cdots = \frac{\partial \text{LE}}{\lambda_i} = 0 \tag{6.9}$$

where i is the number of Lagrange multipliers.

___ *Example 6.3* _____

Solve Example 6.2 using the Lagrange multiplier method.

Solution

From Example 6.2 the objective function is

$$U = x^2 H$$

The equality constraint is

$$b(x) = H + 4x - 96 = 0 \qquad\qquad (6.10)$$

Applying the Lagrange expression

$$LE = x^2 H + \lambda(H + 4x - 96) \qquad\qquad (6.11)$$

and differentiating with respect to x, H, and λ,

$$\frac{\partial LE}{\partial x} = 2xH + 4\lambda = 0 \qquad\qquad (a)$$

$$\frac{\partial LE}{\partial H} = x^2 + \lambda = 0 \qquad\qquad (b)$$

$$\frac{\partial LE}{\partial \lambda} = H + 4x - 96 = 0 \qquad\qquad (c)$$

The solution of Eqs. (a), (b), and (c) gives $x = 16$ in. and $H = 32$ in.; hence, the optimum volume is

$$V = x^2 H = (16)^2(32) = 8192 \text{ in.}^3$$

_____

6.2.3   Search Methods

Search methods are based on examining simultaneous or sequential trial solutions over the entire domain of feasible designs to determine which point is optimal. These methods provide information about a region in which an optimum point is located. However, since these methods are not exact, in general, a rapid rate of convergence is not possible. If the function has a single variable to locate the optimal point in a given interval, single-variable search methods are used. For functions having more than one variable, multiple-variable search methods are used. A simultaneous search examination provides the approximate value of a minimum of a function having several minima in a given interval. However, because of the large number of functions evaluated, considerable computer time is required. For a function having one minimum or maximum (unimodal function), it is more efficient to use either the golden section search method or the Fibonacci search method.

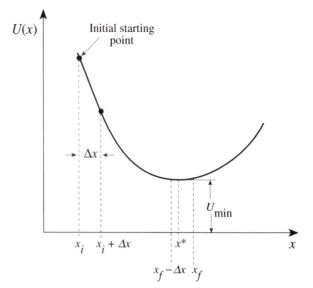

Figure 6.5   Function $U(x)$ with a single variable.

Equal Interval Search   Assume that an objective function $U = f(x)$ is a function of a single variable x and has a single minimum value of U_{min} at x^* as shown in Figure 6.5. In using the equal interval search method, the function $U(x)$ is evaluated at points with equal Δx increments. The calculated values of the function at two successive points, say, x_i and $x_{i+1} = x_i + \Delta x$, are then compared. When an increase in function value at any final point, $x_f = x_{i+1}$, is sensed, then

$$U(x_f - \Delta x) < U(x_f) \tag{6.12}$$

and the minimum value of the function has passed. Thus, lower and upper limits for the interval of uncertainty can be written as

$$x_f - \Delta x < x^* < x_f \tag{6.13}$$

where x^* is the point at which the function $U(x)$ has a minimum. To reduce the interval of uncertainty to an acceptable value, the search process is reversed by a sign change with the increment Δx cut in half. Now the search process is restarted by evaluating the function value at $x_f - \Delta x/2$, and the process is repeated for the next smaller interval of uncertainty [4]. This search process is repeated until the final interval of uncertainty is reduced to a small convergence criterion ε.

—— **Example 6.4** ——

Find a minimum point of the function

$$U(x) = 3x^2 + \frac{1296}{x}$$

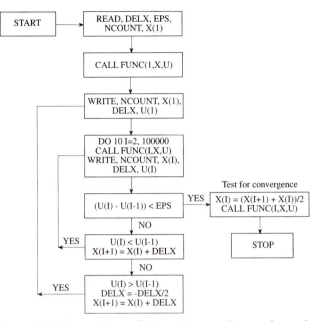

Figure 6.6 The solution of U_{min} using equal interval search method.

Solution:

Following the flowchart given in Figure 6.6 and assuming that $\Delta x = 0.5$ and an initial value of x_i is equal to 1, the value of function U is

$$U(x) = 3(1)^2 + \frac{1296}{1} = 1299.00$$

The initial value of x for the next iteration is

$$x_i + \Delta x = 1.0 + 0.5 = 1.5$$

Then

$$U_{i+1} = 3(1.5)^2 + \frac{1296}{1.5} = 870.75$$

Note from Table 6.1 that an increase in U is sensed at the 12th iteration. Thus, the increment Δx is reduced to one half and the sign is changed. This is shown in Table 6.1 beginning with the 13th iteration. The iteration process is repeated until the solution converges to a minimum value of the function. Based on a convergence criterion of $\varepsilon = 0.001$, the solution to the problem is

$$U_{min} = 324.0002 \text{ at } x = 5.996094$$

TABLE 6.1:   Results of Iterative Calculation
for Equal Interval Search Method

i	x_i	Δx	$U(x)$
1	1.00000	0.50000	1299.00000
—	—	—	—
11	6.00000	0.50000	324.00000
12	6.50000	0.50000	326.13461
13	6.25000	−0.25000	324.54752
14	6.00000	−0.25000	324.00000
15	5.87500	−0.25000	324.57883
16	5.87500	0.12500	324.14264
17	6.00000	0.12500	324.00000
18	6.12500	0.12500	324.13873
19	6.06250	−0.06250	324.03491
—	—	—	—
27	5.98438	−0.01563	324.00220
28	5.99219	0.00781	324.00055
29	6.00000	0.00781	324.00000
	$x = 5.996094$		$U(x) = 324.0002$

Golden Section Search   In the equal interval search method the increment Δx is kept constant until an increase in the function is sensed. The golden section search method provides an alternate procedure in which the increment varies at each search step. This search method finds the minimum value of a given function over a specified interval of length L. The first step of this method is to find the two interior points of the interval at which the function is to be calculated (see Figure 6.7).

Consider the interval to be from the lower limit XL to the upper limit XU. The length of the interval is given by

$$L = XU - XL \tag{6.14}$$

The two interior points x_1 and x_2 are found by using the following equations:

$$x_1 = XL + 0.618L \tag{6.15}$$

$$x_2 = XU - 0.618L \tag{6.16}$$

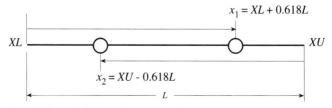

Figure 6.7   Golden section intervals.

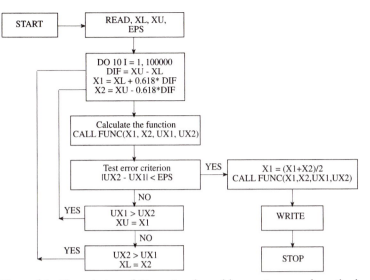

Figure 6.8 The solution of U_{min} using the golden section search method.

As can be seen from the above equations, the interior points at which the function is evaluated are not selected arbitrarily. They are based on the ratio 0.618, known as the *golden ratio.*[2]

Assume that the goal is to find the minimum value of a given function. If the value of the function evaluated at x_1 is larger than the value at x_2, the region to the right of x_1 is eliminated. The new interval is from XL to $XU = x_1$, and the calculations are repeated. If the value calculated at x_2 is larger than the value calculated at x_1, the region to the left of x_2 is eliminated. In this case, the new interval is from $XL = x_2$ to XU. This process is continued until the desired accuracy is obtained. The accuracy is determined by the remaining length of the interval. When the interval length becomes sufficiently small, say $\varepsilon = 0.005$, then the value is, at least, that close to the actual value of the minimum and the iterative process is stopped. The reverse applies to find the maximum of the function. The flow chart for this method is shown in Figure 6.8.

—— *Example 6.5* ————————————————————————

Based on the convergence criterion $\varepsilon = 0.001$ use the golden section search method to find the minimum point of the function

$$U(x) = F(x) = 3x^2 + \frac{1296}{x}$$

in the interval of $1 \le x \le 11$.

[2]G. V. Reklaitis, A. Ravindran, and K. M. Ragsdell, *Engineering Optimization Methods and Applications,* John Wiley & Sons, Inc., New York, 1983.

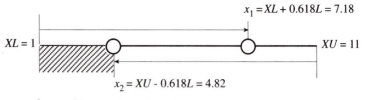

Figure 6.9   Golden section intervals.

Solution

The length of interval L is

$$L = XU - XL$$
$$= 11 - 1 = 10$$

The interior points x_1 and x_2 are

$$x_1 = XL + 0.618L$$
$$= 1 + 0.618(10) = 7.18$$
$$x_2 = XU - 0.618L$$
$$= 11 - 0.618(10) = 4.82$$

Evaluating the function U at interior points x_1 and x_2 yields

$$U(x_1 = 7.18) = 335.15860$$
$$U(x_2 = 4.82) = 338.57687$$

Compare the function values evaluated at x_1 and x_2:

$$U(x_2) > U(x_1)$$

Therefore, as shown in Figure 6.9, the region to the left of x_2 is eliminated and $XL = x_2 = 4.82$. The new interval for the next calculation is from $XL = 4.82$ to $XU = 11$.

The same iteration process is repeated until the solution converges to a minimum value of the function. As shown in Table 6.2, based on a convergence criterion $\varepsilon = 0.001$, the solution to the problem is

$$U_{min} = 324.0000 \text{ at } x = 6.00018$$

TABLE 6.2   Results of Iterative Calculations for the Golden Section Search Method

i	XL	XU	x_1	x_2	UX1	UX2
1	1.00000	11.00000	7.18000	4.82000	335.15860	338.57687
2	4.82000	11.00000	8.63924	7.18076	373.92261	335.17224
3	4.82000	8.63924	7.18029	6.27895	335.16382	324.67960
4	4.82000	7.18029	6.27866	5.72163	324.67819	324.72003
5	5.72163	7.18029	6.62308	6.27884	327.27496	324.67905
6	5.72163	6.62308	6.27873	6.06599	324.67853	324.03891
7	5.72160	6.27873	6.06592	5.93444	324.03882	324.03897
	$x = 6.00018$					$U(x) = 324.0000$

6.2.4 Multivariable Search Method

As discussed previously, in a single variable unconstrained optimization problem a one-dimensional search is required. In most engineering problems the coordinate of the optimum design point is necessary. Hence, to optimize a given mutivariable function, a multidimensional search is required. Although several methods can be used to find the extremes of multivariable functions, the *steepest descent,* one of the simplest methods, is discussed in this section that follows [4].

Steepest Descent Method Suppose an objective function with two variables x_1 and x_2 is given:

$$U = U(x_1, x_2) \tag{6.17}$$

For the given function, U, contour curves of constant U can be obtained by changing values of x_1 and x_2 as shown in Figure 6.10. Point O, shown in this figure, is the optimum point of function U. A_1 is the arbitrary starting point, called the *base point.* Consider the tangent and normal vectors with the unit vectors $\bar{\imath}$ and $\bar{n}$ at point A_1

$$\bar{T} = T \cdot \bar{\imath} \tag{6.18}$$

$$\bar{N} = N \cdot \bar{n} \tag{6.19}$$

The normal vector at point A_1 is called the gradient vector, defined as

$$\bar{N} = \nabla U = \left(\frac{\partial U}{\partial x_1}\right)\bar{I}_1 + \left(\frac{\partial U}{\partial x_2}\right)\bar{I}_2 \tag{6.20}$$

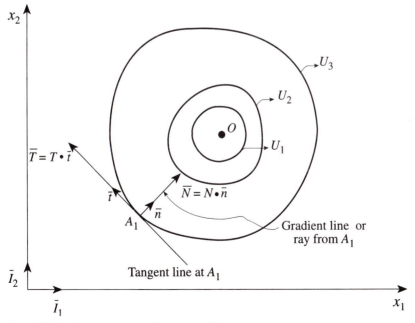

Figure 6.10 Contour curves of constant U.

Thus, the unit vector can be written as

$$\bar{n} = \frac{\left(\frac{\partial U}{\partial x_1}\right)\bar{I}_1 + \left(\frac{\partial U}{\partial x_2}\right)\bar{I}_2}{\sqrt{\left(\frac{\partial U}{\partial x_1}\right)^2 + \left(\frac{\partial U}{\partial x_2}\right)^2}} \tag{6.21}$$

which shows the direction of the gradient as the directional vector of the next step. The steepest search method can be used for multivariable problems with the increment in the gradient direction as shown in Figure 6.11. To move in the direction of the gradient vector, $\bar{N}$, increment Δ is multiplied by the unit vector $\bar{n}$.

$$\bar{\Delta} = \Delta \bar{n} \tag{6.22}$$

or

$$\bar{\Delta} = \left[\frac{\left(\frac{\partial U}{\partial x_1}\right)\Delta\bar{I}_1 + \left(\frac{\partial U}{\partial x_2}\right)\Delta\bar{I}_2}{\sqrt{\left(\frac{\partial U}{\partial x_1}\right)^2 + \left(\frac{\partial U}{\partial x_2}\right)^2}}\right] \tag{6.23}$$

The above equation can be written as

$$\bar{\Delta} = \Delta x_1 \bar{I}_1 + \Delta x_2 \bar{I}_2 \tag{6.24}$$

where

$$\Delta x_1 = \frac{-\left(\frac{\partial U}{\partial x_1}\right)_{A_1} \Delta}{\sqrt{\left(\frac{\partial U}{\partial x_1}\right)^2_{A_1} + \left(\frac{\partial U}{\partial x_2}\right)^2_{A_1}}} \tag{6.25}$$

$$\Delta x_2 = \frac{-\left(\frac{\partial U}{\partial x_2}\right)_{A_1} \Delta}{\sqrt{\left(\frac{\partial U}{\partial x_1}\right)^2_{A_1} + \left(\frac{\partial U}{\partial x_2}\right)^2_{A_1}}} \tag{6.26}$$

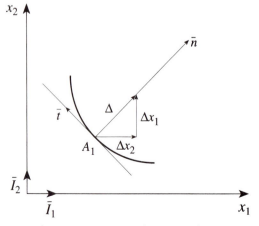

Figure 6.11 Components of step size Δ.

The same Δ and gradient direction are used until an increase in U is sensed. The direction of the gradient is then changed and Δ is reduced to $\Delta/2$. This procedure is repeated until the Δ is reduced to a value corresponding to that of the convergence criterion, ε_1. The next gradient direction (ray) is then determined, and the same steps are repeated until $\partial U/\partial x_1$ or $\partial U/\partial x_2$ is reduced to a value corresponding to that of the convergence criterion ε_2. When x_1 and x_2 reach a minimum or maximum, their values will be equal to the roots of $\partial U/\partial x_1$ and $\partial U/\partial x_2$. Hence, the smaller the partial derivative, $\partial U/\partial x_i$ the closer the x_i values are to the roots. The following example will illustrate the application of this method.

— *Example 6.6*

To illustrate the steepest descent method, consider the function

$$U = F(x) = 3(x_1^2 + x_2^2) + 1296\left(\frac{1}{x_1} + \frac{1}{x_2}\right)$$

Solution

Following the flowchart given in Figure 6.12, choose an arbitrary starting point and increment as

$$x_1 = 8, \qquad x_2 = 1, \qquad \text{and} \qquad \Delta = 1.0$$

Assume a convergence criterion for Δ of $\varepsilon = 0.001$, and a convergence criterion for the ray (gradient line), $\varepsilon = 0.01$. Calculate the function U for x_1 and x_2

$$U = 3(8^2 + 1^2) + 1296\left(\frac{1}{8} + \frac{1}{1}\right) = 1653.00$$

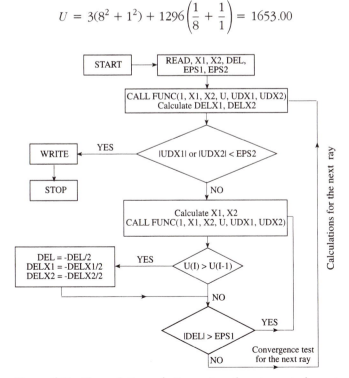

Figure 6.12 The solution of U_{min} using the steepest descent method.

To find the components of the increment, Δ, determine

$$\left(\frac{\partial U}{\partial x_1}\right) = 6x_1 - \frac{1296}{x_1^2}$$

$$= 6(8) - \frac{1296}{8^2} = 27.75$$

$$\left(\frac{\partial U}{\partial x_2}\right) = 6x_2 - \frac{1296}{x_2^2}$$

$$= 6(1) - \frac{1296}{1^2} = -1290$$

From Eqs. 6.25 and 6.26

$$\Delta x_1 = \frac{-\left(\frac{\partial U}{\partial x_1}\right)\Delta}{\sqrt{\left(\frac{\partial U}{\partial x_1}\right)^2 + \left(\frac{\partial U}{\partial x_2}\right)^2}}$$

$$= \frac{-(27.75)(1.0)}{\sqrt{27.75^2 + 1290^2}} = -0.021506$$

$$\Delta x_2 = \frac{-\left(\frac{\partial U}{\partial x_2}\right)\Delta}{\sqrt{\left(\frac{\partial U}{\partial x_2}\right)^2 + \left(\frac{\partial U}{\partial x_2}\right)^2}}$$

$$= \frac{-(-1290)(1.0)}{\sqrt{27.75^2 + 1290^2}} = 0.99977$$

Calculate x_1 and x_2 for the first iteration of the first ray:

$$x_1 = x_1(\text{initial}) + \Delta x_1$$
$$= 8 - 0.021506 = 7.97849$$

$$x_2 = x_2(\text{initial}) + \Delta x_2$$
$$= 1 + 0.99977 = 1.99977$$

Calculate the value of U at the new points x_1 and x_2:

$$U = 3(7.97849^2 + 1.99977^2) + 1296\left(\frac{1}{7.97849} + \frac{1}{1.99977}\right) = 1013.47800$$

Calculations for the next iteration are

$$x_1 = 7.97849 - 0.021506 = 7.95699$$

$$x_2 = 1.99977 + 0.99977 = 2.99954$$

and the new value of the function U is 811.875.

As shown in Table 6.3, at the 6th iteration the values for x_1 and x_2 are 7.87096 and 6.99861, respectively. Note that at this iteration, the value of the function is increased. Thus, for the next iteration, the values of Δ, Δx_1, and Δx_2 are reduced to one-half and the gradient direction sign is changed.

TABLE 6.3. Results of Iterative Calculation for Steepest Descent Method

Ray	i	Δ	x_1	x_2	U
		1.00000	8.00000	1.00000	1653.00000
	1	1.00000	7.97849	1.99977	1013.47800
	2	1.00000	7.95699	2.99954	811.87500
	3	1.00000	7.93548	3.99931	724.27228
	4	1.00000	7.91397	4.99907	685.87415
	5	1.00000	7.89247	5.99884	675.08038
	6	1.00000	7.87096	6.99861	682.63324
	7	−0.50000	7.88171	6.49873	676.91956
1	8	−0.50000	7.89247	5.99884	675.08032
	9	−0.50000	7.90322	5.49896	677.76300
	10	0.25000	7.89784	5.74890	675.80719
	11	0.25000	7.89247	5.99884	675.08032
	12	0.25000	7.88709	6.24879	675.47998
	13	−0.12500	7.88978	6.12382	675.14508
	—	—	—	—	—
	—	—	—	—	—
	36	−0.00195	789175	6.03204	675.07050
	37	0.00098	7.89179	6.03009	675.07056
		1.00000	7.89179	6.03009	675.07056
	1	1.00000	6.89200	6.00979	654.54395
	2	1.00000	5.89220	5.98949	648.10687
2	3	1.00000	4.89241	5.96919	660.71576
	4	−0.50000	5.39231	5.97934	651.57721
	—	—	—	—	—
	—	—	—	—	—
	31	−0.00195	6.00156	5.99171	648.00067
	32	0.00098	6.00351	5.99175	648.00073
		1.00000	6.00351	5.99175	648.00073
	1	−0.50000	5.61280	6.91226	656.24243
3	—	—	—	—	—
	—	—	—	—	—
	29	−0.00195	6.00046	5.99894	648.00000
	30	0.00098	6.00122	5.99714	648.00012
		1.00000	6.00122	5.99714	648.00012
	1	−0.50000	5.60919	6.91710	656.33893
4	—	—	—	—	—
	—	—	—	—	—
	28	−0.00195	5.99969	6.00074	648.00000
	29	0.00098	6.00045	5.99894	648.00006
	NCOUNT = 128		$x_1 = 6.000455$	$x_2 = 5.998939$	$U = 648.0001$

$$\Delta = -\frac{1.0}{2} = -0.500000$$

$$\Delta x_1 = -\left(\frac{-0.021506}{2}\right) = 0.01075$$

$$\Delta x_2 = -\left(\frac{0.99977}{2}\right) = -0.49988$$

Hence,

$$x_1 = 7.87096 + 0.01075 = 7.88171$$
$$x_2 = 6.99861 - 0.49988 = 6.49873$$

and the function U is

$$U = 3(7.88171^2 + 6.49873^2) + 1296\left(\frac{1}{7.88171} + \frac{1}{6.49873}\right) = 676.91956$$

The same iteration procedure is repeated until the increment Δ is equal to the convergence criterion $\varepsilon = 0.001$. As shown in Table 6.3, this occurs at the 37th iteration. The values calculated at this iteration are used as initial values for the next ray's first iteration.

Second ray calculations: The new ray calculations start from the coordinates $x_1 = 7.89179$ and $x_2 = 6.03009$ with $\Delta = 1.00000$. Note that the value of the increment, $\Delta = 1.00000$ remains the same. Calculate the new ray direction as follows:

$$\left(\frac{\partial U}{\partial x_1}\right) = 6x_1 - \frac{1296}{x_1^2}$$

$$= 6(7.89179) - \frac{1296}{7.89179^2} = 26.54160$$

$$\left(\frac{\partial U}{\partial x_2}\right) = 6x_2 - \frac{1296}{x_2^2}$$

$$= 6(6.03009) - \frac{1296}{6.03009^2} = 0.53892$$

Hence,

$$\Delta x_1 = \frac{-\left(\frac{\partial U}{\partial x_1}\right)\Delta}{\sqrt{\left(\frac{\partial U}{\partial x_1}\right)^2 + \left(\frac{\partial U}{\partial x_2}\right)^2}}$$

$$= \frac{-(26.54160)(1.0)}{\sqrt{26.54160^2 + 0.53892^2}} = -0.99979$$

$$\Delta x_2 = \cfrac{-\left(\frac{\partial U}{\partial x_2}\right)\Delta}{\sqrt{\left(\frac{\partial U}{\partial x_2}\right)^2 + \left(\frac{\partial U}{\partial x_2}\right)^2}}$$

$$= \frac{-(0.53892)(1.0)}{\sqrt{26,54160^2 + 0.53892^2}} = -0.020300$$

Calculate x_1 and x_2 for the first iteration

$$x_1 = 7.89179 - 0.99979 = 6.89200$$

$$x_2 = 6.03009 - 0.020300 = 6.00979$$

and

$$U = 3(6.89200^2 + 6.00979^2) + 1296\left(\frac{1}{6.89200} + \frac{1}{6.00979}\right) = 654.54395$$

Iterations for the second ray are repeated until the increment is reduced to the assumed convergence criterion $\varepsilon_1 = 0.001$.

The iteration procedure for each ray is repeated until $\partial U/\partial x_1$ or $\partial U/\partial x_2$ is reduced to a convergence criterion $\varepsilon_2 = 0.01$. As shown in Table 6.3, based on the convergence criterions $\varepsilon_1 = 0.001$ and $\varepsilon_2 = 0.01$, the minimum solution is

$$U_{min} = 648.00006 \quad \text{at} \quad x_1 = 6.00045, \quad x_2 = 5.99894$$

Using elementary calculus, an exact solution of the given problem can also be found as

$$\left(\frac{\partial U}{\partial x_1}\right) = 6x_1 - \frac{1296}{x_1^2} = 0 \Rightarrow x_1 = 6.00000$$

$$\left(\frac{\partial U}{\partial x_2}\right) = 6x_2 - \frac{1296}{x_2^2} = 0 \Rightarrow x_2 = 6.00000$$

Substituting the above determined roots x_1 and x_2 into the equation yields the minimum value of the function

$$U_{min} = 3(6^2 + 6^2) + 1296\left(\frac{1}{6} + \frac{1}{6}\right) = 648.00000$$

The result of the exact solution agrees with the approximate solution found by the steepest descent method.

6.2.5 Linear Programming

The method of linear programming is applicable to a linear objective function subjected to a number of linear constraints. As shown in Figure 6.13, the linear

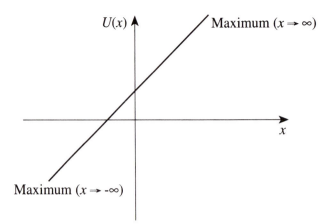

Figure 6.13 The maximum and minimum for an uncon-strained obective function.

objective function $U(x) = a + bx$, without constraints, has a minimum and maximum at $x \Rightarrow -\infty$ and $x \Rightarrow \infty$, respectively.

The maximization problem for a linear objective function subject to linear constraints can be formulated by

1. Optimizing the objective function

$$U(x) = a_1 x_1 + a_2 x_2 + \cdots + a_n x_n \qquad (6.27)$$

2. Setting the linear constraints as

$$h_i(x) = b_{i1} x_1 + b_{i2} x_2 + \cdots + b_{in} x_n \leq, =, \geq C_i \qquad (6.28)$$
$$x_1, x_2, \cdots, x_n \geq 0 \qquad i = 1, 2, \cdots m$$

Where x is the design variable, n is the number of design variables, and a, b, and C are given constants.

Linear programming problems can be solved either analytically or graphically. If there are few unknowns associated with the optimum design, a graphical method can be used to find the optimum points of a given objective function. If the number of unknowns is relatively high, the most common numerical method of solution, *the simplex method*, is used. The graphical method uses the following steps:

1. Identify design variables, objective function, and constraints.

2. Identify the boundaries of the feasible region by using the given constraints. If any inequality constraints are given, use them as equality constraints to find the boundaries of the constraint region so that the feasible region will satisfy simultaneously all the constraints.

3. Plot the objective function to identify the best design point that optimizes the objective function.

▬ *Example 6.7* ▬▬▬▬▬▬▬▬▬▬▬▬▬▬▬▬▬▬▬▬▬▬▬▬▬▬▬▬▬▬▬▬▬▬▬▬▬

Using the graphical method maximize

$$U(x) = 2x_1 + x_2 \tag{6.29}$$

subjected to the following constraints

$$3x_1 + x_2 \geq 30$$
$$x_1 + x_2 \geq 20$$
$$18x_1 + 54x_2 \geq 972$$
$$x_1 \geq 0$$
$$x_2 \geq 0$$

Solution

The two constraints $x_1 \geq 0$ and $x_2 \geq 0$ form the line of boundaries of $x_1 = 0$ and $x_2 = 0$ of the feasible region. After converting the constraint equations to equality constraints

$$3x_1 + x_2 = 30$$
$$x_1 + x_2 = 20$$
$$18x_1 + 54x_2 = 972$$

the feasible region (shaded area) for the optimum solution can be drawn as shown in Figure 6.14. This feasible region simultaneously satisfies all five constraints and represents the area of possible optimum solutions to the design problem. The objective function $U(x) = 2x_1 + x_2$ is also shown on the figure for $U(x) = 8$, $U(x) = 16$, and $U(x) = 32$. As can be observed from the figure, when parallel lines defining the objective function move away from the point of origin, the

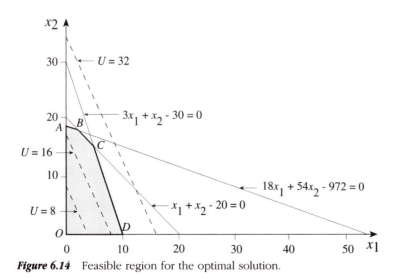

Figure 6.14 Feasible region for the optimal solution.

TABLE 6.4.   Set of Solutions
for $U(x)$

Points	$U(x) = 2x_1 + x_2$
$A\,(0,\ 18)$	18.0
$B\,(3,\ 17)$	23.0
$C\,(5,\ 15)$	25.0
$D\,(10,\ 0)$	20.0
$O\,(0,\ 0)$	0.0

value of the objective function is increased. The objective function for $U(x) = 32$ does not intersect the feasible region boundaries; hence, it should not be included in the optimum design analysis. The set of feasible points lies on the boundaries of the feasible region. The point that maximizes the objective function occurs at the intersection of two or more constraints. Thus, the first step in finding the maximum point is to find the intersection points of pairs of constraints. Next, reduce the intersection points to one point that maximizes the objective function. Table 6.4 shows the coordinates of points A, B, C, D, and O and the corresponding objective function values. The optimum solution is obtained when $x_1 = 5$ and $x_2 = 15$ where the value of the objective function is 25.

6.2.6   Nonlinear Programming Problems

If the objective function or any other constraints that define the optimization problem are nonlinear, the design problem is called a *nonlinear programming* problem. As an example of an optimum design problem that can be solved by using nonlinear programming, consider the beam problem given in Example 6.1. The problem is to find the dimensions of the beam that satisfy strength and geometric constraints and minimize the volume of the beam.

To simplify the problem, assume predetermined values of

$$x_2 = D_2 = 50 \text{ mm}$$
$$x_3 = D_3 = 60 \text{ mm}$$
$$L = 800 \text{ mm}$$

The goal is to determine the dimensions x_1 and x_4 for the minimum volume. The objective function is

$$U(x) = 2\left[\frac{\pi(50^2 - x_1^2)}{4}\right]x_4 + \frac{\pi(60^2 - x_1^2)}{4}(2 \times 800 - 2x_4)$$

Simplifying

$$U(x) = -1728x_4 - 1257x_1^2 + 4523893$$

The problem may be further formulated to find the dimensions, x_1 and x_4, which minimize the objective function, U, and satisfy the following inequality constraints:

1. The beam should resist the maximum load $2F = 16{,}000$ N and the allowable bending stress $\sigma_a = 400$ MPa. Hence, the bending strength inequality constraint for part 1 of the beam is

$$\sigma_b \leq \sigma_a$$

where σ_b, the bending stress due to $2F$, is

$$\sigma_b = \frac{32Fx_4 D_2}{\pi(D_2^4 - x_1^4)}$$

Substituting into the inequality constraint yields

$$\frac{4074367x_4}{6.25 \times 10^6 - x_1^4} \leq 400$$

Similarly, the bending strength inequality constraint for part 2 of the beam is

$$\frac{32FLD_3}{\pi(D_3^4 - x_1^4)} \leq \sigma_a$$

$$\frac{3.9114 \times 10^9}{1.296 \times 10^7 - x_1^4} \leq 400$$

The above equation yields

$$x_1 \leq 42mm$$

2. The inside diameter of the beam x_1 should not be less than 30 mm, and the length x_4 must be equal to or larger than zero. Thus, the additional inequality constraints are

$$x_1 \geq 30\text{mm}$$

$$x_4 \geq 0$$

In summary, the strength and geometric inequality constraints are

$$g_1 \equiv 400 - \frac{4074367x_4}{6.25 \times 10^6 - x_1^4} \geq 0$$

$$g_2 \equiv 42 - x_1 \geq 0$$

$$g_3 \equiv x_1 - 30 \geq 0$$

$$g_4 \equiv x_4 \geq 0$$

A graphical solution for the feasible region of the beam is shown in Figure 6.15. In this example, since g_1 is a nonlinear function of the design variable, x_1, the problem is called a nonlinear programming optimization problem. The designer may also consider an equality constraint such as the length of part 1 of the beam being equal to seven times the inside diameter x_1. Thus, the equality

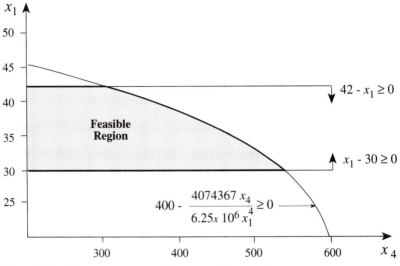

Figure 6.15   The constraint on design variables x_1 and x_4.

constraint is

$$b_1 = x_4 - 7x_1 = 0$$

Figure 6.15 is duplicated as shown in Figure 6.16. Now the feasible region is represented by the dark line. It is interesting to observe that the area of the feasible region is reduced to a line when the equality constraint is included.

6.2.7   Multicriterion Optimization

Almost all real-world problems involve more than one objective. In a multicriterion optimization the aim is to maximize and/or minimize more than one objective

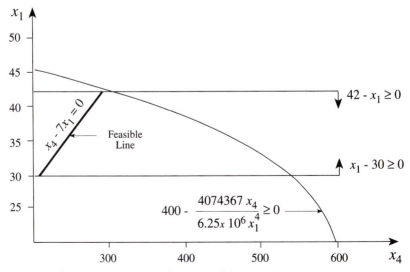

Figure 6.16   The constraint on design variables x_1 and x_4.

function simultaneously. The advantage of this method compared with that of single-criterion optimization is that a set of alternative solutions can be obtained rather than a single solution. In optimizing design problems that involve more than one objective function, the designer may have to

1. Maximize all the objective functions.
2. Minimize all the objective functions.
3. Maximize some and minimize others.

There are many different methods that can be used for multicriterion optimization problems.[3] A simple beam problem can be solved to illustrate a multicriterion optimization. The previously discussed problem can be reformulated as follows. Find the dimensions x_1 and x_2 that satisfy the geometric and strength constraints and minimize the following criteria:

1. Volume of the beam.
2. Static deflection of the beam under the load $2F$.

Both functions are to be minimized. Note that these are contrasting criteria; that is, the best solution for the first criterion yields the worst solution for the second one. More information on multicriterion optimization problems is given by Osyczha and Steuer [10, 11].

BIBLIOGRAPHY

1. Gallagher, R. H. and Zienkiewicz, O. C. *Optimization Structural Design.* John Wiley & Sons, Inc., New York, 1973.
2. Kirsch, U., *Optimum Structural Design,* McGraw-Hill, New York, 1982.
3. Haug, E. J., and Arora, J.S., *Applied Optimal Design,* John Wiley & Sons, Inc., New York, 1979.
4. Johnson, C. Ray, *Optimum Design of Mechanical Elements,* John Wiley & Sons, Inc., New York, 1980.
5. Reklaitis, G. V., Ravindran, A., and Ragsdell, K. M., *Engineering Optimization Methods and Applications.* John Wiley & Sons, Inc., New York, 1983.
6. Morris, A. J., *Foundations of Structural Optimization: A Unified Approach.* John Wiley & Sons, Inc., New York, 1982.
7. Gero, J. S., *Design Optimization.* Academic Press, New York, 1985.
8. Spunt, L., *Optimum Structural Design.* Prentice-Hall, Englewood Cliffs, NJ, 1971.
9. Wilde, D. J., *Optimum Seeking Methods,* Prentice-Hall, Englewoood Cliffs, NJ, 1964.
10. Osyczka, A., *Multicriterion Optimization in Engineering,* Ellis Horwood Ltd., Chichester, West Sussex, England, 1984.
11. Steuer, R. E., *Multiple Criteria Optimization: Theory, Computation, and Application.* John Wiley & Sons, Inc., New York, 1986.

[3] J. L. Cohon, *Multiobjective Programming and Planning.* Academic Press, New York, 1978.

Chapter 7

Statistical Decisions

Engineering statistics deals with a presentation of techniques that are appropriate for making inferences under the conditions of uncertainty that exist in a wide range of enginering activities.

7.1 RANDOM VARIABLE

A function whose values cannot be predicted in a sample space is called a random variable. If a sample is selected in an arbitrary manner from a population, such a sample is said to be a random sample. It should be noted that the word "randomness" is used for the operation of selecting the sample. Rotary machinery, such as turbines and compressors, are designed to operate in a vibration-free manner. However, bearing misalignment, material heterogeneity, and geometric variations, collectively or selectively, cause the rotor axial mass center to be noncoincident with the bearing axis. Such a deviation is termed rotor mass center eccentricity; it causes time-dependent bearing forces to occur in the rotor housing.[1,2] The mass center eccentricity is not a design condition. Instead, the eccentricity is a randomly occurring event (random variable) and differs from rotor to rotor within each machine class. Most of the time it is very difficult, if not impossible, to examine the whole population for a statistical process. More often, a random sampling technique is used instead. For example, the observation obtained by measuring the eccentricity of a rotor is called a population whose sizes are countless. Suppose 20 rotors (random sampling) are selected from a larger (conceivably countless) number of rotors forming a population. Each rotor is discretized into three sections whose

[1] A. Ertas, "The Response of Rotating Machinery to a Random Eccentricity Distribution in the Rotor," *3rd Reunion Academica de Ingenieria Mechanica Conferencias*, San Luis Potosi, Oct.14–18, 1985.

[2] A. Ertas and T. J. Kozik, "Fatigue Loads on the Foundation Due to Turbine Rotor Eccentricity," *Trans. ASME Journal of Energy Resources Technology,* vol. 109, pp. 174–179, 1987.

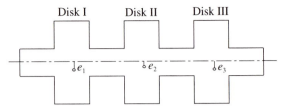

Figure 7.1 Rotor with three discretized disks.

shapes are circular cylinders (disks), as shown in Figure 7.1. Table 7.1 illustrates a typical set of eccentricity measurements, which represents a sample drawn from that population. Rotor eccentricities along the rotor do not remain constant. In general, the locus of the mass centers of the discretized rotor cross sections defines a space curve with respect to the bearing axis. Note that, for simplicity, eccentricities are assumed to be in the same plane.

Figure 7.2 shows the bar charts (frequency histogram) for the data given in Table 7.1. The values plotted along the x axis are the eccentricity measurements of the first, second, and third disk of the 20 rotors, and the number of times the measurements occur (frequency) is plotted along the vertical axis. A distribution is called symmetric, as shown in Figure 7.2a, if the curve can be folded with respect to a vertical axis. A distribution that is not symmetric with respect to a vertical axis is called skewed. Hence, distributions like those shown in Figure 7.2b and 7.2c are called "positively skewed" and "negatively skewed," respectively. In order to learn how the distribution of the eccentricity measurements (data) accumulates, consider another useful description called "cumulative frequency." Table 7.1 is used to construct Table 7.2, which shows a cumulative frequency distribution.

As Figure 7.3 shows, the cumulative frequency curve for the symmetric distribution approaches an "S" shape. Note that the trend of the cumulative frequency curves for the negatively and positively skewed distributions differs from that of the symmetrical distribution.

7.2   MEASURE OF CENTRAL TENDENCY

There are three measures of central tendency that are commonly used to represent the general trend of the eccentricity measurements given in Table 7.1. These are described in the sections that follow.

Table 7.1   Typical Set of Eccentricity Measurements of 20 Rotors

Observations Average Eccentricity e, mm	Frequency of Disk I	Frequency of Disk II	Frequency of Disk III
0.1	1	3	1
0.2	2	6	1
0.3	4	4	2
0.4	6	3	2
0.5	4	2	5
0.6	2	1	6
0.7	1	1	3

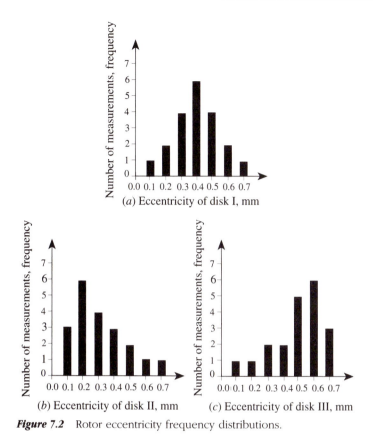

Figure 7.2   Rotor eccentricity frequency distributions.

7.2.1   Mean or Arithmetic Average

The mean is defined as the sum of all of the values of the random sample variables x_i, divided by the size of the random sample, that is,

$$\bar{x} = \frac{\sum_{i=1}^{n} x_i}{n} \tag{7.1}$$

The mean of a random sample is denoted by $\bar{x}$ whereas the mean of a population is denoted by μ. It should be noted that the population mean μ is

Table 7.2   Cumulative Frequence Distributions of Eccentricity Measurements

Eccentricity e, mm	Disk I		Disk II		Disk III	
	Cum. Freq	Cum. Freq. %	Cum. Freq	Cum. Freq. %	Cum. Freq	Cum. Freq. %
0.1	1	5	3	15	1	5
0.2	3	15	9	45	2	10
0.3	7	35	13	65	4	20
0.4	13	65	16	80	6	30
0.5	17	85	18	90	11	55
0.6	19	95	19	95	17	85
0.7	20	100	20	100	20	100

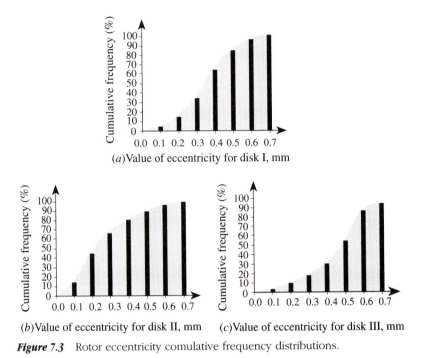

Figure 7.3 Rotor eccentricity comulative frequency distributions.

constant for a particular population, whereas the sample mean $\bar{x}$ can differ from sample to sample.

Suppose the overall mean of the rotors given in Table 7.1 is desired. The individual means of each disk $\bar{x}_1$, $\bar{x}_2$, and $\bar{x}_3$, can be combined by the following formula

$$\overline{X}_{overall} = \frac{n_1\bar{x}_1 + n_2\bar{x}_2 + n_3\bar{x}_3}{n_1 + n_2 + n_3} \qquad (7.2)$$

—— **Example 7.1** ————————————————————————————

Using Table 7.1, estimate the mean eccentricity of each disk and the overall mean of the eccentricity measurements of the 20 rotors.

Solution

The random sample size n is 20. Individual means of each disk are

$$\bar{x}_1 = \frac{\sum_{i=1}^{n} x_i}{n} = \frac{(0.1)(1) + (0.2)(2) + \cdots}{20} = 0.400 \text{ mm}$$

$$\bar{x}_2 = \frac{\sum_{i=1}^{n} x_i}{n} = \frac{(0.1)(3) + (0.2)(6) + \cdots}{20} = 0.310 \text{ mm}$$

$$\bar{x}_3 = \frac{\sum_{i=1}^{n} x_i}{n} = \frac{(0.1)(1) + (0.2)(1) + \cdots}{20} = 0.495 \text{ mm}$$

Combine the individual means of each disk, $\bar{x}_1$, $\bar{x}_2$, $\bar{x}_3$, to determine the overall mean of the eccentricity distribution.

$$\bar{X}_{overall} = \frac{20(0.4) + 20(0.31) + 20(0.495)}{20 + 20 + 20} \approx 0.402 \text{ mm}$$

7.2.2 Median

The median of a set of observations arranged in order of magnitude may be defined as the observation that divides the distribution curve into two equal parts. The value of the median depends on whether the total number of observations is odd or even. If odd, the value of the median is the $[(n/2) + 1]$th-order observation. If even, the value of the median is the average of $(n/2)$th and $[(n/2) + 1]$th-order observation.

Using the same data as in the above example for disk I, the number of observations n is equal to 20. Since the total number of observations is even, we use the value corresponding to the average of $(n/2)$th and $[(n/2) + 1]$th-order to find the median, which is 0.4 mm.

Note that, to find the value of the median this way, the values of the random sample variable should be in increasing or decreasing order of magnitude.

7.2.3 Mode

The last measure of the central tendency is the mode, which is the observation that occurs most frequently. If the same data are used, the mode is equal to 0.4 mm, corresponding to the maximum frequency of 6.

From the foregoing it can be concluded that if the distribution is symmetrical, the value of the *mean*, *median*, and *mode* will be the same. However, as shown in Figure 7.3, in the case of "positively" skewed distributions, the mode is less than the median, and when the mode is larger than the median, the distribution shows a "negative" skew.

7.3 MEASURE OF VARIABILITY

The measure of variations—range, mean deviation, standard deviation, and variance—will now be discussed.

7.3.1 Range

Range, which is the simplest measure of variation (dispersion), is the absolute difference between the largest number and the smallest number in the distribution. To gain a better understanding of the range, the following calculations will be illustrated.

Suppose a set of shafts 25 mm in diameter with an allowable tolerance of 0 to 0.042 mm is needed. As shown in Table 7.3, there are two boxes and the mean value of the tolerances in each box is $\bar{x}_1 = 0.041$ mm and $\bar{x}_2 = 0.045$ mm. Now the question is which box to select. Perhaps the selection will be the first box because the mean value of the tolerance is within the acceptable range. However, when the boxes are opened, it is noted that the number of shafts that can be used is greater inside the second box, even though the mean value of the

Table 7.3   Tolerances of the Shafts in Each Box

Box 1	Box 2
Tolerance, mm	**Tolerance, mm**
0.01	0.03
0.01	0.03
0.02	0.04
0.03	0.04
0.05	0.04
0.05	0.04
0.05	0.05
0.06	0.05
0.06	0.06
0.07	0.07
Mean	Mean
$\overline{x}_1 = 0.041$	$\overline{x}_2 = 0.045$

tolerance in that box is not within the acceptable range. It is obvious that, in this case, the mean does not show the distribution of shaft tolerances inside the boxes. It follows that, in some instances, range can be used to show how far the data in a distribution are from the mean.

▬ *Example 7.2*

By using the data in Table 7.3, calculate the range for the first and second boxes.

Solution

$$\text{Range for the first box} = 0.07 - 0.01 = 0.06 \text{ mm.}$$
$$\text{Range for the second box} = 0.07 - 0.03 = 0.04 \text{ mm.}$$

7.3.2   Mean Deviation

Another measure of variability is the mean deviation, which is the mean of the absolute values of the deviations from the mean. The mean deviation can be calculated by using

$$\text{Mean deviation} = \frac{\sum_{i=1}^{n} |x_i - \overline{x}|}{n} \tag{7.3}$$

7.3.3   Standard Deviation

Although the mean deviation is a useful measure of variability, because of the absolute value involved in calculating the mean deviation, it may not be practical to use. Hence, the standard deviation, which is the most reliable measure of variability, is used in most cases. The following equation is used for computation of the standard deviation of a random sample:

$$S = \sqrt{\frac{\sum_{i=1}^{n}(x_i - \bar{x})^2}{n - 1}} \tag{7.4}$$

where n is the number of observations , and S is the standard deviation of the random sample. To compute the standard deviation of a population, $\bar{x}$ and n are replaced by μ and N in (7.4):

$$\hat{\sigma} = \sqrt{\frac{\sum_{i=1}^{N}(x_i - \mu)^2}{N - 1}} \tag{7.5}$$

where $\hat{\sigma}$ is the standard deviation of a population.

7.3.4 Variance

The square of the standard deviation is another very important quantity known as the variance of a random variable; it shows the spread of the distribution. The variances of the random sample and the population, respectively, are given by

$$S^2 = \frac{\sum_{i=1}^{n}(x_i - \bar{x})^2}{n - 1} \tag{7.6}$$

and

$$\hat{\sigma}^2 = \frac{\sum_{i=1}^{N}(x_i - \mu)^2}{N - 1} \tag{7.7}$$

—— *Example 7.3* ————————————————————————

Compute the sample standard deviation and variance for the data given in Table 7.3.

Solution

$$S_1 = \left[\frac{(0.01 - 0.041)^2 + (0.01 - 0.041)^2 + \cdots (0.07 - 0.041)^2}{10 - 1}\right]^{1/2} = 0.02183 \text{ mm}$$

Hence, $S_1^2 = 0.000477 \text{ mm}^2$.

$$S_2 = \left[\frac{(0.03 - 0.045)^2 + (0.03 - 0.045)^2 + \cdots (0.07 - 0.045)^2}{10 - 1}\right]^{1/2} = 0.01269 \text{ mm}$$

Hence, $S_2^2 = 0.00016 \text{ mm}^2$.

7.3.5 Measure of Skewness

As discussed in the preceeding sections, a distribution that is not symmetric with respect to a vertical axis is said to be skewed. The eccentricity distribution of the rotor disks shown in Figure 7.2*b* is called a positive skew (skewed to the right), and in Figure 7.2*c* the distribution is called a negative skew (skewed to the left). Of course, the easy way to learn whether a distribution is skewed or not is to check whether the mean, mode, and median have the same value. However, skewness of the distribution can be easily determined by the following formula.

$$\text{Skewness} = \frac{3(\text{mean} - \text{median})}{\text{standard deviation}} \qquad (7.8)$$

─── *Example 7.4* ──

Calculate the skewness of Figure 7.2b and c from the data given in Table 7.1.

Solution

Skewness of Figure 7.2b   From Example 7.1, the calculated sample mean for the second disk is equal to 0.310 mm. The standard deviation is

$$S = \sqrt{\frac{\sum_{i=1}^{n}(x_i - \bar{x})^2}{n - 1}} = \sqrt{\frac{3(0.1 - 0.31)^2 + 6(0.2 - 0.31)^2 + \cdots}{20 - 1}} = 0.168 \text{ mm}$$

Therefore

$$\text{Skewness} = \frac{3(\text{mean} - \text{median})}{\text{standard deviation}} = \frac{3(0.310 - 0.300)}{0.168} = 0.179$$

Skewness of Figure 7.2c   From Example 7.1, the calculated sample mean for the third disk is equal to 0.495 mm. Following the same procedure, the standard deviation is found to be $S = 0.167$. Then the skewness of the third disk is

$$\text{Skewness} = \frac{3(\text{mean} - \text{median})}{\text{standard deviation}} = \frac{3(0.495 - 0.500)}{0.167} = -0.090$$

The minus sign indicates that the third disk has negative skewness. From the results of this example, we conclude that if the mean is larger than the median, distribution is positively skewed, and if the mean is smaller than the median, distribution is negatively skewed.

──

7.4   PROBABILITY DISTRIBUTIONS

Some of the well-known probability distributions used in engineering are briefly described in the following discussion.

7.4.1   Discrete Distributions

Binomial Distribution   Binomial distribution is used to describe two possible outcomes, such as a success or a failure, for which the probability remains constant for each trial. The binomial distribution is of great importance in quality control. If the probability of success is designated by p, and failure is designated by q, the probability of x success in n trials is given by

$$p(x) = \binom{n}{x} p^x q^{n-x} \qquad \text{for } x = 0, 1, 2 \cdots n \tag{7.9}$$

where

$\binom{n}{x} = \frac{n!}{x!(n-x)!}$,

x is the number of successes,

$(n - x)$ is the number of failures in n trials.

It can be shown that the mean, standard deviation, and variance of the binomial distribution are[3]

Mean $= n \times p$

Standard Deviation $= \sqrt{n \times p \times q}$

Variance $= n \times p \times q$

—— *Example 7.5* ————————————————————————

Through quality control, it is known that 8 percent of the shafts from a production line are not within the specified tolerances. If samples of 10 shafts are taken, what is the probability of having 2 rejected shafts? Calculate the mean, standard deviation, and variance.

Solution

If the probability of selecting a rejected shaft is called a success p, then the probability of selecting acceptable shafts will be $q = 1 - p$. Hence, $p = 0.08$, and $q = 0.92$.

For sample size $n = 10$, the probability of $x = 2$ rejected shafts can be calculated from (7.9)

$$p(2) = \frac{10!}{2!(10 - 2)!}(0.08)^2(0.92)^{10-2} = 0.1478$$

$$\bar{x} = n \times p = 10 \times 0.08 = 0.8$$

$$S = \sqrt{n \times p \times q} = \sqrt{10 \times 0.08 \times 0.92} = 0.86$$

Variance $= S^2 = 0.736$

Poisson Distribution   This distribution is often used in quality control. Poisson's distribution is an extension of the binomial distribution in which the number of trials n approaches infinity, the probability of success p approaches zero, and the mean value remains finite. Since q is very large compared with p, it can be assumed

---

[3] C. J. Brookes, I. G. Betteley, and S. M. Loxston, *Mathematics and Statistics*, Wiley, Inc., New York, 1966.

as unity. Therefore, the variance of the Possion distribution is equal to the mean value. The probability of x success in n trials is

$$p(x) = \frac{\mu^x}{x!} \exp(-\mu) \tag{7.10}$$

Note that, when random sampling is used, μ should be replaced by the sample mean $\bar{x}$.

▬ *Example 7.6* ─────────────────────────────

Solve example 7.5 by using Poisson's distribution.

Solution

$x = 2$, $n = 10$, $p = 0.08$, then

$$\mu = n \times p = 10 \times 0.08 = 0.8$$

and

$$p(2) = \frac{0.8^2}{2!} \exp(-0.8) = 0.144$$

7.4.2   Important Continuous Life Test Distributions

Weibull Distribution   The Weibull distribution is often used in engineering for predicting the life of machine components. It is considered to be one of the most important prediction methods, since it fits many different machine and electronic component failure distributions. The Weibull distribution was proposed by W. Weibull[4] and applied to ball bearing failure by Lieblein and Zelen.[5]

The Weibull cumulative density function $F(t)$ is

$$F(t) = 1 - exp\left[-\left(\frac{t}{\theta}\right)^b\right] \tag{7.11}$$

The first derivative of 7.11 with respect to t gives the Weibull probability density function $f(t)$ as

$$f(t) = \frac{bt^{b-1}}{\theta^b} \exp\left[-\left(\frac{t}{\theta}\right)^b\right] b > 0, \theta > 0, t \geq 0 \tag{7.12}$$

where
b = shape parameter or Weibull slope

θ = scale parameter

t = time

[4] W. Weibull, "A Statistical Distribution Function of Wide Application," *Journal of Applied Mechanics*, vol. 18, pp. 293–297, 1951.

[5] J. Lieblein and M. Zelen, "Statistical Investigation of the Fatigue Life of Deep-Groove Ball Bearings," *Journal of Res. Nat. Bur. Stand.*, vol. 5, Res. Paper 2719, pp. 273–316, 1956.

For different values of the shape parameter b, the Weibull distribution can be reduced to two special distributions as follows:

Exponential distribution, when $b = 1$

$$F(t) = 1 - \exp\left(-\frac{t}{\theta}\right) \tag{7.13}$$

and

$$f(t) = \frac{1}{\theta} \exp\left(-\frac{t}{\theta}\right) \tag{7.14}$$

Rayleigh distribution, when $b = 2$

$$F(t) = 1 - \exp\left[-\left(\frac{t}{\theta}\right)^2\right] \tag{7.15}$$

and

$$f(t) = \frac{2t}{\theta^2} \exp\left[-\left(\frac{t}{\theta}\right)^2\right] \tag{7.16}$$

Note that in Figure 7.4, for values of $b > 2$, the curve becomes bell shaped (normal distribution).

Gamma Distribution   The gamma distribution is often used in fatigue and wear studies. It was first considered by Gupta and Groll in life test problems.[6] The probability density function $f(t)$ is

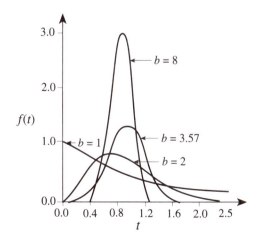

Figure 7.4   The Weibull distribution for different values of the slope b.

---

[6] S. Gupta and P. Groll, "Gamma Distribution in Acceptance Sampling Based on Life Tests," *Journal of the American Statistical Association*, pp. 942–970, 1961.

$$f(t) = \frac{\lambda^\beta t^{\beta-1}}{\Gamma(\beta)} \exp(-\lambda t) \; \lambda > 0, \beta > 0 \text{ and } t \geq 0 \tag{7.17}$$

and the cumulative density function $F(t)$ is

$$F(t) = 1 - \frac{\Gamma(\beta, \lambda t)}{\Gamma(\beta)} \tag{7.18}$$

where λ is the scale factor, β is the shape factor, t is the time, and $\Gamma(\beta)$ is the gamma function and is given as

$$\Gamma(\beta) = \int_0^\infty t^{\beta-1} \exp(-t) dt \tag{7.19}$$

Normal Distribution   The normal distribution (Figure 7.5), which is also called the Gaussian distribution, describes the distribution of many sets of data that occur in engineering. As the number of observations n increases, the binomial distribution approaches the Gaussian distribution. As illustrated, the normal distribution curve is symmetrical about the mean value and has a characteristic bell shape. Hence, the mean, mode, and median have the same value. The equation of the normal curve is given by

$$f(x) = \frac{1}{\hat{\sigma}\sqrt{2\pi}} \exp\left[-\frac{(x-\mu)^2}{2\hat{\sigma}^2}\right] - \infty < x < \infty \tag{7.20}$$

where μ and $\hat{\sigma}$ are the population mean and standard deviation, respectively. The probability density function $f(x)$, has a total area under the bell-shaped curve equal to 1. That is,

$$\int_{-\infty}^\infty f(x) dx = 1 \tag{7.21}$$

If the mean and the standard deviation are known, the normal distribution curve is completely resolved. Figure 7.6 shows the normal probability density curves for different combinations of mean and standard deviations.

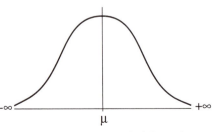

Figure 7.5 Normal probability density curve.

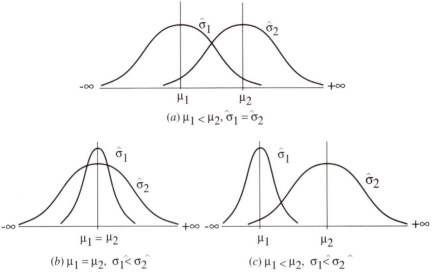

Figure 7.6   Normal probability density curves for different μ and $\hat{\sigma}$.

Equation 7.20 can be integrated to evaluate the probability of certain values of x_1 and x_2 as

$$p(x_1 \le x \le x_2) = \int_{x_1}^{x_2} f(x)dx$$

$$= \frac{1}{\hat{\sigma}\sqrt{2\pi}} \int_{x_1}^{x_2} \exp\left[-\frac{(x-\mu)^2}{2\hat{\sigma}^2}\right] dx \qquad (7.22)$$

As shown in the figures, the normal distribution theoretically ranges from $-\infty$ to ∞. However, approximately 68 percent of the distribution lies within $\mu \mp \hat{\sigma}$, 95 percent lies within $\mu \mp 2\hat{\sigma}$, and 99.7 percent lies within $\mu \mp 3\hat{\sigma}$. Hence, only 0.3 percent lies outside of this range. This is shown in Figure 7.7.

Unit Standard Normal Distribution   Consider the deviation of each value of the normal random variable x from the mean value, which can be expressed as $x - \mu$. Because $\sum(x - \mu) = 0$, the distribution expressed in the aforementioned form will have a zero mean. Transformation of all the observations can be made to a new set

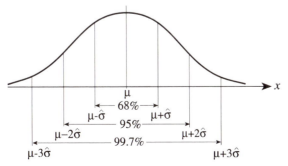

Figure 7.7   Normal distribution percentages lying within $\hat{\sigma}, 2\hat{\sigma}, 3\hat{\sigma}$ of the mean value.

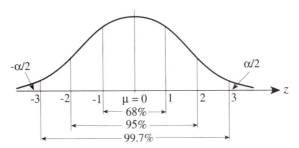

Figure 7.8 Unit normal distributin percentages lying within $\hat{\sigma}, 2\hat{\sigma}, 3\hat{\sigma}$ of the mean value.

of observations by dividing the mean deviation of each random variable x by the standard deviation

$$z = \frac{x - \mu}{\hat{\sigma}}$$

(7.23)

It can be shown that the distributions transformed using (7.23) have a variance equal to 1. Then the new probability density function of the unit standard normal distribution becomes

$$f(z) = \frac{1}{\sqrt{2\pi}} \exp\left(-\frac{z^2}{2}\right)$$

(7.24)

Now Figure 7.7 can be rearranged to the newly transformed form of normal distribution, which has a zero mean and unit standard deviation and variance shown in Figure 7.8. Note that the total area under the curve is equal to 1. In Figure 7.8, $\alpha/2$ is called a two-sided critical region. The values of critical regions corresponding to z values can be obtained from Table C.1 (see Appendix C).

—— *Example* 7.7 ————————————————————————————

A set of pins 10 mm in diameter has been manufactured for use in a conveyor. It is found that the distribution of pin diameter is approximately normal with a mean of 10.05 mm and a standard deviation of 2.1 mm.

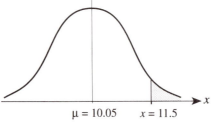

$$\mu = 10.05 \qquad x = 11.5$$

Figure 7.9   Original problem.

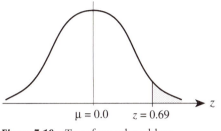

Figure 7.10   Transformed problem.

(a) What percentage of pins picked at random will have a diameter more than 11.5 mm ?

(b) What percentage of pins picked at random will have a diameter between 9.0 and 11.5 mm ?

Solution

(a) To find the probability of picking a pin with a diameter more than 11.5 (area lying to the right of $x = 11.5$), make the transformation of Figure 7.9 into the standard normal form as shown in Figure 7.10

$$z = \frac{x - \mu}{\hat{\sigma}} = \frac{11.5 - 10.05}{2.1} = 0.69$$

From Table C.1 in Appendix C, we obtain the area to the left of z:

$$\text{Area}(z < 0.69) = \frac{1}{\sqrt{2\pi}} \int_{-\infty}^{z} e^{-\frac{x^2}{2}} \, dz = 0.7549$$

Then, the shaded area lying to the right of $z = 0.69$ is

$$\text{Area}(z > 0.69) = 1 - 0.7549 = 0.2451$$

Therefore, the probability of selecting pins with a diameter greater than 11.5 mm is

$$p(x > 11.5) = 0.2451$$

(b) Similarly, Figure 7.11 is transformed into standard normal form as shown in Figure 7.12.

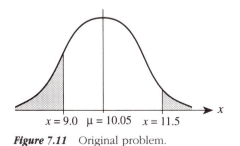

Figure 7.11   Original problem.

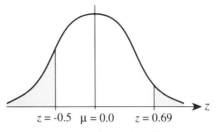

Figure 7.12   Transformed problem.

$$z = \frac{9.0 - 10.05}{2.1} = -0.5$$

The shaded area lying to the left (negative) of $z = -0.5$ is 0.3085, and from part (a) area $(z > 0.69) = 0.2451$. Thus the probability of having a diameter between 9.0 and 11.5 mm is

$$p(9.0 < x > 11.5) = 1 - (0.2451 + 0.3085) = 0.4464$$

7.5   SAMPLING DISTRIBUTIONS

Frequency distributions in statistics based on samples of n observations drawn from a population can be characterized by a sampling distribution. The sampling distribution plays an important role in estimating population parameters by the use of confidence intervals. In the remainder of this chapter some of the important sampling distributions of statistics and their application to problems in statistical inference are discussed.

7.5.1   Sampling Distributions Based on Sample Means and the Central Limit Theorem (Sampling Distribution of the Mean)

As is well known, the degree of confidence in estimating parameters of a population from a sample drawn from it is dependent on the sample size n. It seems intuitively obvious that as the sample size n becomes large, the sample mean $\bar{x}$ approaches the population mean μ. A very important mathematical statistics theorem called the *central limit theorem* states that if a random variable x is normally distributed in the population, then the randomly drawn samples of n observations from this population will also tend toward a normal distribution as the sample size n increases. Hence, the expected value $E\bar{x}$ of the sample mean can be assumed to be equal to the population mean

$$E\bar{x} = \mu \tag{7.25}$$

The variance of the sample mean is

$$\hat{\sigma}_{\bar{x}}^2 = \frac{\hat{\sigma}^2}{n} \tag{7.26}$$

By taking the square root of Eq. 7.26 the standard deviation of the sample mean, which is known as the *standard error of the mean*, is found. This indicates how close the sample mean is to the population mean, and is given as

$$\hat{\sigma}_{\bar{x}} = \frac{\hat{\sigma}}{\sqrt{n}} \tag{7.27}$$

Equation 7.27 can also be written in terms of the unit standard normal distribution $z_{\bar{x}}$ with a variance of 1 and a mean of 0

$$z_{\bar{x}} = \frac{\bar{x} - \mu}{\hat{\sigma}_{\bar{x}}} = \frac{\bar{x} - \mu}{\hat{\sigma}/\sqrt{n}} \tag{7.28}$$

Many frequency distributions experienced in engineering are approximately normal. In general, the normal approximation of the random sample mean is good if $n \geq 30$. This statement is valid even if the population distribution is not normal. If $n < 30$, the approximation is valid only if the population distribution is close to normal.

7.5.2 Student's *t* Distribution

In the previous section, it was assumed that the standard deviation of the population is known. Unfortunately, this is not always true. If the standard deviation of the population from which the random sample drawn is unknown, the standard deviation of a randomly selected sample can be used in Eq. 7.28, providing the sample size $n \geq 30$. However, if the randomly selected sample size n is less than 30, an estimation of the standard deviation of the sample mean will be less dependable. In such cases, assuming that x is normally distributed, the following t distribution is used:

$$t = \frac{\bar{x} - \mu}{s/\sqrt{n}} \tag{7.29}$$

The values of the student distribution for various values of α are given in Table C.2 (see Appendix C).

7.5.3 Chi–Square Distribution (χ^2 Distribution)

Another important sampling distribution used to estimate the population variance $\hat{\sigma}^2$ by using sample variance S^2 is the chi–square distribution.

As shown in Figure 7.13, the form of the chi–square distribution depends on the degrees of freedom ν and approaches a normal distribution as the degrees of freedom increases. The chi-square distribution with $\nu = n - 1$ degrees of freedom is given by the formula

$$\chi^2_{\alpha,\nu} = \frac{(n-1)S^2}{\hat{\sigma}^2} \tag{7.30}$$

It can be seen that because of the square terms, χ^2 can never be negative. The values of the chi-square distribution for various values of $\chi^2_{\alpha,\nu}$ are given in

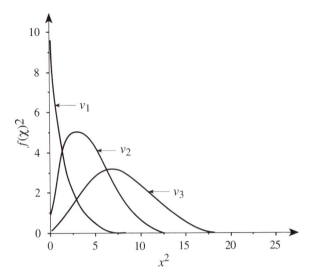

Figure 7.13   The chi-square distribution for different degrees of freedom ($v_3 > v_2 > v_1$).

Table C.3 (see Appendix C). To illustrate the use of Table C.3, assume a chi-square distribution with 18 degrees of freedom, leaving an area of 0.05 to the right, is $\chi^2_{0.05,18} = 28.869$. This is shown graphically in Figure 7.14. A chi-square distribution with a 95 percent level of confidence and 18 degrees of freedom lies between $\chi^2_{0.975,18} = 8.231$ (lower-confidence interval) and $\chi^2_{0.025,18} = 31.526$ (upper-confedence interval). Note that, since the chi-square distribution is not symmetrical, these values are not equal to each other.

7.5.4   F Distribution

The F distribution is used in conjuction with statistical quality control work and is defined as the ratio of two independent sampling distributions drawn from two normal populations with variances $\hat{\sigma}_1$ and $\hat{\sigma}_2$. For sampling distributions n_1 and n_2 with variances S_1^2 and S_2^2, respectively, the F distribution is

$$F_{v_1,v_2} = \frac{S_1^2/\hat{\sigma}_1^2}{S_2^2/\hat{\sigma}_2^2} \tag{7.31}$$

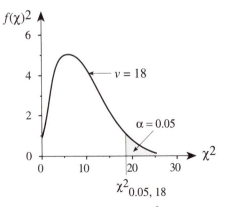

Figure 7.14   Percent point $\chi^2_{0.05,18}$ of the chi-square distribution.

with $v_1 = n_1 - 1$ and $v_2 = n_2 - 1$ degrees of freedom. The values of the F distribution for various combinations of v_1 and v_2 are given in Table C.4 (see Appendix C).

7.6 STATISTICAL INFERENCE

Statistical inference can be divided into two categories: *estimation* and *test of hypotheses*. In this section some methods involving these two areas are discussed and compared.

7.6.1 Estimation

As we have learned, parameters like the population mean μ and the population standard deviation, $\hat{\sigma}$, can be estimated from sample data taken randomly from a population. There are two ways of estimating population parameters, namely point estimates and interval estimates. A point estimate is a single estimated value of a population parameter. For example, the sample mean $\bar{x}$ and variance S^2, which estimate the population mean μ and variance $\hat{\sigma}^2$, respectively, are point estimates. To estimate these parameters in terms of probabilities, an interval estimate is used that is the estimate of a population parameter within a bounded range of values rather than a single estimated value. The accuracy of these parameters depends on how large the randomly selected sample size is. The closer the size of a sample to the population size, the greater the accuracy of the estimated mean and standard deviation of the population. However, selecting a large sample size for estimating these parameters may be impractical. What, then, is the smallest sample size for accurate estimation of the population parameters, and what is the degree of confidence in them? The degree of confidence in the population parameters can be defined by an *interval*, called a *confidence interval*, in which the parameters fall. The limits of this interval are called *confidence limits*.

Confidence Interval Estimation of the Mean: ($\hat{\sigma}$ Is Known)   In the previous sections, the population mean was estimated by using the sample mean $\bar{x}$. However, μ was not assumed to be exactly equal to $\bar{x}$. Since $\bar{x}$ is a random quantity, it may be smaller or greater than the population mean μ, thus introducing the error $(\mu - \bar{x})$. If the population variance $\hat{\sigma}^2$ is known, this error can be defined between two boundary points (two-sided confidence interval) as

$$-z_{\alpha/2}\frac{\hat{\sigma}}{\sqrt{n}} \leq (\mu - \bar{x}) \leq z_{\alpha/2}\frac{\hat{\sigma}}{\sqrt{n}} \tag{7.32}$$

Adding $\bar{x}$ to both sides of the above inequality results in

$$\bar{x} - z_{\alpha/2}\frac{\hat{\sigma}}{\sqrt{n}} \leq \mu \leq \bar{x} + z_{\alpha/2}\frac{\hat{\sigma}}{\sqrt{n}} \tag{7.33}$$

Confidence Interval Estimation of the Mean: ($\hat{\sigma}$ Is Unknown but a Large Sample Size Is Selected)   In most manufacturing applications, the standard deviation of the population is not known. However, if the sample size n is large enough, the distribution of the sample can be assumed to be approximately normal. Consequently, the standard deviation of the population can be estimated by the sample standard deviation S. As discussed in Section 7.5.1, this assumption is valid if the sample size is large enough ($n \geq 30$). In this case, for Eq.7.33, the standard deviation of the population $\hat{\sigma}$ is replaced by the standard deviation of the sample S.

$$\overline{x} - z_{\alpha/2} \frac{S}{\sqrt{n}} \leq \mu \leq \overline{x} + z_{\alpha/2} \frac{S}{\sqrt{n}} \tag{7.34}$$

Confidence Interval Estimation of the Mean: ($\hat{\sigma}$ Is Unknown but a Small Sample Size Is Selected)   In some situations having a sample size $n \geq 30$ can be too expensive. If the distribution of the population is normal, confidence intervals can be computed when the population standard deviation $\hat{\sigma}$ is unknown, and the sample size is less than 30 by using the t distribution

$$\overline{x} - t_{\alpha/2} \frac{S}{\sqrt{n}} \leq \mu \leq \overline{x} + t_{\alpha/2} \frac{S}{\sqrt{n}} \tag{7.35}$$

Confidence Interval Estimation on the Variance of a Normal Distribution   If the standard deviation of a randomly selected sample size n is drawn from a normally distributed population, then, by using a chi-square distribution, a two-sided confidence interval can be calculated by the formula

$$\frac{(n-1)S^2}{\chi^2_{\alpha/2,\nu}} \leq \hat{\sigma} \leq \frac{(n-1)S^2}{\chi^2_{1-\alpha/2,\nu}} \tag{7.36}$$

Confidence Interval Estimation of the Ratio of the Variance of Two Normal Populations
Using the F distribution, discussed in Section 7.5.4, the two-sided confidence interval for the ratio of the two variances $\hat{\sigma}_1^2 / \hat{\sigma}_2^2$ can be calculated from

$$\left(\frac{S_1^2}{S_2^2}\right) F_{1-\alpha/2,\nu_1,\nu_2} \leq \frac{\hat{\sigma}_1^2}{\hat{\sigma}_2^2} \leq \left(\frac{S_1^2}{S_2^2}\right) F_{\alpha/2,\nu_1,\nu_2} \tag{7.37}$$

where the lower tail, $1 - \alpha/2$, of the F distribution is given by

$$F_{1-\alpha/2,\nu_1,\nu_2} = \frac{1}{F_{\alpha/2,\nu_1,\nu_2}}$$

If the confidence interval includes $\hat{\sigma}_1^2 / \hat{\sigma}_2^2 = 1$, it can be concluded that there is no statistical difference in the two different events at a given level of confidence.

Confidence Interval Estimation Based on the Difference in Two Means (Variance Known)
Consider two independent random samples from two populations, x^1 of size n^1 with unknown mean μ_1, and known variance $\hat{\sigma}_1^2$, and x_2 of size n_2 with unknown mean μ_2 and known variance $\hat{\sigma}_2^2$. If the random samples x_1 and x_2 are drawn from normal populations, or sample sizes n_1 and n_2 are both greater than 30, a two-sided $100(1-\alpha)$ percent confidence interval on the difference in means ($\mu_1 - \mu_2$) is computed by

$$(\overline{x}_1 - \overline{x}_2) - z_{\alpha/2} \sqrt{\hat{\sigma}_1^2/n_1 + \hat{\sigma}_2^2/n_2}$$

$$\leq (\mu_1 - \mu_2) \leq (\overline{x}_1 - \overline{x}_2) + z_{\alpha/2} \sqrt{\hat{\sigma}_1^2/n_1 + \hat{\sigma}_2^2/n_2} \tag{7.38}$$

From the above equation, the $100(1-\alpha)$ percent lower and upper confidence intervals on $(\mu_1 - \mu_2)$ can be written

$$\left(\bar{x}_1 - \bar{x}_2\right) - z_{\alpha/2}\sqrt{\hat{\sigma}_1^2/n_1 + \hat{\sigma}_2^2/n_2} \leq (\mu_1 - \mu_2) \tag{7.39}$$

and

$$(\mu_1 - \mu_2) \leq \left(\bar{x}_1 - \bar{x}_2\right) + z_{\alpha/2}\sqrt{\hat{\sigma}_1^2/n_1 + \hat{\sigma}_2^2/n_2} \tag{7.40}$$

respectively. This method is often used to compare the effectiveness of two manufacturing processes.

Confidence Interval Estimation Based on the Difference in Two Means (Variance Unknown)   If the sample size drawn from the normal population is less than 30, the t distribution must be used to compute the confidence interval. To find a $100(1 - \alpha)$ percent confidence interval for the difference in mean $(\mu_1 - \mu_2)$, assume that $\hat{\sigma}_1^2 = \hat{\sigma}_2^2 = \hat{\sigma}^2$. This assumption is often made in comparing manufacturing processes. This unknown common variance, $\hat{\sigma}^2$, can be estimated by using a "combined" or "pooled" estimator. The "pooled" estimator equation is

$$S_p^2 = \frac{(n_1 - 1)S_1^2 + (n_2 - 1)S_2^2}{n_1 + n_2 - 2} \tag{7.41}$$

Therefore, a $100(1 - \alpha)$ percent two-sided confidence interval for the difference in means $(\mu_1 - \mu_2)$ is given by

$$(\bar{x}_1 - \bar{x}_2) - t_{\alpha/2,n_1+n_2-2}S_p\sqrt{1/n_1 + 1/n_2} \leq (\mu_1 - \mu_2)$$

$$\leq (\bar{x}_1 - \bar{x}_2) + t_{\alpha/2,n_1+n_2-2}S_p\sqrt{1/n_1 + 1/n_2} \tag{7.42}$$

For testing the difference in two variables, the above-mentioned test hypothesis is used to calculate the confidence interval. If the confidence interval includes $(\mu_1 - \mu_2 = 0)$, it is concluded that there is no statistical difference in the performance of the two manufacturing processes at a given level of confidence.

7.7   STATISTICAL HYPOTHESIS TESTING

A decision problem in which one of two arguments must be chosen is known as a problem of testing hypotheses. It is often necessary to make a decision about population parameters based on information obtained from the random sample drawn from the population, assuming that the sample and the population have the same distribution. However, if the size of the sample is small, we may not have sufficient evidence for judging that the decision is correct. Hence, the decision must be supplemented by other knowledge. This can be done by postulating a hypothesis and then checking to see whether the statistics of the sample are comparable with the observed results of the population. The test of a statistical hypothesis is a procedure that leads to a decision to *reject* or *accept* the hypothesis under consideration.

In statistical testing, two hypotheses are set: a null hypothesis (initial), denoted H_0, is the hypothesis under test or consideration. The alternative hypothesis is denoted by H_1. After the observation, if the data prove correct with respect to H_0, H_0 is accepted. If the data prove false with respect to H_0, H_0 is rejected. Suppose that the null hypothesis H_0 is that $\mu = \mu_H$. where μ_H is the hypothetical population mean. The alternative hypothesis H_1 could be $\mu > \mu_H$, $\mu < \mu_H$ or $\mu \neq \mu_H$. For example, a drilling company is interested in purchasing drill pipe that, according to the material properties, has a mean yield strength, $\mu_0 = 100$ kpsi, with a standard deviation, $\hat{\sigma}_0 = 10$ kpsi. The drilling company seeks to verify the yield strength of the drill pipe. To do this, a sample of test specimens is prepared and is tested to find the mean yield strength of the sample. On the basis of this sample mean $\bar{x}$ the hypothesis that the true mean is 100 kpsi will be rejected or accepted. Suppose it is found that the sample mean $\bar{x}$ does not differ in any significant manner from the unknown mean yield strength of 100 kpsi, then we state that

$$H_0 : \mu = \mu_H \Longrightarrow \text{the null hypothesis, and}$$

$$H_1 : \mu \neq \mu_H \Longrightarrow \text{the alternative hypothesis.}$$

From the foregoing discussion, it can be understood that the approach is to make a decision about the population parameters based on the information obtained from the sample drawn from it. However, information obtained from the sample may not exactly represent the population parameters, thus, introducing two types of decision errors:

1. A **type I** error is committed if the null hypothesis is rejected when it is true. For example, during finished product evaluation, a type I error is committed if the quality control inspector rejects the product when it is acceptable. This will cause a risk to the producer (producer risk). In a statistical process the probability of rejecting the null hypothesis when it is true (a type I error) is called the level of significance and is denoted by α.

2. A **type II** error is committed if we accept the null hypothesis when it is false. If the quality control inspector accepts the product when it should be rejected, a type II error is committed, denoted by β. This will cause a risk to the consumer (consumer risk).

    Of course, an attempt should be made to minimize the above-mentioned errors. Unfortunately, this may not always be possible. In fact, for a fixed sample size, when the probability of committing a type I error is reduced, the probability of committing a type II error is increased. In general, designers control the probability of making a type I error, and they set the level of α to a desired level, such as 0.01, 0.05, and so forth. There are two ways of minimizing both errors:

    (a) Increase the sample size.

    (b) Fix the level of α (type I error), and specify a critical region that gives the smallest type II error.

In testing a statistical hypothesis, the following steps are carried out:

1. State the hypothesis. The population mean should be equal to some defined hypothetical value μ_H with a given population standard deviation $\hat{\sigma}$. Set

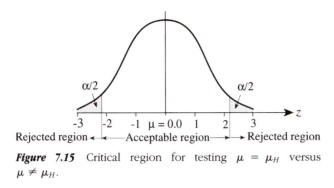

Figure 7.15 Critical region for testing $\mu = \mu_H$ versus $\mu \neq \mu_H$.

- Null hypothesis $\Rightarrow$ H_0: $\mu = \mu_H$
- Alternative hypothesis $\Rightarrow$ H_1: $\mu \neq \mu_H$

2. Decide on the level of significance, α and β.
3. Select the sample size n.
4. Select the statistic to be used in testing the hypothesis. For example,

$$z_{\bar{x}} = \frac{\bar{x} - \mu_H}{\frac{\hat{\sigma}}{\sqrt{n}}} \tag{7.43}$$

5. Define the rejection region at the selected α level of significance as shown in Figure 7.15.
6. Perform the experimental test by using the sample observation, compute $z_{\bar{x}}$, and determine whether it falls in the rejected region.

—— *Example 7.8* ————————————————————————

The Type I Error: Hypothesis Test About the Mean of a Normal Distribution When Standard Deviation Is Known.

Because of severe drilling conditions for a deep drilling operation, a manufacturer of drilling pipes has developed a new welding process to weld the tool joints at the end of the drilling pipe; hence, two drill pipe joints can be connected to each other by the use of tool joints. The quality control manager claims that the mean failure strength and the standard deviation of the drill pipe at the tool joint welding location, as shown in Figure 7.16, is 120 kpsi and 11 kpsi, respectively (normally distributed). To check the claim, a sample of 10 drill pipes is tested. The results are shown in Table 7.4. What is the decision about the hypothesis?

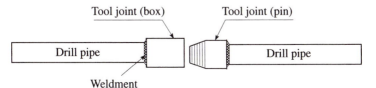

Figure 7.16 Drill pipe tool joint.

Table 7.4   Results of Failure Testing

Test Number	Failure Strength (ksi)
1	120
2	122
3	110
4	105
5	105
6	118
7	104
8	106
9	115
10	115

Solution

Following the procedure outlined in the previous section:

1. Set the null hypothesis.

   - H_0: $\mu = \mu_H = 120$ kpsi with $\hat{\sigma} = 11$ kpsi (assume that the null hypothesis is, in fact, true)
   - H_1: $\mu \neq \mu_H \neq 120$ kpsi (alternative hypothesis)

2. Assume a level of significance, $\alpha = 0.05$ (quality control manager chooses).
3. Size of sample, n=10.
4. Since the standard deviation of the population is known, the sampling distribution of the statistic can be written

$$z_{\bar{x}} = \frac{\bar{x} - \mu_H}{\frac{\hat{\sigma}}{\sqrt{n}}}$$

$$\bar{x} = \frac{\sum x_i}{n} = \frac{120 + 122 \cdots}{10} = 112$$

$$z_{\bar{x}} = \frac{\bar{x} - \mu_H}{\frac{\hat{\sigma}}{\sqrt{n}}} = \frac{112 - 120}{\frac{11}{\sqrt{10}}} = -2.30$$

5. Since the critical region is two sided, dividing the probability equally between two tails ($\alpha/2 = 0.025$), from Table C.1 in Appendix C, the values of z for the critical region are found to be $z = \mp 1.96$; hence,

$$-1.96 < z < 1.96$$

Decision

As shown in Figure 7.17, $z_{\bar{x}} = -2.30$ is not in the acceptable region; therefore, reject the hypothesis H_0 and decide that the mean failure strength is not equal to $\mu_H = 120$ kpsi.

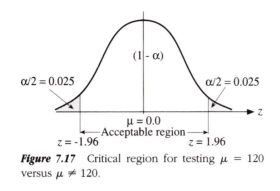

Figure 7.17 Critical region for testing $\mu = 120$ versus $\mu \neq 120$.

—— *Example 7.9* ————————————————————

Type I Error: Hypothesis Test About the Mean of a Normal Distribution When the Standard Deviation is Unknown and the Sample Size is Large ($n \geq 30$).

Electronic components like the capacitors shown in Figure 7.18 are often supported by electrical lead wire through the solder terminals. If this suspended electronic component is subjected to vibration, the electrical lead wire and capacitor will act like a one degree-of-freedom mass–spring system that develops alternating stress loads. Hence, lead wires must be good electrical conductors and must also have high fatigue life. An electronic company has used a new lead wire material and claims that the mean fatigue life of this new electrical lead wire is 6×10^6 cycles. To check whether the mean fatigue life of the new material is equal to the specified value, a random sample of 40 electronic components is tested. The sample mean is $\bar{x} = 5.8 \times 10^6$ cycles, and the standard deviation is $S = 4.8 \times 10^5$ cycles. Determine whether the sample data gives enough evidence to prove that the fatigue life of this new material has been improved.

Solution
1. Set the null hypothesis.

   - $H_0 : \mu = \mu_H = 6 \times 10^6$ cycles (assume that null hypothesis is, in fact, true).

   - $H_1 : \mu \neq \mu_H \neq 6 \times 10^6$ cycles (alternative hypothesis).

2. Assume a level of significance, $\alpha = 0.01$ (two sided).
3. Size of sample, $n = 40$.

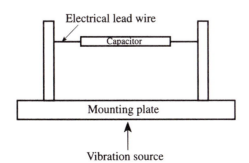

Figure 7.18 Capacitor supported by electrical lead wire.

4. Since the sample size is large enough, the distribution of $\bar{x}$ will be approximately normal. Hence, the standard deviation of the population can be assumed to be equal to the standard deviation of the sample.

$$Z_{\bar{x}} = \frac{\bar{x} - \mu_H}{\frac{S}{\sqrt{n}}} = \frac{5.8 \times 10^6 - 6 \times 10^6}{\frac{4.8 \times 10^5}{\sqrt{40}}} = -2.64$$

5. Assuming the two-sided critical region as shown in Figure 7.19, and $\alpha/2 = 0.005$ from Table C.1 in Appendix C, the value of z for the critical region is found to be $z = \mp 2.575$.

Decision

Since $z_{\bar{x}} = -2.64$ is not in the acceptable region, reject the hypothesis H_0 and decide that there is not sufficient evidence to indicate that the new material increases the fatigue life at a 0.01 level of significance.

▬ *Example 7.10*

The temperature and the color of the exhaust gases t_{eg} of a diesel engine are important parameters of the operating process. A thermocouple is installed in the exhaust manifold near the exhaust valves to measure the exhaust temperature. Consider the two-stroke eight cylinder 4000-hp main diesel engine of a cargo ship. For optimum performance, the engine operator must be sure that the exhaust gas temperature of each of the cylinders is close to the temperature of the others. If, for some reason, some of the cylinder exhaust gas temperatures are higher than others, the cylinders with higher temperatures are assumed to be overloaded. The nonuniform stress distribution over the crank shaft bearing and excessive vibration may result in a possible failure of the machine parts.

Assume that, for the optimum performance of the diesel engine, the mean value of the gas temperature for each cyclinder is 360°C. If the mean exhaust temperature of a cyclinder reaches 370°C, that particular cylinder may be assumed to be overloaded. In both cases, assume that the standard deviation is 5°C (normally distributed). Because of thermal inertia, temperature readings of the gas flow may show changes. Suppose that during engine operation, the operator takes a reading of the exhaust gas temperature of approximately 364°C (average of four readings) from a cylinder. Which of the two following actions should the engine operator take?

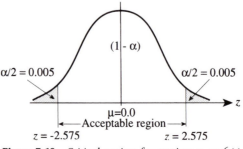

Figure 7.19   Critical region for testing $\mu = 6 \times 10^6$ versus $\mu \neq 6 \times 10^6$.

1. Stop the engine, and start checking the injector valve adjustment.
2. Continue running the engine.

Solution

Several factors may cause the high exhaust temperature of a cylinder. Perhaps the most important one is improper fuel injection valve adjustment for the delivery of the required amount of fuel oil into the cylinder. If the engine is under long duration operation, the injection valves may get out of adjustment. When this situation occurs, the engine operator must take one of the courses of action listed below.

If the operator stops the engine when the engine can still operate normally until the ship reaches port, he or she is making a Type I error. This wrong decision results in loss to the ship company because of down time during the troubleshooting period. Assume that this relative cost is 1 unit and designate it regret r for making a Type I error.

If the operator does not stop the engine when the engine cannot operate normally, he or she is making a Type II error. Making this error also results in loss, in some cases replacement, of very expensive engine parts for not detecting the problem with the engine and system in time. Assume that this relative cost is 6 units and designate it regret R for making a Type II error.

Assume that from past experience, the probability of damage to the engine when any of the cylinders are running with relatively high exhaust gas temperature is 25 percent. Hence, 75 percent of the time engine operation is normal even though some of the cylinders may be experiencing high exhaust gas temperature. Let normal operation of the engine under high exhaust gas temperature be the *null hypothesis*. Table 7.5 shows the probabilities and the regrets of making two different decisions.

$$\text{Expected loss for not stopping the engine} = (0)(0.75) + (6)(0.25) = 1.5$$

$$\text{Expected loss for stopping the engine} = (1)(0.75) + (0)(0.25) = 0.75$$

It can be seen that the expected loss is less for stopping the engine. Hence, stopping is the correct decision. However, consider the statistical decision information as to engine operation. As Figure 7.20 shows, the two temperature distribution curves intersect at a decision-making point corresponding to a temperature of 365°C. To the left of this point, the normal operation curve is above the abnormal operation curve, whereas to the right of this point, the abnormal operation curve is above the normal operation curve. Point 365°C is called the unaffected decision-making point (UDMP), and it can be calculated by taking the average of the two mean values

Table 7.5 Decision Table

	State	
	Normal Operation	**Abnormal Operation**
Decision	$p(\text{normal}) = 0.75$	$p(\text{abnormal}) = 0.25$
Don't stop the engine	$r = 0$	$R = 6$
Stop the engine	$r = 1$	$R = 0$

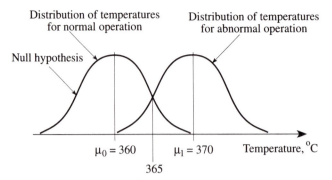

Distribution of temperatures for normal operation

Distribution of temperatures for abnormal operation

Null hypothesis

$\mu_0 = 360$ $\mu_1 = 370$ Temperature, $^\circ$C

365

Figure 7.20 Decision-making curves.

$$\text{UDMP} = \frac{\mu_0 + \mu_1}{2} = \frac{360 + 370}{2} = 365^\circ\text{C}$$

As shown in Figure 7.21, if the temperature is higher than 365°C, the engine operator should stop the engine and start troubleshooting. If the temperature is lower than 365°C, the engine operator should continue operation. Note that, in the above analysis, for both the normal and abnormal cases, equal weight has been given. However, under high exhaust gas temperature operation, 75 percent of the time engine operation is normal and 25 percent of the time engine operation is abnormal. Also, the relative weighting of the regrets is not taken into account. Therefore, the following two steps will be performed to calculate the change in UDMP when the weighting of probabilities and regrets is considered.

1. The change in UDMP when probabilities are taken into consideration is[7]

$$\Delta'_{\text{UDMP}} = \frac{\hat{\sigma}^2}{\mu_1 - \mu_0} \times \ln\frac{P(\text{null hypothesis is true})}{P(\text{null hypothesis is false})}$$

$$\Delta'_{\text{UDMP}} = \frac{5^2}{370 - 360} \times \ln\frac{0.75}{0.25} = 2.75^\circ\text{C}$$

2. The change in UDMP when the regret is taken into consideration is

$$\Delta''_{\text{UDMP}} = \frac{\hat{\sigma}^2}{\mu_1 - \mu_0} \times \ln\frac{P(\text{regrets of type I Error})}{P(\text{regrets of type II Error})}$$

$$\Delta''_{\text{UDMP}} = \frac{5^2}{370 - 360} \times \ln\frac{1}{6} = -4.48^\circ\text{C}$$

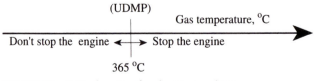

(UDMP)

Gas temperature, $^\circ$C

Don't stop the engine ⟷ Stop the engine

365 °C

Figure 7.21 Critical region for decision making.

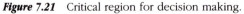

[7]E. A. Robinson, *Statistical Reasoning and Decision Making,* Goose Pond Press, Houston, Tx, 1981.

The net change in the decision-making point is $-4.48 + 2.75 = -1.73°C$, which is a change of 1.73 to the left. Hence, the resulting decision-making point is $365 - 1.73 = 363.27°C$. Now Figure 7.20 can be modified for the new decision-making point, as shown in Figure 7.22. Since the modified decision-making point is 363.27°C and the observed exhaust gas temperature was 364°C, the operator's best decision would be to stop the engine. The probability of a Type I error α can be determined by calculating the area under the normal curve for normal operation to the right of the decision-making point.

$$\alpha = 1 - p\left(z < \frac{x - \mu_0}{\hat{\sigma}/\sqrt{n}}\right) = 1 - p\left(z < \frac{363.27 - 360}{5/\sqrt{4}}\right)$$

$$\alpha = 1 - p(z < 1.308) \approx 0.095$$

The probability of a Type II error β can be determined by calculating the area under the normal curve for abnormal operation to the left of the decision-making point.

$$\beta = p\left(z < \frac{x - \mu_1}{\hat{\sigma}/\sqrt{n}}\right) = p\left(z < \frac{363.27 - 370}{5/\sqrt{4}}\right)$$

$$\beta = p(z < -2.69) = 0.0036$$

These areas are shown in Figure 7.23.

Figure 7.24 shows that, without taking the temperature reading, the probability of normal operation of the engine running under high exhaust gas temperature is 75 percent. However, after the operator took the temperature reading of 364°C it was found that the observed temperature fell in the critical region to the right of the decision-making point of 363.27°C. Hence, the operator's decision should be to

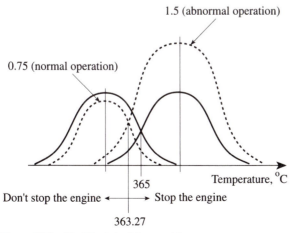

Figure 7.22 Modified decision-making curves.

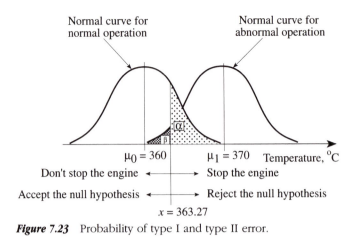

Figure 7.23   Probability of type I and type II error.

reject the null hypothesis H_0 and decide that engine operation should be stopped. From the same tree diagram we observe that the path of probability for a Type I error is 0.071250 with regret 1. The path of probability for a Type II error is 0.0009 with regret 6. Therefore, the expected loss is

$$(0.071250)(1) + (0.0009)(6) = 0.07665$$

The previously calculated smallest loss was 0.75 if the operator did not take the temperature readings. However, by taking the temperature reading it can be seen that the expected loss is reduced by a factor of approximately 10. This example is adapted from Robinson (1981).

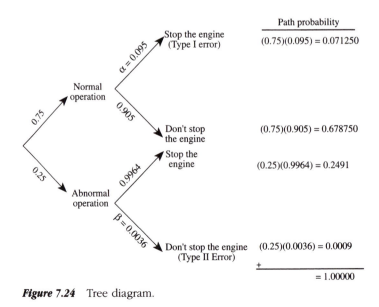

Figure 7.24   Tree diagram.

7.8   STATISTICAL EXPERIMENTAL DESIGN

There is always a need for experimentation to optimize processes. Unfortunately, this is a costly and time-consuming effort; however, by implementing statistical experimental design, cost and time can be controlled. In this section, the up-and-down and Taguchi methods (Orthogonal or Factorial Experiment) of experimentation, which provide the most useful analysis for the least amount of data, are discussed.

7.8.1   Up–and–Down Method

The up–and–down method was developed to obtain data from sensitivity experiments. A sensitivity experiment is one in which only one observation for a given specimen can be made. This means that once a test has been carried out, the specimen is altered so that a valid second statistical result cannot be obtained from the same specimen. This technique is used extensively in various disciplines ranging from engineering to biostatics. Applications include, among others, determining the strength of materials, evaluating the sensitivity of explosives to shock, and psychophysical research dealing with threshold stimuli.[8] To apply this technique in any set of tests, three major conditions must be satisfied:

1. The variant under analysis has to be normally distributed;
2. The standard deviation of the normally distributed variant has to be approximately estimated.
3. The interval between testing levels should be chosen at between 0.5 and 2.0 times the standard deviation.

This technique is discussed in terms of the onset of galling between two surfaces. [9]

Galling depends on a number of factors, including pressure exerted on the specimen, relative surface velocity (rotation), lubrication, surface finish, and specimen material properties. Galling leads to surface distributions, which can be observed as visual manifestations. Since galling is promoted by the above-mentioned factors, the up-and-down technique can be applied by varying one parameter and holding the others constant. For example, the procedure can be performed for varying pressure exerted on the specimen shown in Figure 7.25, at constant speed and with constant materials properties. The first step is to choose the interval for the pressure. The interval must be chosen so that the galling mean pressure (psi) is enveloped. An initial pressure P_0 and a succession of pressures $P_{+1}, P_{+2}, P_{+3}, \ldots$ above P_0 together with a succession $P_{-1}, P_{-2}, P_{-3}, \ldots$ below P_0 is chosen. The first specimen pair is run at pressure P_0 and constant rotational speed. If the posttest inspection reveals that galling has occurred, an x is marked on the data sheet and the second specimen pair is tested at P_{-1}. If galling does not occur, 0 is marked on the data sheet and the second specimen pair is tested at P_{+1}. This implies that any specimen pair would be tested at a pressure immediately above or below that of the previous test depending on whether the specimen revealed *gall* or *no gall*. Using this technique,

---

[8]W. Dixon and F. Massey, *Introduction to Statistical Analysis,* McGraw-Hill, New York, 1957.

[9]A. Ertas et al., "Experimental Investigation of Galling Resistance in OCTG Connections," *ASME Proceedings of the Twelth Annual Energy Resources Technology Congerence and Exhibition,* PD-VOL.29, pp. 15–20, 1990 (permission from ASME).

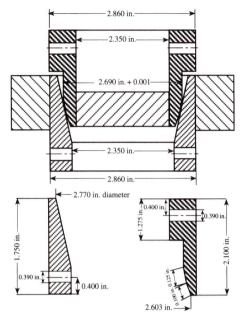

Figure 7.25 Test specimen configuration.

the mean μ and standard deviation $\hat{\sigma}$ of the population can be estimated by the sample mean $\overline{X}$ as

$$\overline{x} = D_0 + I\left(\frac{A}{N} \mp \frac{1}{2}\right) \tag{7.44}$$

where
D_0 is the lowest test level at which the condition under evaluation was recorded
I is the interval between test levels
$+\frac{1}{2}$ is used if analyzing the 0's (no galling condition)
$-\frac{1}{2}$ is used if analyzing the x's (galling condition)

$$N = \sum_{i=0}^{k} n_i \tag{7.45}$$

$$A = \sum_{i=0}^{k} in_i \tag{7.46}$$

where i denotes the number of increments above the lowest testing level at which the symbol under the analysis occurred and n_i denotes the frequency of occurrence of the symbols at the various levels. The standard deviation is calculated from

$$S = 1.620 \times I(\frac{NB - A^2}{N^2} + 0.029) \tag{7.47}$$

where

$$B = \sum_{i=0}^{k} i^2 n_i \tag{7.48}$$

If $(NB - A^2)/N^2$ is less than 0.3 this formula is not valid. The standard deviation of the mean is calculated from

$$S_{\bar{x}} = \frac{6S + I}{7N^{0.5}} \tag{7.49}$$

Equation 7.49 is valid only if I is less than $3S$. For small samples, the approximate confidence limit CL for the population mean can be computed by

$$CL = \bar{x} \mp t_{\nu,\alpha} S_{\bar{x}} \tag{7.50}$$

where ν is the degree-of-freedom, given by

$$\nu = N - 1 \tag{7.51}$$

and t is taken from the Student distribution table.

Example 7.11

Analyze the data given in Table 7.6 by the up-and-down method discussed in this section:

1. Estimate the mean.
2. Estimate the standard deviation.
3. Estimate the standard deviation of the mean.
4. Estimate the 90 percent confidence limits.

Solution

This test series was composed of 19 tests on the L80 tubing specimens shown in Figure 7.25. For this study, a 10-kpsi stress interval with the initial trial at a nominal contact stress of 40.5 kpsi was chosen.

Table 7.6 shows the data obtained in accordance with the up-and-down technique. Testing was initiated at 40.5 kpsi and ranged between 40.5 and 80.5 kpsi. For estimating the population mean and standard deviation the symbol x (galling) or 0 (no galling), which occurs less frequently, is used. From Table 7.6, there are fewer failures (galling) than successes (no galling), so the failures would be used. Since galling did not appear at the 40.5 kpsi stress level, data at this level were eliminated. If the frequencies were equal to each other, either failure x or successes 0 could be chosen. The data obtained from Table 7.6 are tabulated as shown in Table 7.7. In this table, i denotes the number of increments above the 50.5 kpsi

Table 7.6 Up-and-Down Test for L80 Tubing Specimens

Normal Contact Stress, kpsi	1	2	3	4	5	6	7	8	9	10	11	12	13	14	15	16	17	18	19	x	0
80.5					X		X													2	0
70.5				0		0		X		X		X								3	2
60.5			0						0		0		X				X		X	3	3
50.5		0												X		0		0		1	3
40.5	0														0					0	2
																			Total:	9	10

Material . . . L80
Speed = 1.4 rpm
x indicates galling
0 indicates no galling

Table 7.7   Statistical Calculations for L80 Tubing Tests

Nominal Contact Stress (kpsi)	i	n_i	$i \times n_i$	$i^2 \times n_i$
80.5	3	2	6	18
70.5	2	3	6	12
60.5	1	3	3	3
50.5	0	1	0	0
		$N = 9$	$A = 15$	$B = 33$

stress level at which the symbol (x or 0) was recorded, and n_i denotes the frequency of occurrence of the symbols at the various levels. Referring to Table 7.7, the sample mean can now be calculated as

$$\bar{x} = 50.5 + 10\left(\frac{15}{9} - \frac{1}{2}\right) = 62.17 \text{ kpsi}$$

Standard deviation of the sample is

$$S = 1.620 \times 10\left(\frac{9 \times 33 - 15^2}{9^2} + 0.029\right) = 14.87 \text{ kpsi}$$

Standard deviation of the sample mean is

$$S_{\bar{x}} = \frac{6 \times 14.87 + 10}{7 \times 9^{0.5}} = 4.725 \text{ kpsi}$$

and 90 percent confidence limits is

$$90\% CL = 62.17 \mp (1.86)(4.725) = 62.17 \mp 8.79(53.38 \text{ to } 70.96 \text{ kpsi})$$

Note that for $\nu = 9 - 1 = 8$ and the two-sided critical region the value of the t distribution is 1.86 (see Appendix C.2).

To compare the results of the up-and-down technique with other statistical methods, a cumulative frequency distribution was also generated. Four series of tests were conducted. Each test series consisted of 15 replicate tests. These test series were conducted at nominal contact stresses of 51.5, 61.5, 71.5, and 81.5 kpsi, with an interval of 10 kpsi. The results of these test series are plotted on normal probability paper as shown in Figure 7.26. As depicted in the figure, the straight line fits the data reasonably well, indicating that the galling failure has a normal (Gaussian) distribution. As can be seen, 50 percent of the specimens galled at a contact stress of about 67 kpsi. By the theory of normal Gaussian distribution, approximately 68 percent of the distribution lies within $\mu \mp \hat{\sigma}$. The nominal contact stress corresponding to the 16 percent failed level is about 47 kpsi. Subtracting this value from the mean stress of 67 kpsi results in an standard deviation of approximately 20 kpsi.

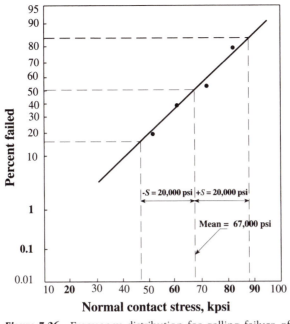

Figure 7.26 Frequency distribution for galling failure of
L 80 specimens

A comparison of the up-and-down test and frequency distribution results is presented in Table 7.8. This table shows that the up-and-down test predicts the mean to within 7 percent of that of the frequency distribution. The standard deviation of the sample, as predicted by the up-and-down method, differs by approximately 25 percent from the frequency distribution. The up-and-down technique can thus be used in predicting the mean, but predicts the other statistical quantities less accurately. As stated before, the interval between testing levels should be set between 0.5 and 2.0 times the standard deviation. The assumed interval (10 kpsi) is 0.5 times the calculated standard deviation of 20 kpsi; thus, the experiment with the up-and-down method satisfies this requirement. The up-and-down technique is an economical and useful method for predicting the mean value. This method enables the experimenter to perform rapid comprehensive parametric studies for the design of experiments.

7.8.2 Taguchi Technique (Orthogonal or Factorial Experiment)

The old way of performing experiments was to vary only one parameter and to hold the others constant and to repeat this procedure for all the parameters involved in the experimental design (full factorial design). Obviously, this method is time consuming and costly. Taguchi designs allow for the measurement of a number of different parameters and their interactions. When this technique is used, a high-

Table 7.8 Comparison Between Results of the Up-and-Down Test and Frequency Distribution

Quantity	Up-and-Down Prediction (kpsi)	Frequency Distribution (kpsi)	Percent Difference
Mean nominal contact stress	62.17	67.0	7%
Standard deviation	14.87	20.0	25%

quality product at a lower cost can be achieved.[10] In this section the Analysis of Variance (ANOVA) and some new terminologies commonly used in the design of experiments are introduced.

ANOVA is the statistical method used to analyze data obtained from an experiment to make necessary decisions. This method is called ANOVA because it takes into account the analysis of variation among the statistical measures, such as the mean. It is important to note that the ANOVA enables an observer to make decisions about a set of observed parameters, which plays an important role in experimental design.

One-Way ANOVA   In experimental design, the parameters that may be varied from trial to trial for an experiment are generally called *factors*, and the value of the factors used in the experiment are referred to as *levels*. Assume that the data shown in Table 7.9 have been collected over a period of three weeks, one week for each concentration setting. This table shows the viscosity of a lithium chloride (LiCl) solution for different concentrations. In this experiment, the variable concentration is the factor, and the three concentration settings are the levels. Thus, this experiment is called a three-level, one-factor design. As can be seen from the table, there are two types of variations in the data:

1. Variations *between* the concentration settings, which was planned to provide data on the effectiveness of the solution with different concentrations of LiCl.

2. Variations *within* each concentration setting (probably due to the humidity and the temperature variations in the lab environment during the each experiment).

Two-Way ANOVA   The two-way ANOVA is introduced by using *temperature* as the second factor, as shown in Table 7.10.

The effect of *temperature* and *concentration* on the viscosity of the LiCl solution can then be evaluated. It is important to note that each data point in Table 7.10 is the mean of the sample taken during a one-week period. For example, viscosity of the LiCl solution for the 30 percent concentration at 80°F is 3.43 percent, which is the mean value of the first column of Table 7.9. This experiment can be simplified by eliminating the third trial (concentration 40 percent at 120°F temperature), thus,

Table 7.9   Viscosity of LiCl Solution at 80°F, c.p.

Day	Concentration, Weight %		
	30%	**35%**	**40%**
M	3.24	5.00	7.84
T	3.32	4.91	7.73
W	3.21	4.82	7.64
Th	3.61	5.13	7.96
F	3.11	4.89	7.54
Sat	3.81	5.48	8.11
Sun	3.73	3.37	7.99

[10]P. J. Ross, *Taguchi Technique for Quality Engineering*, New York, McGraw-Hill, 1988.

Table 7.10   Viscosity of LiCl Solution c.p.

Temperature	Concentration, Weight %		
	30%	**35%**	**40%**
80	3.43	4.80	7.83
100	2.52	3.76	5.76
120	2.05	2.99	4.59

reducing the level of the experiment to two. The two-level factorial experiment is the simplest one for testing the effects of the parameters on performance of the experimental design. This two-level factorial experiment, in which each factor is varied over two levels (*high* level and *low* level), is now considered.

In general, the computation of two-way ANOVA problems can be summarized as shown in Table 7.11. For a two-level experiment, when sample sizes are equal, the following computational formulas are used.

1.   Total sum of squares, SS_T

$$SS_T = SS_Y - C \tag{7.52}$$

where SS_Y is the sum of the squares of each observation Y and C is the correction factor given by

$$C = \frac{T^2}{N} \tag{7.53}$$

where T is the sum of all observations (grand total), and N is the total number of observations.

2.   The sum of squares SS_i for each two-level factor, and the sum of squares for error SS_E can be calculated by

$$SS_i = 2(\text{MEF}_{\text{factor}})^2 \tag{7.54}$$

$$SS_E = SS_T - \left(\sum_{i=1}^{k} SS_i \right) \tag{7.55}$$

where k is the total number of factors and interactions.

Table 7.11   Complete Two-Way ANOVA Summary for Two Factorial Experiments

Source	Degree-of-Freedom (ν)	Sum of Squares (SS_i)	Mean Squares (MS_i)	F Test	Percent of Variance
Factor A					
Factor B					
Interaction between A and B					
Error					

3. The main effect factors and error MEF can be obtained by.

$$\text{MEF} = \frac{\sum Y_i \text{at high level} - \sum Y_i \text{at low level}}{N/2} \tag{7.56}$$

4. The mean square for each factor MS_i and mean square error MS_E are calculcated by

$$MS_i = \frac{SS_i}{\nu_i} \tag{7.57}$$

$$MS_E = \frac{SS_E}{\nu_{\text{error}}} \tag{7.58}$$

where ν is the number of degrees-of-freedom, which is calculated using

$\nu_T = N - 1$

$\nu_i = $ number of levels $- 1$

$\nu_{\text{error}} = (N - 1) - (\sum_{i=1}^{k} \nu_i)$

where ν_T, ν_e, ν_i, and N are the total, error, each factor's degree-of-freedom, and the number of rows in the orthogonal array, respectively.

5. F test for variance comparison: The F test is the ratio of sample variances that provides a decision as to whether there are significant differences in the estimates at known confidence levels. This useful tool is calculated by

$$F_i = \frac{MS_i}{MS_E} \tag{7.59}$$

This ratio is compared with F table values, and if $F_i < F_{\text{Table}}$, it is concluded that the effect of factor i under consideration is not significant.

6. Percent of variance: The percent contribution of factor i to the total variance is determined as

$$PC_i = \frac{100 \left(MS_i - MS_E \right)}{SS_T} \tag{7.60}$$

The percent contribution to error is

$$PC_E = 100 - \left(\sum_{i=1}^{k} PC_i \right) \tag{7.61}$$

Obviously, the total percent contribution should add up to 100 percent.

Two-Level Orthogonal Array (OA) A two level OA[11] consists of 2^k data points where k is the number of factors used in the experiment. A two-level, two factor

---

[11]G. Taguchi, *Introduction to Quality Engineering*, Kraus International Publications, New York, 1987. (Figures 7.27 and 7.28 are reprinted by permission of the Asian productivity organization.)

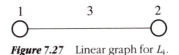

Figure 7.27   Linear graph for L_4.

OA requires $(2)^2 = 4$ data points. This is called an L_4 experiment. A linear graph for an L_4 experiment is shown in Figure 7.27. Nodes 1 and 2 indicate major parameters that will be switched from high (2) to low (1) values, and the line indicates that an interaction is allowed between the major parameters. The L_4 array shown in Table 7.12[11] has four trials and three columns. Factors X_1, X_2, and interaction $X_1 X_2$ between these factors are assigned to columns 1, 2, and 3, respectively. If columns 1 and 2 are acting independently, column 3 is assigned to error. This is shown more clearly in Table 7.13. In this table, "Col." indicates the column in the orthogonal array, and "No." indicates the experiment number. Referring to Table 7.13 and Eq. 7.56, formulas to calculate the main effect factors MEF for an L_4 orthogonal array can be written as

$$\text{MEF}_{X_1} = \left(\frac{Y_3 + Y_4}{2}\right) - \left(\frac{Y_1 + Y_2}{2}\right) \tag{7.62}$$

$$\text{MEF}_{X_2} = \left(\frac{Y_2 + Y_4}{2}\right) - \left(\frac{Y_1 + Y_3}{2}\right) \tag{7.63}$$

$$\text{MEF}_{\text{Error}} = \left(\frac{Y_2 + Y_3}{2}\right) - \left(\frac{Y_1 + Y_4}{2}\right) \tag{7.64}$$

Many other orthogonal arrays are used in experimental design. The most important one is the two level with three factors called the L_8 OA. A linear graph for an L_8 ANOVA is shown in Figure 7.28. The interactions between the factors of the L_8 OA are shown in Tables 7.14 and 7.15, respectively.[11] As shown in Table 7.15, factors X_1, X_2, and X_3 are assigned to columns 1, 2, and 4, respectively. This places the $X_1 X_2$ interaction in column 3, the $X_1 X_3$ interaction in column 5, and the $X_2 X_3$ interaction in column 6. Column 7 is assigned to error accumulation, the effect of which is believed to be the smallest among the factors under consideration. Referring again to Eq. 7.56 and Table 7.15, general formulas for the main effect of factors can be written as follows:

1. Main effect factors for the major parameters

$$\text{MEF}_{X_1} = \frac{(Y_5 + Y_6 + Y_7 + Y_8) - (Y_1 + Y_2 + Y_3 + Y_4)}{4} \tag{7.65}$$

Table 7.12 L_4 Array

Col. →	1	2	3
No. ↓			
1	1	1	1
2	1	2	2
3	2	1	2
4	2	2	1

Table 7.13   Two-Factor L_4 Orthogonal Array

Col. →	Factors, X_i		Error	Observations, Y_i
	1	2	3	
No. ↓				
1	1	1	1	Y_1
2	1	2	2	Y_2
3	2	1	2	Y_3
4	2	2	1	Y_4

$$\text{MEF}_{X_2} = \frac{\left(Y_3 + Y_4 + Y_7 + Y_8\right) - \left(Y_1 + Y_2 + Y_5 + Y_6\right)}{4} \tag{7.66}$$

$$\text{MEF}_{X_3} = \frac{\left(Y_2 + Y_4 + Y_6 + Y_8\right) - \left(Y_1 + Y_3 + Y_5 + Y_7\right)}{4} \tag{7.67}$$

2. Main effect for the interactions, $X_1 X_2$, $X_1 X_3$, and $X_2 X_3$

$$\text{MEF}_{X_1 X_2} = \frac{\left(Y_3 + Y_4 + Y_5 + Y_6\right) - (Y_1 + Y_2 + Y_7 + Y_8)}{4} \tag{7.68}$$

$$\text{MEF}_{X_1 X_3} = \frac{\left(Y_2 + Y_4 + Y_5 + Y_7\right) - \left(Y_1 + Y_3 + Y_6 + Y_8\right)}{4} \tag{7.69}$$

$$\text{MEF}_{X_2 X_3} = \frac{\left(Y_2 + Y_3 + Y_6 + Y_7\right) - \left(Y_1 + Y_4 + Y_5 + Y_8\right)}{4} \tag{7.70}$$

3. Main effect for error

$$\text{MEF}_{\text{Error}} = \frac{\left(Y_2 + Y_3 + Y_5 + Y_8\right) - (Y_1 + Y_4 + Y_6 + Y_7)}{4} \tag{7.71}$$

The following example is a demonstration of this method.

▬ *Example 7.12* ▬▬▬▬▬▬▬▬▬▬▬▬▬▬▬▬▬▬▬▬▬▬▬▬▬▬

Referring to Example 7.11, design a two-level experiment with three factors to determine the relative effects of speed, roughness, and axial load on the galling of a

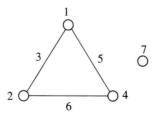

Figure 7.28   Linear graph for L_8.

Table 7.14 L_8 Array

Col. →	1	2	3	4	5	6	7
No. ↓							
1	1	1	1	1	1	1	1
2	1	1	1	2	2	2	2
3	1	2	2	1	1	2	2
4	1	2	2	2	2	1	1
5	2	1	2	1	2	1	2
6	2	1	2	2	1	2	1
7	2	2	1	1	2	2	1
8	2	2	1	2	1	1	2

metal collar. Also evaluate the interaction of the parameters and perform a variance comparison and percent contribution test. Experimental data have been collected,[12] and the results are shown in Table 7.16. Column 1 is the speed with low (1) equated to 1.5 rpm and high (2) equated with 5 rpm. Column 2 is the surface roughness with a low (1) of 75 micro-inches and high (2) of 150 micro-inches. The axial load is shown in column 4 with 60 kpsi low (1) and 100 kpsi high (2). Galling is given a value of 0 for no galling and 10 for severe galling.

Solution

Main effect factor calculations

$$MEF = \frac{\sum Y_i \text{ at high level} - \sum Y_i \text{ at low level}}{N/2}$$

For L_8 orthogonal array, N=8

$$MEF_{speed} = \frac{(6 + 10 + 4 + 8) - (2 + 5 + 0 + 2)}{8/2} = 4.75$$

Table 7.15   Three-Factor L_8 Orthogonal Array

	Ftr. x_1	Ftr. x_2	Int. $x_1 x_2$	Ftr. x_3	Int. $x_1 x_3$	Int. $x_2 x_3$	Err	Obser.
Col. →	1	2	3	4	5	6	7	Y_i
No. ↓								
1	1	1	1	1	1	1	1	Y_1
2	1	1	1	2	2	2	2	Y_2
3	1	2	2	1	1	2	2	Y_3
4	1	2	2	2	2	1	1	Y_4
5	2	1	2	1	2	1	2	Y_5
6	2	1	2	2	1	2	1	Y_6
7	2	2	1	1	2	2	1	Y_7
8	2	2	1	2	1	1	2	Y_8

[12]A. Ertas, H. J. Carper, and W. R. Blackstone, "Development of a Test Machine and Method for Galling Studies," *VII SEM International Congress on Experimental Mechanics*, vol. 2, pp. 1602–1610, 1992.

Table 7.16 L_8 Orthogonal Array for Galling Experiment

Col. →	Speed 1	Rough 2	Sp.×Ro. 3	Load 4	Sp.×Lo. 5	Ro.×Lo. 6	Error 7	Y_i
No. ↓								
1	1	1	1	1	1	1	1	2
2	1	1	1	2	2	2	2	5
3	1	2	2	1	1	2	2	0
4	1	2	2	2	2	1	1	2
5	2	1	2	1	2	1	2	6
6	2	1	2	2	1	2	1	10
7	2	2	1	1	2	2	1	4
8	2	2	1	2	1	1	2	8

$$\text{MEF}_{\text{rough}} = \frac{(0 + 2 + 4 + 8) - (2 + 5 + 6 + 10)}{8/2} = -2.25$$

$$\text{MEF}_{\text{load}} = \frac{(5 + 2 + 10 + 8) - (2 + 0 + 6 + 4)}{8/2} = 3.25$$

$$\text{MEF}_{\text{speed}\times\text{rough}} = \frac{(0 + 2 + 6 + 10) - (2 + 5 + 4 + 8)}{8/2} = -0.25$$

$$\text{MEF}_{\text{speed}\times\text{load}} = \frac{(5 + 2 + 6 + 4) - (2 + 0 + 10 + 8)}{8/2} = -0.75$$

$$\text{MEF}_{\text{rough}\times\text{load}} = \frac{(5 + 0 + 10 + 4) - (2 + 2 + 6 + 8)}{8/2} = 0.25$$

$$\text{MEF}_{\text{error}} = \frac{(5 + 0 + 6 + 8) - (2 + 2 + 10 + 4)}{8/2} = 0.25$$

Analysis of Variance (ANOVA) Table Setup

1. Calculation of degree-of-freedom

$$\nu_T = N - 1 = 8 - 1 = 7, \qquad \nu_{\text{speed}} = \text{number of levels} - 1 = 2 - 1 = 1$$

Similarly,

$$\nu_{\text{rough}} = 1 \qquad \nu_{\text{load}} = 1 \qquad \nu_{\text{speed}\times\text{rough}} = 1 \qquad \nu_{\text{speed}\times\text{load}} = 1$$
$$\nu_{\text{rough}\times\text{load}} = 1$$

$$\nu_{\text{error}} = (N - 1) - \sum_{i=1}^{k} \nu_i$$

$$\nu_{\text{error}} = (8 - 1) - (1 + 1 + 1 + 1 + 1 + 1) = 1$$

2. Calculation of total sum of squares

$$SS_T = SS_y - C$$

$$SS_Y = \sum Y_i^2 = 2^2 + 5^2 + 0^2 + 2^2 + 6^2 + 10^2 + 4^2 + 8^2 = 249$$

$$C = \frac{T^2}{N} \text{ where } N = 8, \text{ and the grand total, } T \text{ is}$$

$$T = 2 + 5 + 0 + 2 + 6 + 10 + 4 + 8 = 37$$

$$C = \frac{37^2}{8} = 171.125$$

$$SS_T = 249 - 171.125 = 77.875$$

3. Calculation of sum of squares

$$SS_i = (N/4)(\text{MEF})^2$$
$$SS_{\text{speed}} = (8/4)(4.75)^2 = 45.125$$
$$SS_{\text{rough}} = (8/4)(-2.25)^2 = 10.125$$
$$SS_{\text{load}} = (8/4)(3.25)^2 = 21.125$$
$$SS_{\text{speed}\times\text{rough}} = (8/4)(-0.25)^2 = 0.125$$
$$SS_{\text{speed}\times\text{load}} = (8/4)(-0.75)^2 = 1.125$$
$$SS_{\text{rough}\times\text{load}} = (8/4)(0.25)^2 = 0.125$$

The sum of squares for error is

$$SS_E = SS_T - \sum_{i=1}^{k} SS_i$$

$$= 77.875 - (45.125 + 10.125 + 21.125 + 0.125 + 1.125 + 0.125)$$

$$= 0.125$$

4. Calculate the mean squares

$$MS_i = \frac{SS_i}{\nu_i}$$

$$MS_{\text{speed}} = \frac{45.125}{1} = 45.125$$

similarly,

$MS_{\text{rough}} = 10.125,$ $\qquad MS_{\text{load}} = 21.125,$ $\qquad MS_{\text{speed}\times\text{rough}} = 0.125,$

$MS_{\text{speed}\times\text{load}} = 1.125,$ $\qquad MS_{\text{rough}\times\text{load}} = 0.125,$ $\qquad MS_E = 0.125$

and
5. Perform the F test

$$F_i = \frac{MS_i}{MS_E}$$

$$F_{\text{speed}} = \frac{45.125}{0.125} = 361$$

similarly,

$$F_{rough} = 81, \qquad F_{load} = 169, \qquad F_{speed \times rough} = 1, \qquad F_{speed \times load} = 9,$$

$$F_{rough \times load} = 1$$

6. Calculate the percent contribution

$$PC_i = \frac{100(MS_i - MS_E)}{SS_T}$$

$$PC_{speed} = \frac{100(45.125 - 0.125)}{77.875} = 57.8\%$$

similarly,

$$PC_{rough} = 12.8\% \qquad PC_{load} = 27.0\%, \qquad PC_{speed \times rough} = 0, \qquad PC_{speed \times load} = 1.3\%,$$

$$PC_{rough \times load} = 0$$

and the percent contribution of error is

$$PC_E = 100 - (\sum_{i=1}^{k} PC_i)$$

$$PC_E = 100 - (57.8 + 27.0 + 12.8 + 1.3) = 1.1\%$$

All the information calculated above can be summarized in the ANOVA table (Table 7.17). When a factor or interaction effect turns out to be small, it is assumed that the factor or interaction has an insignificant effect on the design of the experiment. From this example it is concluded that the factors affecting galling are mostly speed and load. The factor of *roughness* has relatively little effect in this experiment. The total statistical error is 1.1 percent. As can be seen from the ANOVA table, *speed × rough* and *rough × load* have zero effect. The effect of the *speed × load* interaction is insignificant.

Table 7.17 ANOVA Summary for Three r Factorial Galling Experiment

Source	Degrees-of-Freedom (v)	Sum of Squares (SS_i)	Mean Squares (MS_i)	F Test	Percent of Variance
Speed	1	45.125	45.125	361	57.8
Rough	1	10.125	10.125	81	12.8
Load	1	21.125	21.125	169	27.0
Speed × rough	1	0.125	0.125	1	0
Speed × load	1	1.125	1.125	9	1.3
Rough × load	1	0.125	0.125	1	0
Error	1	0.125	0.125		1.1

The percent contribution shows that the factor of *speed* contributes most to the variation observed in this experiment. This agrees with the F test ratios, which also indicate that the factor of *speed* is more significant than the other factors since $F_{speed} = 361$ is higher than the other F ratios. If the calculated value of percent variation is negative, the contribution is assumed to be zero.

── *Example 7.13* ─────────────────────────

Referring to Example 7.12, use a $\alpha = 0.05$ level of significance (95 percent confidence) to test whether there is a significant interaction between the speed and the load.

Solution

The computed value corresponding to the *speed* $\times$ *rough* interaction is

$$F_i = \frac{MS_i}{MS_E}$$

$$F_{speed \times load} = \frac{1.125}{0.125} = 9$$

For a 0.05 level of confidence, the F ratio to look for in the F distribution table is $F_{0.05}(\nu_1 = 1, \nu_2 = 1)$.
From the Table C.4 (see Appendix C)

$$F_{0.05}(\nu_1 = 1, \nu_2 = 1) = 161.4$$

Since the calculated F value of 9 is less than the criterion (F ratio from the table), the effect is not significant at the 0.05 level.

── *Example 7.14* ─────────────────────────

Some factors assigned to an experiment may not be significant. For example, the size of the specimen used in the galling experiment does not affect the galling rate. In this example, the effects of interactions are not significant. In such cases, an alternative approach is to assume that the interactions are minor enough to be placed in a category called residual (error). Thus, only three parameters are needed to do the calculations for the major effects, which simplifies the analysis of the problem significantly. This is called the *lumped interaction model.*

To simplify the analysis of the problem in Example 7.12, assume that the effect of the interactions is insignificant. Using the lumped interaction model shown in Table 7.18, design a two-level, three-factor experiment to determine the relative effects of speed, roughness, and load on the galling of a metal collar. From Example 7.12, the main effect factors are

$$MEF_{speed} = 4.75$$
$$MEF_{rough} = -2.25$$
$$MEF_{load} = 3.25$$

Table 7.18 L_8 Lumped Model for Galling Experiment

Col. →	Speed 1	Roughness 2	Load 3	Y_i
No. ↓				
1	1	1	1	2
2	1	1	2	5
3	1	2	1	0
4	1	2	2	2
5	2	1	1	6
6	2	1	2	10
7	2	2	1	4
8	2	2	2	8

Analysis of Variance:

1. Calculation of degree-of-freedom

$$\nu_T = N - 1 = 8 - 1 = 7, \nu_{\text{speed}} = \text{number of levels} - 1 = 2 - 1 = 1$$

Similarly,

$$\nu_{\text{rough}} = 1, \qquad \nu_{\text{load}} = 1$$

the degree-of-freedom for error is

$$\nu_{\text{error}} = (N - 1) - \sum_{i=1}^{k} \nu_i$$

$$\nu_{\text{error}} = (8 - 1) - (1 + 1 + 1) = 4$$

2. Calculation of the sum of squares, from Example 7.12,

$$SS_{\text{speed}} = 45.125$$

$$SS_{\text{rough}} = 10.125$$

$$SS_{\text{load}} = 21.125$$

3. Total sum of squares from Example (7.12)

$$SS_T = 77.875$$

Error sum of squares, SS_E

$$SS_E = SS_T - \sum_{i=1}^{k} SS_i$$

$$SS_E = 77.875 - (45.125 + 10.125 + 21.125) = 1.5$$

4.   Calculation of mean squares, from Example (7.12),

$$MS_{speed} = 45.125$$

$$MS_{rough} = 10.125$$

$$MS_{load} = 21.125$$

and the error mean square is

$$MS_E = \frac{SS_E}{\nu_{error}} = \frac{1.5}{4} = 0.375$$

5.   Calculate the F test

$$F_i = \frac{MS_i}{MS_E}$$

$$F_{speed} = \frac{45.125}{0.375} = 120$$

similarly,

$$F_{rough} = 27, F_{load} = 56$$

6.   Calculate the percent contribution

$$PC_i = \frac{100(MS_i - MS_E)}{SS_T}$$

$$PC_{speed} = \frac{100(45.125 - 0.375)}{77.875} = 57.5\%$$

similarly,

$$PC_{rough} = 12.52\%, \qquad PC_{total} = 26.69\%$$

7.   The percent contribution of error is

$$PC_E = 100 - \left(\sum_{i=1}^{k} PC_i\right)$$

$$= 100 - (57.5 + 12.52 + 26.65) = 3.33\%$$

Because some of the factors were eliminated from the analysis, the error has increased to 3.33 percent as summarized in Table 7.19.

Table 7.19   ANOVA Summary for Three-Factorial Galling Experiment

Source	Degree-of-Freedom (ν)	Sum of Squares (MS_i)	Mean Squares (SS_i)	F Test	Percent of Variance
Speed	1	45.125	45.125	120	57.5
Rough	1	10.125	10.125	27	12.52
Load	1	21.125	21.125	56	26.65
Error	4	1.5	0.375		3.33

—— *Example 7.15* ——

Referring to Example 7.14, determine the minimum confidence level and the risk α associated with the factor *roughness.*

Solution

The computed value corresponding to *roughness* is

$$F_i = \frac{MS_i}{MS_E}$$

$$F_{\text{rough}} = \frac{10.125}{0.375} = 27$$

The degree-of-freedom for the numerator is $\nu_1 = 1$, and the degree-of-freedom for the denominator is $\nu_2 = 4$. From the *F* distribution table

$$F_{0.1}(\nu_1 = 1, \nu_2 = 4) = 4.54 \qquad \text{(90\% confidence)}$$

$$F_{0.05}(\nu_1 = 1, \nu_2 = 4) = 7.71 \qquad \text{(95\% confidence)}$$

$$F_{0.01}(\nu_1 = 1, \nu_2 = 4) = 21.2 \qquad \text{(99\% confidence)}$$

Since the calculated *F* value 27 is larger than the *F* ratios from the table, the minimum confidence level is 99 percent, and the risk α is 0.01. If the calculated *F* ratio was less than 21.2, for instance 18, it would be concluded that the minimum confidence level was 95 percent.

7.8.3   Regression and Correlation Analysis

Linear Regression   Generally speaking, correlation equations make possible an estimation of one variable from the measurement of another variable. The relationship between variables defined by correlation equations are vital in the design of experiments. Regression is a statistical analysis technique that provides the best curve fit between two or more variables (dependent and independent). As shown in Figure 7.29, x is the independent variable, which can be measured directly, and y is the dependent variable, which is statistically related to x. To find the best linear relationship to estimate y for any measured value of x, an often-used procedure is fitting the data points with a least squares line so that the sum of squares of the vertical distances of the points from the line *AB* is minimum. The line *AB* is said to

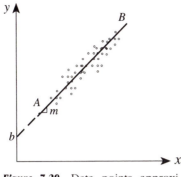

Figure 7.29 Data points approximated by linear regression.

be a "best fitting curve." This procedure is called the least squares method. The error involved with x is due to the accuracy of the observations during the measurement. The dependent variable y error is minimized during this procedure.

If the relationship between y and x is assumed to be linear, the equation of the line of best fit will be in the form

$$y = b + mx \tag{7.72}$$

where constants b and m are the y intercept and the slope of the line AB, respectively. By using the least squares method, the following equations for the constants m and b can be derived:

$$m = \frac{N\sum_{i=1}^{N} x_i y_i - \sum_{i=1}^{N} x_i \sum_{i=1}^{N} y_i}{N\sum_{i=1}^{N} x_i^2 - (\sum_{i=1}^{N} x_i)^2} \tag{7.73}$$

$$b = \frac{\sum_{i=1}^{N} y_i - m\sum_{i=1}^{N} x_i}{N} \tag{7.74}$$

To investigate how close the correlation between two variables is, we consider the *correlation coefficient* r^*, defined by

$$r^* = m\frac{S_x}{S_y} \tag{7.75}$$

where S_x and S_y are the standard deviation of the independent and dependent variables, respectively. The correlation coefficient may vary between -1 and +1. A zero correlation means that there is no relationship between the two variables. As the correlation coefficient approaches ± 1, all the data points tend to coincide with the predicted regression line. A negative correlation coefficient indicates that the two variables are negatively related, that is, the predicted regression line has a negative slope.

Multiple Linear Regression   A more general situation in which y_i observations may depend on more than one independent variable is now considered. For example, the severity of galling of a metal collar is a function of several independent variables. Among them are surface roughness, speed of the applied torque, the applied load, and lubricant. In the previous sections it was noted that the severity of galling changes when the magnitude of the above variables changes. The estimating equation for the linear multiple regression model is given as

$$y = a_0 + a_1 x_1 + a_2 x_2 + \cdots a_n x_n \tag{7.76}$$

where a_0 is a constant and $a_{1 \cdots n}$ are the regression coefficients. Note that where $n = 1$, Eq. (7.76) gives the straight line linear regression model.

7.8.4   Introduction to Statistical Methods Using the SAS Package

Because of the practicality of statistics, statistical methods and data analysis are widely used in every walk of life. Since statistics mainly deals with numbers, computers are naturally the main tools for analysis. To give a better overall picture, large quantities of data are usually collected and analyzed. Computers, being good for number crunching, are used for various data analyses. The SAS package is one of the well-known statistical analysis software packages. The SAS package covers many subjects of data analysis that require substantial effort to master. The SAS package, as presented in this chapter, is used for analyzing linear and nonlinear relationships only.

To use SAS correctly, the user must determine which procedure to use. The user also has to predict the relationship between the dependent and independent variables. The predicted relationship gives rise to the *model* of the relationship. Once the procedure and the model are determined, a short SAS program is written to instruct the SAS package to analyze a set of data according to the required procedures, such as GLM (general linear model), or NLIN (nonlinear regression).

To assist in determining which procedure and model to use, a graph should be plotted showing the relationship between the dependent and the independent variables. For a complex relationship between the dependent and independent variables, the model may have to be modified until an acceptable model is achieved. The parameters to look for in determining whether the model used is good depends on the procedure used for correlation. These important parameters are explained as follows, for NLIN and GLM procedures.

The NLIN Procedure   The NLIN procedure is used to analyze data that are nonlinear, such as data that have a logarithmic relationship. For nonlinear regression models, the least squares method is used. The following four iterative methods are available in the SAS package to obtain the parameter relationship:

1.   Modified Gauss–Newton method.
2.   Marquardt method.
3.   Gradient or steepest-descent method.
4.   Multivariate secant or false position (DUD) method.

Suppose the assumed model is of the form

$$y = \alpha(x_1)^\beta (x_2)^\theta (e)^{\delta x_3} \tag{7.77}$$

where y and x_i are the dependent and independent variables, respectively, and α, β, θ, and δ are the parameters to be estimated. If we use the Gauss-Newton or

Marquardt iterative methods, it is necessary to define the partial derivatives of the model with respect to the various parameters. For example, for the model given in Eq. 7.77, the following partial derivatives must be defined in the analysis:

$$\frac{\partial y}{\partial \alpha} = (x_1)^\beta (x_2)^\theta (e)^{\delta x_3} \tag{7.78}$$

$$\frac{\partial y}{\partial \beta} = \alpha (x_1)^\beta (x_2)^\theta (e)^{\delta x_3} \log x_1 \tag{7.79}$$

$$\frac{\partial y}{\partial \theta} = \alpha (x_1)^\beta (x_2)^\theta (e)^{\delta x_3} \log x_2 \tag{7.80}$$

$$\frac{\partial y}{\partial \delta} = \alpha (x_1)^\beta (x_2)^\theta (e)^{\delta x_3} x_3 \tag{7.81}$$

Note that when the partial derivatives of Eq. 7.77 are evaluated with respect to each parameter, for example, β, all other terms are assumed to be constant except β.

The Gauss–Newton iterative method involves regressing the error on the partial derivatives of the model with respect to the parameters α, β, θ, and δ until the iteration converges. In this program, a DER statement (default for the Gauss–Newton iterative method) must be used for each parameter to be estimated. The expression must be an algebraic representation of the partial derivatives of the model.

The NLIN procedure requires the following information:

1. Models.
2. List of all the variables.
3. Parameter names and their guessed values.
4. Derivatives of the model with respect to each parameter.

The SAS program is executed by typing the SAS *filename*. The filename must have the SAS extension, for example, *filename*.SAS.

Because the NLIN procedure involves derivatives, it not only requires considerable time to compute, it also requires good initial parameter estimates. If the model is very complex, deriving derivatives can be very difficult.

The GLM Procedure   If the model is nonlinear but is primarily logarithmic, the model can be modified by taking the *log* of both sides of the model equation, thus making the model linear. The general model procedure can then be used to determine the parameters with less computational time. Taking the *log* of an equation is also preferred over deriving derivatives. However, care must be taken to ensure that the computed parameters are reconverted to the original logarithmic relationship.

Important SAS Statements

- DATA Statement:$\Longrightarrow$ DATA{*filename*};
  This is the first statement in a SAS job. It creates an SAS data set. Data statements begin with the word DATA followed by a file name and ending

with a semicolon (every SAS statement ends with a semicolon). For example, if the name of the data set chosen is DEN the statement will be

```
DATA DEN;
```

- INPUT Statement: $\Longrightarrow$ INPUT {*variable name*}{*number of column where the data value begins*}{−}{*number of column where the data value ends*}; repeat this for each variable.
  For example, the INPUT file for density-temperature study is written as

```
INPUT T 2-7 DEN 14-18
```

  reads the temperature T in columns 2 to 7 of the data file and it reads the density DEN in columns 14 to 18. The INPUT statement usually follows the small DATA statement. Variable names begin with letters, and a maximum of eight characters can be used.

- INFILE Statement: $\Longrightarrow$ If data values are on the disk or type, include an INFILE statement before the INPUT statement.

- **Procedures:** $\Longrightarrow$ PROC {*procedure name*};

  Procedures are computer programs that read the SAS DATA set, perform various computations, and print the results. For example, if the analysis procedure used is GLM then write

```
PROC GLM;
```

  If the print procedure is to be run, write

```
PROC PRINT;
```
  Note that this step always starts with a PROC statement and is followed by the procedure name.

- MODEL Statement: $\Longrightarrow$ MODEL {*dependents*} {=} {*expression*};
  A model statement defines the prediction equation. For example, to predict the change of density of a solution with temperature, the model statement can be written as

```
MODEL DEN = T;
```

  where DEN and T are the density and temperature of the solution, respectively.
  For more information refer to the *SAS Introductory Guide.*[13]

── *Example 7.16* ──────────────────────────

To learn how to analyze the GLM simple regression (only one independent variable) model, consider the change of density of a LiCl solution with temperature. Establish the regression equation for the data given in Table 7.20.

Solution

The density of the LiCl solution decreases with increase in temperature. Since there is a linear relationship between the independent and dependent variable, the model

---

[13] *SAS Introductory GUIDE,* 1985, 3rd Ed., SAS Institute Inc., Cary, NC.

Table 7.20   Density of
Lithium Chloride Solution

Temperature (°C)	Density (g/ml)
15.55	1.240
21.11	1.238
26.66	1.236
32.22	1.234
37.00	1.232
43.33	1.230

is assumed to be

$$DEN = b + m(T) \qquad (7.82)$$

where
DEN is the density of the lithium chloride,
(T) is the temperature,
m is the gradient (slope), and
b is the intercept of the linear relationship.

DEN is modeled with (T) in the SAS simple GLM procedure as shown below:

```
TITLE 'DENSITY OF LITHIUM CHLORIDE SOLUTION';
OPTIONS LS=72 PS=66 NODATE;
FILENAME IN 'WORK$AREA:[kvkrn]lit.dat';
DATA ONE
INFILE IN;
INPUT T 2-7 WT 14-18;
PROC GLM;
MODEL WT = T;
PROC PRINT;
```

Some of the statements in the above SAS program are explained here:

- TITLE ··· Insert the title for the output printout. The title string must be inserted in quotation marks.
- OPTIONS ··· modifies the printout format. In the above example, LS=80, line size is equal to 80 column; PS=54, page size is equal to 54 lines; and NODATE indicates that data are to be printed.
- FILENAME IN ··· specifies the file name, including its path.
- DATA ONE ··· assigns the data set as data named one.

Output of the SAS program is shown in Table 7.21, and the predicted value of density can be computed by tbe following formula:

$$PREDICTED = INTERCEPT + GRADIENT * TEMPERATURE \qquad (7.83)$$

$$PREDICTED = 1.25 - 0.000364 \times T$$

Table 7.21   SAS Output

Temperature	Actual	Predict
15.55	1.2400	1.2400
21.11	1.2380	1.2380
26.66	1.2360	1.2360
32.22	1.2340	1.2339
37.00	1.2300	1.2299
43.33	1.2300	1.2299
Intercept	. . . $1.25E+00$	
Gradient	. . . $-3.64E-04$	

Example 7.17

Establish the regression equation for the vapor pressure data of a CELD solution ($50\%CaCl_2 + 50\%LiCl$) given in Table 7.22.

Solution

It is easy to see that the vapor pressure increases exponentially with an increase in temperature. This leads us to assume the following relationship between the vapor pressure and the temperature.

$$VP = \alpha * \exp(\beta * T) \tag{7.84}$$

where
VP is the Vapor pressure of the CELD solution.
α is the coefficient of the exponential term.
β is the coefficient of temperature.
(T) is the temperature.

To use the simple regression GLM procedure, must be made linear. If the *log* of both sides of the equation is taken, the following equation is derived.

$$\text{Log}(VP) = \text{Log}(\alpha) + \beta * T \tag{7.85}$$

The value of log(α) represents the intercept, and β represents the gradient. The log(VP) is modeled with temperature in simple regression GLM procedure as

Table 7.22   Vapor Pressure of a CELD Solution

Temperature (°C)	Vapor Pressure (mmHg)
21.11	15.00
26.66	20.00
32.22	27.00
37.77	35.00
43.33	47.00
48.88	63.00
54.44	82.00
60.00	105.00
65.00	134.00

```
TITLE 'VAPOR PRESSURE OF CELD SOLUTION';
OPTIONS LS=72 PS=66 NODATE;
FILENAME IN 'WORK$AREA:[kvkvs]VAP.DAT';
DATA ONE
INFILE IN;
INPUT T 3-7 VP 9-13;
DATA TWO;
SET ONE;
LVP  = LOG(VP);
PROC GLM;
MODEL LVP = T;
PROC PRINT;
```

If the data file is not too large, it can be included in the program, and the above model can be modified as

```
TITLE 'VAPOR PRESSURE OF CELD SOLUTION';
OPTIONS LS=72 PS=66 NODATE;
DATA ONE;
INPUT T 1-5 VP 7-11;
CARDS;
21.11 15.00
26.66 20.00
32.22 27.00
37.77 35.00
43.33 47.00
48.88 63.00
54.44 82.00
60.00 105.0
65.00 134.0
 ;
DATA TWO;
SET ONE;
LVP  = LOG(VP);
PROC GLM;
MODEL LVP = T;
PROC PRINT;
```

Note that, since the data file is in the program, FILENAME IN and FILE IN are eliminated from the second model. Output of the SAS is shown in Table 7.23.

In interpreting the results, the antilog of the intercept must be taken to give the actual value of α in Eq. (7.84). Substituting the intercept and gradient into Eq. (7.85), gives

$$\text{Log}(VP) = 1.67212647 + 0.04996695 \times T$$

Taking the antilog,

$$(VP) = e^{1.67212647} e^{0.04996695 \times T}$$

$$(VP) = \alpha e^{0.04996695 \times T}$$

Table 7.23   SAS Output

Temperature	Actual	Predict
21.11	15.00000	15.28587
26.66	20.00000	20.17099
32.22	27.00000	26.63062
37.00	35.00000	35.14134
43.33	47.00000	46.39513
48.88	63.00000	61.22227
54.44	82.00000	80.82831
60.00	105.00000	106.71305
65.00	134.00000	136.99965
Intercept	...1.67212647	
Gradient	...0.04996695	

calculating $\alpha = e^{1.67212647} = 5.3235$ and rewriting the predicted equation as

$$(VP) = 5.3235 e^{0.04996695 \times T}$$

— Example 7.18

To analyze the performance of a packed tower for a dehumidification process, calculation of the mass and heat transfer coefficients of the packing material used in the tower is necessary. This example is concerned with the interface transfer of heat (gas phase) when air is brought into contact with liquid desiccant (lithium chloride solution). The data shown in Table 7.24 were generated by a computer program that computes the volumetric heat transfer coefficients for a 1-in. Rasching ring. To generate the gas phase heat transfer coefficients H_{GA}, properties of air and liquid desiccant shown in Tables 7.25, and 7.26, respectively, are used.

Gas phase heat transfer in a packed tower depends on air flow G, liquid desiccant flow rate L, and air temperature T_G. Therefore, the independent variables are G, L, and T_G.

The results in Table 7.24 can be presented in graphical form; however, they are presented as a correlated equation. This is most useful in analyzing the performance of the packed tower.

1. Using SAS nonlinear (NLIN) analysis, find the correlation equation of the gas phase heat transfer coefficient in a packed column.

2. Using SAS Linear (GLM) analysis, find the correlation equation of the gas phase heat transfer coefficient in a packed column.

Solution

(a) The correlation model for NLIN analysis can be assumed to be in the form of

$$H_{GA} = \alpha (L)^\beta (G)^\theta e^{(\delta T_G)} \tag{7.86}$$

where H_{GA} is the gas phase heat transfer coefficient, and α, β, θ and δ, are the parameters that need to be determined.

Table 7.24   Lithium Chloride Lit. Data File

Air Temperature (°C)	Air Flow Rate (kg/m²)	Liquid Flow Rate (kg/m²)	H_{GA} (w/m³ °C)
$0.40000E+02$	$0.30000E+00$	$0.70000E+00$	$0.16834E+04$
$0.40000E+02$	$0.30000E+00$	$0.17000E+00$	$0.16834E+04$
$0.40000E+02$	$0.50000E+00$	$0.70000E+01$	$0.23344E+04$
$0.40000E+02$	$0.50000E+00$	$0.17000E+00$	$0.26961E+04$
$0.40000E+02$	$0.70000E+00$	$0.70000E+00$	$0.28953E+04$
$0.40000E+02$	$0.30000E+00$	$0.17000E+00$	$0.33439E+04$
...	...	...	...
...	...	...	...
$0.45000E+02$	$0.30000E+00$	$0.70000E+00$	$0.16905E+04$
$0.45000E+02$	$0.30000E+00$	$0.17000E+00$	$0.19524E+04$
$0.45000E+02$	$0.50000E+00$	$0.70000E+00$	$0.23442E+04$
$0.45000E+02$	$0.50000E+00$	$0.17000E+00$	$0.27074E+04$
$0.45000E+02$	$0.70000E+00$	$0.70000E+00$	$0.29075E+04$
$0.45000E+02$	$0.70000E+00$	$0.17000E+00$	$0.33580E+04$
...	...	...	...
...	...	...	...
$0.50000E+02$	$0.30000E+00$	$0.70000E+00$	$0.16977E+04$
$0.50000E+02$	$0.30000E+00$	$0.17000E+00$	$0.19607E+04$
$0.50000E+02$	$0.50000E+00$	$0.70000E+00$	$0.23542E+04$
$0.50000E+02$	$0.50000E+00$	$0.17000E+00$	$0.27189E+04$
$0.50000E+02$	$0.70000E+00$	$0.70000E+00$	$0.29198E+04$
$0.50000E+02$	$0.70000E+00$	$0.17000E+00$	$0.33722E+04$
...	...	...	...
...	...	...	...

Differentiating H_{GA} with respect to α, β, θ, and δ, respectively, we have the following equations (note that $\alpha = A$, $\beta = B$, $\theta = D$, and $\delta = F$) :

$$DER.A = L^B(G)^D(e)^{FT_G}$$

$$DER.B = AL^B(G)^D(e)^{FT_G} Log(L)$$

$$DER.D = AL^B(G)^D(e)^{FT_G} Log(G)$$

$$DER.F = AL^B(G)^D(e)^{FT_G} T_G$$

These equations are used in the NLIN SAS procedure as follows:

```
TITLE 'GAS PHASE HEAT  TRANSFER COEFFICIENT:HGA';
OPTIONS LS=80 PS=54 NODATE;
```

Table 7.25   Air Data File

Temperature (°C)	High Capacity (J/kg K)	Viscosity (kg/m s)	Prandtl Number	Density (kg/m³)
40.0	1005.3	$1.9192E-05$	0.7190	1.1272
45.0	1005.9	$1.322E-05$	0.7190	1.1095
50.0	1006.0	$1.9545E-05$	0.7190	1.0924

Table 7.26   Lithium Chloride Data File

Temper-ature (°C)	Concen-tration (W%)	Density (kg/m³)	Viscosity (kg/m s)	Mol. Weight (kg/kmol)	Heat Capacity (J/kg K)	Heat Condition (J/ms K)
32.22	32.5	1187.0	0.003441	42.20	0.547	2720.9
37.77	35.0	1202.0	0.003765	42.40	0.547	2720.9
43.33	37.5	1216.0	0.004079	42.40	0.547	2720.9

```
FILENAME IN 'WORK$AREA:[kvkrn]LIT.DAT';
DATA ONE
INFILE IN;
INPUT TG 3-13 G 15-25  L27-37 HG 39 49;
DATA TWO;
SET ONE;
PROC NLIN METHOD=MARQUARDT CONVERGE=0.0000000001;
PARMS A=1.9887E+3 B=2.4093E-1 D=6.64E-1 F=9.1742E-4;
HGA=A*L**B*G**D*EXP(F*TG);
MODEL HG=HGA;
*   ... DERIVATIVES ...
DER.A=L**B*G**D*EXP(F*TG);
DER.B=A*L**B*G**D*EXP(F*TG)*LOG(L);
DER.D=A*L**B*G**D*EXP(F*TG)*LOG(G);
DER.F=A*L**B*G**D*EXP(F*TG)*TG;
PROC PRINT;
```

Some of the statements in the above SAS program are explained here:

- DATA TWO ··· assigns a new data file, in this case, DATA TWO.
- SET ONE ··· creates an SAS data set from another SAS data set.
- PROC NLIN METHOD=MARQUARDT CONVERGE=0.000000001 ··· specifies the procedure to be used by SAS, the method of regression, and the convergence criteria.
- PARMS $A = 1.9887E + 3$ $B = 2.4093E - 1$ $D = 6.64E - 1$ $F = 9.1742 E - 4$; ··· assigns the initial guessed values for each parameter.
- * ··· an asterisk is used for comment lines.

As mentioned previously, each statement must end with a semicolon; otherwise, SAS assumes that the statement continues on the next line. Statements can begin from any column, but the position of the statement relative to others is important, since SAS runs from the top to the bottom of the program. Output of the SAS is shown in Tables 7.27 and 7.28. In the output, the phrase "convergence criteria is met" will apear when the 95 percent confidence interval is reached. This is shown in Figure 7.30 for parameter A. Otherwise, the program should be run with the different initial guesses, or the model should be changed.

The correlation equation can be written by substituting the estimated values of output into Eq. 7.86 as

$$H_{GA} = 3726.72(L)^{0.162}(G)^{0.64}e^{0.00084 T_G} \tag{7.87}$$

Table 7.27   Lithium Chloride
SAS Output

H_{GA} Calculated	Estimated H_{GA} Nonlinear
1683.40002	1683.54285
1944.19995	1943.79749
2334.39990	2334.57251
2696.10010	2695.46802
2895.30005	2895.53833
$\cdots$	$\cdots$
$\cdots$	$\cdots$
$\cdots$	$\cdots$
1960.69995	1960.25293
2354.19995	2354.33594
2718.89990	2718.28687
2919.80005	2820.05078
3372.19995	3371.45410
Intercept	$\ldots 8.22326087$
TG	$\ldots 0.00084385$
LG	$\ldots 0.64000008$
LL	$\ldots 0.16233595$

(b) In the case of a multiple regression GLM SAS procedure, consider the same assumed model equation

$$H_{GA} = \alpha(L)^{\beta}(G)^{\theta} e^{(\delta T_G)} \qquad (7.88)$$

or

$$H_{GA} = A(L)^{B}(G)^{D} e^{(FT_G)} \qquad (7.89)$$

Taking the log of the assumed model equation:

$$LOG(H_{GA}) = LOG(A) + B * LOG(L) + D * LOG(G) + F * T_G \qquad (7.90)$$

where $LOG(A)$ is the intercept of the linear equation. B, D, and F are the coefficients of $LOG(L)$, $LOG(G)$, and T_G, respectively. This procedure makes the model linear. The values of B, D, and F are then determined by the SAS GLM procedures. The value of $LOG(A)$ is also determined by the SAS GLM procedure, from which the actual value of parameter A can be calculated by taking the antilog of the value of $LOG(A)$.

The following SAS program is for the GLM procedure for the same model used in the NLIN SAS procedure.

TAble 7.28   Lithium Chloride SAS Output (continued)

Parameter	Estimate	Asymptotic STD. Er.	Asymptotic 95% Low. Con. Int.	Up. Con. Int.
A	3726.71799	0.06696065	3726.58349	3726.85248
B	0.16233	0.00000364	0.16233	0.16235
D	0.63999	0.00000519	0.63999	0.64001
F	0.00084	0.00000039	0.00084	0.00084

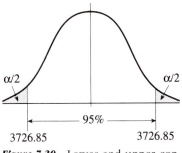

α/2   α/2

— 95% —

3726.85   3726.85

Figure 7.30   Lower and upper confidence intervals for SAS output.

```
TITLE 'GAS PHASE HEAT  TRANSFER COEFFICIENT:HGA';
OPTIONS LS=72 PS=66 NODATE;
FILENAME IN 'WORK$AREA:[kvkrn]LIT.DAT';
DATA ONE
INFILE IN;
INPUT TG 3-13 G 15-25 L 27-37 HG 39-49;
DATA TWO;
SET ONE;
LG = LOG(G);
LL = LOG(L);
LHGA = LOG(HG);
PROC GLM;
MODEL LHGA = TG LG LL;
PROC PRINT;
```

The above SAS program has similar statements to those that have been explained earlier, except for a few lines. For DATA TWO, the log was taken for *G*, *L*, and *HG*. Also, the model has multiple independent variables, *TG*, *LG*, and *LL*.

The results of the GLM method are presented in Table 7.29. From the results of the output:

- Intercept LOG(*A*) is equal to 8.22326087
- Coefficient of *LL* is $B = \beta$ and is equal to 0.16233595
- Coefficient of *LG* is $D = \theta$ and is equal to 0.64000008
- Coefficient of T_G is $F = \delta$ and is equal to 0.00084385

Substituting the estimated parameters into Eq. 7.91

$$LOG(H_{GA}) = 8.22326087 + 0.16233595 * LOG(L)$$

$$+ 0.64000008 * LOG(G) + 0.00084385 * T_G \qquad (7.91)$$

Taking the antilog of both sides of the above equation yields the correlation equation for the gas phase heat transfer coefficient.

$$H_{GA} = 3726.63(L)^{0.162}(G)^{0.64}e^{0.00084T_G} \qquad (7.92)$$

Table 7.29   Lithium chloride SAS output

H_{GA} Calculated	Estimated H_{GA} Nonlinear	Estimated H_{GA} Linear
1683.40002	1683.54285	1683.34082
1944.19995	1943.79749	1943.56433
2334.39990	2334.57251	2334.29224
2696.10010	2695.46802	2695.14478
2895.30005	2895.53833	2895.19092
. . .	. . .	. . .
. . .	. . .	. . .
. . .	. . .	. . .
1960.69995	1960.25293	1959.95898
2354.19995	2354.33594	2353.98291
2718.89990	2718.28687	2717.87915
2919.80005	2920.05078	2919.61279
3372.19995	3371.45410	3370.94873
Intercept	. . . 8.22326087	
TG	. . . 0.00084385	
LG	. . . 0.64000008	
LL	. . . 0.16233595	

From the above example, it is interesting to note that the estimated parameters are identical by both methods. However, the computer took far less time to determine the parameters using the NLIN procedure.

BIBLIOGRAPHY

1. DIMITRIS, N. and CHORAFAS, P. E., *Statistical Processes and Reliability Engineering.* Van Nostrand Reinhold, New York, 1966.

2. RICHMOND, B. S., *Statistical Analysis.* The Ronald Press Company, New York, 1964.

3. WALPOLE, E. R. and MYERS, H. R., *Probability and Statistics for Engineers and Scientists.* Macmillan, New York, 1978.

4. EDWARDS, L. A., *Statistical Analysis.* Rinehart & Co., Inc., New York, 1958.

5. SIMON, E. L., *An Engineer's Manual of Statistical Methods.* Wiley, New York, 1941.

6. DHILLON, S. B., *Reliability Engineering in Systems Design and Operation.* Van Nostrand Reinhold, New York, 1983.

7. NYLANDER, E. J., "Statistical Distributions in Reliability," *Transactions on Reliability and Quality Control*, vol. RQC-11, pp.43–53, 1961.

8. TAGUCHI, G., *System of Experimental Design,* Vol. 2, 2nd ed. UNIPUB/Kraus International Publications, New York, 1988.

9. TAGUCHI, G., *Introduction to Quality Engineering.* UNIPUB/Kraus International Publications, New York, 1987.

10. HICKS, C. R., *Fundamental Concepts in the Design of Experiments.* 3rd ed. CBS College Publishing, New York, 1982.

11. HINES, W. W. and MONTGOMERY, D. C., *Probability and Statistics in Engineering and Management Science,* 3rd ed. Wiley, New York, 1990.

12. ROBINSON, E. A., *Statistical Reasoning and Decision Making.* Goose Pond Press, Houston, Tx, 1981.

Chapter 8

Design for Reliability

A chain is only as strong as the weakest link.

8.1 INTRODUCTION

This chapter introduces the concept of reliability as applied to engineering design. Based on the failure rates of systems, devices, and components, reliability is discussed from a statistical point of view. Understanding of the principles of reliability is an essential ingredient in designing systems and is one of the essential elements of modern engineering design.

Reliability is generally defined as the probability that a system, device, or component will successfully perform for:

1. A given range of operating conditions.
2. A specific environmental condition.
3. A prescribed economic survival time.

It is interesting to note that high performance expectation from a component, in general, implies that the component is heavily loaded, which affects the operational survival time. Therefore, care must be taken by the designer to achieve a proper balance between performance and reliability so that an adequate safety factor is provided. Reliability is not the only design parameter that affects performance. Reliability predictions are especially useful when used in conjuction with maintainability, availability, dependability, optimization, and cost parameters. The interconnection between these concepts is depicted by Figure 1.9. Consideration of maintainability in the design process results in a system that can be maintained effectively within given time constraints. Availability is an important parameter in complex systems. Having

285

high reliability does not ensure that the system will be operational (available) when needed. Thus, it is essential that the system be capable of being repaired easily and inexpensively when it fails. It is evident that the availability of a system, device, or component is a function of its reliability and maintainability. Dependability of a component is another important design parameter that provides a measure of the system condition combining its reliability and maintainability. To determine the importance of these attributes to the system design, it is necessary to examine the mathematical relationships involved.

8.2   DEFINITIONS AND PROBABILITY LAWS

Any event identified as a subset of a sample space has a probability of occurrence ranging between zero and one. This can be explained by use of the Venn diagram, as shown in Figure 8.1 The area of the square designated by S represents the sample space, which is equal to one. The circular area designated by $P(A)$ represents the probability of event A. For instance, when we make a statement that domestic cars will probably accumulate 100,000 miles before the engine starts to burn oil, *burning engine oil* is called the *event*. Considering this event, if there is no chance of burning engine oil before the car reaches 100,000 miles, the probability of *burning engine oil* is equal to zero, which is shown as

$$P(\text{event}) = 0 \tag{8.1}$$

If there is a 100 percent chance that the engine will burn oil before the car reaches 100,000 miles, the probability of this event is

$$P(\text{event}) = 1 \tag{8.2}$$

Then the probability of an event cannot be less than zero or greater than one and must satisfy the following inequality:

$$0 \leq P(\text{event}) \leq 1 \tag{8.3}$$

Events can be classified as either independent or conditional events.

8.2.1   Independent Events

When the outcome of one event does not depend on another events, this event is called an *independent event*. For example, in tossing two die, the probability of the

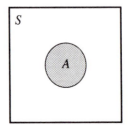

Figure 8.1   Venn diagram for an event.

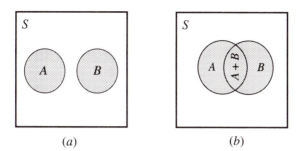

Figure 8.2 Venn diagram for (a) mutually exclusive events, (b) events that are not mutually exclusive.

sum of the numbers obtained does not depend on the previous toss. Probability laws for independent events are as follows.

Multiplication Law   If A and B are independent events, the joint probability of both A and B, $P(A$ and $B)$, is equal to the product of the probabilities of the individual events.

$$P(A \text{ and } B) = P(A)P(B) = P(AB) \tag{8.4}$$

Addition Law   If A and B are independent events and are not mutually exclusive, the probability of occurrence of, at least, one of the two events is the sum of their individual probabilities less the probability of their joint occurrence.

$$P(A \text{ or } B) = P(A + B) = P(A) + P(B) - P(AB) \tag{8.5}$$

A special case of the addition law is when A and B are mutually exclusive events. Then, the probability of occurrence of A or B equals the sum of the individual probabilities

$$P(A \text{ or } B) = P(A + B) = P(A) + P(B) \tag{8.6}$$

Note that, since events A and B cannot occur at the same time, $P(AB) = 0$. Both cases are shown in Figure 8.2.

▬ *Example 8.1* ▬▬▬▬▬▬▬▬▬▬▬▬▬▬▬▬▬▬▬▬▬▬▬▬▬▬▬▬▬

Suppose that 60 percent of engine component failures are due to fatigue and 30 percent are due to corrosion. If the probability of failure from both fatigue and corrosion is 20 percent, determine the probability of failure due to fatigue or corrosion.

Solution

Let A denote the event, failure due to fatigue, and B denote the event, failure due to corrosion. Then $P(A) = 0.60$, $P(B) = 0.30$, $P(AB) = 0.20$. Since $P(AB)$ is not equal to zero, these are nonmutually exclusive events; hence, the probability of failure due to fatigue or corrosion, $P(A$ or $B)$, can be calculated by using Eq. 8.5

$$P(A \text{ or } B) = P(A) + P(B) - P(AB) = 0.60 + 0.30 - 0.20 = 0.70$$

8.2.2   Conditional Events

Probabilities determined for dependent events are called conditional probabilities. For example, consider a bag containing 10 colored identical pens, which are drawn randomly. Three of the pens are yellow, three are blue, and four are black. The probability that the second pen drawn will be a certain color depends on the color of the first pen drawn. For any events A and B in sample space S, the conditional probability of B, given A, is denoted by $P(B|A)$, and is defined by the equation

$$P(B|A) = \frac{P(A \text{ and } B)}{P(A)} \tag{8.7}$$

8.2.3   Bayes' Theorem

Bayes' theorem is an extension of the law of conditional probability which states that, if an event B is dependent on any of the mutually exclusive events A_1, A_2, ...A_k that partition sample space S, and if event B has occurred, then the probability that event A_k has also occurred is defined by Bayes' theorem as

$$P(A_k|B) = \frac{P(A_k)P(B|A_k)}{\sum_{i=1}^{n} P(A_i)P(B|A_i)} \tag{8.8}$$

—— *Example 8.2* ——————————————————————

Suppose a company manufactures the following three types of engine valves:

A_1 = regular valves

A_2 = valves are hard-faced for maximum wear resistance

A_3 = aluminum coated valves to increase resistance to oxidation

Let B denote the event, failure due to fatigue. Suppose it is known that 60 percent of the company production is regular valves, 28 percent hard-faced valves, and 12 percent aluminum-coated valves. Assume that 50 percent of the regular valves, 10 percent of the hard-faced valves, and 14 percent of the aluminum-coated valves fail by fatigue. If it is observed that an engine valve fails by fatigue, what is the probability that the valve is hard faced?

Solution

$P(A_1) = 0.60$, $P(A_2) = 0.28$, $P(A_3) = 0.12$, $P(B|A_1) = 0.50$, $P(B|A_2) = 0.10$, $P(B|A_3) = 0.14$. From Eq. 8.8, we have $P(\text{hard-faced valves}|\text{failure due to fatigue}) = P(A_2|B)$.

$$P(A_2|B) = \frac{P(B|A_2)P(A_2)}{P(A_1)P(B|A_1) + P(A_2)P(B|A_2) + P(A_3)P(B|A_3)}$$

$$P(A_2|B) = \frac{0.28 \times 0.10}{0.60 \times 0.50 + 0.28 \times 0.10 + 0.12 \times 0.14} = 0.081$$

That is, there is only an 8.1 percent chance that the failed valve is hard faced.

8.3 BASIC RELIABILITY EQUATION

The failure cumulative density function of a component at time t is defined by

$$F(t) = \int_0^t f(t)dt \tag{8.9}$$

where $f(t)$ is the failure probability density function. By definition, reliability is the probability of no failure, which is given at time t by

$$R(t) = 1 - F(t) = 1 - \int_0^t f(t)dt \tag{8.10}$$

Properties of the probability density function yield

$$R(t = 0) = 1 \tag{8.11}$$
$$R(\infty) = 0 \tag{8.12}$$

After differentiating, Eq. 8.10 can be rearranged to give the failure density function in terms of the reliability

$$f(t) = -\frac{d}{dt}R(t) \tag{8.13}$$

The hazard rate or instantaneous failure rate is the conditional probability that the component has not failed by time t and can be obtained by

$$b(t) = \lim_{\Delta t \to 0} \frac{F(t + \Delta t) - F(t)}{\Delta t} \frac{1}{R(t)} = \frac{f(t)}{R(t)} \tag{8.14}$$

Substituting Eq. 8.13

$$b(t) = -\frac{1}{R(t)}\frac{dR(t)}{dt} \tag{8.15}$$

Furthermore, the above equation can be rewritten as

$$b(t) = -\frac{d}{dt}\ln R(t) \tag{8.16}$$

Integrating both sides of Eq. 8.16 yields

$$\ln R(t) = -\int_0^t b(t)dt \tag{8.17}$$

and, finally, reliability can be expressed as

$$R(t) = \exp\left[-\int_0^t b(t)dt\right] \tag{8.18}$$

Note that $b(t)$ is the time-dependent failure rate. For the constant failure rate of a component, which is so often noted in reliabilty, $b(t)$ is set to λ:

$$b(t) = \lambda \tag{8.19}$$

Substituting Eq. 8.19 in Eq. 8.18 leads to

$$R(t) = \exp(-\lambda t) \tag{8.20}$$

In Eq. 8.20, λ is the *failure rate*, also referred to as the *hazard* or *instantaneous failure rate*, which predicts the number of failures of a system, device, or component that has occurred over a particular length of time. The practical features of Eq. 8.20 is that it is simple and can be applied any time other than after component failures. Only an exponential distribution has this special feature.

8.3.1 Mean Time to Failure (MTTF)

Mean time to failure (MTTF) is the average time that a nonrepairable system, device, or component will operate before experiencing a failure. From the basic probability definitions, the mean value of a random continuous variable x is given by

$$E(x) = \int_{-\infty}^{\infty} xf(t)dx \tag{8.21}$$

Hence, for time to failure, at $t \geq 0$

$$E(t) = \text{MTTF} = \int_{0}^{\infty} tf(t)dt \tag{8.22}$$

and it can be shown that

$$\text{MTTF} = \int_{0}^{\infty} R(t)dt \tag{8.23}$$

8.4   PROBABILITY DISTRIBUTION FUNCTIONS USED IN RELIABILITY

This section briefly discusses selected probability distributions for finding the failure rate of a single component.

8.4.1 Bathtub Hazard Distribution

The failure probability density function of the bathtub distribution for a component is defined by

$$f(t) = b\theta(\theta t)^{b-1} \exp\left\{-\left[e^{(\theta b)^b} - \theta t^b - 1\right]\right\} \text{ for } b > 0, \theta > 0, t \geq 0 \tag{8.24}$$

where t is the time, θ is the scale parameter, and b is the shape parameter. A component reliability function $R(t)$ can be obtained by substituting Eq. 8.24 into Eq. 8.10

$$R(t) = \exp\left\{-\left[e^{(\theta t)^b} - 1\right]\right\} \tag{8.25}$$

By utilizing Eq. 8.14 with Eq. 8.24 and Eq. 8.25, the hazard rate can be determined:

$$h(t) = b\theta (\theta)^{b-1} e^{(\theta t)^b} \tag{8.26}$$

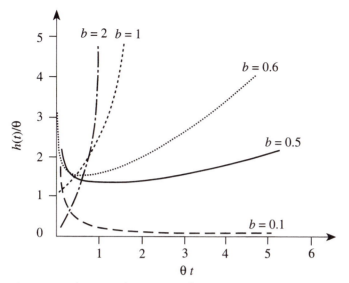

Figure 8.3 The Hazard rate curves for $\theta=1$.

Typical hazard rate curves are shown in Figure 8.3 for different values of b at the constant value of $\theta = 1$. Experience indicates that the failure behavior of many components follow the bathtub hazard rate function. As shown in Figure 8.4, at $b = 0.5$ the hazard rate function gives a typical statistical curve of the change of failure rate during the operating life of a component. Since the curve is similar to that of a bathtub, it is known as the *bathtub curve*. This curve represents the failure of electronic components that have three different life stages. Consider a population of new components. In the first stage of operation the failure rate will be high but will decrease with time. Failure at this stage is called *infant mortality* and is attributed to design or manufacturing defects and poor quality control. At the end of the first stage, the defective components are eliminated. During the second stage, called the *useful life period*, the remaining components will have a fairly constant failure rate for some extended period of time. The cause of failure during this stage is not readily evident, and failures are assumed to be random. Finally, the

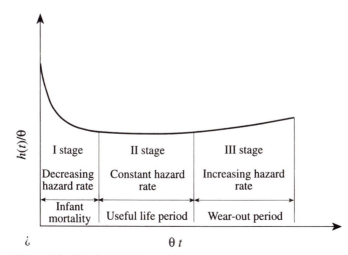

Figure 8.4 The bathtub hazard rate curve.

last stage is referred to as the *wear-out* stage. During this period of time, the rate of failure begins to increase because the component has already completed its useful operating life. In general, critical equipment and components are replaced before they enter the *wear-out* period.

Mechanical component failure rates follow the hazard rate curve for $b = 0.6$ in Figure 8.3. The hazard rate curve with $b = 0.1$ depicts computer software failure. It is clear from this curve that the wear-out period does not exist. Most software failure occurs in the first stage because of undetected errors in the computer program. A special case of hazard rate distribution is the extreme value distribution at $b = 1$, which has proved to be very useful in reliability engineering.[1] Often in a reliability study of a component, the second stage of the bathtub hazard rate curve is used.

8.4.2 Exponential Distribution

This is the most simple and widely used distribution in reliability engineering. The failure probability density function $f(t)$ of a component is given by

$$f(t) = \frac{1}{\theta}\exp\left(-\frac{t}{\theta}\right), \quad t > 0 \tag{8.27}$$

Let $\lambda = 1/\theta$, then

$$f(t) = \lambda e^{-\lambda t} \tag{8.28}$$

Substituting Eq. 8.28 into Eq. 8.10 and integrating yields the reliability function

$$R(t) = e^{-\lambda t} \tag{8.29}$$

Referring to Eq. 8.14, λ is the constant failure rate of a component. Substituting Eq. 8.29 into Eq. 8.23 and integrating leads a component's *mean time to failure* as

$$\text{MTTF} = \int_0^\infty e^{-\lambda t}\,dt = \frac{1}{\lambda} \tag{8.30}$$

—— *Example 8.3* ————————————————————————————————

Assume that the constant failure rate of an electronic plotter is 0.00055 failure/hour. Determine the following:

1.   The reliability of the plotter for 450 hours of operation.
2.   The mean time to failure (MTTF) of the plotter.

Solution

1.   Using Eq. 8.29 yields

$$R(450) = e^{-(0.00055)(450)} = e^{-0.2475} = 0.7808$$

[1]E. J. Gumbel, *Statistics of Extremes,* Columbia University Press, New York, 1958.

2. Using Eq. 8.30

$$\text{MTTF} = \frac{1}{\lambda} = \frac{1}{0.00055} = 1818.18 \text{ hr}$$

8.4.3 Weibull Distribution

Another life-predicting distribution is the Weibull distribution. The failure probability density function $f(t)$ of a component is given by

$$f(t) = \frac{bt^{b-1}}{\theta^b} \exp\left[-\left(\frac{t}{\theta}\right)^b\right] \quad \text{for} \quad b > 0, \, \theta > 0, t \geq 0 \tag{8.31}$$

where t is the time, b is the shape parameter, and θ is the scale parameter. Substituting Eq. 8.31 into Eq. 8.10 and integrating yields the component reliability function $R(t)$ as

$$R(t) = \exp\left[-\left(\frac{t}{\theta}\right)^b\right] \tag{8.32}$$

By utilizing Eq. 8.14 with Eq. 8.31 and Eq. 8.32, the hazard rate can be determined.

$$h(t) = \frac{bt^{b-1}}{\theta^b} \tag{8.33}$$

At $\theta = 1$, the curves of Eq. 8.33 for various values of b are shown in Figure 8.5. Note that when $b = 1$, the Weibull hazard function gives a constant failure rate and, when $b = 2$, it provides the Rayleigh hazard rate function, which is a linearly increasing hazard rate function. Substituting Eq. 8.32 into Eq. 8.23 and integrating yields the

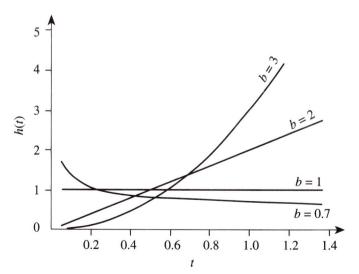

Figure 8.5 The Weibull hazard rate function for $\theta=1$.

MTTF of a component

$$\text{MTTF} = \theta\Gamma\left(\frac{1}{b} + 1\right)$$

(8.34)

where Γ is the gamma function given as

$$\Gamma(\beta) = \int_0^\infty t^{\beta-1}e^{-t}dt \quad \text{for} \quad \beta > 0$$

(8.35)

Values of Γ corresponding to β can be obtained from Figure 8.6.

── **Example 8.4** ───────────────────────────────────

Boiler failures that cause reduced or total loss of propulsion power on a group of ships during a period of two years fits well with the Weibull distribution of the following form[2]

$$R(t) = \exp\left[-\left(\frac{t}{1609.8}\right)^{0.8941}\right]$$

(8.36)

Calculate the following:

1. The reliability for 1000 hours of operation
2. The mean time to failure (MTTF)
3. The hazard rate

Solution

1. From Eq. 8.36, we obtain

$$R(t) = \exp\left[-\left(\frac{1000}{1609.8}\right)^{0.8941}\right]$$

$$R(t) = 0.52$$

2. Mean time to failure is

$$\text{MTTF} = \theta\Gamma\left(\frac{1}{b} + 1\right) = 1609.8\Gamma\left(\frac{1}{0.8941} + 1\right)$$

$$\text{MTTF} = 1609.8\Gamma(2.12) = 1609.8 \times 1.031 = 1660 \text{ h}$$

3. The hazard rate is

$$b(t) = \frac{bt^{b-1}}{\theta^b} = \frac{(0.8941)(1000)^{0.8941-1}}{(1609.8)^{0.8941}}$$

$$b(t) = 0.00058 \text{ failure}/\text{h}$$

Note that the value of $\Gamma(2.12) = 1.031$ can be obtained from Figure 8.6.

───

[2]C. Boe and O. J. Tveit, "Reliability Engineering and Safety at Sea," *IEEE Trans. on Reliability*, vol. R-23, no. 3, 1974.

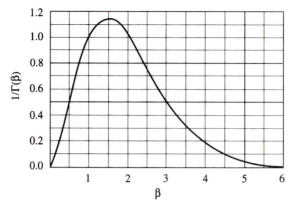

Figure 8.6   The Gamma function.

8.5   BASIC SYSTEM RELIABILITY

An important problem in system design is to make sure that a system will operate or will be available when it is required to function. This becomes an important design criterion, especially when the system involves many elements and a number of design constraints. Performance of a system with n elements depends on the performance of each and every component. The system reliability can be defined in terms of the reliability of the elements of the sytem. In this section the reliability of multicomponent systems that are connected in series, parallel, and in more complex configurations is discussed.

8.5.1   Series System

The block diagram of a k component series is shown in Figure 8.7. Each block represents an element or component. Let R_1 be the reliability of component 1, R_2 the reliability of component 2, and so on. The arrangement shown in Figure 8.7 implies that all of the components must work to ensure system success. Since the system fails even if one element fails, the reliability of the system R_s is given by the product of the component reliabilities

$$R_s = \prod_{k=1}^{n} R_k = R_1 R_2 R_3 \ldots R_n \tag{8.37}$$

where R_k denotes the reliability of component k. If the component reliabilities are assumed to be identical, Eq. 8.37 becomes

$$R_s = R^k \tag{8.38}$$

where R is the reliability of the component.

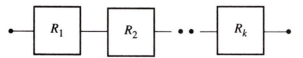

Figure 8.7   The component block diagram of a series system.

Figure 8.8   The Block diagram for radio station.

Example 8.5

Assume that a high school student who has a limited knowledge of electronics and radio decides to operate a small radio station called PUTV (pump up the volume). He buys used Army equipment to put his radio station together. Suppose that the system consists of three main components having reliabilities 0.63, 0.72, and 0.86. For proper system operation he was told that the three components must be connected in series, as shown in Figure 8.8. Determine the reliability of the radio station.

Solution

Using Eq. 8.37, the reliability of the radio station is

$$R_s = \prod_{k=1}^{n} R_k = R_1 R_2 R_3 = (0.63)(0.72)(0.86) = 0.39$$

8.5.2   Parallel Systems

When components of a system can be connected in parallel, the system will only fail if all of its components fail. The block diagram of a k-component parallel system is shown in Figure 8.9. The parallel system reliability R_p can be obtained by using

$$R_p = 1 - \prod_{k=1}^{n}(1 - R_k) \tag{8.39}$$

For components with the same reliability, Eq. 8.39 becomes

$$R_p = 1 - (1 - R_k)^k \tag{8.40}$$

For parallel systems, the system reliability increases as the number of components

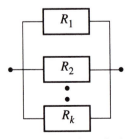

Figure 8.9   The Block diagram of a parallel system.

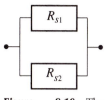

Figure   8.10 The Component   block diagram for a radio station.

increases. Admittedly, increasing the number of parallel components increases the initial cost, weight, and volume of the system as well as the maintenance requirements. Therefore, the number of components must be carefully optimized.

—— *Example 8.6* ———

Note that in Example 8.5, the student bought used equipment; hence, the expected system reliability is low. To increase the reliability of the system, the student decides to buy a complete backup system. Determine the increase in system reliability.

Solution

The student used a simple approach to interconnect the two systems as shown in Figure 8.10. Since this is a parallel configuration, the total system reliability is

$$R_p = 1 - (1 - R_k)^k = 1 - (1 - 0.39)^2 = 0.63$$

The increase in reliability ΔR is

$$\Delta R = 0.63 - 0.39 = 0.24$$

8.5.3   Multistage System with Parallel Redundancy

Figure 8.11 shows a multistage system with parallel redundancy. Many industrial processes, such as power plants, can be represented as multistage systems. As shown, the system consists of n stages in series where the components have identical reliabilities and are connected in parallel at each stage. The system

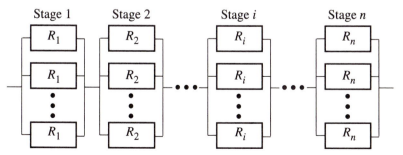

Figure 8.11   The block diagram of a multistage system with parallel redundancy.

reliability is the product of reliabilities of each stage. Using Eq. 8.37 and 8.40, the reliability of a multistage system with parallel redundancy is determined as

$$R_{sp} = \prod_{k=1}^{n} \left[1 - (1 - R_k)^{Y_k} \right]$$

(8.41)

where Y_k is the number of components in each stage, and R_k is the common reliability of the components.

___ *Example 8.7* _____

In Example 8.6, the student increased the reliability of the system. However, it needs to be increased further. The student has a limited budget and cannot affort to buy another system to increase the system reliability further. An electronics technician is contacted for other solutions to increase the reliability of the radio station. The technician recommends going to a multistage system with parallel redundancy, as shown in Figure 8.12. Determine the reliability of the system.

Solution

The reliability of the new system is

$$R_{sp} = \prod_{k=1}^{n} \left[1 - (1 - R_k)^{Y_k} \right]$$
$$= \left[1 - (1 - 0.63)^2 \right] \left[1 - (1 - 0.72)^2 \right] \left[1 - (1 - 0.86)^2 \right] = 0.78$$

The result shows that proper combination of the components can improve the system reliability.

_____

8.5.4   Parametric Method in System Reliability Evaluation

In 1972 Banerjee and Rajamani proposed a parametric method to evaluate the reliability of complex systems.[3] Calculation of complex system reliability by using the classical approach becomes very difficult and requires long computational time. However, by using the parametric method, the reliabilty of complex systems can be

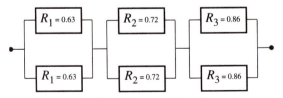

Figure 8.12   The block diagram for radio station.

[3]© 197-IEEE, S. K. Banerjee and K. Rajamani, "Parametric Representation of Probability in Two-Dimensions-A New Approach in System Reliability Evaluation," *IEEE Trans. on Reliability*, vol. R-21, pp. 56–60, 1972.

easily evaluated. This method introduces a new parameter, ϕ, defined by

$$\phi = \frac{1 - R}{R} \tag{8.42}$$

where R is the reliability of a component.

Series System   Substitution of Eq. 8.42 into Eq. 8.37 yields

$$(1 + \phi_s) = \prod_{k=1}^{n}(1 + \phi_k) \tag{8.43}$$

Consider a two element system, $n = 2$, where the system parameter is

$$\phi_s = \phi_1 + \phi_2 + \phi_1\phi_2 \tag{8.44}$$

Assuming $\phi \ll 1$, the higher order terms of Eq. 8.44 can be neglected, and the corresponding parameter for the series system ϕ_s can be approximated as

$$\phi_s = \sum_{k=1}^{n} \phi_k \tag{8.45}$$

Once the parameter ϕ_s is known, the system reliability is calculated by using Eq. 8.46:

$$R_S = \frac{1}{1 + \phi_s} \tag{8.46}$$

Parallel Systems   Substitution of Eq. 8.42 into Eq. 8.39 yields

$$\frac{\phi_p}{1 + \phi_p} = \prod_{k=1}^{n} \frac{\phi_k}{1 + \phi_k} \tag{8.47}$$

For small values of ϕ, the $(1 + \phi)$ term can be approximated as unity. Hence, the parameter, ϕ_p for a parallel system becomes

$$\phi_p = \prod_{k=1}^{n} \phi_k \tag{8.48}$$

Similarly, after calculating ϕ_p, Eq. 8.46 can be used to determine the system reliability.

Multistage System with Parallel Redundancy   Since the value of the parameter is the same for all components in a stage, Eq. 8.48 can be written as

$$\phi_p = \phi_k^{Y_k} \tag{8.49}$$

where Y_k is the number of components in k^{th} stage. The parameter for the series parallel system shown in Figure 8.11 can be determined by modifying Eq. 8.45.

$$\phi_{sp} = \sum_{k=1}^{n} \phi_k^{Y_k} \tag{8.50}$$

Complex systems reliability can be determined by using the parametric method with the simpler equations. Note that the parametric method assumes that $\phi \ll 1$; however, when R is less than 0.5, $\phi \approx 1$. Hence, the range of ϕ is an important criterion to consider in determining the error of calculations using the parametric method. The parametric method can be used to determine the system reliability for systems having components with high reliability.

▬ Example 8.8 ▬

Repeat Example 8.5 using the parametric approach.

Solution

Parameters, ϕ_1, ϕ_2, and ϕ_3, for the first, second, and third components, respectively, are

$$\phi_1 = \frac{1 - R_1}{R_1} = \frac{1 - 0.63}{0.63} = 0.587$$

$$\phi_2 = \frac{1 - R_2}{R_2} = \frac{1 - 0.72}{0.72} = 0.389$$

$$\phi_3 = \frac{1 - R_3}{R_3} = \frac{1 - 0.86}{0.86} = 0.163$$

The system parameter ϕ_s is

$$\phi_s = \phi_1 + \phi_2 + \phi_3 = 0.587 + 0.389 + 0.163 = 1.139$$

The system reliability can now be obtained:

$$R_s = \frac{1}{1 + \phi_s} = \frac{1}{1 + 1.139} = 0.47$$

The error between the method used in Example 8.5 and this example is $0.47 - 0.39 = 0.08$. If the reliabilty of system components are relatively high, this error will be negligible.

8.6 OPTIMIZATION OF SYSTEM RELIABILITY

As shown in Figure 8.11, the overall reliability of a series system can be increased by providing redundancy at each stage. As mentioned before, increasing the number of components at each stage increases the cost and weight of the system. To design a system with high performance requires compromises between reliability and cost or weight. The problem is to optimize the reliability with respect to cost or weight by a reasonable trade-off among the various components at each stage. In this section three cases of reliability optimization of a multistage parallel redundant system with different linear constraints, using a parametric approach, are discussed:[4]

1. Maximum reliability for a given cost constraint.
2. Minimum cost for a given reliability constraint.
3. Maximum reliability for given cost and weight constraints.

[4]© 197-IEEE, S. K. Banerjee and K. Rajamani, "Optimization of System Reliability Using a Parametric Approach," *IEEE Trans. on Reliability*, vol. R-22, pp. 35–39, 1973.

To optimize the redundant system mentioned above, Eq. 8.50 is used. This equation is simple in structure but requires the minimization of ϕ instead of the maximization of R, as in the classical method. Hence, use of the parametric method to optimize the reliability of a complex system provides considerable simplification and time reduction in computations.

8.6.1 Reliability Optimization for a Given Cost Constraint

Consider a system where the number of stages and component reliabilities are given. It is desirable that the total cost of the system not exceed a given cost limit C_L. Thus

$$C_L \geq \sum_{k=1}^{n} C_k Y_k \tag{8.51}$$

where C_k is the cost of one component. The constraint function can be written as

$$\psi = \sum_{k=1}^{n} C_k Y_k - C_L = 0 \tag{8.52}$$

Since reliability is subject to optimization, the objective function U is defined by

$$U = \sum_{k=1}^{n} \phi_k^{Y_k} \tag{8.53}$$

The Lagrange expression, $LE = U + \lambda\psi$, for cost optimization is

$$LE = \sum_{k=1}^{n} \phi_k^{Y_k} + \lambda\left(\sum_{k=1}^{n} C_k Y_k - C_L\right) \tag{8.54}$$

To optimize the number of components Y_i in the ith stage, as shown in Figure 8.11, differentiate the Lagrange equation with respect to Y_i:

$$\frac{\partial LE}{\partial Y_i} = \ln\phi_i \phi_i^{Y_i} + \lambda C_i = 0 \tag{8.55}$$

Since only one stage is considered, the summation in Eq. 8.55 is deleted. The above equation can be written in the form

$$Y_i = a_i \ln\lambda + b_i \tag{8.56}$$

where

$$a_i = \frac{1}{\ln\phi_i} \tag{8.57}$$

$$b_i = \frac{\ln K_i}{\ln\phi_i} \tag{8.58}$$

$$K_i = -\frac{C_i}{\ln\phi_i} \tag{8.59}$$

By using Eq. 8.51 and 8.56, the expression for the Lagrange multiplier λ can be obtained in the following form:

$$\lambda = e^s \tag{8.60}$$

where

$$s = \frac{C_L - \sum_{k=1}^{n} C_k b_k}{\sum_{k=1}^{n} C_k a_k} \tag{8.61}$$

To obtain a sequence of numbers that will optimize the reliability for the given cost limit, it is assumed that a sequence $\bar{n} = (n_1, n_2 \cdots, n_i, \cdots, n_n)$ is produced for the system reliability of $R(\bar{n})$ and cost $C(\bar{n})$. Let there be another series $\bar{m} = (m_1, m_2, m_3, \cdots, m_i, \cdots, m_n)$ where $R(\bar{m}) > R(\bar{n})$ and $C(\bar{m}) > C(\bar{n})$. The sequence $\bar{m}$ dominates sequence $\bar{n}$, provided that the cost constraint is satisfied. A sequence of undominated numbers, $\bar{n}$, can be obtained from the smallest n_i, which will satisfy the following inequality[5]

$$\frac{\phi_i^{Y_i}}{\left[1 + \phi_i\right]^{(1+Y_i)}} < \lambda C_i \tag{8.62}$$

By calculating λ from Eq. 8.60, the smallest numbers that will satisfy the above inequality can be obtained. By changing the value of λ in decrements, a different sequence that corresponds to each λ can be obtained. By choosing proper decrements for λ, optimum component numbers for each stage that satisfy the cost constraint can be obtained by using the iteration procedure shown in the flowchart of (Figure 8.13). As can be seen from the flowchart, the value of λ changes with the cost and the reliability of the element. The decrements of λ should be chosen for different sets of data. An interactive Fortran program is given in Appendix D that solves the above problem.

8.6.2 Cost Minimization for a Given Reliability Constraint

Consider the case where the specific system should achieve a certain reliability requirement while keeping the lowest possible price. In this situation, the objective function that should be minimized is

$$U = \sum_{k=1}^{n} C_k Y_k \tag{8.63}$$

The constraint equation that should be satisfied is

$$\phi_L \leq \sum_{k=1}^{n} \phi_k^{Y_k} \tag{8.64}$$

where the limiting value of ϕ_L is given by

$$\phi_L = \frac{1 - R_L}{R_L} \tag{8.65}$$

[5]R. E. Barlow and F. Proschan, *Mathematical Theory of Reliability,* John Wiley & Sons, Inc., 1965.

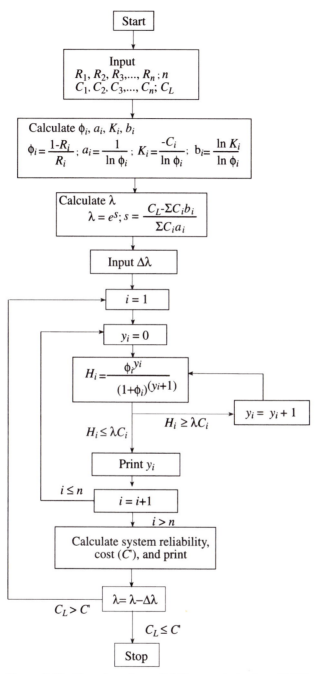

Figure 8.13 Flowchart for reliability maximization (©197-
IEEE, S.K. Banerjee and K. Rajamani, "Optimization of Reliability Using a Parametric Approach," *IEEE trans. on Reliability*,
vol. R-22, pp. 35-39, 1973).

where R_L is the reliability limit. The cost constraint function can be written as

$$\psi = \sum_{k=1}^{n} \phi_k^{Y_k} - \phi_L \tag{8.66}$$

The Lagrangian equation can be written as

$$LE = \sum_{k=1}^{n} C_k Y_k + \lambda \left(\sum_{k=1}^{n} \phi_k^{Y_k} - \phi_L \right) \tag{8.67}$$

To optimize the number of components Y_i in the ith stage, differentiate the Lagrangian equation with respect to Y_i

$$\frac{\partial LE}{\partial Y_i} = C_i + \lambda \ln \phi_i \phi_i^{Y_i} = 0 \tag{8.68}$$

for $i = 1$, Eq. 8.68 becomes

$$\lambda \ln \phi_1 \phi_1^{Y_1} = -C_1 \tag{8.69}$$

Eliminating λ from Eq. 8.68 and 8.69, the resulting equation is

$$\phi_i^{Y_i} = K_i \phi_1^{Y_1} \tag{8.70}$$

where

$$K_i = \frac{\ln \phi_1 C_i}{\ln \phi_i C_1} \tag{8.71}$$

From Eq. 8.69 and 8.70, λ can be calculated as

$$\lambda = -\frac{C_1}{S \ln \phi_1} \tag{8.72}$$

where

$$S = \frac{\phi_L}{K} \tag{8.73}$$

$$K = \sum_{k=1}^{n} K_k \tag{8.74}$$

From the theory of undominated modes, a sequence of numbers can be generated by finding the smallest number that will satisfy the following inequality (Banerjee and Rajamani, 1973)

$$\lambda \frac{\phi_i^{Y_i}}{(1 + \phi_i)^{(1+Y_i)}} < C_i \tag{8.75}$$

To determine the optimum cost for a given reliability constraint, an interactive Fortran program given in Appendix D is used. The algorithm of the program is shown in Figure 8.14.

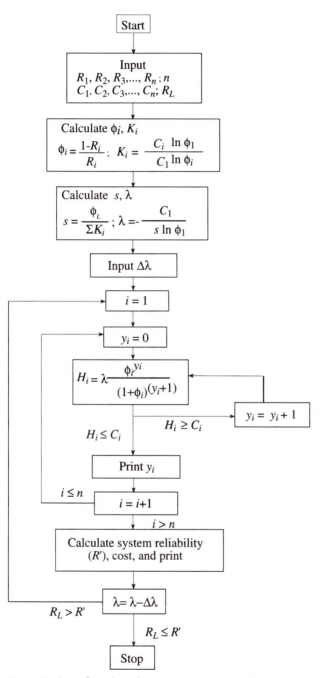

Figure 8.14 A flowchart for cost minimization (© 197-IEEE, S. K. Banerjee and K. Rajamani, "Optimization of System Reliability Using a Parametric Approach," *IEEE Trans. on Reliability,* vol. R-22, pp. 35–39, 1973.).

8.6.3   Optimization Reliability for a Given Cost and Weight Constraint

Consider a system similar to that of Section 8.6.2 except that the system is subjected to two linear constraints. The objective function that should be minimized is

$$U = \sum_{k=1}^{n} \phi_k^{Y_k} \tag{8.76}$$

The weight and the cost of the system are both linear functions of the number of elements and are used as constraint functions. The total weight and cost of the system should not exceed a given weight limit W_L and a cost limit C_L:

$$C_L \geq \sum_{k=1}^{n} C_k Y_k \tag{8.77}$$

$$W_L \geq \sum_{k=1}^{n} W_k Y_k \tag{8.78}$$

where W_k is the weight of a component at the kth stage. The constraint functions can be written as

$$\psi_1 = \sum_{k=1}^{n} C_k Y_k - C_L = 0 \tag{8.79}$$

$$\psi_2 = \sum_{k=1}^{n} W_k Y_k - W_L = 0 \tag{8.80}$$

Then the Lagrangian equation can be expressed as

$$LE = \sum_{k=1}^{n} C_k Y_k + \lambda \left(\sum_{k=1}^{n} \phi_k^{Y_k} - \phi_L \right) \tag{8.81}$$

$$LE = \sum_{k=1}^{n} \phi_k^{Y_k} + \lambda_1 \left(\sum_{k=1}^{n} C_k Y_k - C_L \right) + \lambda_2 \left(\sum_{k=1}^{n} W_k Y_k - W_L \right) \tag{8.82}$$

differentiating with respect to Y_i gives

$$\frac{\partial LE}{\partial Y_i} = \ln \phi_i \phi_i^{Y_i} + \lambda_1 C_i + \lambda_2 W_i = 0 \tag{8.83}$$

Rearranging Eq. 8.83 yields

$$Y_i = \frac{1}{\ln \phi_i} \left[\ln(a_i \lambda_1 + b_i \lambda_2) \right] \tag{8.84}$$

where

$$a_i = -\frac{C_i}{\ln \phi_i} \tag{8.85}$$

$$b_i = -\frac{W_i}{\ln \phi_i} \tag{8.86}$$

Differentiating the Lagrange equation with respect to λ_1 and λ_2, respectively,

$$\frac{\partial LE}{\partial \lambda_1} = \sum_{k=1}^{n} C_k Y_k - C_L = 0 \tag{8.87}$$

$$\frac{\partial LE}{\partial \lambda_2} = \sum_{k=1}^{n} W_k Y_k - W_L = 0 \tag{8.88}$$

Substituting Eq. 8.84 into Eq. 8.87, and Eq. 8.88,

$$C_L = -\sum_{k=1}^{n} a_k \left[\ln(a_k \lambda_1 + b_k \lambda_2) \right] \tag{8.89}$$

$$W_L = -\sum_{k=1}^{n} b_k \left[\ln(a_k \lambda_1 + b_k \lambda_2) \right] \tag{8.90}$$

An interactive Fortran program to solve the preceding problem is given in Appendix D. The algorithm of the program is shown in Figure 8.15. The nonlinear simultaneous equations are solved by the Newton–Raphson method. The above equations can be easily evaluated for systems with three or more linear constraints by simply introducing a λ_i for each constraint.

To determine λ_1 and λ_2, the above two nonlinear equations should be solved simultaneously. Approximate values of λ are sufficient to start the iterations. Again, a sequence of undominated modes can be generated from the smallest numbers that will satisfy the following inequality (Bannerjee and Rajamani, 1973)

$$\frac{\phi_i^{Y_i}}{\left[1 + \phi_i \right]^{(1+Y_i)}} < \lambda(g_1 C_i + g_2 W_i) \tag{8.91}$$

where
$$\lambda = \lambda_1 + \lambda_2 \tag{8.92}$$
$$g_1 = \frac{\lambda_1}{\lambda} \tag{8.93}$$
$$g_2 = \frac{\lambda_2}{\lambda} \tag{8.94}$$

—— *Example 8.9* ——————————————————————

The result of Example 8.7 shows that the reliability of the radio station was increased to 0.78. It is evident that this reliability is far from satisfactory to minimize maintenance problems. If the student has $40 to buy more equipment for his station, maximize the reliability for a three-stage redundant system. As shown in Table 8.1, it is assumed that the cost of components at stages 1, 2, and 3 is $2, $5, and $7, respectively. The reliability of components at the first, second, and third stages is 0.63, 0.72, and 0.86, respectively.

Solution

Referring to the flowchart in Figure 8.13, the calculations are as follows:

Step 1. Calculate ϕ_i, a_i, b_i, and K_i. The results are shown in Table 8.2.

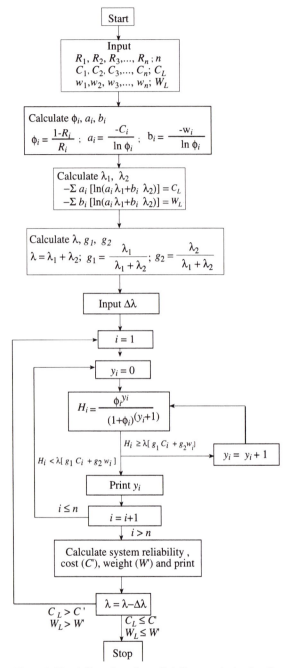

Figure 8.15   A flowchart for reliability maximization for multiple constraints (© 197-IEEE, S. K. Banerjee and K. Rajamani, "Optimization of System Reliability Using a Parametric Approach," *IEEE Trans. on Reliability,* vol. R-22, pp. 35–39, 1973.).

Table 8.1   Radio Station Model

Stage i	Cost C_i	Reliability R_i
1	2	0.63
2	5	0.72
3	7	0.86
Constraint	$C_L = 40$	

Step 2.   Calculate the summations:

$$\sum_{k=1}^{n} C_k b_k = -19.0022$$

$$\sum_{k=1}^{n} C_k a_k = -12.9080$$

and determine S for the cost limit $C_L = 40$,

$$S = \frac{C_L - \sum_{k=1}^{3} C_k b_k}{\sum_{k=1}^{3} C_k a_k} = \frac{40 - (-19.0022)}{-12.9080} = -4.57099$$

Then, λ is

$$\lambda = e^{S} = e^{-4.57099} = 1.034786 \times 10^{-2}$$

Step 3.   Assume that the decrement $\Delta \lambda = 0.001$ and start the iteration with $\lambda_1 = 1.034786 \times 10^{-2}$. The number of components for the first stage is determined as follows:

(a) For $i = 1$, the inequality equation can be written as

$$\frac{\phi_1^{Y_1}}{\left[1 + \phi_1\right]^{(1+Y_1)}} < \lambda_1 C_1 \tag{8.95}$$

Table 8.2   Results of Parameters

Stage i	ϕ_i	a_i	b_i	K_i
1	0.5873	−1.8789	−2.4874	3.7578
2	0.3888	−1.0588	−1.7646	5.2940
3	0.1628	−0.5508	−0.7435	3.8561

The right-hand side of (8.95) is

$$\lambda_1 C_1 = (1.034786 \times 10^{-2})(2) = 2.069572 \times 10^{-2}$$

For the inequality to be satisfied the left-hand side (LHS) of Eq. 8.95 must be less than the right-hand side (RHS). The LHS of the equation for the initial iteration $Y_1 = 0$ is

$$\frac{\phi_1^0}{(1 + \phi_1)^{(1+0)}} = \frac{1}{1 + \phi_1} = \frac{1}{1 + 0.5873} = 0.6300$$

Since this is larger than the LHS, increment Y_1 by 1 and recalculate the LHS:

$$\frac{\phi_1^1}{(1 + \phi_1)^{(1+1)}} = \frac{\phi_1^1}{(1 + \phi_1)^2} = \frac{0.5873}{(1 + 0.5873)^2} = 0.2331$$

When $Y_1 = 4$, the RHS is equal to 1.18807×10^{-2}, which satisfies the inequality equation. Thus, the number of components in the first stage is 4. Similarly, this same calculation can be performed for stages 2 and 3. Using $\lambda = 1.034786 \times 10^{-2}$, the number of components for stages 2 and 3 can be found to be 3 and 2, respectively.

(b) Calculate the system reliability

$$R_{sp} = \prod_{k=1}^{n} \left[1 - (-R_k)^{Y_k} \right]$$

$$R_{sp} = \left[1 - (1 - 0.63)^4 \right]\left[1 - (1 - 0.72)^3 \right]\left[1 - (1 - 0.86)^2 \right] = 0.9409$$

and the system cost C_s is

$$C_s = \sum_{k=1}^{n} C_k Y_k = (2 \times 4) + (5 \times 3) + (7 \times 2) = \$37$$

Since the calculated systems cost is less than the cost limit, the same iteration must be repeated for a new λ.

$$\lambda_2 = \lambda_1 - \Delta\lambda = 1.034786 \times 10^{-2} - 0.001 = 0.0093478$$

When the same calculation procedure is repeated with $\lambda_2 = 0.0093478$, the same solution is found as with $\lambda_1 = 1.034786 \times 10^{-2}$. If $\Delta\lambda$ is selected to be very small, repeating solutions may occur as in this example. It is safe to select $\Delta\lambda$ small in order not to miss the correct optimum solution. The only drawback is computation time. A good rule of thumb to select $\Delta\lambda = 0.2\lambda_1$. In this example, for $\lambda_2 = 0.0093478$, $\lambda_3 = 0.0083478$, $\lambda_4 = 0.0073478$, and $\lambda_5 = 0.0063478$, this same solutions are obtained. For $\lambda_6 = 0.0063478 - 0.001 = 0.0053478$, the RHS of the inequality is

$$\lambda_6 C_1 = (0.0053478)(2) = 0.0106956$$

Now, start the iteration with $Y_1 = 0$, and repeat the foregoing solution procedure. The correct optimum solution with the cost constraint $C_L = 40$ is then obtained.

The number of components is 5, 3, and 2 for the first, second, and third stages, respectively. The reliability of this system is

$$R_{sp} = \prod_{k=1}^{n} \left[1 - (1 - R_k)^{Y_k} \right]$$

$$= \left[1 - (1 - 0.63)^5 \right] \left[1 - (1 - 0.72)^3 \right] \left[1 - (1 - 0.86)^2 \right] = 0.9522$$

and the system cost C_s is

$$C_s = \sum_{k=1}^{n} C_k Y_k = (2 \times 5) + (5 \times 3) + (7 \times 2) = \$39$$

8.7 MAINTAINABILITY

During the design and development phase, emphasis is also given to the maintainability requirements of equipment. Maintainability may be defined as "the probability that a device will be restored to operational effectiveness within a given period of time when the maintenance action is performed in accordance with prescribed procedures."[6] Since no equipment can be designed with absolute reliability, it is important that it be repaired within a specified time constraint with optimum ease and economy. However, the designer must be careful not to improve maintainability by introducing maintenance requirements that reduce reliability. To achieve satisfactory maintainability, the following maintenance design guidelines should be considered by the designer:[7]

1. Minimize the number and complexity of maintenance tasks by using a simple design that includes optimum interchangeability and standardized equipment.
2. Permit rapid and positive identification of multifunctions, defective parts, or assembly.
3. Minimize the need for high degrees of personnel skill and training.
4. Minimize the special tools and test equipment necessary to perform maintenance tasks.
5. Provide optimum accessibility in all systems, equipment, and components so that they can be reached easily for maintenance.
6. Provide maximum safety for both equipment and personnel.
7. Minimize the required maintenance time to ensure availability of the system to satisfy operational demands.
8. Maintain the lowest life-cycle cost possible.

[6]A. S. Goldman and T. B. Slattery, *Maintainability, A Major Element in Systems Effectiveness,* John Wiley & Sons, Inc., New York, 1964.

[7]V. J. Taylor, "Weapon System Design and Supportability, A Function of Failure Prediction," *IRE Trans. Reliability and Quality Control,* RQC-11, pp. 13–17, 1962.

8.7.1 Maintenance

Maintenance tasks can be classified into two main categories:

1. *Preventive maintenance* is performed periodically to keep equipment in a satisfactory operational condition. The objective of preventive maintenance is to keep the equipment failure rates from increasing above the design levels. This can be achieved by checking all equipment on a scheduled periodic basis. For example, consider a diesel engine system of a cargo ship. All the components, such as oil pumps, cooling water pumps, injectors, valves, and the like, included in the operating system, should be routinely checked in accordance with the maintenance schedule. Thus, maintenace, such as inspection, adjustments, cleaning, calibration, and replacing components are performed. If preventive maintenance is not performed in accordance with the schedule, the company may have to bear the cost of an unexpected shutdown of the main engine due to component failure. Preventive maintenance is especially important in this circumstance because if the main engine of a ship is not operational at sea during bad weather conditions, neither the safety of the ship nor the crew can be assured.

2. *Corrective maintenance* is concerned with restoring equipment as soon as possible to an acceptable operating condition after a failure has occurred. This maintenance can be achieved by replacing, repairing, adjusting, or calibrating the failure source that has caused interruption or breakdown of the system.

It is impossible to design equipment with absolute reliability so that failures are not incurred. In almost every system, components are designed for limited life and thus the consideration of corrective maintenance is essential for any program. However, if equipment is unusually costly and difficult to repair, it may complete its entire life cycle without maintenance (e.g., satellites, etc.).

The maintainability function $M(t)$ for any given distribution can be expressed as

$$M(t) = \int_0^t f(t)\,dt \tag{8.96}$$

where $f(t)$ is the repair time probability density function and t is the allowable repair time constraint. Assuming the exponential probability density function, and substituting in Eq. 8.96, the corrective maintainability function $M(t)$ can be found:

$$M(t) = 1 - \exp(-\mu t) \tag{8.97}$$

where μ is the constant repair rate of a component given as

$$\mu = \frac{1}{\text{MTTR}} \tag{8.98}$$

where MTTR is the mean time to repair a particular component after a malfunction has occurred.

Example 8.10

Suppose that an electronic component is designed for the space shuttle. Assume that during launch, a maximum 0.65-hour delay can be allowed for repair in the event of component failure. If a contractor's design reveals that the component constant repair rate μ is 1.4 repairs/hour, what is the probability of achieving maintenance success (adapted from reference [1])?

Solution

For a repair time constraint of 0.65 hour, the probability of maintenance success is

$$M(t) = 1 - \exp(-\mu t) = 1 - \exp\left[-(0.65)(1.4)\right] = 0.5974 \qquad (8.99)$$

and the MTTR becomes

$$\text{MTTR} = \frac{1}{\mu} = \frac{1}{1.4} = 0.714 \qquad (8.100)$$

The result shows that the contractor's design presents only a 59.7 percent probability of meeting the specified time constraint for repair of a random failure.

With a design change, if the contractor repair rate is increased to 2.8 repairs/hour the MTTR becomes

$$\text{MTTR} = \frac{1}{\mu} = \frac{1}{2.8} = 0.357 \qquad (8.101)$$

The new maintenance success probability within the time constraint of 0.65 hour is

$$M(t) = 1 - \exp(-\mu t) = 1 - \exp\left[-(0.65)(2.8)\right] = 0.8379 \qquad (8.102)$$

The contractor is now able to offer an 83.8 percent probability of maintenance success. For any repair time constraint, the probability of achieving maintenance success depends on assuring a low MTTR. To the customer, a high $M(t)$ represents a greater probability of savings in maintenance expense and a substantial step toward establishing system effectiveness.

8.8 AVAILABILITY

Availability is a measure of the impingement of failures on the operation of a system and thus indicates how often a system will be operational (available) when needed. Availability is an important design parameter that primarily depends on reliability, the effects of failures, and maintainability. In system design, it is essential to ensure that the system can be maintained easily and inexpensively.

Availability can be stated as the probability that a particular component will be capable of operation for a predicted time duration t. In reliability and main-tainability studies the terms steady-state availabilitiy and instantaneous availabilty are commonly used. Steady-state availability A is defined as the proportion of total operation time that a component is available for use. That is,

$$A = \frac{\text{operating time}}{\text{operating time} + \text{down time}}$$ (8.103)

An alternative form of the availability equation is

$$A = \frac{\text{MTBF}}{\text{MTBF} + \text{MTTR}}$$ (8.104)

where *mean time between failures* (MTBF) is the mean number of failures in a given time for a repairable system, device, or component. For example, if the total operating time of a component is 12,000 hours and 8 failures occur, the mean time between failures is

$$\text{MTBF} = \frac{12000}{8} = 1500$$

If the constant failure rate, $\lambda = (\text{MTBF})^{-1}$ and the constant repair rate $\mu = (\text{MTTR})^{-1}$ are given, the availability equation can be rearranged as

$$A = \frac{\mu}{\lambda + \mu}$$ (8.105)

The time-dependent instantaneous availability is defined as the probability that at time t the system will be available. A single-system instantaneous availability $A(t)$ at a given time t is given by

$$A(t) = \frac{\mu}{\lambda + \mu} + \frac{\lambda}{\lambda + \mu}\exp\left[-(\lambda + \mu)t\right]$$ (8.106)

Note that, as t goes to infinity, Eq. 8.106 approaches the stationary steady-state availablitiy. Equation 8.106 reveals that, as the repair rate μ increases, the probability of maintenance success increases and availability increases.

The product law can be used with the individual component availabilities to calculate the steady-state system availability for repairable component installation in series as

$$A_{ss} = \prod_{k=1}^{n} A_k$$

Similarly, steady-state availablity for a repairable parallel component installation is given by

$$A_{ps} = 1 - \prod_{k=1}^{n} \left(1 - A_k\right)$$

_____ **Example 8.11** _____

If a diesel generator has constant failure and repair rates of 0.002 failure/hour and 0.015 repair/hour, respectively, calculate the steady-state availability of the diesel generator.

Solution

For $\mu = 0.015$ and $\lambda = 0.002$, the availability is

$$A = \frac{\mu}{\lambda + \mu} = \frac{0.015}{0.002 + 0.015} = 0.882$$

In this example, if the repaired diesel generator is as good as new, compute the availability of the diesel generator at time $t = 65$ hours. The time-dependent instantaneous availability is

$$\begin{aligned}
A(t = 65) &= \frac{\mu}{\lambda + \mu} + \frac{\lambda}{\lambda + \mu} \exp\left[-(\lambda + \mu)t\right] \\
&= \frac{0.015}{0.002 + 0.015} + \frac{0.002}{0.002 + 0.015} \exp\left[-(0.002 + 0.015)(65)\right] \\
&= 0.921
\end{aligned}$$

In other words, 92.1 percent of the time, the diesel generator will be available for operation after 65 hours of usage.

There is an intimate relation between cost and availability that affects design performance. Figure 8.16 shows how the design performance changes with respect to constant cost profiles and availability.[8] As can be seen from this figure, there are numerous design solutions for different levels of performance. For example, for constant availability, there are three possible solutions varying from low cost with low performance to high cost with high performance. Cost profile 2 may be preferred because it gives fair performance with reasonable cost. The decision maker should be aware of the trade-off between performance and availability in selecting systems between A and B on curve 2. If the choice is a system with high

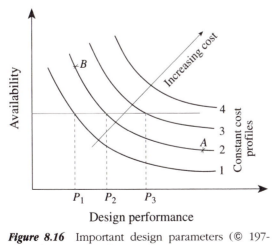

Figure 8.16   Important design parameters (© 197-IEEE, from H. David).

---

[8]© 197-IEEE, H. David, "Availability-Concept and Definations," *Proceedings, Annual Reliability and Maintainability*, pp.482–490, 1976.

availability, the system will have low performance. It is clear from the figure that a system with high design performance and excellent availability is expensive. In some instances it is very difficult to make a decision on which situation is best. Before the designer makes a choice, he or she has to study the information and the resulting alternative solutions from the trade-off analysis.

8.9   DEPENDABILITY

Dependability of a component is another important design parameter that provides a single measure of system condition(s), combining reliability and maintainability. Dependability can be defined as the probability that a component either does not fail or fails and will be repaired within an allowable time interval t. The concept of dependability is significant because it permits cost trade-offs between reliability and maintainablity. Dependability can be determined by using Eq. 8.107[9]

$$D(t) = 1 - \left(\frac{1}{d-1}\right)(e^{-\lambda t})\left[1 - e^{-(d-1)\lambda t}\right] \tag{8.107}$$

where d is the component dependability ratio given as

$$d = \frac{\mu}{\lambda} \tag{8.108}$$

Equation 8.105 can be modified to establish the relationship between availability and dependability as

$$A = \frac{\mu}{\mu + \lambda} = \frac{\mu/\lambda}{1 + \mu/\lambda} = \frac{d}{1 + d} \tag{8.109}$$

Component availability versus the dependability ratio is shown in Figure 8.17. As shown in the figure, the component dependability ratio d increases rapidly above the availability level of 0.9 and decreases rapidly for values of availability less than 0.1. These extremes define the regions of maximum sensitivity to change of the dependability ratio d.

For a given value of d, the minimum value of $D(t)$ in relation to λt can be obtained by taking the derivative of Eq. 8.107 with respect to λt

$$\frac{\partial[D(t)]}{\partial(\lambda t)} = 0 = \left(\frac{e^{-\lambda t} - e^{-d\lambda t}}{d - 1} - e^{-d\lambda t}\right) \tag{8.110}$$

Solving for λt yields

$$\lambda t = \frac{\ln d}{d - 1} \tag{8.111}$$

Substituting λt in Eq. 8.107 the minimum obtainable $D(t)$ is

$$D(t)_{\min} = 1 - \left(\frac{1}{d - 1}\right)\left(e^{-\ln d/d - 1} - e^{-d\ln d/d - 1}\right) \tag{8.112}$$

[9]© 197-IEEE, J. G. Wohl, "System Operational Readiness and Equipment Dependability," *IEEE Trans. Reliability*, vol. R-15, no. 1, pp. 1–6, 1966.

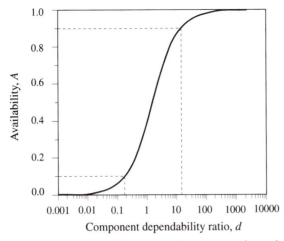

Figure 8.17 Component availability versus dependability ratio (© 197-IEEE, from J. G. Wohl).

Figure 8.18 shows the component dependability $D(t)$ versus λt for different values of the component dependability ratio d. As shown in the figure, dependability $D(t)$ increases with increasing dependability ratio d. This figure also shows that minimum dependability, $D(t)_{min}$ increases with increasing d. The concepts of dependability and dependability ratio are important design parameters. However, it is important to remember that reliability, maintainability, availability, and dependability are very closely interrelated and their effects on design performance and effectiveness should be considered as a whole. For more information on dependability, refer to references [2] and [3].

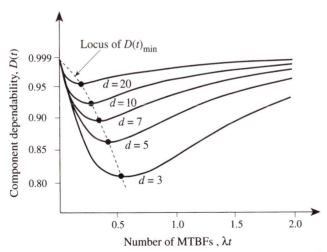

Figure 8.18 Component dependability $D(t)$ versus dependability ratio d (©197-IEEE, adapted from J. G. Whol).

8.10   FAULT TREE ANALYSIS

Fault tree analysis (FTA) is a commonly used technique for describing system reliability and safety. FTA gives a quantitative interpretation of failure consequences and provides an objective basis for the following:

1. Analyzing system design.
2. Justifying system changes.
3. Performing trade-off studies.
4. Analyzing common mode failures.
5. Demonstrating compliance.

The following basic steps are generally present in a fault tree analysis:

1. System definition.
2. Construction of the fault tree.
3. Qualitative evaluation of the fault tree.
4. Quantitative evaluation of the fault tree.
5. Recommendation for corrective action.

The construction of a fault tree starts with the identification of an undesirable event (system failure) called the *top event*. The *top event* is linked to basic failures, called *primary events*, by logic gates and event statements. These event statements and gates are represented by symbols, some of which are given in Table 8.3.

Table 8.3   Fault Tree Symbols

Symbols	Definitions
	Resultant event: represents the fault event above the gates, which is a result of the combination of other fault events.
	Basic fault event: represents a fault where the failure probability can be driven from empirical data. It is the limit of resolution of the fault tree.
	House event: revents a basic event that is expected to occur during the system operation.
	Transfer symbol: transfer-in and transfer-out triangles used to transfer fault tree from one part to another.
	AND gate: output fault event occurs only when all the input fault events occur simultaneously.
	OR gate: output fault event occur when one or more input fault events occur.

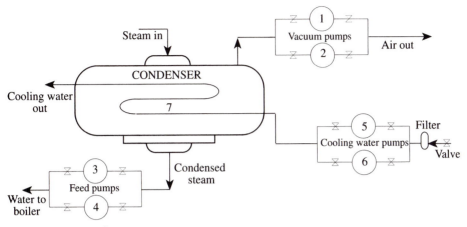

Figure 8.19   A condenser system.

The following example demonstrates fault tree construction by using some of the symbols included in Table 8.3.

As shown in Figure 8.19, a condenser is used to convert the exhaust steam of an engine into water so that it can be reused in the boiler. It also creates a partial vacuum at the engine exhaust through the use of a vacuum pump and thereby increases the efficiency of the engine. To condense the steam, cold water is circulated through the condensing tubes by the use of a circulating pump. Finally, condensed exhaust steam is pumped by a feed-water pump to a boiler.

The system, as shown in the figure, is composed of series and parallel subsystems. As the first step, a block diagram can be constructed by breaking the system down into series and parallel subsystem configurations, as shown in Figure 8.20. It is now relatively easy to develop a fault tree network representing the condenser system.

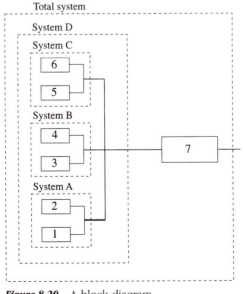

Figure 8.20   A block diagram.

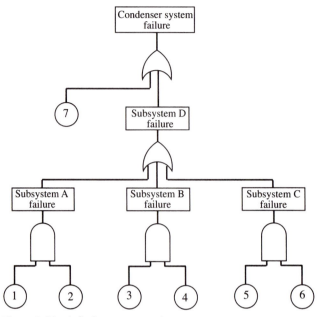

Figure 8.21 A fault tree network.

Several conditions can cause the failure of a condenser system. The most important one is the loss of the vacuum in the condenser. Other problems include the failure of pumps, dirty filters, leaking gaskets, and high boiler pressure. Consider the fault tree network for the condenser system shown in Figure 8.21. Each block in the figure represents an independent mechanical component and is assigned a number. The *top event* is the condenser system failure that may result from several prior faults (events) represented by the numbered circles. An *Or* gate is used because a failure may occur when any one of two events, 7, or system D failure, occur. Note that three *and* gates are used because failure occurs only if pumps 1 and 2, 3 and 4, or 5 and 6 fail simultaneously.

Although the concepts of fault tree network development appear to be simple, fault tree analysis for complex systems requires a thorough understanding of the system being analyzed and the techniques of fault tree development.

8.11   PROBABILISTIC DESIGN

The performance of a mechanical component depends on the variability of its internal parameters. These parameters include material properties, tolerances, and environmental and time-dependent effects, such as external loads. Figure 8.22 shows the relationship between a performance characteristic of a mechanical component and a parameter P.[10] For example, the life of an airplane depends on the external loads to which the airplane will be subjected during its useful lifetime. In this example, the parameter is the external load, which depends on certain uncontrollable factors such as weather conditions, flying habits of pilots,

[10]R. E. Mesloh and D. G. Mark, "Variability Analysis of Precision Mechanical Devices," *SAE Third Annual Aerospace Reliability and Maintainability Conf.*, pp. 576–588, 1964.

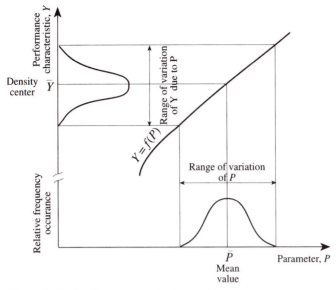

Figure 8.22 Performance variations with respect to parameter P (reprinted with permission from SAE Technical paper 640617, © 1964, Society of Automotive Engineers, Inc.).

and the like. It is clear from Figure 8.22 that as the parameter values change the performance characteristics change in response. Note that at the design center of Y, the parameter has its mean value. Frequency distribution of the parameter indicates the variability that can be expected in P during the life of the component. The variability in the performance characteristic can be obtained by simply projecting the frequency distribution of P up to the curve $Y = f(P)$ and over to the Y axis as shown in the figure. The performance distribution depends on the parameter distribution and will have the same shape as the parameter distribution. The relationship between the performance and the parameter provides important and extremely useful information to the engineer in finalizing a design.

8.11.1 Safety Factor

A simple definition for safety factor is the ratio of strength to stress due to external loadings

$$n = \frac{S}{\sigma} \tag{8.113}$$

or the ratio of yield strength to actual working stress

$$n = \frac{S_y}{\sigma_{\text{working}}} \tag{8.114}$$

However, uncertainties will cause a variation of strength ΔS and of stress $\Delta \sigma$; therefore, the lowest possible strength can be written as[11]

$$S_{\text{min}} = S - \Delta S \tag{8.115}$$

[11]K. Dimitri and C, David, "Designing a specified Reliability Directly into a Component," *SAE Third Annual Aerospace Reliability and Maintainability Conf.*, paper 640617, pp. 546–565, 1964.

and the highest possible stress is

$$\sigma_{max} = \sigma + \Delta\sigma \tag{8.116}$$

For no failure the following inequality must be satisfied

$$S_{min} \geq \sigma_{max} \tag{8.117}$$

or

$$S\left[1 - \frac{\Delta S}{S}\right] \geq \sigma\left[1 + \frac{\Delta\sigma}{\sigma}\right] \tag{8.118}$$

For conservative design, the minimum safety factor can be obtained from the above equation as

$$n = \frac{S}{\sigma} = \frac{1 + \frac{\Delta\sigma}{\sigma}}{1 - \frac{\Delta S}{S}} \tag{8.119}$$

For example, if the maximum variations of stress and strength are 20 percent and 25 percent, respectively, the minimum safety factor is

$$n = \frac{1 + 0.20}{1 - 0.25} = 1.6 \tag{8.120}$$

8.11.2   Interference Model

The failure probability of a component can be determined by plotting the stress and strength probability density curves as shown in Figure 8.23. Whenever there is an overlap between two distributions and a mean difference in probability less than zero results, a component failure is predicted. The mean difference μ_D is given by

$$\mu_D = \mu_S - \mu_\sigma \tag{8.121}$$

where μ_S and μ_σ are the mean strength and mean stress, respectively. The difference

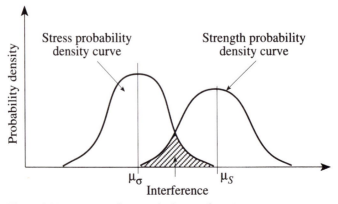

Figure 8.23   Stress and strength density functions.

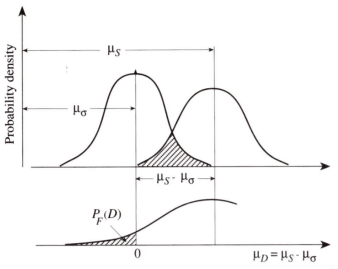

Figure 8.24   Difference diagram for normal distribution (reprinted with permission from SAE Technical paper 640616, © 1964, Society of Automotive Engineers, Inc.).

standard deviation $\hat{\sigma}_D$ can be expressed as

$$\hat{\sigma}_D = \sqrt{\hat{\sigma}_S^2 + \hat{\sigma}_\sigma^2} \qquad (8.122)$$

where $\hat{\sigma}_S$ and $\hat{\sigma}_\sigma$ are the standard deviations for the strength and stress, respectively.

When the stress and strength are represented by normal distributions, the probability of failure is the probability of having a negative difference between the stress and strength. This probablity can be shown to be the negative tail end of a difference distribution, as shown in Figure 8.24. The probability of failure $P_F(D)$ is equal to the area of the negative tail and is given as[12]

$$P_F(D) = \int_{-\infty}^{0} f_F(D)d\mu_D \qquad (8.123)$$

where $f_F(D)$ is the probability density of the difference. For a normal distribution, the negative tail area can be determined by using

$$z = \frac{\mu_D}{\hat{\sigma}_D} \qquad (8.124)$$

After calculating the value of z, the probability of failure (area) is obtained from Table C.1 (Appendix C). Once the probability of failure is known the component reliability can be determined using Eq. 8.125:

$$R = 1 - P_F(D) \qquad (8.125)$$

—— ***Example 8.12*** ——

The data collected from an airplane engine structural member are the following (this example has been reprinted with permission from SAE Technical paper 640579, Society of Automotive Engineers, Inc. [6]):

[12]M. J. Bratt, G. Reethof, and G. W. Weber, "A Model for Time Varying and Interfering Stress-Strength Probability Density Distributions with Consideration for Failure Incidence and Property Degradation," *SAE Third Annual Aerospace Reliability and Maintainability Conf.* paper 640616, pp. 566–575, 1964.

1. The mean estimated stress at flight conditions as measured by simulated conditions in the test cell and flight test are

    (a) mean thermal stress, $\mu_{\sigma t}$ = +35,500 psi

    (b) mean maneuver stress, $\mu_{\sigma m}$ = +1,750 psi

    (c) mean residual stress, $\mu_{\sigma r}$ = −11,200 psi

2. The standard deviation determined from test data, manufacturing tolerances and technical requirements are

    (a) $\hat{\sigma}_{thermal}$ = +2,650 psi

    (b) $\hat{\sigma}_{maneuver}$ = +437 psi

    (c) $\hat{\sigma}_{residual}$ = 667 psi

3. The mean material strength (from material data) is μ_S = 38,500 psi.

4. The standard deviation of the strength (from material data) is $\hat{\sigma}_S$ = 3,500 psi.

Using the above given data, calculate the probability of failure of the engine structural member.

Solution

The net mean stress μ_σ in the structural member is

$$\mu_\sigma = 35,500 + 1,750 - 11,200 = 26,050 \text{ psi}$$

The mean difference is

$$\mu_D = \mu_S - \mu_\sigma = 38,500 - 26,050 = 12,450 \text{ psi}$$

The combined standard deviation of the average stress is

$$\hat{\sigma}_\sigma = \sqrt{2,650^2 + 437^2 + 667^2} = 2,767 \text{ psi}$$

The difference standard deviation is

$$\hat{\sigma}_D = \sqrt{2,767^2 + 3,500^2} = 4,462 \text{ psi}$$

Then the z value is

$$z = \frac{\mu_D}{\hat{\sigma}_D} = \frac{12,450}{4,462} = 2.79 \text{ psi}$$

From Table C.1 of Appendix C, the failure probability $P_F(D)$ = 0.0026 can be obtained. Reliability of the structure can be calculated using Eq. 8.125:

$$R = 1 - 0.0026 = 0.9974$$

8.12   WORST CASE DESIGN

As discussed previously, because of both external loads and material property variabilities the choice of a meaningful safety factor to increase the reliability of a design can be complicated. To overcome this difficulty, one may chose an unrealistically high safety factor to exceed the desirable confidence level. However, such a decision will result in expensive and overdesigned components and the presence of excessive material that may actually weaken the component and cause poor performance.

Parameters that control the design performance are related in a manner that causes variations in one to cancel or increase the effect of variations in the other. Manufacturing tolerances or material property variations are examples. The poorest combination of parameters results in the most reliable component design, called the *worst case design*.

Based on the theory of variances, the worst case design can be defined as

$$d\phi = \left(\frac{\partial\phi}{\partial x}\right)dx + \left(\frac{\partial\phi}{\partial y}\right)dy \tag{8.126}$$

where ϕ is the function of two independent variables x and y. The following example illustrates the use of Eq. 8.126:

── *Example 8.13* ──────────────────────────

The maximum axial stress σ in a pressure vessel (shown in Figure 8.25) due to bending can be approximated as[13]

$$\sigma = \mp\frac{138KQL}{R^2t} \tag{8.127}$$

where
K = constant which varies with the ratio A/L
L = pressure vessel length, ft
Q = load on the supports, lb
R = radius of the pressure vessel, ft
t = thickness of the pressure vessel, in.

If the allowable stress is 10,000 psi, A/L = 0.1, L = 40 ft ($\mp$0.2 ft), and R = 4 ft ($\mp$0.1 ft), determine the thickness of the pressure vessel for the worst-case design.

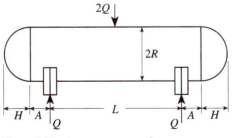

Figure 8.25   A pressure vessel.

[13]Permission from ASME, L. P. Zick, "Stresses in Large Horizontal Cylinder Pressure Vessels on Two Saddle Supports," *ASME Pressure Vessel and Piping Design and Analysis*, vol. 2, pp. 959–970, 1972.

Table 8.4   Cases for Thickness Calculations

Case	dL	dR	dt	$t(0.0207 \mp dt)$
1	0.2	0.1	−0.000931	0.01977
2	−0.2	0.1	−0.001139	0.01956
3	0.2	−0.1	0.001139	0.02184
4	−0.2	−0.1	0.000931	0.02163

Solution

If A/L = 0.1, K can be approximated as 0.6.[13] To simplify Eq. 8.127 further, assume Q = 1 lb, then the thickness is

$$t = \frac{138KL}{\sigma R^2} = \frac{(138)(0.6)(L)}{(10,000)(R^2)} = \frac{0.00828L}{R^2} \tag{8.128}$$

To calculate the nominal thickness, substitute L = 40 and R = 4 ft

$$t = \frac{(0.00828)(40)}{4^2} = 0.0207 \text{in.} \tag{8.129}$$

As can be seen from Eq. 8.126, thickness corresponds to the function ϕ and length and radius are the parameters under consideration. Taking the partial derivative of Eq. 8.128 with respect to L yields

$$\frac{\partial t}{\partial L} = \frac{0.00828}{R^2} = \frac{0.00828}{4^2} = 0.00052$$

and the partial derivative with respect to R is

$$\frac{\partial t}{\partial R} = \frac{-0.01656L}{R^3} = \frac{(-0.0248)(40)}{4^3} = -0.01035$$

The total variance equation is

$$dt = \left(\frac{\partial t}{\partial L}\right) dL + \left(\frac{\partial t}{\partial R}\right) dR$$

$$= (0.00052)(\mp 0.2) + (-0.01035)(\mp 0.1)$$

The possible thickness calculations are shown in Table 8.4. The results show that case 3 yields the greatest wall thickness. Hence, t = 0.02184 in. would be the thickness for the worst case design (adapted from reference [5] pp. 101–102).

BIBLIOGRAPHY

1. BARBER, C. F., "Expanding Maintainability Concept and Techniques," *IEEE Trans. Reliabilty*, vol. R-16, no. 1, pp. 5–9, 1967.

2. WOHL, J. G., "System Operational Readiness and Equipment Dependability," *IEEE Trans. Reliability*, vol. R-15, no. 1, pp.1–6, 1966.

3. PETERSON, E. L., "Maintainability Application to System Effectiveness Quantification," *IEEE Trans. Reliability*, vol. R-20, no. 1, pp. 3–7, 1971.

4. WILLIAM, J. M. F., "Bayes' Equation, Reliability, and Multiple Hypothesis Testing," *IEEE Trans. Reliability*, vol. R-21, no. 3, pp.136–139, 1972.

5. KIVENSON, G., *Durability and Reliability in Engineering Design,* Hayden Book Co., New York, 1971.

6. FREBERG, D. D. and SPECTOR, R. B., "Reliability Analysis and Prediction for Turbojet Engines–Results versus Needs," *SAE Third Annual Aerospace Reliability and Maintainability Conf.*, Paper 640579, pp. 253–262, 1964.

7. DHILLON, B. S., *Quality Control, Reliability, and Engineering Design.* Marcel Dekker, New York, 1985.

8. LEWIS, E. E., *Introduction to Reliability Engineering.* Wiley, Sons, New York, 1987.

9. HENLY, J. E., *Reliability Engineering and Risk Assessment.* Prentice-Hall, Englewood Cliffs, NJ, 1981.

10. DHILLON, B. S., *Mechanical Reliability: Theory, Models and Applications.* AIAA Education Series, SW, Washington, D. C., 1988.

11. BILLINTON, R. and ALLAN, N. R., *Reliability Evaluation of Engineering Systems.* Plenum Press, New York, 1983.

Chapter 9

Safety and Environmental Protection

Only within the moment of time represented by the present century has one species—man—acquired significant power to alter the nature of his world. . . . and the early mornings are strangely silent where once they were filled with the beauty of bird song. **Rachel Carson**

9.1 OUR CONTAMINATED ENVIRONMENT

The condition of the environment and what can be done to protect it in the future ranks high among the concerns of Americans as the twenty-first century approaches. Rachel Carson's *Silent Spring*, from which the above quote was taken, awakened America's environmental conscience when it was published in 1962. Judging from the continued degradation in the environment that has occurred during the intervening years, awakening to the existence of the problem and being concerned is not enough; there must be a widespread acceptance of the facts that correction of ecological abuse in the past and protection of the environment for the future are of the highest priority and could quite possibly determine the future nature of man's existence on this planet. There must be a willingness to make hard choices and to accept the changes in life-styles that will inevitably result.

The seriousness of the environmental problem was dramatized by events that occurred in the 1980s. The Chernobyl nuclear accident in Soviet Russia emphasized the importance of proper preplanning and worker training in protecting the environment (and surrounding population) from contamination by nuclear radiation as well as the significance of atmospheric conditions in making an event that occurred in a remote part of the world (the Ukraine) into a problem affecting all of Europe. The U.S. Department of Energy (DOE) decided to end its practice of paying environmental fines and other penalties incurred by contractors running nuclear weapons plants. Correcting environmental problems plaguing these plants would reportedly

329

require billions of dollars.[1] Long Island's infamous floating barge of garbage focused attention on the dwindling number of landfill areas in the United States as it cruised the seas for weeks trying to find a place to unload its unwanted cargo. Despite the $11.2 billion that has been spent to clean up hazardous waste, only 14 of the 802 national priority list sites were considered decontaminated as the 1980s ended; another 22,000 sites suspected to be contaminated were waiting to be placed on the priority list. Discarded syringes, vials of contaminated blood, and other medical debris that washed in with the tide closed beaches along the eastern seaboard and brought the following assessment from New York City's health commissioner, "The planet is telling us we can't treat it this way any more."[2]

Drought in the U.S. farmbelt during the late 1980s was blamed by many on climatic changes brought on by a worldwide warming trend attributed to the *greenhouse effect*, a buildup in the heat-trapping shroud of gasses that surrounds the earth, resulting from the cumulative effect of burning fossil fuels coupled with reduced terrestrial carbon dioxide-absorbing capability. During the 1980s researchers detected depletion of the ozone layer over both of the earth's poles. The ozone layer is the veil of gas above the earth that prevents most of the sun's harmful ultraviolet rays from reaching the earth's surface. Scientists predict that the increased amount of ultraviolet radiation reaching the earth due to ozone layer depletion will cause an increase in the incidence of human skin cancer and increased eye damage.[3] Acid rain, which includes both sulfuric and nitric acids, caused significant environmental problems in the northeastern United States and Canada during this period. Sulfuric acid rain is a result of the combustion of high-sulfur coal in utility plants whereas nitric acid rain is caused by both stationary and motor vehicle fuel combustion sources. The sulfuric and nitric oxides produced by the combustion processes combine with moisture in the air and return to earth in the form of precipitation to contaminate lakes, streams, and forests in locations remote from the original source of the combustion process itself. In 1987, 13 percent of all U. S. communities violated Environmental Protection Agency (EPA) water quality standards, and it was estimated that one quarter of America's lakes and rivers were contaminated.[4] This may be understated, for the writer remembers that every creek, pond, river, and lake was a safe swimming hole in rural north Texas during the 1940s. Today, there are few holes that could be considered safe for swimming, even less that could be used as a safe source of drinking water.

The quality of the mantle of air that surrounds the earth has been degraded to the extent that warnings are issued in many cities when contamination levels reach the danger zone. Joggers are warned about jogging at times of the day when smog levels are high, and many metropolitan areas in the world have enacted motor vehicle and other industrial emission controls in an attempt to lower air pollution levels. More than 100 cities in the United States routinely exceed safe ozone levels as a result of smog buildup. In Mexico City, 20 million people live in an atmosphere so murky that the sun is obscured, so poisonous that school is sometimes delayed until late morning when the air clears.[5]

[1]"Agency to Stop Paying for Nuclear Weapons Plant Negligence," *Lubbock Avalanche Journal*, Thursday, December 28, 1989.

[2]"The Planet Speaks," *Commonweal*, vol. 115, pp. 451–452, 1988.

[3]"Our Threatened Planet," *Maclean's*, vol. 101.2, pp. 38–43, September 5, 1988.

[4]"The Muck Stops Here: How We Can Halt the Trashing of America," *Glamour*, vol. 86, pg. 72, Feb. 1988.

[5]"Can Man Save This Fragile Earth?" *National Geographic*, vol. 174, pp. 765–945, Dec. 1988.

It seems clear that future generations on this planet will be faced with environmental problems, life-style modifications, and restrictions largely influenced by actions taken during the 1990s. An example of the kinds of actions being taken is a vehicle user's fee recently considered in Stockholm, Sweden. In an effort to reduce carbon monoxide, nitrogen oxides, and hydrocarbons produced by motor vehicles, a fee of $50 a month was proposed for motorists to drive on Stockholm roads. If this action proved inadequate in discouraging traffic, authorities planned to increase the fee to an amount that would generate the desired result.[6] In the Los Angeles basin, which has the most stringent air pollution restrictions of any metropolitan area in the world, a major effort is underway to promote the use of motor vehicles using alternative fuels. In 1988, the South Coast Air Quality Management District, which regulates air pollution in the Los Angeles basin, initiated a five-year, $30 million program to fund methanol and natural gas vehicle projects.

As stewards of the earth's ecology, this generation has both a moral and ethical obligation to protect the earth's environment so that it passes on to future generations without further degradation. The responsibility of a steward implies the relationship of an agent, managing the resource (the earth's ecology) for the real owner (the future generation). To be good stewards, the generation of the 1990s must find the knowledge necessary for understanding the earth's ecology, the management ability to cope with problems on a global scale, and the degree of commitment required to accept the changed behavior patterns necessary for a solution to this threat. The extremely complex interactions of the atmosphere, oceans, land, and human activities must be understood well enough to develop models that can predict the consequences of man's actions so that sound decisions can be made concerning the future of planet earth.

9.2   THE ENGINEER AND THE ENVIRONMENT

The foregoing summary paints a bleak picture for the future of this planet. The problems described are not easily solved technically, and the scope, which is global in many cases, adds enormous complication. Water flows and the winds blow without regard for distance or boundaries. Pesticides applied to crops in California are consumed in Dallas and Peoria. Soil-ravaging practices used in Iowa impoverish the nation. A nuclear accident in Russia spreads radioactive material over a wide area of Europe. Clearing of rain forests in Brazil contributes to the greenhouse effect on a global scale. Coal-burning electrical utility power plants in the U.S. Midwest cause acid rain in Canada. The list of ecological disasters-in-the-making (or already here) is long, and most proposed remedies are intricate, hugely expensive, and politically difficult.[2] Environmental problems are basically research and development challenges of a different order. They can and must be solved by scientists and engineers working in concert with political entities that can enact the necessary legislation, obtain the required international cooperation, and provide the needed funding, directly or indirectly. Engineers, as representatives of the profession most often identified with advancing technology, must accept some responsibility for the present state of the environment. As individuals, members of professional groups, employees, managers, and company owners, engineers

---

[6]"An Attack on Air Pollution," *Parade Magazine, Lubbock Avalanche Journal*, Jan. 14, 1990.

can shoulder this responsibility by supporting actions that lead to the restoration of ecological balance to planet earth. As concerned individuals, engineers can buy fuel efficient or alternate-fueled vehicles, cut back on air conditioning and heating, avoid overpackaged goods, pass up plastics, start recycling, plant trees and other vegetation, and be willing to sacrifice and suffer some inconvenience. As professionals, engineers must begin to apply concepts from the second law of thermodynamics to new product and process development, namely, the effect on the surroundings must be taken into account. If the process rejects anything to the environment that is not biodegradable (in a reasonable period of time), is not recyclable in some economically feasible manner, or has some measurable and long-lasting effect on the surroundings that cannot be accommodated by the environment's reservoir characteristics (which are increasingly limited), it should be seriously reexamined, redesigned or, as a last resort, abandoned in favor of some alternate process that may have less (apparent) return on investment. The environment can no longer be considered an infinite and forgiving reservoir in which chemical discharges, toxic material dumping, and harmful stack vapors can be deposited based on the lack of a measurable effect on the immediate surroundings. Managing the environment is a global problem that cannot be based on monitoring and control at the local level only. Engineers will have to play a key role in providing the requisite technology for understanding these global problems and in implementing workable solutions.

9.3 THE ENVIRONMENTAL PROTECTION AGENCY

In 1969, Congress began to consider a whole new range of legislation to protect the nation's environment. Two extremely important laws came out of these deliberations—the National Environmental Policy Act (NEPA) and the Clean Air Act. These laws took effect in 1970, and shortly thereafter the Environmental Protection Agency (EPA) was established within the Executive Branch of the federal government by reorganization, largely from pieces of several other agencies. Section 101 of the NEPA instructs the federal government to use "all practical means and measures . . . to create and maintain conditions under which man and nature can exist in productive harmony, and fulfill the social and economic and other requirements of present and future generations of Americans." The legal link between the NEPA and the EPA was forged by Section 309 of the Clean Air Act, which made the EPA a clearinghouse for reviewing proposed actions by other federal agencies that might adversely affect the environment. It further required that if the EPA determined that any federal agency proposal was considered unsatisfactory from the standpoint of public health or environmental quality, the administrator of the EPA was to make that fact known to the public and was to refer the matter to the Council on Environmental Quality (CEQ), which is the office that oversees the Executive Branch's implementation of NEPA. Although the CEQ and the EPA are the two primary environmental entities within the Executive Branch, they are charged with completely different missions. The CEQ oversees the implementation of the NEPA, but the EPA is the nation's primary environmental regulatory agency, responsible for cleaning up and maintaining the environment. Through its NEPA and Section 309 review responsibilities, the EPA is generally regarded as the nation's "environmental watchdog," which is the role Congress envisioned for the agency when it was established in 1970.

In spite of the many ecological problems that exist today, the EPA has made considerable progress in removing several specific contaminants from the air and water. The pesticides of the 1960s, such as DDT, have been largely eliminated and significant progress has been made in reducing air contaminants such as sulfur dioxide, nitrogen dioxide, particulates, carbon monoxide, and lead. The problems faced by the EPA in achieving even greater success are compounded by the changing life-styles in America. Population growth in large metropolitan areas and along the nation's coasts tend to increase air, water, and ground pollution problems in these locations. The ever-increasing production of plastic products is adding enormously to the nation's solid-waste disposal burden. Since plastic tends to be bulky, it consumes more and more of the nation's dwindling landfill area and, since it is largely nonbiodegradable, these landfill areas cannot be recovered (see Figure 9.1). Americans generate approximately $4\frac{1}{2}$ lb per person of waste every day. If this is multiplied by 280 million, a real appreciation for the problem of finding adequate landfill areas begins to evolve. Proliferation in the number and types of chemicals produced and distributed in America results in increased pollution and adds contaminants to the environment for which the hazards have not been fully characterized or understood. It is obvious that the EPA cannot solve these problems without the cooperation of industry and the help of every U.S. citizen in being willing to accept changes in life-style, to begin recycling, and to begin seriously conserving resources.

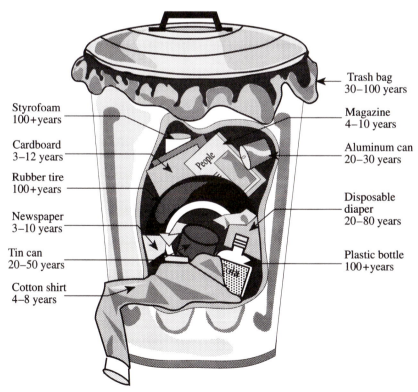

Figure 9.1 Typical trash and rate of decomposition (from *Lubbock Avalanche Journal*, March 6, 1990).

The EPA's goals for the future are based on a coordinated strategy with other government programs to achieve the maximum affordable reduction of the most significant risks. To accomplish this, understanding of the risks and the associated chemical, physical, and meteorological processes must be increased, and the public must become better informed. The EPA plans to accomplish this part of its mission by informing the public as to the scope and severity of the environmental challenges ahead and by convincing them that they are part of the pollution problem as well as the solution. Increasing responsibilities and capabilities of the various states will also be recognized. Technical support is to be improved by increasing the training and technical assistance provided in line with the EPA's statutory responsibility for setting standards and ensuring compliance. Finding solutions to environmental problems such as chlorofluorocarbons threatening the stratospheric ozone layer and global emissions of carbon dioxide contributing to the gradual warming of the earth's atmosphere will require new ways of cooperating with other nations. The Montreal protocol for protection of the stratosphere signed in 1989 provides hope that nations will be capable of resolving their economic and political differences in the interest of protecting a shared environment. Pollution prevention will continue to be emphasized in future EPA programs, including reduction in the amount of waste from homes and industry, restriction of toxic chemicals in places where they might enter drinking water supplies or endanger fish and wildlife, banning the disposal of untreated wastes on land when it threatens human health or the environment, restricting the harmful development of wetlands, and continuing the promotion of farming practices that will prevent agricultural chemicals from contaminating ground and surface waters.

The EPA has initiated a long-range planning program to determine what kind of environment the nation wants in the future and the best strategies to achieve it. This will require tremendous vision, considering the tough choices that must be made between environmental and competing social goals. Creativity and hard-nosed realism will be required to design systematic, coordinated program strategies and to follow through with the dedication necessary to achieve the desired end result, a quality environment coupled with a life-style that continues to be challenging, rewarding, and not diminished in any significant, long-term way from that enjoyed at the present.[7]

9.4 GROUNDWATER CONTAMINATION

Groundwater is one of the most important natural resources in the United States, and degradation of its quality could have a major effect on the welfare of the nation. Groundwater is the source of drinking water for more than one half of the nation's population and essentially all of its rural population. In 1985 it was the source of approximately 40 percent of the nation's public water supply, 33 percent of the water used for irrigation, and 17 percent of the freshwater used by industries who supply their own water. Groundwater is also the source of about 40 percent of the average annual streamflow in the United States, except during drought periods when groundwater discharge provides nearly all of the basic streamflow.[8] Because of this

---

[7]"Environmental Progress and Challenges: EPA's Update," U.S. Environmental Protection Agency, EPA 230-07-88-033, August 1988.

[8]National Water Summary 1986, U.S. Geological Survey Water-Supply Paper 2325, 1988.

interaction between aquifers and streams, contamination of groundwater may well result in contamination of a stream.

Groundwater contamination is not a new problem. Early investigations of groundwater contamination are abundant in scientific literature. The classic work of Dr. John Snow in 1854 linked the contamination of wells by cholera to seepage from earth privy vaults before the microorganisms responsible for the disease were identified.[9] In 1959, a European publication cited 60 cases in which groundwater had become contaminated with petroleum products.[10] Predicting the extent of contaminated zones is difficult, however, owing to several factors, including the variety of waste materials, their range in toxicity and adverse effects, the variable pattern of waste disposal and accidental release of contaminants in the ground, the unpredictable behavior of each contaminant in the soil, water, and rock environment, and the wide range of geologic and hydrologic conditions encountered.[11] In addition to these problems, many potentially hazardous contaminants are colorless, odorless, and tasteless, and are therefore difficult to detect. Many of the synthetic organic chemicals require sophisticated, expensive sampling and analytical techniques for accurate identification and quantification.

For the nation as a whole, whether contamination of groundwater is a serious problem at the present time (early 1990s) depends on who is making the assesment. Locally severe problems do exist and incipient conditions are present that may result in wide-scale contamination of the nation's groundwater in the future. Although naturally occurring constituents, such as nitrates, and human-induced substances, such as synthetic organic chemicals, are frequently detected in groundwater, their concentrations usually do not exceed the maximum allowables for concentrations in drinking water.[8] Contamination of groundwater falls into two categories related to the source(s) of the contamination (i.e., point and broad scale). In specific localities, high concentrations of toxic metals, organic chemicals, and petroleum products have been detected in groundwater associated with point sources, such as waste-disposal sites, storage-tank leaks, and hazardous chemical spills. Larger areas have also been identified where contamination is often found in shallow wells. These areas are usually associated with broad-scale sources of contamination, such as agricultural activities or concentrations of domestic waste disposal (septic system) in urban centers. Although the degree of contamination of potable groundwater in the United States at present is considered minimal, data, especially concerning the occurrence of synthetic organic chemicals and toxic substances, are considered inadequate to determine the full extent of the contamination problem or to define trends in groundwater quality[8].

New EPA regulations for municipal landfills designed to curtail groundwater pollution are scheduled to be implemented during the early 1990s. The cost of compliance with these proposed regulations may triple the cost of waste disposed by landfill operations and will probably result in the termination of many small operations. In one region in the State of Texas where 130 landfills were inspected in the past by State Health Department officials, it is expected that only a dozen or

[9]S. C. Prescott, and M. P. Horwood, *Sedgwick's Principles of Sanitary Science and Public Health*, New York, Macmillan, 1935.

[10]Michels et al., "Expert Opinion on Questions of Protection of Aquifers Against Contamination of Ground Water," *Bundesministerium für Atomkernenergie and Wasserwirtschaft*, Bad Godesberg, 1959.

[11]H. E. LeGrand, "Patterns of Contaminated Zones of Water in the Ground," *Water Resources Research*, vol. 1, no. 1, pp. 83–95, 1965.

so will remain open after the new regulations go into effect.[12] The new regulations stipulate the following:

- All garbage trenches dug after the regulations take effect must be double lined and have a leachate collection system between the liners for catching liquid waste infiltration.
- Filled garbage trenches must be closed, covered, and monitored for 30 years after closure.
- A cover must be placed on top of the garbage each day.
- Groundwater monitoring at landfill sites must be conducted.
- A system for controlling methane gas and other potentially explosive gases must be developed.

The new regulations will have the greatest effect on type 2 and type 3 landfills, those that accommodate less than 5000 and less than 1500 people, respectively. Most of the type 1 landfills, those that service more than 5000 people, are already meeting many of the new requirements and thus will not be impacted severely. In effect, the new regulations will essentially result in elimination of type 2 and type 3 landfills. Figure 9.2 shows an example of a municipal landfill under the proposed criteria.

Efforts have been made in recent years to assess the degree of occurrence of organic chemicals in groundwater supplies. A survey conducted by the EPA provided information on the frequency with which volatile organic compounds (VOCs) were

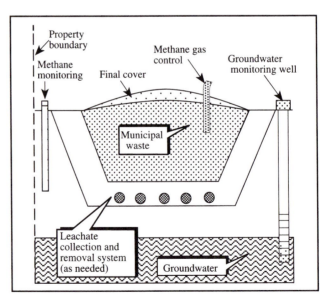

Figure 9.2 Example of municipal landfill under proposed criteria (from *Lubbock Avalanche Journal*, March 4, 1990).

---

[12]"Area Soon May be Down in the Dumps—New Landfill Regulations to Carry Staggering Costs," *Lubbock Avalanche Journal*, March 4, 1990.

detected in 466 randomly selected public groundwater supply systems.[13] One or more VOCs were detected in 16.8 percent of the small systems and 28.0 percent of the large systems sampled. The two VOCs found most often in this survey were trichloroethylene (TCE) and tetrachloroethylene (PCE), both biodegradable only over long periods of time. Two or more VOCs were found in 6.8 percent of the small systems and 13.4 percent of the large systems. Most groundwater contamination incidents involve substances released at or near the surface. As a result, it is shallow groundwater, that is affected initially by contaminant releases. Shallow groundwater is thus considered more susceptible to surface sources of contamination than is the deeper ground water.

One significant source of groundwater contamination in the United States is the many millions of underground liquid storage tanks located across the nation. It has been estimated that as many as 100,000 of the 3.5 million existing underground storage tanks are leaking.[14] In view of the large number of automotive service stations utilizing underground storage tanks for fuels and lubricating oils, it is very apparent that a serious environmental threat exists that must be dealt with quickly and effectively. Accordingly, legislation has been passed at both the federal and state levels that specifies requirements for existing and future underground tank installations. In recognition of the enormous cost associated with detecting the extent and level of contamination, as well as site restoration procedures, states have made funding available to assist in site remediation.

The primary ways in which groundwater contamination occurs are the following:[15]

1. ***Infiltration:*** This is the most common groundwater contamination mechanism. Water that has fallen to earth slowly infiltrates the soil through pore spaces in the soil matrix. As the water moves downward under the influence of gravity, it dissolves materials, including contaminates, with which it comes in contact, forming leachate. The leachate will continue to migrate downward until the saturation zone is reached, after which horizontal and vertical spreading of the contaminates in the leachate will occur primarily in the direction of the groundwater flow, as shown in Figure 9.3.

2. ***Direct Migration:*** Contaminants can migrate directly into groundwater from below-ground sources, such as storage tanks and pipelines that lie within the saturation zone. Significant contamination can occur because of the continually saturated conditions. Storage sites and landfills excavated to near the water table may permit direct contact of contaminants with the groundwater. Vertical leakage through seals around well casings or through improperly abandoned wells may also allow direct entry of contaminants into the groundwater system.

3. ***Interaquifer Exchange:*** Contaminated groundwater can mix with uncontaminated groundwater through a process known as interaquifer exchange, in which one water-bearing unit "communicates" hydraulically with another.

[13]J. J. Westrick, J. W. Mello, and R. F. Thomas, "The Ground Water Supply Survey: Summary of Volatile Organic Contaminant Occurrence Data," U. S. Environmental Protection Agency, Office of Drinking Water, Cincinnati, OH, 1983.

[14]S. Niaki and J. A. Broucious, "Underground Leak Detection Methods, A State of The Art Review," NTIS PB86-137155.2, 1986.

[15]*Handbook on Ground Water*, Environmental Protection Agency Research Laboratory EPA/625/016, March 1987.

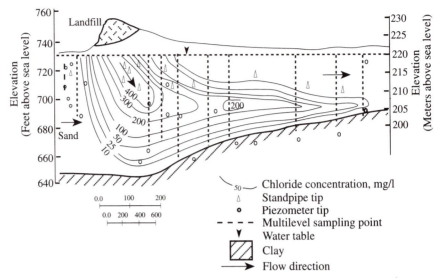

Figure 9.3 Plume of leachate migrating from a sanitary landfill on a sandy aquifer using contours of chloride concentration (from R. A. Freeze and J. A. Cherry, *Groundwater*, © 1979, p 437. Adapted by permission of Prentice Hall, Englewood Cliffs, NJ, 1979).

This is most common in bedrock aquifers where a well penetrates more than one water-bearing formation. When the well is not being pumped, water will move from the formation with the greatest potential to formations with less potential. If the formation with the greatest potential contains contaminated water, the quality in another formation can be degraded. Figure 9.4 shows an improperly abandoned well with a corroded casing

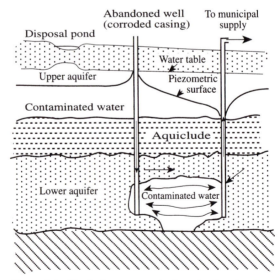

Figure 9.4 Vertical movement of contaminants along old, abandoned, or improperly constructed well (from M. Deutsch, "Incidents of Chromium Contamination of Ground Water in Michigan," U.S. Department of Health, Education and Welfare, Cincinnati, OH, 1961).

that was originally used to tap a lower uncontaminated aquifer but now allows migration from an overlying contaminated zone to communicate directly with the lower aquifer. The pumping of a nearby well tapping the lower aquifer creates a downward gradient between the two water-bearing zones allowing contaminated water to migrate through the lower aquifer to the pumping well.

4. ***Recharge from Surface Water:*** Normally, groundwater moves toward surface water bodies. Occasionally, however, the hydraulic gradient is such that surface water has a higher potential than groundwater (such as during flood stages), causing reversal in flow. Contaminants in the surface water can then enter the groundwater system.

Three processes govern the migration of chemical constituents in groundwater:

1. ***Advection:*** movement caused by the flow of groundwater. The hydraulic conductivity of a geologic formation depends on a variety of physical factors including porosity, particle size and distribution, the shape of the particles, particle arrangement (packing), and secondary features, such as fracturing and dissolution. In general, for unconsolidated porous materials, hydraulic conductivity values vary with particle size. Fine-grained, clayey materials exhibit lower values of hydraulic conductivity, whereas coarse-grained sandy materials normally exhibit higher conductivities.

2. ***Dispersion:*** movement caused by the irregular mixing of waters during advection. In porous materials, the pores possess different sizes, shapes, and orientations. Similar to stream flow, a velocity distribution exists within the pore spaces such that the rate of movement is greater in the center of the pore than at the edges. Therefore, in saturated flow through these materials, velocities vary widely across any single pore and between pores. As a result, a miscible fluid will spread gradually to occupy an ever-increasing portion of the flow field when it is introduced into a flow system. This mixing phenomenon is known as dispersion.

3. ***Retardation:*** chemical and physical mechanisms that occur during advection. Four general mechanisms can retard or slow the movement of chemical constituents in groundwater: dilution, filtration, chemical reaction, and transformation.

Despite increasing awareness that some of the nation's groundwater is contaminated with a variety of toxic metals, synthetic organic chemicals, radionuclides, and pesticides, public policy in this area is still in the formative stages. In spite of increasing efforts by state and federal agencies to protect groundwater, the extent of contamination is likely to appear to increase during the 1990s because more agencies will be searching for contamination, and they will be using increasingly sensitive detection methods.[8] Past excesses and ignorance that (later) result in groundwater contamination will undoubtedly come to light and will add to the staggering projected cost of cleaning up the nation's groundwater. The present groundwater situation may be satisfactory, but it is probably just a matter of time.

9.5 SOIL AND GROUNDWATER RESTORATION

A number of techniques are available to contain pollutants and to treat soil and groundwater to clean up contamination. These techniques range from removal of the polluted material and physical, chemical, or biological treatment on the surface, to physical containment and in situ treatment with chemicals or microbes. When selecting soil and groundwater cleanup methods for a particular site, the nature and location of the release, the soil type and geologic conditions, and the required degree of remediation must be evaluated. Liability, cost, period of time necessary to complete the remediation procedure, and any federal or state compliance requirements must also be considered.

The purpose of removing contaminated soil and groundwater associated with a plume of contamination is to treat and/or relocate the wastes to a better engineered and controlled site. The costs of excavation, transportation and new site preparation, soil removal, and reburial are often excessive for a large site and, thus, this method is used only as a last resort, or in instances of severe pollution where the cost is not significant compared with the importance of the resource being protected. In some cases, removal and reburial in an approved facility is simply transferring a problem from one location to another. Excavation is a technique often used for tank releases when the area of contamination is restricted and well defined. For these cases, excavation costs are usually low compared with other remediation methods, time required for cleanup is short, equipment is generally available, and a wide variety of contaminants can be remediated.

Low-temperature thermal stripping is a process in which soil is excavated and placed in a mobile unit designed to provide heat and drive off contaminants. Heat is usually supplied indirectly as the soil passes through the unit. Contaminants released from the soil are treated in an afterburner, in an activated carbon adsorption system, or condensed for recovery. Soil is treated on site using this method and usually can be accomplished in a relatively short time period. The cost of thermal stripping is higher than landfilling, however, and some contaminants cannot be successfully treated because of the low temperature.[16]

Soil gas extraction, or venting, is a technique frequently used to clean up VOCs released from storage tanks. This process involves extraction of the contaminate vapor from the soil by using wells and vacuum pumps. Volatile compounds are removed from the area between the soil particles by applying negative pressure to screened wells located in the vadose zone. This process does not require removal of the soil from the site and thus offers certain advantages over extraction. However, cleanup can often take lengthy periods of time, and soils like heavy clays cannot be readily treated using this method.

___ *Example 9.1* _____

A spill has been detected from an underground gasoline storage tank and abatement procedures have been implemented to prevent any additional leakage. A plan is required that will allow site cleanup in a period of approximately 6 months.

To determine the extent of contamination, organic vapor readings were taken in the basements of all surrounding buildings and utility tunnels in the immediate

---

[16] *Guidance Manual for Leaking Petroleum Storage Tanks in Texas*, Texas Water Commission, January 1990.

area. Venting was initiated in areas where organic vapors were detected, to prevent fire or explosion.

A hydrogeologic study was then performed to determine the depth to the water table, the local soil strata, and the hydraulic gradient. From these data likely paths of contaminant migration were determined. All water wells in the plume migration path were then sampled for both floating and dissolved hydrocarbons. Surface water near the spill site was also sampled for contamination. Monitoring wells were then drilled, and soil borings were taken at locations radiating out from the spill site. Soil and water samples from these wells were analyzed in a certified laboratory to determine the level of contaminate concentrations. Organic vapor readings were also taken during the drilling process. The resulting data were then assimilated to define the extent of the contamination from the spill. The following determinations were made from these data.

1. No free-floating hydrocarbons were found on the water table.

2. The gasoline contamination was weathered and located vertically in a 2-ft depth region of homogeneous sandy soil.

3. The horizontal migration was contained in a circle of a 36-ft radius with the center located directly below the source of the spill.

4. The total recoverable hydrocarbon concentration was determined to be 1000 ppm.

The volume of hydrocarbon contamination within the soil can be approximated as follows:

The density ρ of sandy soils can be approximated as 1.5 g/cm^3. The volume of contaminated soil is

$$V = hA \tag{9.1}$$

where V is the volume of soil, h is the height of zone, and A is the horizontal area of the zone. Therefore,

$$V = 2\text{ft.} \times \pi 36^2 = 8143 \text{ ft}^3$$

The mass m_s of contaminated soil is

$$m_s = \rho V \tag{9.2}$$
$$= (1.5 \text{ g/cm}^3)(8143 \text{ ft}^3)(1728 \text{ in.}^3/\text{ft}^3)(16.387 \text{ cm}^3/\text{in.}^3)$$
$$= 345.88(10)^6 \text{ g}$$

The mass of the contaminant m_c can now be found by using the concentration of the contaminant, which was found to be 1000 ppm. For every part of soil there is 0.001 parts of contaminant; thus,

$$m_c = 0.001 m_s \tag{9.3}$$
$$m_c = 0.001(345.88 \times 10^6 \text{ g})$$
$$m_c = 345.88 \text{ kg}$$

The desired cleanup period is determined to be 6 months, which requires an acceptable contaminate removal rate R_{acc} of

$$R_{acc} = \frac{345.88 \text{ kg}}{180 \text{ days}}$$

$$R_{acc} = 1.922 \text{ kg/day}$$

It must now be determined whether this removal rate is attainable. Equation 9.4[17] can be used to determine the estimated contaminant vapor concentration (C_{est}) based on the physical characteristics of the various chemical species.

$$C_{est} = \sum_i \frac{x_i P_i^v m_{w,i}}{R_u T} \text{ mg/liter} \tag{9.4}$$

where

x_i = mole fraction of constituent
P_i^v = pure component vapor pressure, atm
$m_{w,i}$ = molecular weight of component, mg/mol
R_u = universal molar gas constant, l-atm/mol K
T = absolute temperature of soil, K

Based on the chemical composition of weathered gasoline shown in Table 9.1 and the physical properties of these constituents shown in Table 9.2, C_{est} is determined to be 220 mg/liter. It should be noted that this estimate is valid for vapor concentration at the beginning of venting when the removal rate is the greatest. Contaminant concentration declines with time because of changes in composition, lowered contaminant residual levels, and increased resistance to diffusion. This leads to the conclusion that there is a practical limit to the amount of contaminant that can be removed by venting alone. Thus, venting should be considered as only one of the processes used to clean up a contaminated site.

TABLE 9.1: Composition of Weathered Gasoline

Component Number	Chemical Formula	$M_{w,i}$ (g)	Initial Mass Fraction	Initial Mole Fraction
1. Propane	C_3H_8	44.1	0.0000	0.0000
2. Isobutane	C_4H_{10}	58.1	0.0000	0.0000
3. *n*-Butane	C_4H_{10}	58.1	0.0000	0.0000
4. *trans*-2-Butene	C_4H_8	56.1	0.0000	0.0000
5. *cis*-2-Butene	C_4H_8	56.1	0.0000	0.0000
6. 3-Methyl-1-butene	C_5H_{10}	70.1	0.0000	0.0000
7. Isopentane	C_5H_{12}	72.2	0.0200	0.0269
8. 1-Pentane	C_5H_{10}	70.1	0.0000	0.0000
9. 2-Methyl-1-butene	C_5H_{10}	70.1	0.0000	0.0000
10. 2-Methyl-1,3-butadiene	C_5H_8	68.1	0.0000	0.0000
11. *n*-Pentane	C_5H_{12}	72.2	0.0114	0.0169
12. *trans*-2-Pentene	C_5H_{10}	70.1	0.0000	0.0000
13. 2-Methyl-2-butene	C_5H_{10}	70.1	0.0000	0.0000
14. 3-Methyl-1,2-butadiene	C_5H_8	68.1	0.0000	0.0000

[17]P. C. Johnson, C. C. Stanley, M. W. Kemblowski, D. L. Byers, and J. D. Colthart, "A Practical Approach to the Design, Operation and Monitoring of In Situ Soil-Venting Systems," *Ground Water Monitoring Review*, Spring 1990, pp. 159–178.

TABLE 9.1:   (Continued)

Component Number	Chemical Formula	$M_{w,i}$ (g)	Initial Mass Fraction	Initial Mole Fraction
15. 3,3-Dimethyl-1-butene	C_6H_{12}	84.2	0.0000	0.0000
16. Cyclopentane	C_5H_{10}	70.1	0.0000	0.0000
17. 3-Metyl-1-pentene	C_6H_{12}	84.2	0.0000	0.0000
18. 2,3-Dimethylbutane	C_6H_{14}	86.2	0.0600	0.0744
19. 2-Methylpentane	C_6H_{14}	86.2	0.0000	0.0000
20. 3-Methylpentane	C_6H_{14}	86.2	0.0000	0.0000
21. *n*-Hexane	C_6H_{14}	86.2	0.0370	0.0459
22. Methylcyclopentane	C_6H_{12}	84.2	0.0000	0.0000
23. 2,2-Dimethylpetane	C_7H_{16}	100.2	0.0000	0.0000
24. Benzene	C_6H_6	78.1	0.0100	0.0137
25. Cyclohexane	C_6H_{12}	84.2	0.0000	0.0000
26. 2,3-Dimethylpentane	C_7H_{16}	100.2	0.1020	0.1088
27. 3-Metylhexane	C_7H_{16}	100.2	0.0000	0.0000
28. 3-Ethylpentane	C_7H_{16}	100.2	0.0000	0.0000
29. 2,2,4-Trimethylpentane	C_8H_{18}	114.2	0.0000	0.0000
30. *n*-Heptane	C_7H_{16}	100.2	0.0800	0.0853
31. Methylcyclohexane	C_7H_{14}	98.2	0.0000	0.0000
32. 2,2-Dimethylhexane	C_7H_{18}	114.2	0.0000	0.0000
33. Toluene	C_7H_8	92.1	0.1048	0.1216
34. 2,3,4-Trimethylpentane	C_8H_{18}	114.2	0.0000	0.0000
35. 2-Methylheptane	C_8H_{18}	114.2	0.0500	0.0468
36. 3-Methylheptane	C_8H_{18}	114.2	0.0000	0.0000
37. *n*-Octane	C_8H_{18}	114.2	0.0500	0.0468
38. 2,4,4-Trimethylhexane	C_9H_{20}	128.3	0.0000	0.0000
39. 2,2-Dimethylheptane	C_9H_{20}	128.3	0.0000	0.0000
40. *p*-Xylene	C_8H_{10}	106.2	0.1239	0.1247
41. *m*-Xylene	C_9H_{10}	106.2	0.0000	0.0000
42. 3,3,4-Trimethylhexane	C_9H_{20}	128.3	0.0250	0.0208
43. *o*-Xylene	C_8H_{10}	106.2	0.0000	0.0000
44. 2,2,4-Trimethylheptane	$C_{10}H_{22}$	142.3	0.0000	0.0000
45. 3,3,5-Trimethylheptane	$C_{10}H_{22}$	142.3	0.0250	0.0188
46. *n*-Propylbenzene	C_9H_{12}	120.2	0.0829	0.0737
47. 2,3,4-Trimethylheptane	$C_{10}H_{22}$	142.3	0.0000	0.0000
48. 1,3,5-Trimethylbenzene	C_9H_{12}	120.2	0.0250	0.0222
49. 1,2,4-Trimethylbenzene	C_9H_{12}	120.2	0.0250	0.0222
50. Methylpropylbenzene	$C_{10}H_{14}$	134.2	0.0373	0.0297
51. Dimethylbenzene	$C_{10}H_{14}$	134.2	0.0400	0.0319
52. 1,2,4,5-Tetramethylbenzene	$C_{10}H_{14}$	134.2	0.0400	0.0319
53. 1,2,3,4-Tetramethylbenzene	$C_{10}H_{14}$	134.2	0.0000	0.0000
54. 1,2,4-Trimethyl-5-ethylbenzene	$C_{11}H_{16}$	148.2	0.0000	0.0000
55. *n*-Dodecane	$C_{12}H_{26}$	170.3	0.0288	0.0181
56. Napthalene	$C_{10}H_8$	128.2	0.0100	0.0083
57. *n*-Hexylbenzene	$C_{12}H_{20}$	162.3	0.0119	0.0078
58. Methylnaphthalene	$C_{11}H_{10}$	142.2	0.0000	0.0000
Total			1.0000	1.0000

From P. C. Johnson, et al. "A Practical Approach to the Design, Operation, and a Monitoring of In-Situ Soil Venting Systems," GWMR, Spring 1990, Copyright © 1990.

TABLE 9.2: Physical Properties of Regular Gasoline Constituents

Component Number	P_i^v (20°C, atm)	T_B (1 atm, °C)	S_i (20°C, mg/l)	k_{ow}
1. Propane	8.500	−42	62	73
2. Isobutane	2.930	−12	49	537
3. *n*-Butane	2.110	−1	61	946
4. *Trans*-2-Butene	1.970	1	430	204
5. *cis*-2-Butene	1.790	4	430	204
6. 3-Methyl-1-butene	0.960	21	130	708
7. Isopentene	0.780	28	48	1,862
8. 1-Pentene	0.700	30	148	710
9. 2-Methyl-1-butene	0.670	31	155	525
10. 2-Methyl-1,3-butadiene	0.650	34	642	323
11. *n*-Pentane	0.570	36	40	2,511
12. *trans*-2-Pentene	0.530	36	203	708
13. 2-Methyl-2-butene	0.510	38	155	525
14. 3-Methyl-1,2-butadiene	0.460	41	1,230	148
15. 3,3-Dimethyl-1-butane	0.470	41	23	1,350
16. Cyclopentene	0.350	50	158	871
17. 3-Metyl-1-pentene	0.290	54	56	1,820
18. 2,3-Dimethylbutane	0.260	57	20	4,786
19. 2-Methylpentane	0.210	60	14	6,457
20. 3-Methylpentane	0.200	64	13	6,457
21. *n*-Hexane	0.160	69	13	8,710
22. Methylcyclopentane	0.150	72	42	2,239
23. 2,2-Dimethylpentane	0.110	79	4.4	16,600
24. Benzene	0.100	80	1,780	135
25. Cyclohexane	0.100	81	55	3,236
26. 2,3-Dimethylpentane	0.072	90	5.3	16,600
27. 3-Metylhexane	0.064	92	3.2	22,400
28. 3-Ethylpentane	0.060	94	3.2	22,400
29. 2,2,4-Trimethylpentane	0.051	99	2.2	42,660
30. *n*-Heptane	0.046	98	3	30,000
31. Methylcyclohexane	0.048	101	14	11,220
32. 2,2-Dimethylhexane	0.035	107	1.5	57,544
33. Toluene	0.029	111	515	490
34. 2,3,4-Trimethylpentane	0.028	114	1.8	42,658
35. 2-Methylheptane	0.021	116	0.9	77,625
36. 3-Methylheptane	0.020	115	0.8	77,625
37. *n*-Octane	0.014	126	0.7	104,700
38. 2,4,4-Trimethylhexane	0.013	131	1.4	147,911
39. 2,2-Dimethylheptane	0.011	133	0.3	199,526
40. *p*-Xylene	0.0086	138	198	1,413
41. *m*-Xylene	0.0080	139	162	1,585
42. 3,3,4-Trimethylhexane	0.0073	140	1.4	147,911
43. *o*-Xylene	0.0066	144	175	589
44. 2,2,4-Trimethylheptane	0.0053	149	0.8	389,000
45. 3,5-Trimethylheptane	0.0037	156	0.8	389,000
46. *n*-Propylbenzene	0.0033	159	60	4,786
47. 2,3,4-Trimethylheptane	0.0031	160	0.8	389,000
48. 1,3,5-Trimethylbenzene	0.0024	165	73	12,883
49. 1,2,4-Trimethylbenzene	0.0019	169	57	12,883

TABLE 9.2: (Continued)

Component Number	P_i^v (20°C, atm)	T_B (1 atm, °C)	S_i (20°C, mg/l)	k_{ow}
50. Methylpropylbenzene	0.0010	182	6.8	33,884
51. Dimethylbenzene	0.0007	190	21	44,668
52. 1,2,4,5-Tetramethylbenzene	0.00046	196	3.5	12,883
53. 1,2,3,4-Tetramethylbenzene	0.00033	205	21	12,883
54. 1,2,4-Trimethyl-5-ethylbenzene	0.00029	210	7	204,000
55. *n*-Dodecane	0.0004	216	0.004	1,537
56. Napthalene	0.00014	218	33	1,738
57. *n*-Hexylbenzene	0.00010	226	1.3	309,000
58. Methylnaphthalene	0.000054	241	27	7,943

From P. C. Johnson, et al. "A Practical Approach to the Design, Operation, and a Monitoring of In-Situ Soil Venting Systems," GWMR, Spring 1990, Copyright © 1990.

A realistic vapor flowrate must now be determined. Equation 9.5[17] can be used to estimate a range of flowrates as a function of soil permeability (k). For sandy soils the permeability varies from 1 to 10 darcy (approximately 10^{-8} cm^2).

$$1 \text{ Darcy} = \frac{\frac{(1 \text{ centipoise})(1 \text{ cm}^3/\text{s})}{1 \text{ cm}^2}}{1 \text{ atm/cm}} = 0.987(\mu m)^2 = 0.987 \times 10^{-8} \text{ cm}^2$$

$$\frac{Q}{H} = \frac{\pi \left(\frac{k}{\mu}\right)(P_w)\left[1 - \left(\frac{P_{atm}}{P_w}\right)^2\right]}{\ln \frac{R_w}{R_i}} \tag{9.5}$$

$$= \frac{\pi \left(\frac{10^{-8}}{1.8 \times 10^{-4}}\right)(0.9)(1.01 \times 10^6)\left[1 - \left(\frac{1}{0.9}\right)^2\right]}{\ln \frac{0.0508}{12}}$$

$$= 6.810 \text{ to } 68.10 \text{ cm}^2/\text{s}$$

$$Q = 6.810(60.96) = 415 \text{ to } 4150 \text{ cm}^3/\text{s}$$

$$= 24.9 \text{ to } 249 \text{ liters/min}$$

where

k = soil permeability to air flow, darcy
μ = viscosity of air, 1.8×10^{-4} g/cm-s
H = height of well screen interval, 60.96 cm (2 ft)
P_w = absolute pressure at extraction well, typically 0.9 to 0.95 atm
P_{atm} = 1.0 atm(1.01×10^6 g/cm-s^2)
R_w = radius of extraction well, assume 5.08 cm
R_i = radius of influence of vapor extraction well, cm

R_i can be assumed to be 12 m (40 ft) without significant loss of accuracy[17]. The estimated removal rate R_{est} can now be determined using Eq. 9.6.

$$R_{est} = C_{est}Q \tag{9.6}$$

$$R_{est} = 220 \text{ mg/liter}(24.9 \text{ liters/min})(1 \text{ kg}/10^6 \text{ mg})(1440 \text{ min/day})$$

$$R_{est} = 7.9 \text{ kg/day to } 79 \text{ kg/day}$$

where Q is the vapor flowrate, l/min. Since the acceptable removal rate, $R_{acc} = 1.922$ kg/day, the range for the actual removal rate appears to be more than adequate.

States have established target cleanup goals for contaminated groundwater. A maximum allowable residual of 100 ppm total petroleum hydrocarbon has been adopted by many states. For this example this corresponds to removal of 90 percent of the original 1000 ppm contamination level. From Figure 9.5, which plots the maximum predicted removal rate for a weathered gasoline against the volume of air drawn through the contaminated zone per unit mass of contaminant, it can be seen that approximately 100 liters of vapor per gram of contaminant must pass through the soil to remove 90 percent of the contaminant. Recalling that $m_c = 345.88$ kg, the quantity of vapor that must pass through the soil is $345,880$ g(100 liters/g) $= 345.88 \times 10^5$ liters.

$$Q = \frac{(345.88 \times 10^5\, l)(1\text{m}^3/1000l)(35.3ft^3/\text{m}^3)}{180\ \text{days}\ (1\text{day}/1440\text{m})} = 4.71\ \text{cfm}$$

Since this value is within the level of attainable flowrates, cleanup can be accomplished within the planned six month period.

Hydrodynamic controls are often employed to isolate a plume of contamination from the normal groundwater flow regime to prevent the plume from moving into a well field, another aquifer, or surface water. Isolation of the contaminated plume is accomplished when uncontaminated groundwater is circulated around the plume in the opposite direction to the natural groundwater flow. The circulated zone creates a groundwater (hydrodynamic) barrier around the plume. Groundwater upgradient of the plume will flow around the circulated zone while groundwater downgradient will be essentially unaffected.[15]

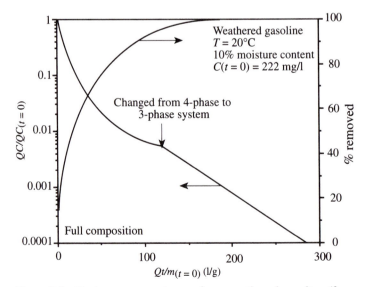

Figure 9.5   Maximum removal rates for a weathered gasoline (from P. C. Johnson, C. C. Stanley, M. W. Kemblowski, D. L. Byers, and J. D. Colthart, "A Practical Approach to the Design, Operation and Monitoring of In Situ Soil-Venting Systems," GWMR, Spring 1990, copyright© 1990.)

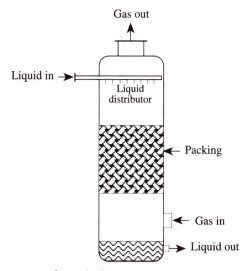

Figure 9.6   Packed tower.

Withdrawal and treatment of contaminated groundwater is one of the most often used processes for cleaning up aquifers. The type of contamination and the cost associated with treatment determine what specific treatment technology will be used. There are three broad types of treatment possibilities: (1) physical, which includes adsorption, density separation, filtration, reverse osmosis, air and steam stripping, and incineration, (2) chemical, which includes precipitation, oxidation/reduction, ion exchange, neutralization, and wet air oxidation, and (3) biological, which includes activated sludge, aerated surface impoundments, land treatment, anaerobic digestion, trickling filters, and rotating biological discs.[15]

The wide variety of devices that may be used to remove contaminants from water by transfer to air include diffused aeration, coke tray aerators, cross-flow towers, and countercurrent packed towers. In the packed tower, water containing the contaminant flows down through the packing while the air flows upward and is exhausted through the top to the atmosphere or to emission control devices (Figure 9.6). The kinetic theory of gases states that molecules of dissolved gases can readily move between the gas and liquid phases. Consequently, if water contains a volatile contaminant in excess of its equilibrium level, the contaminant will move from the liquid phase (water) to the gas phase (air) until equilibrium is reached. If the air in contact with the water is continuously replenished with fresh, contaminant-free air, eventually all of the contaminant will be removed from the solution. This is the basic operating principal of air-stripping processes. The objective in the design of air-stripping equipment is to maximize the rate of this mass transfer process at minimum cost.

Another method of cleaning up contamination in groundwater has been proposed by researchers at the Texas Agricultural Experiment Station, College Station, TX. In this approach, aquatic bacteria with a taste for toxic waste is used to break down one of the most common pollutants, TCE. TCE is an industrial solvent used in degreasing operations as well as in dry cleaning procedures, refrigerants, fumigants, and most septic tank cleaning fluids. As indicated previously, TCE is a contaminant in many public drinking water supplies and has been detected in samples from a large number of water wells across the United States. The proposed method, which is depicted in Figure 9.7, uses a treatment zone to clean the water like

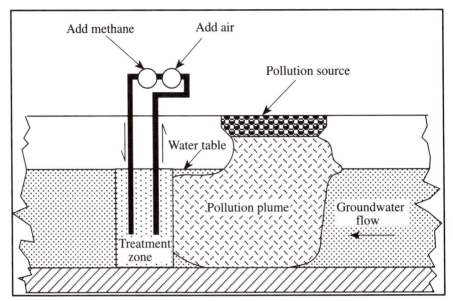

Figure 9.7 Cleaning groundwater (from *Lubbock Avalanche Journal,* July 1, 1990).

a filter. The filtering bacterium used in the process must be activated by using oxygen and methane as nutrients, which cause the bacteria to grow and to produce the enzyme that breaks down TCE. Bioremediation has some advantages over air stripping and other similar treatments, since it destroys the chemical pollutant. With bioremediation, TCE is converted into microbial cells and carbon dioxide, whereas with air stripping and some other techniques, the pollutant is merely removed from the water and added to the air, which may require subsequent scrubbing processes before release.[18]

9.6 DESIGN OF A PACKED TOWER

The basic design of packed towers has been well documented in the literature.[19,20] The packed tower offers a cost-effective solution for many scrubbing (change in composition of the air) and stripping (change in composition of the water) operations. The packed tower facilitates high mass transfer rates by incorporating a packing material with a large surface area for the water and air to interact and a configuration that enhances turbulence in the water stream to constantly expose fresh water surfaces to the air. Packings should also have large void areas to minimize pressure drop through the tower. Large-size packing materials are less expensive per unit volume and allow the use of smaller diameter towers for a given water flowrate. Packing materials of smaller size provide greater mass transfer coefficients and, therefore, allow the use of shorter towers. When the degree of contaminant removal is small, larger packing use is appropriate,

[18]"New Method Devised to Clean Up Contamination in Groundwater," *Lubbock Avalanche Journal,* July 1, 1990.

[19]J. S. Eckert, "Design Techniques for Sizing Packed Towers," *Chemical Engineering Progress,* vol. 57, no. 9, pp. 54–58, 1961.

[20]M. C. Kavanaugh and R. R. Trussel, "Design of Aeration Towers to Strip Volatile Contamininant from Drinking Water." Reprinted from *Journal of American Water Works Association,* vol. 72, no. 12, pp. 684–692, (Dec 1980), by permission. Copyright ©1980, American Water Works Association.

TABLE 9.3   Mass Transfer Coefficients

| Packing Type | $K_G a$ for a Nominal Packing Size $kmol/s\ m^3 atm \times 10^{-2}$ | | | | |
	1 in.	1.5 in.	2 in.	3 in.	3.5 in.
Super Intalox	1.43		0.94	0.62	
Pall rings	1.10	0.98	0.85		0.62
Tellerettes			0.85		
Maspak			0.78	0.89	
Heil-Pak			1.25	0.53	
Raschig rings (ceramic)	1.03	0.85	0.74		0.62
Berl saddles (ceramic)	1.11	0.80	0.85		

From M. C. Kavanaugh and R. R. Trussel, "Design of Aeration Towers to Strip Volatile Contaminant from Drinking Water." Reprinted from *Journal of American Water Works Association*, vol. 72, no. 12, pp. 684–692, December 1980. Copyright ©1980, American Water Works Association.

whereas for high removal rates smaller size packing may be more economical. As a general rule, the ratio of tower diameter to packing size should be at least 15:1 to avoid poor liquid distribution caused by the tendency of the water flowrate to be greater near the walls of the tower.[21] Other desirable features of packing materials include low weight, being chemically inert to fluids being processed, structural strength for easy handling, and low cost. Table 9.3 lists several of the commonly used packing types with corresponding values of $K_G a$, which is the product of the overall gas phase mass transfer coefficient and the interfacial surface area per volume of packing over which the mass transfer occurs. For towers in which the mass transfer is controlled by the liquid phase resistance, which is the case for most water treatment air stripping operations, the coefficient of interest is K_L, the overall liquid phase mass transfer coefficient. After selecting a suitable packing from Table 9.3, the product $K_L a$ can be approximated by using Eq. 9.7,

$$K_L a = \frac{K_G a H_A}{C_o} \tag{9.7}$$

where H_A = Henry's constant for solute A, and C_o = molar density of water which is 55.6 kmol/m³. When more accurate values for $K_L a$ are required, as in an actual design situation, pilot studies should be conducted. For the purpose of gaining a general understanding of the principles involved in designing packed towers, which is one of the purposes of this chapter, the use of equation 9.7 will provide satisfactory values for $K_L a$. For most contaminants in wastewater and public water supplies, pilot studies should be performed to develop accurate mass transfer data.

Henry's law states that the mass of a slightly soluble gas that dissolves in a definite mass of liquid at a given temperature is very nearly directly proportional to the partial pressure of the gas in the air above the liquid. Henry's constant thus provides a measure of the equilibrium concentration of the contaminant in the air-stripping process. The larger Henry's constant, the greater the equilibrium concentration of the contaminant in the air and the easier the contaminant is to remove from the liquid. For compounds with high values of H, as with slightly soluble gases (e.g., O_2) and volatile liquids (e.g., trichloroethylene), the mass transfer rate depends primarily on the liquid phase resistance. For contaminants with small

[21]R. E. Treybal, *Mass Transfer Operations*, McGraw-Hill, New York, 1980.

TABLE 9.4:   Henry's Constants for Selected Compounds[a]

Compound	Formula	Henry's Constant (atm)
Vinyl chloride[b]	CH_2CHCl	3.55×10^3
Oxygen	O_2	4.3×10^4
Nitrogen	N_2	8.6×10^4
Methane	CH_4	3.8×10^4
Ozone	O_3	3.9×10^3
Toxaphene[b]	$C_{10}H_{10}Cl_8^C$	3.5×10^3
Carbon dioxide	CO_2	1.51×10^3
Carbon tetrachloride[b]	CCl_4	1.29×10^3
Tetrachloroethylene[b]	C_2Cl_4	1.1×10^3
Trichloroethylene[b]	$CCHCl_3$	5.5×10^2
Hydrogen sulfide	H_2S	5.15×10^2
Chloromethane[b]	CH_3Cl	4.8×10^2
1,1,1-Trichloroethane[b]	CCH_3Cl_3	4.0×10^2
Toluene[b]	$C_6H_5CH_3$	3.4×10^2 (25°C)
1,2,4-Trimethylbenzene[b]	$C_6H_3(CH_3)_3$	3.53×10^2 (25°C)
Benzene[b]	C_6H_4	2.4×10^2
1,4-Dichlorobenzene[b]	$C_6H_4Cl_2$	1.9×10^2
Chloroform[b]	$CHCl_3$	1.7×10^2
1,2-Dichloroethane[b]	CH_3CHCl_2	61
1,1,2-Trichloroethane[b]	CCH_3Cl_3	43
Sulfide dioxide	SO_2	38
Bromoform[b]	$CHBr_3$	35
Ammonia	NH_3	0.7600
Pentachlorophenol[b]	$C_6(OH)Cl_3$	0.1200
Dieldrin[b]		0.0094

[a]Temperature 20°C except where noted otherwise.
[b]Computed from water solutiblity data and partial pressure of pure liquid at specified temperature.
[c]Synthetic; approximate chemical formula.

M. C. Kavanaugh and R. R. Trussel, "Design of Aeration Towers to Strop Volatile Contamininant from Drinking Water." Reprinted from *Journal of American Water Works Association,* vol. 72, no. 12, pp. 684–692, (Dec 1980), by permission. Copyright ©1980, American Water Works Association.

values of H, characteristic of soluble gases or low volatility liquids, the mass transfer will often be controlled by the gas phase resistance. Table 9.4 gives values of Henry's constant for various compounds of interest at 20°C. For other temperatures a correction can be applied based on the enthalpy change caused by dissolution of the contaminant in water.

In designing a packed tower for stripping and scrubbing operations involving air and water, the following important design variables are considered:

1.   Air-to-water flowrate ratio

2.   Diameter of the tower

3.   Height of the packing

As with most design processes there is more than one solution that will satisfy the design requirements. Removal of a specific contaminant can be accomplished at various air-to-liquid flowrate ratios and at different tower diameters and heights. It is thus important to use an iterative design approach providing several different design options from which the most cost effective can be selected. The optimum

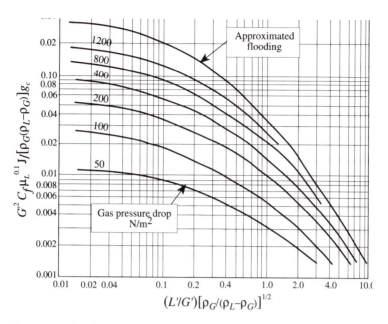

Figure 9.8   Flooding and pressure drop in random-packed towers (from R. E. Treybal, *Mass Transfer Operations*, McGraw-Hill, New York, 1980).

design will be that with the least total cost, including the initial capital cost as well as the operating cost over the life of the system. The time required to complete the cleanup operation must be considered when one determines the overall cost.

9.6.1   Sizing the Tower Diameter

The first step in determining the diameter of a packed tower is to assume values of the stripping factor R and to calculate the corresponding gas to liquid flowrate ratio G/L using Eq. 9.8.

$$G/L = \frac{RP_T}{H} \qquad (9.8)$$

where
  G = gas flowrate (kmol/m²s)
  L = liquid flowrate (kmol/m²s)
  P_T = the ambient pressure (atm)
  H = Henry's constant for the contaminant (atm)
  R = stripping factor (dimensionless)

The flowrate ratio can be converted to SI units (kg/m²s) by multiplying the mol ratio by the gas and liquid molecular weights; thus,

$$\frac{G'}{L'} = (\frac{G}{L})(\frac{MW_{air}}{MW_{water}}) \qquad (9.9)$$

After the G'/L' ratio has been determined, Figure 9.8 can be used to establish values for the gas and liquid flowrates. To obtain the value for the abscissa in Figure 9.8, evaluate the following expression

$$\frac{L'}{G'}\left(\frac{\rho_G}{\rho_L - \rho_G}\right)^{1/2} \tag{9.10}$$

where

ρ_G = density of the gas
ρ_L = density of the liquid

For a given flowrate the pressure drop of the air rising countercurrent to the water in a packed tower increases approximately in proportion to the square of the gas velocity. At high gas flowrates, entrainment of the liquid may occur, causing a sudden increase in the gas pressure drop. When the liquid flowrate increases at a fixed gas flowrate, the tower may fill with liquid, a condition known as flooding. The tower should be designed to operate at conditions well below flooding. According to Treybal[21], most stripping towers are designed for gas pressure drops of 200 to 400 N/m^2 per meter of packing depth (0.25 to 0.50 in. H$_2$O/ft). From the calculated values for the abscissa and a selected pressure drop, values for the ordinate from Figure 9.8 can now be obtained and Equation 9.11 can be solved for values of G':

$$G' = \left[\frac{\rho_G(\rho_L - \rho_G)g_c}{C_f \mu_L^{0.1} J} \times \text{ordinate value}\right]^{1/2} \tag{9.11}$$

where

μ_L = liquid viscosity
g_c = gravity correction factor (for SI units g_c = 1,
      for English units g_c = 4.18 × 10^8)
C_f = packing factor (from Table 9.5)
J = conversion factor (for SI units J = 1.0,
     for English units J = 1.502)

The water flowrate L' can now be determined by using Eq. 9.10, and the abscissa from Figure 9.8. The tower diameter ϕ can then be obtained by solving Eq. 9.12:

$$\phi = \left[\frac{4}{\pi}\frac{Q_L \rho_L}{L'}\right]^{1/2} \tag{9.12}$$

where Q_L = volumetric liquid flowrate.

TABLE 9.5   Packing Factors C_f for Common Plastic Packings and Raschig Rings

| | C_f for Normal Packing Size Shown | | | | |
Packing Type	1 in.	1.5 in.	2 in.	3 in.	3.5 in.
Super Intalox	33		21	16	
Pall rings	52	32	25		16
Tellerettes	40		20		
Maspak			32	20	
Heil-Pak	45		18	15	
Raschig rings (ceramic)	155	95	65	37	
Berl saddles (ceramic)	110	65	45		

From J. S. Eckert, "How Tower Packings Behave," *Chemical Engineering,* vol. 82, no. 70, April 4, 1975.

9.6.2   Determining the Height of the Tower

The height of the packing Z required to achieve the desired removal of contaminant is the product of the height of a transfer unit (HTU), and the number of transfer units (NTU).

$$Z = (HTU)(NTU) \tag{9.13}$$

The HTU is inversely proportional to the product of the overall liquid phase mass transfer coefficient and the interfacial area $K_L a$ and is a function of the efficiency of mass transfer from water to air:

$$HTU = \frac{L}{(K_L a)(C_o)} \tag{9.14}$$

The NTU provides a measure of the difficulty of removing the contaminant from the water and can be calculated by using

$$NTU = \frac{R}{R-1} \ln \frac{\frac{X_{in}}{X_{out}}(R-1) + 1}{R} \tag{9.15}$$

where X_{in} and X_{out} are the concentrations of the compound to be removed (percent).

▬ Example 9.2 ▬▬▬▬▬▬▬▬▬▬▬▬▬▬▬▬▬▬▬▬▬▬▬▬▬▬▬▬▬▬▬

Assume, after chlorination, a 50-liter/s surface water supply contains 200 $\mu g/$liter of chloroform. The effluent concentration must be reduced to 40 $\mu g/$liter. Considering the water and air temperature to be 20°C, determine the dimensions of the packed tower.
Assume:

- Stripping factor, $R = 3$.
- Allowable pressure drop in tower is 300 N$/$m²$-m$.

Solution

Selection of the packing material: For many years, the most popular packings were Raschig rings and Berl saddles, but these have been almost entirely replaced by Pall rings and Super Intalox made from plastic or ceramics. For this design problem, 2 in. Super Intalox is selected because of its low cost and high mass transfer efficiency. From Table 9.5, we find the packing factor $C_f = 21$.

1. Compute the number of transfer units NTU for $R = 3$.
    The desired removal efficiency η_e is

$$\eta_e = 1 - \frac{40}{200} = 80\%$$

Number of transfer units,

$$NTU = \left(\frac{R}{R-1}\right)\ln\left[\frac{\frac{X_{in}}{X_{out}}(R-1)+1}{R}\right]$$

$$= \left(\frac{3}{3-1}\right)\ln\left[\frac{\frac{200}{40}(3-1)+1}{3}\right] = 1.95$$

2. Determine the allowable airflow at 20°C and P_T=1 atm for 300 N/m²−m gas pressure drop.
   From Table 9.4, Henry's constant for $CHCl_3$ is, $H = 170$ atm. Hence,

$$\frac{H}{P_T} = \frac{170}{1} = 170$$

Knowing
$$\rho_{air} = 1.205 \text{ kg/m}^3,$$
$$MW_{air} = 28.8 \text{ kg/kmol},$$
$$\rho_{water} = 998 \text{ kg/m}^3,$$
$$MW_{water} = 18.02 \text{ kg/kmol}$$

$$R = \frac{H}{P_T}\frac{G}{L} = 170\frac{G}{L}$$

or,

$$R = 170\frac{28.8}{18.02}\frac{G'}{L'} = 272\frac{G'}{L'}$$

Then, for $R = 3$, we find $G'/L' = 0.011$.
   Figure 9.8 can be used to obtain the value of G' and L'. The abscissa of Figure 9.8 is

$$\frac{L'}{G'}\left(\frac{\rho_G}{\rho_L-\rho_G}\right)^{1/2} = 90.56\left(\frac{1.205}{998-1.205}\right)^{1/2} = 3.149$$

For 3.149 and 300 N/m²(m) pressure drop, from Figure 9.8, the ordinate value is obtained:

$$\frac{G'^2 C_f \mu_L^{0.1} J}{\rho_G(\rho_L-\rho_G)g_c} = 0.0042$$

For SI units $J = 1$, and $g_c = 1$, and $\mu_L = 0.001$ kg/m-s. The airflow rate is found by using Eq. 9.11:

$$G' = \left[\frac{0.0042 \times 1.205 \times (998-1.205)}{21 \times (0.001)^{0.1}}\right]^{1/2} = 0.69 \text{ kg/m}^2 - s$$

Hence,

$$L' = 90.56G' = 90.56 \times 0.69 = 62.70 \text{ kg/m}^2 - s$$

3. Determine the tower diameter ϕ.
   Using Eq. 9.12, the tower diameter is

$$\phi = \left[\frac{4}{\pi} \frac{Q_L \rho_L}{L'} \right]^{1/2}$$

$$\phi = \left[\frac{4}{\pi} \frac{0.050 \times 998}{62.70} \right]^{1/2} = 1.013 \text{ m}$$

4. Compute the height of the packed tower.
   Since the contaminant is being removed from the liquid (water), to determine HTU, the liquid phase mass transfer coefficient K_L controls the process. Estimate $K_L a$ using Eq. 9.7:

$$K_L a = \frac{K_G a H_A}{C_0} = \frac{0.0094(170)}{55.6} = 0.029 \text{ s}^{-1}$$

$$L = \frac{L'}{MW_{water}} = \frac{62.77}{18.02} = 3.48 \text{ kmol/m}^2 \text{ s}$$

then

$$HTU = \frac{L}{K_L a C_o} = \frac{3.48}{0.029 \times 55.6} = 2.16 \text{ m}$$

5. Compute the height of tower.

$$Z = (HTU)(NTU) = (2.16)(1.95) = 4.2 \text{ m}$$

6. Select the appropriate tower dimensions.
   A nominal size tower diameter and height that is readily available should be selected. A reasonable safety factor should also be included. For this application a reasonable choice might be $z = 4.88$ m (16.0 ft) and $\phi = 1.22$ m (4.0 ft). To optimize the tower design, an iterative procedure for various values of stripping factor R is recommended.

9.7 AIR POLLUTION AND TOXIC CHEMICAL EXPOSURE

Air pollution has been a major concern in the United States since the mid-1950s. This concern was translated into law with the enactment of the 1970 Clean Air Act, and through subsequent amendments to that act in the years 1977 and 1990. Most states have developed regulatory, enforcement, and administrative programs to reduce air pollution. These include controls on pollution from industries and utilities as well as motor vehicle emission inspection programs. These state programs have been augmented by federal regulations and control measures for reducing emissions from new motor vehicles and certain industrial sources. The U. S. EPA is responsible for coordinating and approving most state air pollution plans and programs. The EPA also provides substantial technical assistance to states which ranges from information on new air pollution control technology to studies of the health effects of air pollution. The overall goal of these programs is to ensure that national air quality standards (NAAQS) are achieved and maintained for major air pollutants, including lead, carbon monoxide, ozone, nitrogen oxides, sulfur dioxide, and particulates.

Much public concern has been focused on the harmful effects that acid rain has on fish and other wildlife, lakes, forests, crops, and on man-made objects, such as buildings and statues. Certain aspects of the acid rain phenomenon are generally accepted by the scientific community, but many other causes and effects are not well understood. The geographic range of damage from acid rain, the rate at which acidification takes place, and the combination of pollutants that are involved are uncertain. For example, most attention has been focused on the contribution of sulfur emissions to acid deposition, but other pollutants (including oxides of nitrogen) are also contributers.

Another major air quality issue involves the transport of ozone and its precursors. This chiefly involves a process whereby emissions of VOCs—ozone precursors—are transported over long distances from the source of the release. This phenomenon has complicated the development of control strategies for some areas of the country and has also compounded the difficulties of achieving air quality standards for ozone. This is a particularly difficult problem for some of the larger urban areas of the country.[22]

Approximately 5000 air monitors across the United States measure and record air pollution levels. During the period from 1981 through 1990 the pollution levels for the six pollutants listed above decreased according to EPA's annual urban air quality trends report.[23] Lead levels decreased 85 percent during the 10-year period, ambient carbon dioxide levels declined 29 percent, sulfur dioxide ambient levels fell 24 percent, ambient smog levels dropped 10 percent, and particulate (dust, dirt, soot) levels decreased 3 percent. The report also points out the magnitude of the air quality problem that still exists. In 1990, more than 74 million Americans lived in areas where the air was considered to be unhealthy. Approximately 22 million people lived in areas exceeding the carbon monoxide standard, and almost 19 million resided in areas where the particulate standard was exceeded. Smog was still the greatest problem, with 63 million people living in areas exceeding the standard for this contaminant.

Hazardous pollutants also pose serious but primarily localized health problems. This includes hazards to the general public from exposure to refuse and other discarded materials from chemical, metallurgical, agricultural, and other industrial processes as well as exposures in the workplace. The EPA, the Occupational Safety and Health Administration (OSHA), and the Consumer Product Safety Commission (CPSC) share the responsibility for issuing and enforcing regulations in this area based on research performed by other government agencies, including the National Institute for Occupational Health (NIOSH), the National Institute of Environmental Health Sciences (NIEHS), the National Cancer Institute (NCI), and the Food and Drug Administration (FDA). Although it is probably impossible to determine the magnitude of the toxic chemical risk, it is recognized as a major and growing public health problem. Residents of dwellings in close proximity to hazardous waste landfills are in danger of exposure by direct contact or by inhalation of dusts, fumes, or vapors. Employees working in an environment where chemical

---

[22]"Trends in the Quality of the Nation's Air," U.S. Environmental Protection Agency, January 1985.

[23]"National Air Quality and Emissions Trends Report," U.S. Environmental Protection Agency, April 1991.

and other industrial processes are ongoing can be exposed by accident, through ignorance or carelessness. The difficulty in defining the magnitude of the problem is hampered by the long latent period that exists between exposure and the onset of chemically induced disease as well as the relative newness of the science of environmental toxicology.[24]

OSHA has published a document entitled "Air Contaminants—Permissible Exposure Limits" that provides information and recommendations relating to permissible exposure limits (PELs), chemical and physical properties, health hazard information, respiratory protection, and personnel protection and sanitation practices for 600 substances.[25] This is a small portion of the many thousands of chemicals in the environment, however, and OSHA is in the process of developing generic limits that will simultaneously cover many substances. Substance limits in this OSHA document are provided for direct exposures involving chemical absorption through the skin as well as respiratory exposure wherein chemicals enter the body through the breathing process. Compliance with these limits in the workplace will protect workers against a wide variety of health effects that could cause material impairment of health or functional capacity.

9.7.1 Particulate Contamination

Particulate contamination contributes to both occupational health and nuisance problems in the workplace. Accordingly, OSHA has promulgated allowable limits that are applicable to various chemical, mineral, and grain dusts for both health concerns and as workplace irritants. These requirements are specified in U.S. Department of Labor report on air contaminants, OSHA 3112. Certain dusts also pose explosive hazards and must be controlled for this reason as well. The U.S. Department of Agriculture has published several documents relating to the problems and prevention of grain dust explosions. OSHA PELs for grain dust are less than the minimum explosive concentrations (e.g., the PEL for grain dust is given as 10 mg/m^3, whereas the minimum explosive concentration for corn dust is 40 g/m^3).

The rate of dust generation during the handling of a material is a function of the material, the material handling rate, and the relative air velocity near the surface of the material. Dust particles less than 125 μm have surface irregularities that engender Van der Waals surface charges that cause the particles to adhere to each other and to other surfaces.[26] When minimum entrainment velocities are reached, these bonds are broken and the dust is released into the air. If material-handling rates and relative air velocity near the material surface are both held constant, the dust generated by the system will remain essentially constant for each operation.

Dust cloud permanence is related to the settling velocities of the particles, air velocity, turbulence, and flow path. The settling velocity of dust is the terminal velocity that the particle reaches when its drag force is equal to the force of gravity acting on the particle. The terminal settling velocity of a particle as derived by

---

[24]"Health Effects of Toxic Pollution: A Report from the Surgeon General," U.S. Department of Health and Human Services, Serial No. 96-15, Aug. 1980.

[25]"Air Contaminants-Permissible Exposure Limits," Title 29 Code of Federal Regulations Part 1910.1000, 1989.

[26]Review of Literature Related to Engineering Aspects of Grain Dust Explosions, U.S. Department of Agriculture Publication 1375, 1979.

Stokes is

$$V_{ts} = \sqrt{\frac{4\rho_p d_p g}{3 C_D \rho_g}}$$ (9.16)

where
ρ_p = density of the particle
ρ_g = density of the gas (air)
C_D = drag coefficient for the particle
d_p = aerodynamic diameter of the particle

where the particle drag coefficient C_D is given by

$$C_D = \frac{4}{3} \frac{\rho_p d_p g}{\rho_g V_{ts}^2}$$ (9.17)

The solution of V_{ts} from Eqs. 9.16 and 9.17 requires an iterative method. An alternative approach is to introduce the Reynolds number squared:

$$R_e^2 = \left(\frac{\rho_g V_{ts} d_p}{\eta} \right)^2$$ (9.18)

To eliminate the velocity-squared term in Eq. 9.17 use

$$C_D R_e^2 = \frac{4}{3} \frac{d_p^3 \rho_p \rho_g g}{\eta^2}$$ (9.19)

where η is the viscosity of the air.

Now $C_D R_e^2$ can be calculated directly. Equation 9.16 is valid for spherical particles and particles with Reynolds numbers less than 1.0. For small particles (size less than 125 μm) the Reynolds number is in the range of $0.05 < R_e < 4.0$. For this range, an empirical equation was developed by Davies,[27] which permits the direct calculation of V_{ts}^* using $C_D R_e^2$:

$$V_{ts}^* = \frac{\eta}{\rho_g d_p} \left[\frac{C_D R_e^2}{24} - 2.3363 \times 10^{-4} (C_D R_e^2)^2 \right.$$
$$\left. + 2.0154 \times 10^{-6} (C_d R_e^2)^3 - 6.9105 \times 10^{-9} (C_D R_e^2)^4 \right]$$ (9.20)

The aerodynamic diameter of the particle d_p can be calculated by using

$$d_p = d_e \left(\frac{\rho_p}{\rho_0 \chi} \right)^{1/2}$$ (9.21)

where
ρ_0 = unit density (1.0 g/cm^3)
χ = shape correction factor

[27]C. N. Davies. "Particle Fluid Interaction," *Journal of Aerosol Sciences*, vol. 10, pp. 477–513, 1979.

TABLE 9.6 Dynamic Shape Factors

Shape	Shape Factor (χ)
Geometric shapes	
Sphere	1.00
Cube	1.08
Cylinder ($L/D = 4$)	
Axis horizontal	1.32
Axis vertical	1.07
Cluster of spheres	
2 chain	1.12
3 chain	1.27
4 compact	1.15
4 chain	1.32
4 compact	1.17
Dusts	
Bituminous coal	1.05–1.11
Quartz	1.36
Sand	1.57
Talc	2.04

Reprinted with permission from C. N. Davis, "Particle Fluid Interaction," *Journal of Aerosol Science,* vol. 10, pp. 477–513, copyright ©1979, Pergamon Press Ltd.

In Eq. 9.21, d_e is the equivalent diameter of a sphere of the same volume as an irregular particle, and it can be experimentally determined. The shape parameter χ can be obtained from Table 9.6.[28] The following example demonstrates how the dust concentration in a work area can be determined by applying Eq. 9.20 to ascertain the quantity of the dust generated that settles out of the air.

—— *Example 9.3* ——

Lubbock State School, a school for mentally and physically handicapped students, contacted the Texas Tech University Mechanical Engineering Department in June 1991 regarding a problem they had encountered with the generation of dust during the operation of a gravel bagging system used in vocational training. The gravel bagging system raises gravel from an underground gravel pit through use of a bucket/belt elevator system to a hopper that releases measured amounts of gravel into bags. Large quantities of dust were being produced by the system, raising the dust concentration in the room to a level intolerable to clients working in the area and thought to be in excess of the OSHA limit for nuisance dust in the workplace.

    On-site inspection of the bagging operation by a student group interested in the project led to the conclusion that the dust generation was occurring at four different gravel transfer points (Figure 9.9). Standard practices for controlling dust include wetting, ventilation, and the use of fabric dust arresters. An air inlet and exhaust fan had been previously installed in the room to remove the dust but had proved to be ineffective. Based on the on-site inspection the student group decided to undertake the project. The objective established was to develop a dust suppression system to reduce the dust concentration below OSHA nuisance dust requirements.

[28]W. C. Hinds, *Aerosol Technology*, John Wiley & Sons, New York, 1982.

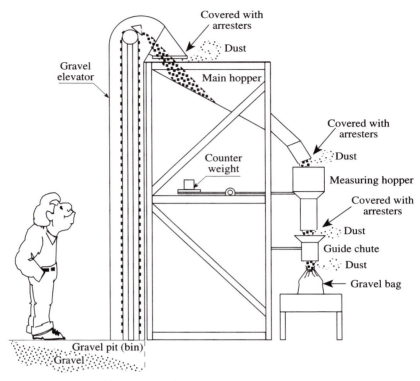

Figure 9.9   A gravel bagging machine.

The concentration of dust in the room is strongly influenced by the terminal settling velocity. To calculate the terminal settling velocity V_{ts}^* for a range of particle sizes, the following parametric values were used:[28]

$$\rho_g = 1.054 \text{ g/cm}^3$$
$$\rho_p = 2.7 \text{ g/cm}^3$$
$$\chi = 1.57$$
$$g = 981 \text{ cm/s}^2$$
$$\eta = 1.81 \times 10^{-4} \text{ g/cm−s}$$

Assuming an equivalent particle diameter, $d_e = 1 \ \mu$m, the aerodynamic diameter of the particle is

$$d_p = d_e \left(\frac{\rho_p}{\rho_0 \chi} \right)^{1/2}$$

$$d_p = 1 \left(\frac{2.7}{1 \times 1.57} \right)^{1/2} = 1.311 \ \mu\text{m} = 1.311 \times 10^{-4} \text{cm}$$

and from Equation 9.19

$$C_D R_e^2 = \frac{4}{3} \frac{d_p^3 \rho_p \rho_g g}{\eta^2}$$

TABLE 9.7   Settling Velocity of Gravel Dust

Equivalent Particle Diameter d_e (μm)	Aerodynamic Particle Diameter d_p(μm)	Davies Settling Velocity V_{ts}^*(cm/s)
1	1.31	0.014
2	2.62	0.056
3	3.93	0.126
4	5.25	0.233
5	6.56	0.349
6	7.87	0.503
7	9.18	0.684
8	10.49	0.893
9	11.80	1.130
10	13.11	1.395
15	19.67	3.137
20	26.23	5.524
25	32.78	8.540
30	39.34	12.112

$$C_D R_e^2 = \frac{4}{3} \frac{(1.311 \times 10^{-4})^3 (2.7)(1.054 \times 10^{-3})(981)}{(1.81 \times 10^{-4})^2}$$

$$C_D R_e^2 = 2.560136 \times 10^{-4}$$

Using $C_D R_e^2 = 2.560136 \times 10^{-4}$ in Eq. 9.20 gives a settling velocity of 0.014 cm/s. Similar calculations can be performed for other equivalent particle diameter to obtain Table 9.7 for the gravel dust at Lubbock State School.

The dust concentration in the bagging room can be determined by assuming that the dust concentration at any point in time in the room is equal to the dust concentration exhausted out of the room through the ventilation system. This assumption allows the concentration of dust in the room to be calculated knowing the total effective mass of dust suspended in the air and the total effective volume of air. The dust concentration is

$$C = \frac{M_e}{V_e} \qquad (9.22)$$

where
   C = dust concentration, mg/m³
   M_e = effective mass of dust
   V_e = effective volume of air

The total effective mass of dust M_e in the room is the initial suspended dust mass M_0 in the room plus the dust generated during the operation of the bagging machine M_{in} minus the dust that continously settles out of the air M_s during the time of operation. That is,

$$M_e = M_0 + M_{in} - M_s \qquad (9.23)$$

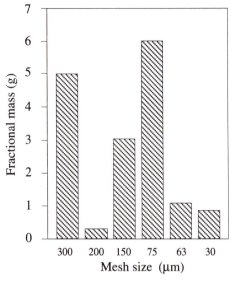

Figure 9.10 Gravel size distribution.

Since the dust generated in the room during time t includes the dust exiting the room via the ventilation system, it can be formulated as

$$M_{in} = \int_0^t \dot{M}_{in}\, dt = \dot{M}_{in} t \tag{9.24}$$

where $\dot{M}_{in}$ is the constant rate of mass of dust generated by the system.

The mass rate of dust generated was estimated by fractionating a sample of the gravel. An American Soil Testing Machine using soil-fractionating sieves was used. The mass distribution for dust particle sizes from 300 to 30 μm is shown in Figure 9.10. The mass of particles smaller than 30 μm was determined to be 1 gram for each gravel bag filled. Since the bag processing rate is 1 per minute, this gives a dust generation rate of 1 g/min for particles less than 30 μm in size. The vertical component of velocity for air moving through the bagging room was measured to be 10 cm/s. Since the settling velocities for particles 30 μm and larger in size is greater than the vertical air velocity component (see Table 9.7), particles in these size ranges will tend to settle and not contribute to airborne dust contamination within the room. This assumes that personnel working in the bagging room will not be breathing air through which the larger particles are settling (e.g., when the dust generation source is at a level below the worker's head). The mass rate of dust generated, M_{in}, minus the mass rate of the dust that settled, M_s, was thus estimated to be equal to 1 g/min.

In Eq. 9.22, the total effective volume V_e is the volume of the room plus the volume of air exiting the room during time t and can be defined as

$$V_e = V_0 + \int_0^t \dot{Q}\, dt = V_0 + \dot{Q}t \tag{9.25}$$

where V_0 is the volume of the room and $\dot{Q}$ is the ventilation flowrate. Equation 9.22

can be rewritten in the form

$$C = \frac{M_0 + \dot{M}_{in}t}{V_0 + \dot{Q}t} \qquad (9.26)$$

The air flowrate of the ventilation system was measured by using a hot-wire anemometer. The air flowrate in the ventilation system was found to be 56.6 m^3/min. The volume of the bagging room was measured to be 142.7 m^3. Knowing the mass rate of dust generated, $\dot{M}_{in}$, the ventilation flowrate, $\dot{Q}$, volume of the room, V_0, and the initial dust mass, M_0, the dust concentration inside the bagging room can now be estimated.

$$C = \frac{M_0 + \dot{M}_{in}t}{V_0 + \dot{Q}t}$$

$$C = \frac{0 + (1)(4 \times 60)}{142.7 + (56.6)(4 \times 60)} = 0.0175 \ g/m^3$$

$$C = 17.5 \ mg/m^3$$

The permissible eight-hour exposure limit (PEL) for inert or nuisance dust is 5 mg/m^3.[29] Particles less than 10 μm are considered to be respirable; thus, particles larger than this size are not hazardous to the health.[28]

The bagging machine has four hoppers that generate the dust. It was assumed that each hopper contributed an equal portion of the total dust produced by the gravel machine. With one hopper covered with a fabric dust arrestor, $\dot{M}_{in}$ = 0.75 g/min, with two hoppers, $\dot{M}_{in}$ = 0.5 g/min, and so on. Figure 9.11 shows the effect of eliminating individual hopper sources on the dust concentration in the

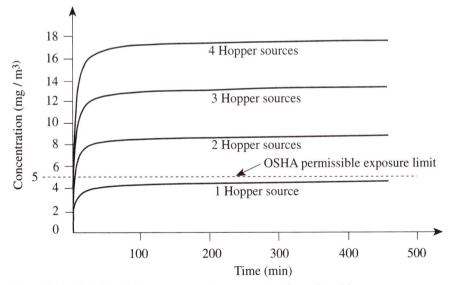

Figure 9.11   Calculated dust concentration versus time for reduced hopper sources.

[29]Contaminants-Permisible Exposure Limits, OSHA 3112, 1989.

room. It can be seen that reducing dust generation sources to one reduces the dust concentration below the PEL for nuisance dust.

Installation and Testing

The dust suppression method selected was to use fabric dust arrestors at three of the four dust generation points (see Figure 9.9) in the gravel bagging system. The fabric material selected was nylon-reinforced plastic, which was flexible, durable, and impermeable to dust. The main hopper was enclosed by using velcro around the elevator discharge and a draw rope around the rim of the hopper. This enclosure included a large zipper to allow quick access to the main hopper in case of system clogging. The transfer point between the exit chute of the main hopper and the inlet of the measuring hopper was covered using Velcro™ hook and loop fasteners at the top and a draw rope at the bottom. The transfer point between the measuring hopper and the guide chute was also covered by using these fasteners at each end.

After installation the air quality in the room was tested using a cascade inertial impactor type sampler which simulates the deposition of particles within the human respiratory system. Air is drawn through the sampler at the rate of 1 cfm by a small vacuum pump. The flowrate can be adjusted by using a bleed-off valve. The flowrate at the beginning and end of each test was measured using a Rockwell gas meter. As air enters the sampler, it is drawn through a series of seven orifices and impaction filters. The incoming particles are deposited at each stage according to size and weight.

The normal bagging rate is one bag per minute, which corresponds to a rate of 600 bags per eight-hour day. To test the system, a sleeve was fitted to the guide chute. The sleeve guided the exiting gravel into a wheelbarrow, allowing multiple bagging operations to be simulated before dumping was necessary. During the simulation of the bagging operation, three air sample tests were conducted for periods of approximately one hour each. The sample time was recorded, and the preweighed collection filters were removed for subsequent gravimetric analysis after each test. Air-borne particulate concentrations and size distributions were then determined. The concentration in the room was found to be 1.94 mg/m^3, which is less than the OSHA Permissible Exposure Limits for nuisance dust (5 mg/m^3).

Conclusion

The objectives of this project were satisfied by installation of the fabric dust arrestors. The dust arrestors require no maintenence, are easily removed, and are very durable. The dust concentration level in the bagging room was reduced to less than one half of the maximum level of nuisance dust recommended by OSHA. The operation of the bagging machine was only minimally affected by installation of the arrestors. The customer was pleased with the installation.

9.8 AIR CONTAMINATION FROM HAZARDOUS LIQUID SPILLS

The sector downwind of the source of a toxic air contaminant release point is defined as the hazard corridor. Hazardous liquid vapor, released into the atmosphere from near ground level, will expand and mix with the air. As the vapor is carried away by the wind, its density decreases as mixing occurs in a process referred to as turbulent or eddy diffusion. The rate of this diffusion is dependent on complex motions of the air, in both the horizontal and vertical planes. High wind speeds produce a high dilution factor, since the volume of air into which the vapor is introduced is large. A fluctuating wind direction will spread the vapor horizontally about the axis of the mean wind direction. The vapor will also be diffused in the vertical direction, and this is related to the thermal structure of the lowest part of the atmosphere. Terrain, trees, buildings, and diurnal considerations all affect these motions, making accurate predictions difficult. For these reasons, predictions are often made in statistical terms. Experimental evidence shows that three meteorological parameters need to be considered in determining the rate of atmospheric dispersion of small toxic chemical spills.

1. The mean wind speed $\bar{V}$ as an indicator of the downwind plume travel from a continuous source.

2. The standard deviation of wind direction σ_A which is an indicator of the lateral rate of mixing.

3. The temperature gradient ΔT which is an indicator of the vertical rate of mixing.

For decreasing temperatures with increasing altitude, the atmosphere is considered unstable, and for increasing temperatures with increasing altitude, the atmosphere is considered stable. An unstable atmosphere is characterized by significant vertical and horizontal mixing and large contaminant concentration reduction, whereas a stable atmosphere implies little mixing and little reduction in contaminant concentration.

For toxic chemical spills the normalized peak concentration C_p/Q at distance x is generally the quantity desired. The regression equation for this expression usually takes the form

$$\frac{C_P}{Q} = Kx^a \bar{V}^b \sigma_A^c (\Delta T + t)^d \qquad (9.27)$$

where K, a, b, c, and d are parameters of fit and t is a constant added to ΔT to avoid raising a negative number to a power.[30]

The Ocean Breeze/Dry Gulch (OB/DG) dispersion model is used extensively by the USAF Air Weather Service for predicting the hazard zone resulting from accidental toxic chemical spills. This is an empirical and statistical model derived from the Ocean Breeze, Dry Gulch, and Prairie Grass experiments initiated during the late 1950s. The OB/DG model is based on data from three different geographical locations with significant variations in terrain, vegetation, and meteorological conditions. The equation used with this model is an adaptation of the regression equation (Eq. 9.27), which relates the hazard distance to the source strength and the temperature difference between heights of 6 and 54 ft above the ground.[31]

$$\frac{C_P}{Q} = 0.000175x^{-1.95}(\Delta T + 10)^{4.92} \qquad (9.28)$$

---

[30]"Predicting Diffusion of Atmospheric Contaminants by Consideration of Turbulent Characteristics of WSMR," Frank V. Hansen, Atmospheric Sciences Laboratory, White Sands Missile Range, ECOM-5170, January 1968.

[31]"The Development of a Dispersion Model to Replace the Ocean Breeze–Dry Gulch Model," Bruce A. Kunkel, *Atmosphere Transport and Diffusion Modelling*, S&EPS Workshop, June 1985.

where

C_P = concentration at distance x,g/m^3
Q = source strength, g/s
ΔT = temperature difference between 54 and 6 ft above ground, F°
x = downwind distance, m

Note that the mean wind speed $\bar{V}$ and the standard deviation of wind direction σ_A are not included in Eq. 9.28. It has been found that the contribution of $\bar{V}$ to predictions made by using Eq. 9.28 are not significant.[32,33] For estimates of the width of the hazard corridor, the lateral movement of the plume must be accounted for, and σ_A must be included in the calculation. The width of the corridor can be estimated using W = $8\sigma_A$ with 95 percent confidence that this value will not be exceeded.[30]

To calculate the hazard corridor length for a particular propellant, Eq. 9.28 can be rearranged and the units simplified to yield

$$x\text{(in feet)} = 2.62(\frac{C_P}{Q})^{-0.513}(\Delta T + 10)^{2.53} \qquad (9.29)$$

and

$$x\text{(in feet)} = 2.20(\frac{C_P}{Q})^{-0.513}(\Delta T + 10)^{2.53} \qquad (9.30)$$

Equations 9.29 and 9.30 are valid for a continuous point source (large spills) for two well-known aerospace propellants, nitrogen tetroxide (N_2O_4) and unsymmetrical dimethylhydrazine (UDMH). For instantaneous releases (small spills) of these propellants, the following equations hold:

$$x\text{(in feet)} = 4.23(\frac{C_P}{Q})^{-0.513}(\Delta T + 10)^{2.53} \qquad (9.31)$$

and

$$x\text{(in feet)} = 3.55(\frac{C_P}{Q})^{-0.513}(\Delta T + 10)^{2.53} \qquad (9.32)$$

For other fuels or oxidizers, the constants of Eqs. 9.29, 9.30, 9.31, and 9.32 must be modified accordingly.

Determining the length of the hazard corridor resulting from a spill also requires an estimate of the source strength Q, which is the quantity of gas or vapor released into the atmosphere per unit time. For the liquid propellants mentioned above, a spill results in the formation of a pool, which then evaporates into the air. Estimating the source strength is a complex calculation involving Reynolds' number, the molar mass velocity of the air, the molecular weight and vapor pressure of the spilled liquid, the equilibrium temperature of the liquid pool, and a stability index that is a function of the atmospheric temperature profile and the surface roughness. The Ille and Springer model is probably the most comprehensive source strength model, since it includes an energy balance computation for pool temperature and also uses a stability index.[30] The U.S. EPA has also published information on chemical spills that accounts for source strength by providing tabular data for specific chemicals based on the boiling point, vapor pressure, and temperature of the spilled chemical.[34] Hazard corridor length is then determined as a function of the *level of concern* and *rate of release*, which have to do with the health hazard to personnel and the rate at which the chemical is released from containment. Because of the complexity and site-peculiar nature of source strength estimates, they are often based on experimental data. Vaporization rate data have been previously reported by NASA for nitrogen tetroxide and several members of the hydrazine family.[35] These values are as follows:

1. N_2O_4 (NO_2 Vapor)—0.22 to 0.50 lb/ft^2 min with the larger rate occurring at the start of the spill.

2. UDMH—0.25 lb/ft^2 min for large spills and 0.065 lb/ft^2 min for small spills (1 gal or less).

3. MMH—0.089 lb/ft^2 min for large spills and 0.021 lb/ft^2 min for small spills (1 gal or less).

4. N_2H_4—0.025 lb/ft^2 min for large spills and 0.005 lb/ft^2 min for small spills (1 gal or less).

For elevated releases, the basic dispersion equation (generalized Gaussian plume equation)[36] is used:

$$C(x, y, z, H) = \frac{Q}{2\pi\sigma_z\sigma_y\mu} \exp\left[-\frac{1}{2}\left(\frac{y}{\sigma_y}\right)^2\right]\left\{\left[\exp-\frac{1}{2}\left(\frac{z-H}{\sigma_z}\right)^2\right]\right.$$

$$\left.+\left[\exp-\frac{1}{2}\left(\frac{z+H}{\sigma_z}\right)^2\right]\right\} \tag{9.33}$$

where
Q = the source strength
C = the concentration
μ = the wind speed
x = the distance in the direction of the mean wind
y = the cross wind distance
z = the height above the ground
H = the effective height of emission
σ_y = the standard deviation in the cross wind direction of the plume concentration distribution.
σ_z = the standard deviation in the vertical of the plume concentration distribution.

[34]"Technical Guidance for Hazards Analysis," U.S. EPA, Federal Emergency Management Agency, December 1987.
[35]"Propellant Evaporation Rate Study," NASA Report PP-516, 1969.
[36]D. B. Turner,"Workbook of Atmospheric Dispersion Estimates," Air Resources Field Research Office, ESSA, U.S. Department of Health and Human Services, Cincinnati, OH, 1967.

The effective height of emission is dependent on the height of the vent stack and the plume rise. Plume rise is a function of the temperature of the effluent at the top of the stack, the exit velocity of the effluent, stack diameter, air temperature, barometric pressure, wind speed at the top of the stack, and atmospheric stability. Other factors being equal, the higher the effluent temperature relative to the free air temperature, the greater will be the plume rise. Plume rise is also directly proportional to stack diameter.

—— *Example 9.4* ——

A test facility involved in testing small rocket engines using the propellants nitrogen tetroxide (N_2O_4) and unsymmetrical dimethylhydrazine (UDMH) is attempting to establish operational criteria concerning routine and emergency spills. The test facility is to be enclosed by a perimeter fence and, since activities outside of this boundary are not under the control of facility personnel, operating procedures must be conducted in a manner that will ensure that vapor concentration levels at the boundary fence do not exceed allowables. Accordingly, the fence location relative to the facility in which propellant operations are conducted must be such that the time-weighted average threshold limit value (TLV-TWA) and ceiling limits (TLV-C) established by OSHA and EPA for these two contaminants are not exceeded.

TLV-TWA is the threshold limit value–time weighted average concentration for a normal 8-hour workday and a 40-hour workweek, to which nearly all workers may be repeatedly exposed, day after day, without adverse effect. The threshold limit value–short-term exposure limit (TLV-STEL) is the concentration to which workers can be exposed continuously for a short period of time without suffering from (1) irritation, (2) chronic or irreversible tissue damage, or (3) narcosis of sufficient degree to increase the likelihood of accidental injury, impair self-rescue, or materially reduce work efficiency, provided that the daily TLV-TWA is not exceeded. It is not a separate independent exposure limit; rather, it supplements the time-weighted average (TWA) limit where there are recognized acute effects from a substance whose toxic effects are primarily of a chronic nature. STELs are recommended only where toxic effects have been reported from high short-term exposures in humans or animals. A STEL is defined as a 15-minute time-weighted average exposure that should not be exceeded at any time during the workday even if the 8-hour time-weighted average is within the TLV. Exposures at the STEL should not be longer than 15 minutes and should not be repeated more than four times per day. There should be at least 60 minutes between successive exposures at the STEL. An averaging period other than 15 minutes may be recommended when this is warranted by observed biological effects. The threshold limit value ceiling (TLV-C) is the concentration that should not be exceeded during any part of the working exposure. In conventional industrial hygiene practice, if instantaneous monitoring is not feasible, then the TLV-C can be assessed by sampling over a 15-minute period, except for those substances that may cause immediate irritation with exceedingly short exposures. For some substances (e.g., irritant gases), only one category, the TLV ceiling, may be relevant. For other substances, either two or three categories may be relevant, depending on their physiologic action. It is important to observe that if any one of these three TLVs is exceeded, a potential hazard from that substance is presumed to exist. The limit for the propellants used at this test facility are as follows:

UDMH: TWA = 0.5 ppm,   Ceiling = 100 ppm

N₂O₄: TWA = 3 ppm,   Ceiling = 30 ppm

Diffusion prediction equations are used to make calculations for contaminate levels. It should be emphasized that diffusion predictions are estimates subject to many random atmospheric motions and are, therefore, to be used as the best available guideline—not as absolute values. Equations 9.29, 9.30, 9.31, and 9.32 are generally reliable within a factor of two for hazard corridor lengths of a few hundred meters and within a factor of three for hazard corridor lengths of a few hundred kilometers.

Source Strength

The source strength is determined using the following rationale,

1. For large oxidizer spills (emergency spills) the vaporization rate is 30 percent of the total quantity spilled in 10 minutes or 3 percent/min, but not less than 12 lb/min.

2. For oxidizer spills of 1 gallon or less, the total quantity is assumed to evaporate in 1 minute.

3. For bulk fuel spills (emergency) the source strength is 0.25 lb/min−ft² of exposed surface area for UDMH. This source strength is based on a 10-minute spill evaporation time.

4. For fuel spills of 1 gallon or less, the source strength is 0.065 lb/min−ft² of UDMH. One gallon will cover an area of approximately 50 ft².

Operations Involving Nitrogen Tetroxide (N_2O_4)

For routine (instantaneous) ground releases the use of Eq. 9.31 allows the determination of hazard corridor lengths for all source strength and lapse rates of interest. Based on the size of the spill anticipated (which is a function of the type of operation planned), the measured lapse rate (ΔT), and the acceptable contaminate level, the hazard corridor length is determined. For a C_p = 3 ppm, Q = 3 lb/min and a $\Delta T = 0°F$:

$$x = 4.23(\frac{C_P}{Q})^{-0.513}(\Delta T + 10)^{2.53}$$

$$x = 4.23(3/3)^{-0.513}(0 + 10)^{2.53}$$

$$x = 1434 \text{ ft}$$

For emergency (continuous) ground releases, Eq. 9.29 is used. Assuming C_p = 30 ppm, Q = 12 lb/min, and $\Delta T = 0°F$,

$$x = 2.62(\frac{C_P}{Q})^{-0.513}(\Delta T + 10)^{2.53}$$

$$x = 2.62(30/12)^{-0.513}(0 + 10)^{2.53}$$

$$x = 555 \text{ ft}$$

Operations Involving Unsymmetrical Dimethylhydrazine (UDMH)

For routine (instantaneous) ground releases Eq. 9.32 is used. For a $C_p = 0.5$ ppm, $Q = 3$ lb/min, and $\Delta T = 0°F$,

$$x = 3.55(\frac{C_p}{Q})^{-0.513}(\Delta T + 10)^{2.53}$$

$$x = 3.55(0.5/3)^{-0.513}(0 + 10)^{2.53}$$

$$x = 3016 \text{ ft}$$

For emergency (continuous) ground releases, Eq. 9.30 is used. For a $C_p = 100$ ppm, an exposed surface area of 100 ft², and $\Delta T = 0°F$,

$$x = 2.20(\frac{C_p}{Q})^{-0.513}(\Delta T + 10)^{2.53}$$

$$x = 2.20(100/25)^{-0.513}(0 + 10)^{2.53}$$

$$x = 366 \text{ ft}$$

9.9   OCCUPATIONAL SAFETY AND HEALTH

The Occupational Safety and Health Act was signed into law on December 29, 1970. This law is enforced by OSHA, an agency of the Department of Labor. The purpose of the act is to assure working men and women safe and healthful working conditions and to preserve the nation's human resources. The act applies to every employer engaged in business affecting commerce. It does not apply to the self-employed or to workers to the extent they are covered by other federal safety and health laws. It is estimated that the act covers six million workplaces employing 75 million workers.[37]

The job safety and health standards included in this act provide rules for the elimination or avoidance of hazards proven by research and experience to be harmful to personal safety and health. Standards are enforced by OSHA compliance officers who can conduct inspections at any workplace covered by the act. Employers can require the inspector to provide a warrant. Representatives of employers and employees accompany the inspector during the inspection, which is normally scheduled according to priorities established by OSHA. However, employees can request an inspection if an imminent danger exists that could cause death or serious physical harm. When a violation is found, a written citation is issued requiring abatement within an appropriate time period. Depending on the seriousness of the violation and whether it is willful or repeated, a penalty can be assessed up to $10,000 for each violation. Any employer who fails to correct a violation for which a citation has been issued can be penalized up to $1000 per day for each day the hazard persists beyond the abatement day specified in the citation. A willful violation by an employer that results in the death of a worker is punishable, on conviction, by a fine of up to $10,000 or imprisonment for up to six months, or both. Criminal penalties are also included in the act for making false statements or giving unauthorized advance notice of an OSHA inspection.[38]

[37] *OSHA Reference Book*, U. S. Dept. of Labor, OSHA 3081, 1985.
[38] *All About OSHA*, U. S. Dept. of Labor, OSHA 2056, 1985.

The act encourages states to assume the fullest responsibility for the administration and enforcement of their job safety and health laws through the vehicle of an approved state plan. Up to 50 percent federal funding is provided for these state programs. Once a state receives approval from OSHA to operate its own safety and health plan, OSHA continues to monitor the state's performance and can withdraw approval if the state should fail to continue to meet the criteria established for state plans. In FY 1984, 25 states and territories conducted 110,135 inspections, with a staffing level of 1065 compliance officers. Federal grants to states with safety and health plans totaled approximately $53 million in FY 1985. The act also includes provisions for conducting training and education programs designed to assist employers and employees in avoiding unsafe or unhealthy work practices.[37] In 1979, NIOSH sponsored an Engineering Control Technology Workshop, which concluded that there was a critical need to include occupational safety and health (OS&H) in the education of engineers. In 1980, the Public Health Service issued a report entitled "Promoting Health/Preventing Disease: Objectives for the Nation." This report includes a major objective: "By 1990, at least 70 percent of all graduate engineers should be skilled in the design of plants and processes that incorporate OS&H control technologies." Accordingly, NIOSH initiated Project SHAPE to encourage improvements in engineering practice, education, and research. This project has included a series of workshops to further these goals as well as to establish an engineering school faculty network to ensure that any educational resource material developed by NIOSH is made available to prospective users.

9.9.1 Occupational Safety and Health in Design

The National Safety Council[39] has recommended fundamental guidelines for designers to ensure adequate safety and health for products and processes. These guidelines are listed below in order of descending effectiveness. These rules are inclusive, and as many as possible should be applied, within the constraints of manpower, cost, and schedule.

1. Eliminate hazards by changing the design, the materials used, or the maintenance procedures.

2. Control hazards by capturing, enclosing, or guarding at the source of the hazard.

3. Train personnel to be cognizant of hazards and to follow safe procedures to avoid them.

4. Provide instructions and warnings in documentation and post them in appropriate locations.

5. Anticipate credible abuse and misuse and take appropriate action to minimize the consequences.

6. Provide appropriate personal protective equipment and establish procedures to ensure that it is used as required.

Engineers must be able to identify hazards associated with their designs and to quantify the relative severity and likelihood of occurrence. Safety hazards normally result in accidents that occur over a relatively brief period of time and for which the acute effects are readily apparent. The effects of health hazards, on the other hand, may not be apparent for some time, often months or years, but the results can be

---

[39] *Accident Prevention Manual*, 9th Ed., National Safety Council, Chicago, 1986.

just as devastating.[40] Several techniques have been proposed as aids in the process of recognizing, quantifying, and reducing hazards. Haddon's 10 rules comprise one of the more widely recognized strategies:[41]

1. Prevent the creation of the hazard (e.g., prevent the production of hazardous and nonbiodegradable chemicals).

2. Reduce the magnitude of the hazard (e.g., reduce the amount of lead in gasoline).

3. Eliminate hazards that already exist (e.g., ban the use of chlorofluorocarbons).

4. Change the rate of distribution of a hazard (e.g., control the rate of venting a hazardous propellant).

5. Separate the hazard from that which is being protected (e.g., store flammable materials at isolated locations).

6. Separate the hazard from that which is being protected by imposing a barrier (e.g., separate fuel and oxidizer storage areas by using berms or other barriers).

7. Modify basic qualities of the hazard (e.g., use breakaway roadside poles).

8. Make the item to be protected more resistant to damage from the hazard (e.g., use fabric materials in aircraft that do not create toxic fumes when combusted).

9. Counter the damage already done by the hazard (e.g., move people out of a contaminated area).

10. Stabilize, repair, and rehabilitate the object of the damage (e.g., rebuild after a fire).

Although the above guidelines and rules do not encompass every possible safety consideration in a design project, they do provide a checklist against which the design can be evaluated and modified as necessary. The designer must develop the habit of constantly evaluating the design for safety, considering not only the design itself but the personnel involved in fabricating the product, using the procedure, and in maintaining and repairing the product or system as well as the end user or purchaser. As indicated previously, developing the manufacturing processes as well as the maintenance and operating procedures early during the design process will assist in revealing safety problems at a time when corrective action can be taken at minimum cost. Assigning selected design team members, supported by appropriate specialists, to this task will usually pay off in the long run.

9.10 THE SAFETY ENGINEER

The organizational position of the safety engineer in many industry and government operations generates an inherent dichotomy, a sort of organizational "Catch 22" situation. Since the responsibility of safety engineers often involves making

---

[40]H. Gage, "Integrating Safety and Health into M. E. Capstone Design Courses," *ASEE Southwest Regional Conf.*, Texas Tech University, Lubbock, TX, 1989.

[41]W. Haddon, "The Basic Strategies for Reducing Damage from Hazards of All Kinds," *Hazard Prevention*, Sept./Oct 1980.

decisions that slow production, delay testing, and otherwise (apparently) impede the furtherance of the organization's overall assigned mission, they are normally assigned organizationally to a staff function at managment level. This places the safety engineer in the position of being an outsider to the production organization on which his or her decisions often have the greatest impact and, unfortunately, often results in poor communication, mutual suspicion, and general uncooperativeness between the two, with only grudging compliance with safety decisions on the part of the production organization. The justification for this organizational approach is that the safety engineer must be free of coercion in making decisions and, therefore, must not be under the influence of the production group, which might tend to interpret safety considerations by their effect on production levels or schedule. The schism that often exists between the element of the organization that is trying to meet schedules, production quotas, and other commitments and the safety group is a serious issue and can have an enormous impact on the overall effectiveness of the company or government organization involved. The writer's experience in association with a relatively small test organization adjacent to a very large government facility provides a good example of the effect that this schism can have on productivity and responsiveness. In this case the small test organization was frequently requested to perform testing for elements of the large government facility that could have been easily accomplished by test organizations within the government facility. However, the unrealistic, uncompromising attitude and "high-handed" approach adopted by the safety group within the government facility often worked to delay proposed test effort significantly and unnecessarily, and caused test organizations to seek more responsive and cooperative support.

It is often stated that the best organizational approach in the world will not compensate for poor quality employees and, in contrast, if an organization has good, dedicated employees that feel an ownership responsibility for their work, the work will be accomplished effectively in spite of a poor organizational approach. If this can be applied to the position of safety engineer, then it should be possible to overcome the difficulties referred to above, but the individual(s) assigned the safety engineering responsibility must possess an extraordinary amount of common sense, be the kind of person(s) that other employees enjoy working with, and exhibit a level of technical knowledge that will earn the respect of those around him or her. Most importantly, the person assigned as safety engineer must be sensitive to the overall needs and goals of the organization and must render decisions that further these goals in a safe manner rather than enforcing blanket compliance with broad safety edicts that are narrowly interpreted and applied. The logic of safety decisions must be recognized by the workers affected to be accepted and complied with willingly.

The responsibilities of safety engineers vary widely but generally include the preparation of safety procedures and other safety-related documents, safety training, issuance and control of personal safety equipment, consultation and advice on safety aspects of new or changed processes, and inspection of ongoing work to ensure that appropriate safety requirements are being complied with. Safety engineers often work closely with reliability and quality assurance personnel and, in some organizations, these two functions are placed in the same group. The safety engineer's interface with the production-level working group is of paramount importance, however, and cooperation and trust at this level may be enhanced by having a separate safety engineering function.

Engineers in all disciplines and in every workplace have a responsibility to ensure that the products of their labors incorporate an in-depth consideration of safety. The design of products and processes should consider approaches that will ensure safe manufacture, maintenance, use (including alternate uses), and retirement on completion of useful life. Decisions concerning the safety features necessary (or desirable) during fabrication, assembly, maintenance, and first use are necessarily driven by economic considerations, including potential liability as well as initial product cost. Design accommodations for potential failures involving safety during alternate (or second) product use and for ultimate product retirement are more difficult to justify but, nevertheless, should be considered and adopted or abandoned by conscious decision, after evaluation. Society in the United States is very litigious, and engineers need to be especially aware of this in product design. Unfortunately, it is no longer the simple questions of gross negligence, intent, or even ignorance that determines fault. If the accident was *foreseeable* in the view of the jury, sizable awards, sometimes driving companies into bankruptcy, are often made, even when the product was misused and the fault lies entirely with the user. Further complicating this issue, the definition of the term *foreseeable* is a moving target that changes with time. What may have been *unforeseeable* during the design phase and on the day of the accident may be considered very *foreseeable* on the following Monday morning, or on the day of the trial. In spite of this seemingly "no-win situation," engineers must attempt to foresee potential safety problems with their designs and, for the credible occurrences, must provide some solution. In case of later litigation, if the product safety evaluation has been accomplished adequately and the judicial process properly recognizes the distinction between *foreseeable* and *credible*, the product's safe design should be substantiated.

9.11 CLOSURE

It should be obvious that the material presented in this chapter constitutes only a cursory treatment of the very broad subject of safety and environmental protection. Some understanding of the relative roles of the EPA and OSHA is important; thus, a brief description of the responsibilities and authority of these organizations has been included. Since the pollution problem involves air, land, and water, some limited discussion of the principal considerations affecting these has also been included. However, the pollution of the earth's oceans and rivers, which is certainly a serious problem, is not discussed at all. The examples and design problems also represent only a fraction of the type of pollution abatement techniques in use or under development, but it is hoped that they will serve to introduce the student to this technology and pique his or her interest to delve into this subject in greater depth. There is a vast amount of material available on the subjects mentioned in this chapter, and there are certainly enough challenges to provide a very fertile field for future engineers.

BIBLIOGRAPHY

1. BALL, W. B., JONES, M. D., and KAVANAUGH, M. C., "Mass Transfer of Volatile Organic Compounds in Packed Tower Aeration," *Journal WPCF*, vol. 56, no. 2, pp. 127–136, Feb. 1984.

2. KAVANAUGH, M. C. and TRUSSEL, R. C., "Design of Aeration Towers to Strip Volatile Contaminants from Drinking Water," *Journal AWWA*, pp. 684–692, Dec. 1980.

3. PERRY & CHILTON, Chemical Engineers' Handbook, 4th Edition, McGraw-Hill, New York.

4. WILKE, C. R. and CHANG, PIN, "Correlation of Diffusion Coefficients in Dilute Solutions," *AICHE Journal*, pp. 264–270, June 1955.

5. National Water Summary 1986, U.S. Geological Survey Water Supply Paper 2325, 1988.

6. *Handbook on Ground Water*, Environmental Protection Agency Research Laboratory EPA/625/016, March 1987.

7. TREYBAL, R. E., *Mass Transfer Operations*, McGraw-Hill, New York, 1980.

8. *National Air Quality and Emissions Trends Report*, U.S. Environmental Protection Agency, April 1991.

9. Air Contaminants—Permissible Exposure Limits, Title 29 Code of Federal Regulations, Part 1910-1000, 1989.

10. *Accident Prevention Manual*, 9th Edition, National Safety Council, Chicago, 1986.

Chapter 10

Engineering Ethics

ETHICS: The science or doctrine of the sources, principles, sanctions, and ideals of human conduct and character; the science of the morally right.

10.1 ETHICS IN INDUSTRY

During the past two decades U.S. industry has experienced a veritable explosion of interest in the development and teaching of corporate ethics. According to the Ethics Resource Center in Washington, DC, a 1979 study of Fortune 500 companies and the top 150 service firms in the United States revealed that 73 percent had written standards of ethics, half of which were five years old or less. A decade later, better than 9 out of every 10 companies from 2000 surveyed reported having standards of ethical conduct. In addition, about 11 percent of these companies provide ombudsmen, with whom employees can discuss ethical concerns and other problems.[1] Only since the mid-1970s have articles on ethics been commonly found in the more well-known engineering journals. Today, it is increasingly common to find articles on ethics in journals published by professional societies like the American Society of Mechanical Engineers, the American Society of Chemical Engineers, the American Society of Civil Engineers, and the Institute of Electrical and Electronics Engineers.[2] This is a significant change from the situation in 1959 when the writer, who was then a new first-level supervisor for a large aerospace company, attempted to get the company to establish a policy pertaining to employees accepting vendor lunches. At that time, very few companies had any published ethical standards, and most were reluctant to establish standards even when requested to do so by their own employees. Needless to say, the company refused to establish a policy regarding vendor lunches and, to keep from going on record, failed to respond to the request in any manner.

---

[1]"Teaching Ethics Takes Off in U. S. Firms," *Engineering Times,* June 1989.
[2]M. W. Martin and R. Schinzinger, *Ethics in Engineering,* 2nd ed., McGraw-Hill, New York, 1989.

In today's environment, ethics is clearly recognized by leading companies as a critical element in business and policy decisions and as a valuable tool in resolving difficult management problems. However, training in corporate ethics is not restricted to just those employees that have wide authority to make decisions for the company. Increasingly, industry is recognizing that all employees, regardless of position or assignment, need to maintain high standards of conduct on the job for the company's performance to be sustained and enhanced. Company performance depends on the actions of its employees, acting alone or in concert with other employees, and these actions are strongly influenced by the employee's perception of the company and his or her knowledge and understanding of the company's policies and goals. Techniques used by industry to provide employee instruction in corporate ethics vary widely. One method that seems to be effective is based on the use of case studies concerning problems the firm has faced in the past and how to anticipate, recognize, and avoid situations involving questionable ethics in like situations. Training programs can also include specific issues and company policies having to do with ethical questions. If the company has a policy in regard to the acceptance of lunches and other gratuities from vendors, this can be specifically spelled out and discussed during the training sessions.

Effective and ongoing communication with employees is the key to a successful corporate ethics program. General Dynamics Corporation has instituted a Standards of Business Ethics and Conduct Booklet as part of their corporate ethics program. These standards alert employees to occasions where misconduct might occur because of ignorance, misunderstanding, or a mistaken notion of what is expected. They provide simple, explicit guidelines to follow to avoid the fact, and sometimes even the appearance, of misconduct. Additional methods of employee communication used in this program include the use of videotapes featuring top management and other employees, stories on the program in internal employee news publications, posters advertising particular features of the program, and annual (and other) reports that include informative articles on the program. Twenty-nine ethics hotlines are provided throughout the company with a toll-free 800 number to the corporate office. This gives employees a readily available channel to ask questions, seek advice, voice concerns, or raise allegations. In 1987, the director of the ethics program at General Dynamics received 5760 communications from employees, 56 percent of these came over the hotlines. The goals of this program are to support individual employees in their daily business conduct, to enhance the administrative performance of the company in basic business relationships, and to help build the bond of trust between the company and its customers, suppliers, employees, shareholders, and the communities in which it functions.[3]

10.2   ETHICS AND THE UNIVERSITY

In spite of the growth of industrial ethics programs, there remains considerable controversy about the effectiveness of teaching ethics in universities. Many people believe that a person's ethical outlook is established at a young age under the influence of family, church, friends, and elementary and secondary school teachers, and that this outlook cannot be changed or influenced significantly at the young adult stage, the university entry age level. It is obvious from the emphasis being placed on ethics in universities that people involved in education believe that a

---

[3]"Ethics Program Update," General Dynamics Corporation, 1988.

person's basic value system can be influenced and reinforced, however marginally, by providing an appropriate environment, one that encourages strong beliefs about honesty, personal integrity, fair play, and wholesome competition. The belief is that university students can be educated to deal more effectively with ethical dilemmas by being exposing to these situations in a classroom environment where they can be discussed freely with both peers and professors. Case studies involving situations typical of those encountered by professional engineers in industry, government, and private practice can also be used to demonstrate ethical principles and applicable codes. With this type of training the student should learn how to recognize situations in which ethical difficulties often arise at an early stage so that an alternative course of action can be devised. The student will also become familiar with the particular code of ethics applicable to his or her engineering discipline and how this code is enforced and applied.

The Accreditation Board for Engineering and Technology (ABET), which is the officially recognized organization for accrediting university programs in the United States, makes the following statement in regard to curriculum requirements: "An understanding of the ethical, social, economic, and safety considerations in engineering practice is essential for a successful engineering career. Course work may be provided for this purpose, but as a minimum it should be the responsibility of the engineering faculty to infuse professional concepts into all engineering course work."[4] Thus, it is a requirement for universities that wish to maintain and improve their accreditation standing to include material on ethics in their engineering programs. This can be accomplished by integrating ethical concepts into lectures, problems, and design projects, including a study of ethics as a separate topic within an upper level course and/or developing a formal course in professionalism with specific attention to the role of ethics. ABET has developed guidelines to assist engineering institutions in their efforts to include ethics and professionalism in engineering curricula. These guidelines suggest that the following areas of instruction be included in any formal course on ethics and professionalism: an introduction, which should include justification for the course; some material on philosophical ethics, including several of the major ethical theories; concepts of professionalism, which should emphasize the importance of registration; codes of ethics, including their purpose and form; and pertinent case studies, which should emphasize real-life ethical dilemmas.[5]

10.3   THE FOUNDATION OF ETHICS

A technological revolution has occurred during the twentieth century. With this technological revolution has come a social revolution and an understanding of the interrelationship of these two is important to the subject of engineering ethics. At the beginning of this century the horse-and-buggy was the primary means of personal transportation; it seems that the twentieth century will end with the space shuttle making its way to a space station orbiting the earth. Possibly a select few astronauts will be planning a trip to another planet. At the beginning of this century there were no automobiles, no airplanes, no radios, no telephones, no television, no spacecraft, and no atom bombs. Most people had traveled within only a few miles of

---

[4]Accreditation Board for Engineering and Technology 1988 Annual Report, pp. 110–111.

[5]"Guide on Professionalism and Ethics in Engineering Curricula," Accreditation Board for Engineering and Technology, July 1989.

their homes and to contemplate man traveling to the moon was incomprehensible. The United States was isolationist in outlook, and the primary concerns were with basics such as food and shelter, friends, and loved ones. In small towns the local hardware store represented the state of technology, and on hot summer afternoons the visit of the local ice wagon (and later truck) on its daily delivery route was the high point. Even in the larger cities, the attention of industry was focused primarily on domestic needs and social life was largely neighborhood oriented. Today, there are few people who have not traveled to at least one other country, commuting 50 miles a day to work is considered quite normal, concerns and interests are global, and most people barely know their neighbors well enough to speak to them. Our problems have more to do with substance abuse, pollution, greenhouse effects, ozone layers, crime, abortion, and disintegration of the family than with having enough food on the table or with the sick neighbor down the street. The enormous technological changes that have occurred have had dramatic and lasting effects on personal values. A good argument could be made that value systems have failed to keep pace with the technological and sociological changes that have occurred in this century.

Ethics has its roots in secular and religious philosophy. Evidence of a moral value system in secular society extends back into recorded history to the Egyptian, Babylonian, Greek and Roman empires. Hammurabi, king of Babylon in the eighteenth century B.C., established the following building code:

> If a builder has built a house for a man and has not made the work sound, and the house which he has built has fallen down and so caused the death of the householder, that builder shall be put to death. If it causes the death of the householder's son, they shall put that builder's son to death. If it causes the death of the householder's slave, he shall give slave for slave to the householder. If it destroys property he shall replace anything it has destroyed; and because he has not made sound the house which he has built and it has fallen down, he shall rebuild the house which has fallen down from his own property. If a builder has built a house for a man and does not make his work perfect and the wall bulges, that builder shall put that wall into sound condition at his own cost[2].

It is doubtful that early Babylon had to deal with many unethical builders. The Greek philosopher Socrates, the fifth century B.C., also contributed to the development of secular ethics with his dialogues on virtue, justice, and piety. Socrates equated virtue with knowledge of one's true self, holding that no one knowingly does wrong. Aristotle (384–322 B.C.) transformed the common Greek word for habit, *ethikos,* into the modern concept of ethics.[6] To Aristotle these habits were the ways in which individuals work to fulfill their potential. Later philosophers like John Locke (1632–1704), Immanuel Kant (1724–1804), and John Stuart Mill (1806–1873) further developed the field of secular ethics to include concepts and theories about morality that have to do with maximizing good, duties to and respect for other people, and human rights. Contemporary philosophers have further defined and developed these ideas. Alasdair MacIntyre suggests that belief in a goal or "end" that transcends immediate gratification is the key to moral conduct. He emphasizes

---

[6]J. L. Gottlieb, "How We Should Think about Ethics in Business," *Stanford Business School Magazine,* February 1988.

the "internal good' such as the creation of useful and safe products over "external goods" such as fame and prestige.[7]

The religions of the world have provided much of our moral understanding and ethical development. Even religions that do not include belief in God in their tenets, such as Buddhism, Confucianism, and Taoism, emphasize a *right path* from which is derived a code of ethics.[2] In the United States, Judeo/Christian beliefs are held by most people, and ethical value systems are largely based on writings in the Bible. The Old Testament includes a significant amount of text concerning the way a person should live. The most notable text in this regard is the Ten Commandments, Exodus 20:3–17. The New Testament also has much to say about value systems and ethics. Possibly the most widely quoted verse from this portion of the Bible is the statement of Jesus in Matthew 7:12 which is paraphrased as the Golden Rule:

> Therefore all things whatsoever ye would that men should do to you, do ye even so to them.(KJV)

It can be seen from the above that the principal difference in secular and Judeo/Christian ethics is that secular ethics are defined by man and as such, are subject to change and interpretation, depending on the interpreter and the time in which he or she lives, whereas Judeo/Christian ethics are based on God's word, the Bible, which does not change but is subject to interpretation (Figure 10.1). The former is transitory or situational, and the latter is permanent or absolute. In spite of these significant differences, secular and Judeo/Christian ethics seem to have, at least, one of the same goals—that of striving to give meaning to a person's life through morally right living. Both seem to based on the premise that people are not driven to moral behavior, but decide to behave morally for the sake of a cause to which they are committed, for a person they love, or for the sake of their God.[8] If each person has a unique cause or mission in life, then there are

Figure 10.1 And man has been interpreting it since. (From National Review, June 16, 1989, p 36.)

[7]MacIntyre, *After Virtue*, Notre Dame University Press, Notre Dame, IN, 1984.
[8]V. E. Frankle, *Man's Search for Meaning*, Washington Square Press, Inc., New York, 1963, p. 158.

concrete tasks and assignments that must be carried out to accomplish this mission. By this premise each individual is unique and irreplaceable. Each individual's path is traversed only once and cannot be repeated. This is strong encouragement indeed for applying high ethical standards to all relationships and dealings as individuals and as engineers and for quickly rectifying mistakes made along the way.

10.4   ETHICS IN ENGINEERING

On October 10, 1973, Spiro T. Agnew resigned as Vice-President of the United States amidst charges of bribery and tax evasion related to his previous position as County Executive of Baltimore County, Maryland. Agnew was a civil engineer and lawyer and as County Executive during the years 1962–1966 he had authority to award contracts for public works projects to architect/engineering firms. In exercising that authority he benefited from a lucrative kickback scheme in which certain firms were given special consideration in receiving contracts for public works projects if they made secret payments in return[2]. Several architect/engineering firms were involved in this illegal kickback scheme, so engineers cannot justify Agnew's conduct on the basis that he was a lawyer or politician. Although this may be the most notable incident in recent times of illegal and unethical conduct by engineers, it is certainly not unique. However, it does present a problem as to what can be done about intentional ethical misconduct in the engineering profession. A few lectures or even a course on ethics will not keep a person with the wrong goal from being tempted to bend, or break, the rules of acceptable moral and legal conduct. If the goal in an individual's life is to amass a large amount of money, great power, or enormous prestige, one will probably be susceptible to shortcuts and get-rich schemes. If, on the other hand, a person's goal is to build a career and reputation as a good engineer and trustworthy individual, such schemes will have little appeal. In all probability the only deterrent to people that intentionally act in an unethical and illegal manner is to increase the likelihood of apprehension and to make the penalties severe. This can be achieved when illegal acts are involved, but when the act is unethical, but not illegal, the responsibility for meting out the appropriate punishment falls on the profession itself, which is poorly equipped to deal with such occurrences.

Breach of ethical conduct can also occur because of ignorance. We live in a complicated and often confusing world in which the rules governing professional conduct are constantly expanding and being reinterpreted. Although industry is rapidly correcting this problem, there are many working engineers that are not aware of company ethical policies and guidelines, nor are they familiar with professional codes of ethics that apply to their particular engineering discipline. This is a problem that can be solved simply by providing adequate training at the university level augmented by further, and more specific, training as the graduating engineer enters industry. Fundamentals of the codes as well as techniques and strategies for handling ethical decisions and challenges can be taught in the university environment. Case studies can be used to clarify the principles covered in the codes and to analyze ethical dilemmas in which moral principles are in conflict or are ambiguous. Needs, such as finding a match between the graduating student's ethical value system and that of potential employers can be discussed. Managing the situation in which one's values are tested by peers, subordinates, or superiors is an especially fertile field for discussion and analysis in the university classroom.

ABET has defined engineering as "...the profession in which a knowledge of the mathematical and natural sciences gained by study, experience and practice is applied with judgment to develop ways to utilize, economically, the materials and forces of nature for the benefit of mankind."[9] It is with the words "with judgment" that engineers have the greatest ethical difficulty. This is not a question of knowing or unwitting violation of ethical codes and moral values, but one in which a choice must be made between options, all of which have varying degrees of ethical value and liability. Judgment is one of those attributes that varies from individual to individual and, on a larger scale, from company to company. That an individual's judgment is strongly affected by his or her ethical value system, as well as education and experience, seems to be self-evident. Corporations, as well as individual engineers, are increasingly being called on to justify and defend their judgment in regard to technical decisions as a result of structural/equipment failures and accidents. The Space Shuttle *Challenger* accident, the DC10 cargo door failures, and the Ford Pinto gasoline tank fires are all notable examples of situations in which judgment and ethical practices have been questioned. The ethical dilemma presented by each of these examples is that in all cases the decision maker was advised as to the possible outcome of the decision but considered the extremely low likelihood of failure, and the attendant consequences, to be favorable to the alternative. The corporation's financial loss associated with making the recommended change(s) was undoubtedly uppermost in the mind of the decision maker in these examples, but other factors invariably affect decisions of this magnitude, consciously or unconsciously. The personal cost to the decision maker in terms of lost prestige and advancement potential as a result of trying to obtain concurrence for such a costly decision could well have been involved, as well as the ethical trade-off between responsibility to the using public for safety vis-á-vis financial management responsibility to the employer.

It is in this area of "judgment" that a lot of second guessing is encountered, especially after an incident has occurred. Nothing is as clear as hindsight. This is not to excuse faulty judgment but is pointed out to emphasize the importance of full and open discussion of all the pertinent facts and ethical consequences, and their appropriate weighting, when making significant decisions. It is extremely important that decision makers make decisions on the basis of a full and complete discussion of the facts and that they not be influenced by extraneous matters, such as personal ego, interpersonal relationships, customer/contractor relationships and authority, history of the program, or how hard the decision may be to sell. In the case of the *Challenger* accident, the decision-making process failed, not due to the process itself, but because of the extraneous factors that interfered with making a decision based on the facts. The customer/contractor relationship obviously got in the way of appropriate consideration of the facts. Success in previous launches led the customer, NASA, to believe that the launch could be successfully accomplished in spite of the engineering data that indicated otherwise. NASA management ego was probably one of the factors that led to the decision to overrule the recommendations of the engineers close to the problem. The NASA project managers at the Marshall Space Flight Center did not accept the contractor's initial technical decision as being valid and, using intimidation, persuaded the contractor to consider the "management" consequences. Contractor management was overwhelmed by the

[9]Accreditation Board for Engineering and Technology 1988 Annual Report, p. 108.

threatening conduct of the customer and subsequently agreed with NASA over the recommendations of their own engineers, a sad commentary on engineering decision making and ethical conduct. *Technical decisions must be made on their own merit–personal feelings have no place in this process and, when they are involved, will almost always result in less than the best decisions.*

What can be accomplished in the university environment to strengthen the ethical judgment of students? Is it possible to infuse values that have not been developed earlier by the student's family, religious institutions, friends, and teachers? Probably not, but the university educational experience will have an influence on the shape and structure of the student's value system, and it is the university's responsibility to ensure that this influence is positive. The student should experience strong and positive value reinforcement in all encounters with faculty and university administration. A value-reinforcing climate should be nurtured in all campus facilities and functions, including dormatories, recreation centers, student newspapers, and student organizations. Finally, some strengthening and modifying of value systems can be realized by using classroom techniques that encourage students to test their own value systems. This can be achieved by using case studies that present ethical dilemmas that demonstrate credible engineering problems.

10.5   LEGAL RESPONSIBILITIES OF ENGINEERS

Law and the legal profession dominate modern American society to an unprecedented extent. The impact of law and our legal system falls on engineers to, at least, as great an extent as any other profession or occupation.[10]

First, one may not practice the profession of engineering, at least, in most states, without satisfying certain legal requirements of training, education, examination, and experience. Second, there are numerous federal, state, and local laws, regulations, and ordinances (such as building codes) that govern the work that engineers do. Third, the work of engineers is almost always performed, and disputes resolved, pursuant to some contractual agreement, which is itself governed by various provisions of the law. Finally, engineers cannot practice their profession without remaining aware at all times of the potential of liability for mistakes in judgment, intentional wrongdoing, or even inadvertent mistakes.

The laws of each state of the United States require that an engineer be registered before being allowed to practice professional engineering in that state. Typically, although state laws vary on the requirements, a person must satisfy requirements of education and experience and pass an examination to be registered as a professional engineer. A person who practices professional engineering without satisfying state registration requirements is subject to various penalties and fines. Registration laws and requirements vary from state to state, and registration in one state does not give the engineer the right to act as a professional engineer in another state. However, many states have reciprocal agreements under which registration is simplified if the engineer is already registered in another state.

---

[10]This section relies on R. C. Vaughn, *Legal Aspects of Engineering,* Kendall/Hunt Publishing, Dubuque, IA, 1977; J. Sweet, *Legal Aspects of Architecture, Engineering, and the Construction Process,* West Publishing, St. Paul, MN, 1977; and J. Miller, *Architect/Engineer Liability: A Growth Period,* American Bar Association, Chicago, IL, 1984.

Because engineering is now considered to be a profession, it is, like other professions, said to be self-regulating. This phrase usually means that the members of the profession act together to ensure that other members of the profession act ethically and competently. Professions typically organize associations and societies that, in turn, adopt codes of ethics to govern professional conduct. This self-regulation is most often informal and based on peer pressure and reputation within the profession. However, many states have incorporated professional codes of ethics into the law governing the registration of engineers. In those states, violation of the code of ethics can subject the professional to legal penalties up to the revocation of the license to practice the profession.

Apart from direct regulation from the state, the law impacts on engineers greatly through the imposition of civil liability. There has been a tremendous increase in the number of lawsuits filed against engineers in the last 20 to 30 years. There are two basic ways in which an engineer may become liable to another party or parties. The English common law, which is the historical basis for the American legal system, distinguished between duties owed as result of mutual consent, that is, by contract, and duties owed by all persons to act with reasonable care, skill, and diligence so as not to injure other persons or property. The law governing liability imposed in the absence of a contract is known as torts.

Contracts are a critically important aspect of the legal rights of engineers. Virtually all engineering work is done pursuant to some contract. Most often that contract is in writing and includes detailed language and many provisions. Any time a party, including an engineer, enters into a contract, his or her rights and responsibilities are defined with reference to the language of the contract. No one should ever sign a written contract without reading the entire contract and understanding its terms. In the case of most written contracts, courts are not sympathetic to claims that a person has not read a contract he or she signed or that the person did not understand the contract's language.

Ordinarily, a breach of a provision of a contract entitles another party to the contract to recover money damages. That is done by filing a civil lawsuit for breach of contract. Most construction contracts to which engineers are parties, however, contain a provision that all disputes arising under the contract will be submitted to arbitration. Arbitration is a procedure in which a dispute is submitted to an arbitrator or panel of arbitrators, selected by the parties, and whose decision on the matter is final and binding. Arbitration has several advantages over litigation: it usually saves significant time and cost, can be kept private and confidential if the parties so desire, and the decision can be made by a person or persons who have more knowledge and experience in the field than do judges and juries. In most states, and in all contracts that involve interstate commerce, the courts will enforce an agreement to arbitrate and will enforce the decision of the arbitrator.

In contrast to contract law, under the law of negligence, persons may be held liable even in the absence of a prior contractual agreement with another party. In general, persons owe a duty to other persons to exercise reasonable care in going about their business so as not to injure other persons. If someone acts negligently or carelessly, that person will be liable to anyone who is injured as a foreseeable consequence of the negligent act. The term "malpractice" is used to refer to negligence by professionals. Malpractice is commonly defined as a dereliction of professional duty or a failure of professional skill or learning that results in injury, loss, or damage. All professionals owe a duty to their clients and to the public to exercise skill and ability sufficient to enable one to perform professional services at least ordinarily and reasonably well. However, undertaking professional services does not imply or warrant a satisfactory result. An error of judgment is not necessarily evidence of a lack of skill or care. The question of liability in malpractice is ultimately a question of performing as a reasonable professional would perform under the same or similar circumstances. Often, in addition, allowance is made for the locality where the defendant professional practices. Thus, the question of liability often becomes a question of the credibility of expert witnesses on the issue of how a reasonable engineer would have performed under the same circumstances.

A relatively recent development in the law of torts is the imposition of liability even in the absence of negligence or other fault, known as strict liability. Manufacturers of products can be held liable if a defective product causes injury, even in the absence of proof of negligence in the design or manufacture of the product. The law of strict liability varies from state to state, and its impact on engineers is not certain, but engineers should be aware of the possibility of strict liability.

10.6   CODES OF ETHICS

All of the major professions in the United States have adopted codes of ethics to provide guidance and support for their profession and its membership. These codes deal with common problems like competency, confidentiality, and conflicts of interest and, generally, are more beneficial in the guidance that they provide than they are in a negative, or disciplinary, sense. Engineering codes of ethics encourage ethical conduct and provide guidance concerning the obligations of engineers. Codes provide support to engineers seeking to act ethically in situations that involve conflict with their superiors and serve as a deterrent to unethical conduct. Codes also enhance the profession's public image and generally encourage professionalism.

There are many codes of ethics within the profession of engineering; generally each separate discipline has adopted its own code. However, all of these codes include the basic principles incorporated in the code of ethics published by the National Society of Professional Engineers (NSPE), which is a voluntary organization of engineering professionals dedicated to serving the public and the profession. NSPE is the umbrella organization for the state professional engineering societies and, as such, represents all of the individual engineers registered to practice under the various state registration laws:

> NSPE strives to insure the application of engineering knowledge and skills in the public interest and to foster public understanding of the role of engineering in society. Further, the NSPE promotes the professional, social and economic interests of the engineering profession and its individual members.

One of the goals of NSPE is to provide "professionwide leadership on selected professional issues of importance to all segments of engineering." One of the objectives under this goal is "to seek common acceptance, understanding and enforcement of ethical standards."[11]

It is pursuant to this objective that NSPE has adopted and promulgated a Code of Ethics for Engineers. The NSPE Code of Ethics establishes standards of conduct

---

[11] *Texas Society of Professional Engineers Chapter/State Reference Handbook*, p. III-2, 1986.

Figure 10.2 Trying to eliminate activities unfavorable to the engineering profession.(From *National Review*, Sept. 29, 1989, p. 36)

for all engineers. It is enforceable only as applied to members through the action of the state societies. Each state society has adopted the Code as the state society's code, with minor changes in some states. The NSPE issues interpretations of the Code in particular fact circumstances through its Board of Ethical Review (Figure 10.2)[12]. As indicated previously, in some states the principal elements of the NSPE Code have become state law through the enactment of laws concerning professional engineering registration within the state. In these cases, disciplinary action can be considerably more severe.

<hr>

EXERCISE 10.1

<hr>

Jack Smithfield Case Study

<hr>

NOTE: This case is fictional and does not describe an actual event or situation. It was provided by the General Dynamics, Ft. Worth, TX, Division, and was prepared to facilitate discussion of important management issues, not to demonstrate the proper or improper handling of a business situation.

Jack Smithfield, the division general manager at General Dynamics (GD), La Jolla, CA, division, whistled contentedly as he opened the crisp, new envelope at the top of his mail file. Marked "Personal," the innocuous looking envelope was to command his time for the balance of the day. "Jack!" began the yellow "sticky-note" attached at a skewed angle atop the thick document inside. Glancing quickly at the bottom of the note, Jack saw that the document came from one of the division's best marketing managers, Peter Franklin. "Attached you will find an interesting document that came into my possession recently. It is not a 'classified document' which might run afoul of the DoD's current campaign. I know you will make good use of it in preparing our bid. And don't worry, the copy is genuine."

Thumbing through the substantial document, Jack immediately realized it was a series of cost figures for Southcorp Enterprises, GD's main competitor

[12]R. A. Gorlin, "Codes of Professional Responsibility," Bureau of National Affairs, Inc., 1986.

for the advanced tactical fighter contract currently up for production bids. GD and Southcorp had both produced prototypes under Navy contracts and now sought the production contract. Although the Navy might require a second producer down the road, the winner of this first contract would get the lion's share of the production. Winning the contract was top priority for GD as a whole as well as for Jack's division. As clearly as Jack could tell, the cost figures were those prepared by Southcorp for inclusion in their bid.

Jack immediately wondered where Peter got the document. He suspected that service program officers were not above passing such a document to help out their favored contractor. He also suspected that documents flowed all too freely around Washington, especially for offers of "reciprocal aid." These and other disquieting thoughts raced through his mind.

Peter's note said nothing about the source of the document. He assumed, or maybe hoped, Peter was not paying cash for such information, but he wondered what type of "reciprocal aid" might have been involved or how else the document may have been obtained. Jack could immediately tell the document was going to be a goldmine of information for the preparation of their own bid, which was due in just two weeks. But he also wondered whether the recent ethics flap and GD's commitment to ethics, reemphasized in mid-1985, had any bearing on the division's possession and use of the document.

Competitive analysis and data gathering were certainly important parts of doing business these days. Just three months ago, Jack himself had given a strong talk to the division's top 100 managers on the importance of getting reliable intelligence on competitors' capabilities and plans. After all, competitor analysis was all the rage in the corporate strategy literature, as well as within GD itself.

To reassure himself, Jack called in his executive assistant and his marketing vice-president to discuss the document. The marketing vice-president could immediately see its usefulness and argued: "Jack, you know this happens all the time. Documents get passed around so freely that it is all part of the game. Our competitors know we are trying to learn as much about them as they are about us. I would assume Southcorp has various of our documents too. Don't worry about it."

Jack's administrative assistant also commented on the value of the document but raised some questions that made Jack uncomfortable. "Suppose Peter did something 'wrong' to get this thing—couldn't it blow up in our faces? Can GD afford another scandal? And even if he didn't buy it, is it any different if he promised to do his source a favor some other time?"

"Come on," said the marketing vice-president, "let's not get carried away. I agree with you that obtaining something like this might be wrong, but Peter is not dumb, and he wouldn't do something wrong. And, as for favors, the business system runs on favors. We are always helping each other out with useful information. That's the only reason this capitalist system runs so efficiently."

"I have another concern," said the executive assistant. "We've been hiring outside consultants to give us briefings on our competitors and their plans. I have wondered where they get all that juicy data, particularly the stuff on costs. Are we some kind of accomplice if they use questionable means to get that data we pay so highly for?" "Are you talking about that market research

I have been buying?" said the marketing vice-president. "Why Jack, you even slapped that fellow on the back last week and said 'good job'—remember that, Jack? We couldn't do our job in marketing without that kind of research. It is the consultant's responsibility to stay on the up and up, not ours."

The marketing vice-president addressed the executive assistant, "I think you have entirely too narrow a view on competitive information. Whenever we hire someone from government or from another company, one of the benefits we get is the knowledge of the government's plans or the general capabilities of their former employer. With military or government types, we can't have them represent us before their former groups—I know that. But no one has ever said it is unethical to hire such people for their general knowledge and information. That's why I have been having my marketing staff debrief many of the people we hire from competitors and the government."

"Well, I have some questions about that too," said the executive assistant. "Shouldn't we be giving our people some guidance about what types of information we can 'mine' from new hires and what we can't? Isn't all this too close to espionage for comfort?"

Smithfield ushered the two rancorous executives out of his office and sat down to ponder what he should do about the document still sitting on the center of his desk. The debate between his two executives unnerved him a bit. He knew he had a decision to make concerning the document—but now he wondered whether his own recent comments on the need to do good competitor analysis contributed to Peter's efforts to secure and send along this document. Should he then see and use his handling of this incident as an opportunity to give some clearer direction to his marketing organization about the limits of competitor intelligence gathering? If so, what kind of signal did he want to give?

Jack knew an Ethics Steering Committee had been formed as part of the new GD ethics program, but he also knew they had held a meeting last week and would not meet again for another three and one-half weeks. Besides, as a hardened business veteran, he had some questions about how realistic a decision he might get from a committee with so many corporate staff types on it. The committee had burned him last week by not letting him make a contribution to a retirement gift for a distinguished general, as innocent a gesture as he could imagine, particularly when Southcorp and two other defense contractors were already contributing. He still had not made the embarrassing call to renege on the verbal commitment he had made on that one. Maybe it was just better to decide for himself what to do with this document.

Questions

1. What should Jack do with the document and why?

2. Does it matter if the document is classified?

3. Does it matter if the document is proprietary? What obligation does a firm have regarding its own proprietary information? Does Southcorp's negligence (if any) remove the proprietary considerations?

4. Does it matter how Peter got the document? What if it was obtained from a disgruntled Southcorp employee, from a consultant, or from the customer? What if the document was found in a bar?

5. What if it wasn't a document—what if the information was overheard? Does it matter? Why? Suppose it was overheard on a bus, or in a car by a consultant. Does it matter if the information was paid for or free? Or if it came from a disgruntled Southcorp employee?

6. What other obligations does Jack have? Why? Should he tell the customer, Southcorp? Someone in General Dynamics? Should he disqualify himself, Peter, and others from work on the proposal? Should he discipline Peter, or was he responsible for Peter's conduct?

7. What is proper conduct in gathering information? Did Jack give encouragement to get information? Should division policies and practices be developed? Corporate policies and practices? How would you initiate such policies?

8. What about the policy of "mining" new employees hired from competitors and government? Is it ethical to hire these people for the information they possess?

10.7 CODE RULES AND INTERPRETATIONS

The references given in the following sections are taken directly from the NSPE Code of Ethics for Engineers and the particular code paragraphs are identified in parenthesis. Some liberty has been taken in the selection of the particular Code sections quoted, but an attempt has been made to address those elements considered to be of paramount concern to students about to become engineers. The cases used to demonstrate the various sections of the Code have been provided by NSPE, and the NSPE Board of Ethical Review (BER) Case Numbers are identified by referenced footnote. It is recommended that cases be thoroughly discussed relative to the Code sections quoted and that the BER's decisions be consulted only after discussion is completed. The complete Code of Ethics for Engineers is included at the end of this chapter and it is recommended that all students using this textbook familiarize themselves with this document.

Case Study 10.1

A. Participation in Production of Unsafe Equipment

Engineers of Company A prepared plans and specifications for machinery to be used in a manufacturing process, and Company A turned them over to Company B for production. The engineers of Company B in reviewing the plans and specifications came to the conclusion that they included certain miscalculations and technical deficiencies of a nature that the final product might be unsuitable for the purposes of the ultimate users, and that the equipment, if built according to the original plans and specifications, might endanger the lives of persons in proximity to it. The engineers of Company B called the matter to the attention of appropriate officials of their employer who, in turn, advised Company A of the concern expressed by the engineers of Company B. Company A replied that its engineers believed that the design and specifications for the equipment were adequate and safe and that Company B should proceed to build the equipment as designed and specified. The officials of Company B instructed its engineers to proceed with the work.

Question: What are the ethical obligations of the engineers of Company B under the stated circumstances?[13]

B. Whistle-blowing

Engineer A is employed by a large industrial company that engages in substantial work on defense projects. Engineer A's assigned duties relate to the work of subcontractors, including the review of the adequacy and acceptability of the plans for material provided by subcontractors. In the course of this work, Engineer A advised his superiors by memoranda of the problems he found with certain submissions of one of the subcontractors and urged management to reject such work and to require the subcontractors to correct the deficiencies he outlined. Management rejected the comments of Engineer A, particularly his proposal that the work of a particular subcontractor be redesigned because of Engineer A's claim that the subcontractor's submission represented excessive cost and time delays.

After the exchange of further memoranda between Engineer A and his management superiors, and continued disagreement between Engineer A and management on the issues he raised, management placed a critical memorandum in his personnel file, and subsequently placed him on three months' probation, with the further notation that if his job performance did not improve, he would be terminated.

Engineer A has continued to insist that his employer had an obligation to ensure that subcontractors deliver equipment according to the specifications, as he interprets same, and thereby to save substantial defense expenditures. He has requested an ethical review and determination of the propriety of his course of action and the degree of ethical responsibility of engineers in such circumstances.

Question: Does Engineer A have an ethical obligation, or an ethical right, to continue his efforts to secure change in the policy of his employer under these circumstances, or to report his concerns to proper authority?[14]

References

1. (II.1.a) Engineers shall at all times recognize that their primary obligation is to protect the safety, health, property and welfare of the public. If their professional judgement is overruled under circumstances where safety, health, property or welfare of the public are endangered, they shall notify their employer or client and such other authority as may be appropriate.

2. (II.1.b) Engineers shall approve only those engineering documents that are safe for the public health, property and welfare in conformity with accepted standards.

---

[13]NSPE Board of Ethical Review, Case No. 65–12, Vol. II, p. 35, 1966.
[14]NSPE Board of Ethical Review, Case No. 82–5, Vol. VI, p. 23, 1989.

3. (II.2.b)Engineers shall not complete, sign, or seal plans and/or specifications that are not of a design safe to the public health and welfare and in conformity with accepted engineering standards. If the client or employer insists on such unprofessional conduct, they shall notify the proper authorities and withdraw from further service on the project.

Case Study 10.2

Confidentiality of Engineering Report

Engineer A offers a homeowner inspection service, whereby he undertakes to perform an engineering inspection of residential property for prospective purchasers. Following the inspection, Engineer A renders a written report to the prospective purchaser. Engineer A performed this service for a client (husband and wife) for a fee and prepared a one page written report, concluding that the residence under consideration was in generally good condition requiring no major repairs, but noting several items needing attention. Engineer A submitted his report to the client showing that a carbon copy of it was sent to the real estate firm handling the sale of the residence. The client objected that such action prejudiced their interests by lessening their bargaining position with the owners of the residence. They also complained that Engineer A acted unethically in submitting a copy of the report to any others who had not been a party to the agreement for the inspection services.

Question: Did Engineer A act unethically in submitting a copy of the home inspection report to the real estate firm representing the owners?[15]

References

1. (II.1.c)Engineers shall not reveal facts, data, or information obtained in a professional capacity without the prior consent of the client or employer except as authorized by law or this code.
2. (II.4)Engineers shall act in professional matters for each employer or client as faithful agents or trustees.

Case Study 10.3

Signing and Sealing Plans Not Prepared by an Engineer

Engineer A is the chief engineer in a large engineering firm, and affixes his seal to some of the plans prepared by registered engineers working under his general direction who do not affix their seals to their plans. At times, Engineer A also seals plans prepared by nonregistered, graduate engineers working

---

[15]NSPE Board of Ethical Review, Case No. 82–2, Vol. VI, p. 15, 1989.

under his general supervision. Because of the size of the organization and the large number of projects being designed at one time, Engineer A finds it impossible to give a detailed review or check of the design. He believes he is ethically and legally correct in not doing so because of his confidence in the ability of those he has hired and who are working under his general direction and supervision. By general direction and supervision, Engineer A means that he is involved in helping to establish the concept, the design requirements, and review elements of the design or project status as the project progresses. Engineer A is consulted about technical questions, and he provides answers and directions in these matters.

Question: Is it ethical for Engineer A to seal plans that have not been prepared by him, or which he has not checked and reviewed in detail?[16]

References:

1. (II.2.a) Engineers shall undertake assignments only when qualified by education or experience in the specific technical fields involved.

2. (II.2.b) Engineers shall not affix their signatures to any plans or documents dealing with subject matter in which they lack competence, nor to any plan or document not prepared under their direction and control.

3. (II.2.c) Engineers may accept assignments and assume responsibility for coordination of an entire project and sign and seal the engineering documents for the entire project, provided that each technical segment is signed and sealed by the qualified engineers who prepared the segment.

Case Study 10.4

Public Criticism of Bridge Safety

Engineer A, a renowned structural engineer, is hired for a nominal sum by a large city newspaper to visit the site of a state bridge contruction project, which has had a troubled history of construction delays, cost increases, and litigation primarily as a result of several well-publicized, on-site accidents. Recently the state highway department has announced the date for the opening of the bridge. State engineers have been proceeding with repairs based on a specific schedule.

Engineer A visits the bridge and performs a one-day visual observation. Her report identifies, in very general terms, potential problems and proposes additional testing and other possible engineering solutions. Thereafter, in a series of feature articles based on information gleaned from Engineer A's report, the newspaper alleges that the bridge has major safety problems that jeopardize its successful completion date. Allegations of misconduct and incompetence are made against the project engineers and the contractors as

[16]NSPE Board of Ethical Review, Case No. 86–2, Vol. VI, p. 76, 1989.

well as the state highway department. During an investigation by the state, Engineer A states that her report was intended merely to identify what she viewed were potential problems with the safety of the bridge and was not intended to be conclusive as to the safety of the bridge.

Question: Was it ethical for Engineer A to agree to perform an investigation for the newspaper in the manner stated?[17]

References:

1. (II.3.a) Engineers shall be objective and truthful in professional reports, statements or testimony. They shall include all relevant and pertinent information in such reports, statements, or testimony.

2. (II.3.b) Engineers may express publicly a professional opinion on technical subjects only when that opinion is founded on adequate knowledge of the facts and competence in the subject matter.

3. (II.3.c) Engineers shall issue no statements, criticisms or arguments on technical matters that are inspired or paid for by interested parties, unless they have prefaced their comments by explicitly identifying the interested parties on whose behalf they are speaking, and by revealing the existence of any interest the engineers may have in the matters.

4. (III.2.a) Engineers shall seek opportunities to be of constructive service in civic affairs and work for the advancement of the safety, health, and well-being of their community.

5. (III.3.a) Engineers shall avoid the use of statements containing a material misrepresentation of fact or omitting a material fact necessary to keep statements from being misleading or intended or likely to create an unjustified expectation; statements containing an opinion as to the quality of the Engineer's services; or statements intended or likely to attract clients by the use of showmanship, puffery, or self-laudation, including the use of slogans, jingles, or sensational language or format.

Case Study 10.5

Engineer's Disclosure of Potential Conflict of Interest

Engineer A is retained by the state to perform certain feasibility studies relating to a possible highway spur. The state is considering the possibility of constructing the highway spur through an area that is adjacent to a residential community in which Engieer A's residence is located. After learning of the proposed location for the spur, Engineer A discloses to the state the fact that his residental property may be affected by the new spur and fully discloses the potential conflict with the state. The state does not object to Engineer A's performing the work. Engineer A proceeds with his feasibility study and

---

[17]NSPE Board of Ethical Review, Case No. 88–7, Vol. VI, p. 117, 1989.

ultimately recommends that the spur be constructed. The highway spur is constructed.

Question: Was it ethical for Engineer A to perform the feasibility study despite the fact that his land may be affected thereby?[18]

References:

1. (II.4) Engineers shall act in professional matters for each employer or client as faithful agents or trustees.

2. (II.4.a) Engineers shall disclose all known or potential conflicts of interest to their employers or clients by promptly informing them of any business association, interest, or other circumstances that could infuence or appear to influence their judgment or the quality of their services.

Case Study 10.6

Objectivity of Engineer Retained as Expert

Engineer A is a forensic engineer. He is hired as a consultant by Attorney Z to provide an engineering and safety report and courtroom testimony in support of a plaintiff in a personal injury case. Following Engineer A's review and analysis, Engineer A determines that he cannot provide an engineering and safety analysis report favorable to the plaintiff because the results of the report would have to suggest that the plaintiff and not the defendant was at fault in the case. Engineer A's services are terminated and his fee is paid in full. Thereafter, Attorney X, representing the defendant in the case, learns of the circumstances relating to Engineer A's unwillingness to provide a report in support of Attorney Z's case and seeks to retain Engineer A to provide an independent and separate engineering and safety analysis report. Engineer A agrees to provide the report.

Question: Was it ethical for Engineer A to agree to provide a separate engineering and safety analysis report?[19]

References:

1. (II.1.c) Engineers shall not reveal facts, data, or information obtained in a professional capacity without the prior consent of the client or employer except as authorized or required by law or this Code.

2. (II.3.a) Engineers shall be objective and truthful in professional reports, statements, or testimony. They shall include all relevant and pertinent information in such reports, statements, or testimony.

3. (II.3.b) Engineers shall not accept compensation, financial or other-wise, from more than one party for services on the same project, or for

---

[18] NSPE Board of Ethical Review, Case No.85–6, Vol. VI, p. 70, 1989.
[19] NSPE Board of Ethical Review, Case No. 85–4, Vol. VI, p. 64, 1989

services pertaining to the same project, unless the circumstances are fully disclosed to, and agreed to, by all interested parties.

4. (III.4.b) Engineers shall not, without the consent of all interested parties, participate in or represent an adversary interest in connection with a specific project or proceeding in which the engineer has gained particular specialized knowledge on behalf of a former client or employer.

Case Study 10.7

Gifts to Engineer

Engineers A, B, and C are principals or employees of a consulting engineering firm that does an extensive amount of design work for private developers. The engineers are involved in recommending to the developers a list of contractors and suppliers to be considered for selection on a bidding list for construction of the projects. Usually, the contractors and suppliers recommended by the engineers for the selected bidding list obtain most of the contracts from the developers. Over a period of years, the officers of the contractors or suppliers developed a close business and personal relationship with the engineers of the firm.

From time to time, at holidays or on birthdays of the engineers with whom they dealt, the contractors and suppliers would give Engineers A, B, and C personal gifts of substantial value, such as home furnishings, recreational equipment, gardening equipment, and the like.

Question: Was it ethical for Engineers A, B, and C to accept gifts from the contractors and suppliers?[20]

Reference

1. (II.4.c) Engineers shall not solicit or accept financial or other valuable consideration, directly or indirectly, from contractors, their agents, or other parties in connection with work for employers or clients for which they are responsible.

2. (II.5.b) Engineers shall not offer, give, solicit, or receive, either directly or indirectly, any political contribution in an amount intended to influence the award of a contract by public authority, or which may be reasonably construed by the public of having the effect or intent to influence the award of a contract. They shall not pay a commission, percentage, or brokerage fee in order to secure work except to a bona fide employee or bona fide established commercial or marketing agencies retained by them.

3. (III.5.b) Engineers shall not accept commissions or allowances, directly or indirectly, from contractors or other parties dealing with clients or

[20]NSPE Board of Ethical Review, Case No.81–4, Vol. VI, p. 7, 1989.

employers of the Engineer in connection with work for which the Engineer is responsible.

Case Study 10.8

Conflict of Interest—Consultant to a Goverment Body; Member of a Local Authority

Engineer A, who is in full-time private practice, is retained by the county as county engineer for a stipulated monthly fee. His duties include reviewing plans and construction drawings to determine whether they meet county requirements, and making recommendations to local developers, county commissions, and the planning and zoning board. In addition, Engineer A is retained by the city as city engineer for a stipulated annual fee. His duties include making recommendations to the city council concerning the approval of completed engineering work. Engineer A also serves as project administrator for the county airport authority and, as such, is responsible for formulating a plan for the continued development of an airport industrial park. Finally, Engineer A is administrator of the city block grant program and, as such, oversees engineering work on various projects. Engineer A has been retained as a consultant by several private firms to help develop city and county project proposals.

Questions

1. May Engineer A, who serves as city engineer and county engineer for a retainer fee, provide engineering services in a private capacity to the city or county?

2. May Engineer A, who serves as a member of local boards or commissions that sometimes require the services of engineers, provide services through his private firm to those boards and commissions?

3. May Engineer A, who serves as city engineer and county engineer for a retainer fee, provide approval or render judgment on behalf of the city and county relative to projects on which Engineer A has furnished services through a private client?[21]

References

1. (II.4.d) Engineers in public service as members, advisors, or employees of a governmental body or department shall not participate in decisions with respect to professional services solicited or provided by them or their organizations in private or public engineering practice.

[21]NSPE Board of Ethical Review, Case No.82-4, Vol. VI, p. 19, 1989.

2. (II.4.e) Engineers shall not solicit or accept a professional contract from a government body on which a principal or officer of their organization serves as a member.

Case Study 10.9

Conflict of Interest—Duty of Loyalty of Terminated Employed Engineer to Employer; Misleading Brochure

Engineer A worked for Engineer B. On November 15, 1982, Engineer B notified Engineer A that Engineer B was going to terminate Engineer A because of lack of work. Engineer A thereupon notified clients of Engineer B that Engineer A was planning to start another engineering firm and would appreciate being considered for future work. Meanwhile, Engineer A continued to work for Engineer B for several additional months after the November termination notice. During that period, Engineer B distributed a previously printed brochure listing Engineer A as one of Engineer B's key employees, and continued to use the previously printed brochure with Engineer A's name in it well after Engineer B did in fact terminate Engineer A.

Questions

1. Was it ethical for Engineer A to notify clients of Engineer B that Engineer A was planning to start a firm and would appreciate being considered for future work while still in the employ of Engineer B?

2. Was it ethical for Engineer B to distribute a brochure listing Engineer A as a key employee in view of the fact that Engineer B had given Engineer A a notice of termination?

3. Was it ethical for Engineer B to distribute a brochure listing Engineer A as a key employee after Engineer A's actual termination?[22]

References

1. (II. 4) Engineers shall act in professional matters for each employer or client as faithful agents or trustees.

2. (II.5.a) Engineers shall not falsify or permit misrepresentation of their, or their associates', academic or professional qualifications. They shall not misrepresent or exaggerate their degree of responsibility in or for the subject matter of prior assignments. Brochures or other presentations incident at the solicitation of employment shall not misrepresent pertinent facts concerning employers, employees, associates, joint ventures or past accomplishments with the intent and purpose of enhancing their qualifications and their work.

3. (III.3.a) Engineers shall avoid the use of statements containing a material misrepresentation of fact or omitting a material fact necessary to keep

---

[22]NSPE Board of Ethical Review, Case No.83-1, Vol. VI, p. 30, 1989.

statements from being misleading; statements intended or likely to create an unjustified expectation; statements containing prediction of future success; statements containing an opinion as to the quality of the Engineers' services; or statements intended or likely to attract clients by the use of showmanship, puffery, or self-laudation, including the use of slogans, jingles, or sensational language or format.

4. (III.4.a) Engineers in the employ of others shall not without the consent of all interested parties enter promotional efforts or negotiations for work or make arrangements for other employment as a principal or to practice in connection with a specific project for which the Engineer has gained particular and specialized knowledge.

5. (III.7) Engineers shall not compete unfairly with other engineers by attempting to obtain employment or advancement or professional engagements by taking advantage of a salaried position, by criticizing other engineers, or by other improper or questionable methods.

Case Study 10.10

A. Political Contributions—Solicitation by Retained Consultant

Richard Roe, P.E., had a continuing series of engineering contracts with a three-member county board. While working under this arrangement, Roe in several successive elections headed a special campaign soliticitation committee which raised substantial amounts of money for the political campaigns of two incumbent county commissioners. Following their reelection in each case, Roe continued to receive engineering assignments from the county board. The political committee headed by Roe operated in strict accordance with applicable state and local laws governing political contributions.

Question: Was it ethical for Roe to continue to accept engineering assignments from the county board having engaged in raising funds for incumbent members of the county board?[23]

B. Payment of Fee to Landscape Architect Above true Value of Working in Order to Receive Leads

A local landscape architect, through a network of contacts, is able to locate engineering projects throughout the state. The landscape architect contacts Engineer A and proposes to refer these clients to Engineer A in return for a fee over and above the value of the landscaping work that the landscape architect would presumably perform on these jobs. Generally, little landscaping work is required on the project. Engineer A accepts the proposal.

Question: Was it ethical for Engineer A to accept the landscape architect's proposal to refer clients to Engineer A in return for a fee over and above the

[23]NSPE Board of Ethical Review, Case No.76-12, Vol. V, p. 23, 1981.

value of the landscape work that the landscape architect would presumably perform on each of the projects?[24]

References

1. (II.5) Engineers shall avoid improper solicitation of professional employment.

2. (II.5.b) Engineers shall not offer, give, solicit or receive, either directly or indirectly, any political contribution in an amount intended to influence the award of a contract by a public authority, or which may be reasonably construed by the public of having the effect or intent to influence the award of a contract. They shall not offer any gift, or other valuable consideration in order to secure work. They shall not pay a commission, percentage or brokerage fee in order to secure work except to a bona fide employee or bona fide established commercial or marketing agencies retained by them

Case Study 10.11

Engineer's Dispute with Client over Design

Client hired Engineer A to design a particular project. Engineer A develops what he believes to be the best design and meets with the client to discuss the design. After discussing the design plans and specifications, the client and Engineer A are involved in a dispute concerning the ultimate success of the project. The client believes Engineer A's design is too large and complex and seeks a simpler solution to the project. Engineer A believes a simpler solution will not achieve the result and could endanger the public. The client demands that Engineer A deliver over to him the drawings so that he can present them to Engineer B to assist Engineer B in completing the project to his liking. The client is willing to pay for the drawings, plans, specifications, and preparation but will not pay until Engineer A delivers over the drawings. Engineer A refused to deliver the drawings.

Question: Would it be ethical for Engineer A to deliver the plans and specifications to the client?[25]

References

1. (II.1.a) Engineers shall at all times recognize that their primary obligation is to protect the safety, health, property, and welfare of the public. If their professional judgment is overruled under circumstances where the safety, health, property, or welfare of the public are endangered, they shall notify their employer or client and such other authority as may be appropriate.

[24]NSPE Board of Ethical Review, Case No.83-5, Vol. VI, p. 41, 1989.
[25]NSPE Board of Ethical Review, Case No.84–4, Vol. VI, p. 51, 1989.

2. (II.1.b) Engineers shall advise their clients or employers when they believe a project will not be successful.

3. (II.1.e) Engineers having knowledge of any alleged violation of the Code shall cooperate with the proper authorities in furnishing such information or assistance as may be required.

Case Study 10.12

Participation in Protest Action as Part of a Political Campaign

Engineer A is a candidate for the state legislature from a district in which there is a substantial percentage of unskilled workers who are represented by a union. In a particular plant where many of these employees work, the third worker in a year was killed recently in an industrial accident. After many discussions between workers and management, the workers set up a picket line to protest what they claim are unsafe working conditions and alleged management indifference to employee safety. During the political campaign Engineer A visits the picket site and participates without having visited the plant to investigate the specific conditions of the previous accident. With cameras focused on her, Engineer A holds up a placard that accuses the company of callous disregard for the workers and then joins the protesting employees in the picket line.

Question: Was it unethical for Engineer A to accuse the company of callous disregard for the workers at the plant?[26]

References

1. (II.3) Engineers shall issue public statements only in an objective and truthful manner.

2. (II.1.e) Engineers shall not actively participate in strikes, picket lines, or other collective coercive action.

3. (II.1.f) Engineers shall avoid any act tending to promote their own interest at the expense of the dignity and integrity of the profession.

4. (III.2.a) Engineers shall seek opportunities to be of constructive service in civic affairs and work for the advancement of the safety, health, and well-being of their community.

Case Study 10.13

Demand for Promotion Based on Transfer of Client

Engineer A, a vice president of a broad-scope engineering firm that engaged in international engineering work through a wholly owned subsidiary firm,

[26]NSPE Board of Ethical Review, Case No.84–6, Vol. VI, p. 55, 1989.

was placed in charge of the subsidiary company. His responsibilities included the development of new business. He spent several years in that capacity and while overseas developed personal contacts with foreign agencies and their representatives. Engineer A was moderately successful in these endeavors, and in due course reported that he had arranged a very large and desirable contract with attractive profit potential. In accordance with the practice in the foreign country, the subsidiary firm was entitled to a substantial advance against its fee, but the parent firm was required to post a form of security for the advance through the furnishing of a letter of credit payable to the foreign agency in the event the firm defaulted. That was arranged and an irrevocable letter of credit was issued, and work proceeded under the contract with the subsidiary firm.

Subsequently, Engineer A told the parent firm that he demanded that he be promoted to presidency of the parent firm, given a commensurate increase in salary and other benefits, and be awarded a percentage of the profits flowing from the ongoing contract. If the parent firm refused his demands, he said, he was in a position to, and would, have the contract taken away from the subsidiary firm and awarded to another firm that he would establish. The officers of the parent firm were convinced that, because of the personal relationship of Engineer A to the officials of the foreign agency, he could, in fact, have the contract terminated and awarded to another firm, in which case the parent firm would risk a severe financial loss through the process of termination of work in progress and its recovery of the letter of credit, as well as the profit potential from the assignment.

Question: Was Engineer A ethical in making his demands under these circumstances?[27]

References

1. (II.4) Engineers shall act in professional matters for each employer or client as faithful agents or trustees.

2. (III.1) Engineers shall be guided in all their professional relations by the highest standards of integrity.

3. (III.1.f) Engineers shall avoid any act tending to promote their own interest at the expense of the dignity and integrity of the profession.

4. (III.4.a) Engineers in the employ of others shall not, without consent of all interested parties, enter promotional efforts or negotiations for work or make arrangements for other employment as a principal or to practice in connection with a specific project for which the Engineer has gained particular and specialized knowledge.

5. (III.4.b) Engineers shall not, without the consent of all interested parties, participate in or represent an adversary interest in connection with a specific project or proceeding in which the Engineer has gained particular specialized knowledge on behalf of a former client or employer.

6. (III.5) Engineers shall not be influenced in their professional duties by conflicting interest.

7. (III.7) Engineers shall not compete unfairly with other engineers by attempting to obtain employment or advancement or professional en-

---

[27]NSPE Board of Ethical Review, Case No.81-3, Vol. VI, p. 35, 1989.

gagements by taking advantage of a salaried position, by criticizing other engineers, or by other improper or questionable methods.

Case Study 10.14

Advertising Services of Engineering Staff

Smith & Jones, Inc., is an engineering and construction company. It carries an advertisement in various magazines with a heading "RENT-AN-ENGINEER." The text of the advertisement explains that Smith & Jones, Inc., offers its surplus engineering capacity on a rental basis, claiming that those using the services of such engineers benefit from having the services of experienced engineering personnel without increasing the permanent work force of the employer who utilizes the offer of Smith & Jones, Inc. The advertisement then lists various engineering disciplines that are available for rental and invites inquiries for details of its "Rent-an-Engineer" plan.

Question: Is the advertisement of Smith & Jones, Inc., a violation of the Code of Ethics?[28]

References

1. (III.3) Engineers shall avoid all conduct or practice which is likely to discredit the profession or deceive the public.

2. (III.3.a) Engineers shall avoid the use of statements containing a material misrepresentation of fact or omitting a material fact necessary to keep statements from being misleading; statements intended or likely to create an unjustified expectation; statements containing prediction of future success; statements containing an opinion as to the quality of the Engineer's services; or statements intended or likely to attract clients by the use of showmanship, puffery, or self-laudation, including the use of slogans, jingles, or sensational language or format.

Case Study 10.15

Joint Authorship of Paper

Engineers A and B are faculty members at a major university. As part of the requirement for obtaining tenure at the university, both Engineers A and B are required to author articles for publication in scholarly and technical journals. During Engineer A's years as a graduate student he had developed a paper that was never published and that forms the basis of what he thinks would be an excellent article for publication in a journal. Engineer A discusses his idea with Engineer B, and they agree to collaborate in developing the article. Engineer

---

[28]NSPE Board of Ethical Review, Case No.84-2, Vol.VI, p. 47, 1989.

A, the principal author, rewrites the article, bringing it up to date. Engineer B's contributions are minimal. Engineer A agrees to include Engineer B's name as coauthor of the article as a favor to enhance Engineer B's chances of obtaining tenure. The article is ultimately accepted and published in a refereed journal.

Questions

1. Was it ethical for Engineer A to use a paper he developed at an earlier time as the basis for an updated article?

2. Was it ethical for Engineer B to accept credit for developmeant of the article?

3. Was it ethical for Engineer A to include Engineer B as a coauthor of the article?[29]

References

1. (III.1) Engineers shall be guided in all their professional relations by the highest standards of integrity.

2. (III.3.c) Consistent with the foregoing, engineers may prepare articles for the lay or technical press, but such articles shall not imply credit to the author for work performed by others.

Case Study 10.16

City Engineer Seeking to Retain Employees of Engineering Firm Independent of Their Firm

The city of West Eastville requests proposals from various consulting engineers for a major job it is planning. Engineer A, a principal in a large engineering firm in West Eastville, decides to have his firm submit a proposal. Engineer A asks three engineers on his staff (Engineers X, Y, and Z) to develop a proposal for the firm. Engineers X, Y, and Z develop a proposal that is ultimately submitted to the city. Soon thereafter, the city learns that Engineers X, Y, and Z are the engineers who actually developed the proposal for the firm. A city official approaches Engineers X, Y, and Z and asks if they would agree to work as consultants, independent of Engineer A's firm. Engineers X, Y, and Z disclose the facts to Engineer A, resign from the firm, and enter into negotiations with the city.

Question: Was it unethical for Engineers X, Y, and Z to agree to a contract for consulting services independent of Engineer A's firm? [30]

References

1. (II.4) Engineers shall act in professional matters for each employer or client as faithful agents or trustees.

---

[29]NSPE Board of Ethical Review, Case No.85-1, Vol. VI, p. 57, 1989.
[30]NSPE Board of Ethical Review, Case No.86-5, Vol. VI, p. 83, 1989.

2. (III.4.a) Engineers in the employ of others shall not without the consent of all interested parties enter promotional efforts or negotiations for work or make arrangements for other employment as a principal or to practice in connection with a specific project for which the engineer has gained particular or specialized knowledge.

3. (III.7) Engineers shall not compete unfairly with other engineers by attempting to obtain employment or advancement or professional engagements by taking advantage of a salaried position, by criticizing other engineers, or by other improper or questionable methods.

Case Study 10.17

Payment for Employment

Engineer A, a recent engineering graduate seeking employment, had a direct offer from Company X for a position in its sales department, and at the same time had an offer from Company Z through an employment agency for a position in its design division. Engineer A was attracted to the second offer for work more to his liking, but the proposed salary was $2000 a year less than that offered by Company X. The employment agency told Engineer A that Company Z would not increase the amount of the proposed starting salary during the initial year of employment because of its salary system applicable to all newly hired engineers, but that the employment agency would, from its own funds, pay Engineer A an "acceptance bonus" of $2000 if he accepted the offer of Company Z. Engineer A has inquired whether it would be ethical for him to accept the bonus arrangement.

Question: Is it ethical for an engineer to accept a bonus payment from an employment agency as an inducement to accept employment with a particular employer?[31]

References

1. (III.6.a) Engineers shall not accept remuneration from either an employee or employment agency for giving employment.

Case Study 10.18

A Consulting Fee—Fee Based on Savings to Client

Engineer A, a specialist in utility systems, offers to industrial clients a service consisting of a technical evaluation of the client's use of utility services (electric power, gas, telephone, and the like) including, where appropriate,

[31] NSPE Board of Ethical Review, Case No.81-2, Vol. VI, p. 3, 1989.

recommendations for changes in the utility facilities and systems, methods of payment for such utilities, study of pertinent rating schedules, discussions with utility suppliers on rate charges, and renegotiation (where found applicable) of rate schedules forming the basis of charges to the client. Engineer A is compensated for these services solely on the basis of a percentage of money saved by the client for utility costs.

Question: Is it ethical for Engineer A to be compensated solely on the basis of a percentage savings to his clients? [32]

B. Contingency Fees

A city council adopted a resolution tentatively approving the construction of five parking lots to be built in the central business district. The city contacts Engineer A and requests that Engineer A make studies of the proposed project, including field investigations, data collection, and environmental impact reports. The city proposed that compensation for those services be contingent on approval of a general construction fund by the state agency responsible for funding the construction. Engineer A agrees to perform the services requested.

Question: Was Engineer A's performance of the requested services under the above facts a violation of the Code of Ethics? [33]

References

1. (III.7.a) Engineers shall not request, propose, or accept a professional commission on a contingent basis under circumstances in which their professional judgment may be compromised.

Case Study 10.19

Criticizing Other Engineers' Work

In a southwestern state legislature, various bills involving water supply, flood control, and production of electric power are awaiting action. The question of how to achieve the bill's goals most efficiently and economically has been debated within the legislature and in public forums for several years. A state legislative committee on public works calls a hearing to receive comments and recommendations on the various proposals. Engineer A, representing the state power commission, testifies that from an engineering standpoint, her team's studies point to a series of low dams as the most efficient solution. Engineer B, representing a private power company, testifies that according to his engineering analysis, a single high dam would produce the same results both faster and for less money. Both engineering witnesses submit voluminous

---

[32]NSPE Board of Ethical Review, Case No.73–4, Vol. IV, p. 49, 1976.
[33]NSPE Board of Ethical Review, Case No.83–2, Vol. VI, p. 33, 1989.

engineering data in support of their positions and do not hesitate to criticize the other's analysis and findings.

Question: Is it ethical for Engineers A and B to criticize each other's analysis and findings?[34]

References

1. (III.8) Engineers shall not attempt to injure, maliciously or falsely, directly or indirectly, the professional reputation, prospects, practice or employment of other engineers, nor indiscriminately criticize other engineer's work. Engineers who believe others are guilty of unethical or illegal practice shall present such information to the proper authorities for action.

Case Study 10.20

Using Technical Proposal of Another Without Consent

Engineer B submitted a proposal to a county council following an interview concerning a project. The proposal included technical information and data that the council requested as a basis for the selection. Smith, a staff member of the council, made Engineer B's proposal available to Engineer A. Engineer A used Engineer B's proposal without Engineer B's consent in developing another proposal, which was subsequently submitted to the council. The extent to which Engineer A used Engineer B's information and data is in dispute between the parties.

Question: Was it unethical for Engineer A to use Engineer B's proposal without Engineer B's consent in order for Engineer A to develop a proposal which Engineer A subsequently submitted to the council?[35]

References

1. (III.7) Engineers shall not compete unfairly with other engineers by attempting to obtain employment or advancement or professional engagement by taking advantage of a salaried position, by criticizing other engineers, or by other improper or questionable methods.

2. (III.10) Engineers shall give credit for engineering to those whom credit is due, and will recognize the proprietary interests of others.

3. (III.10.a) Engineers shall, whenever possible, name the person or persons who may be individually responsible for designs, inventions, writings, or other accomplishments.

---

[34]NSPE Board of Ethical Review, Case No.63–6, Vol. I (not available).
[35]NSPE Board of Ethical Review, Case No.83-3, Vol. VI, p. 36, 1989.

Case Study 10.21

Engineer's Duty to Report Data Relating to Research

Engineer A is performing graduate research at a major university. As part of the requirement for Engineer A to complete her graduate research and obtain her advanced degree, Engineer A is required to develop a research report. In line with developing the report, Engineer A compiles a vast amount of data pertaining to the subject of her report. The vast majority of the data strongly supports Engineer A's conclusion as well as prior conclusions developed by others. However, a few aspects of the data are at variance and not fully consistent with the conclusions contained in Engineer A's report. Convinced of the soundness of her report, and concerned that inclusion of the ambiguous data will detract from and distort the essential thrust of the report, Engineer A decides to omit references to the ambiguous data in the report.

Question: Was it unethical for Engineer A to fail to include reference to the unsubstantiative data in her report?[36]

References

1. (II.3.a) Engineers shall be objective and truthful in professional reports, statements, or testimony. They shall include all relevant and pertinent information in such reports, statements, or testimony.

2. (III.3.a) Engineers shall avoid the use of statements containing a material misrepresentation of fact or omitting a material fact necessary to keep statements from being misleading; statements intended or likely to create an unjustified expectation; statements containing prediction of future success; statements containing an opinion as to the quality of the engineers' services; or statements intended or likely to attract clients by the use of showmanship, puffery, or self-laudation, including the use of slogans, jingles, or sensational language or format.

3. (III.11) Engineers shall cooperate in extending the effectiveness of the profession by interchanging information and experience with other engineers and students, and will endeavor to provide opportunity for the professional development and advancement of engineers under their supervision.

10.8   CODE OF ETHICS FOR ENGINEERS

Engineering is an important and learned profession. The members of the profession recognize that their work has a direct and vital impact on the quality of life for all people. Accordingly, the services provided by engineers require honesty, impartiality, fairness and equity, and must be dedicated to the protection of the public health, safety, and welfare. In the practice of their profession, engineers must perform under a standard of professional behavior that requires adherence to the

---

[36]NSPE Board of Ethical Review, Case No. 85-5, Vol. VI, p. 67, 1989.

highest principles of ethical conduct on behalf of the public, clients, employers, and the profession.

- I.   Engineers, in the fulfillment of their professional duties, shall:

  1. Hold paramount the safety, health, and welfare of the public in the performance of their professional duties.

  2. Perform services only in areas of their competence.

  3. Issue public statements only in an objective and truthful manner.

  4. Act in professional matters for each employer or client as faithful agents or trustees.

  5. Avoid improper solicitation of professional employment.

- II. Rules of practice

  1. Engineers shall hold paramount the safety, health, and welfare of the public in the performance of their professional duties.

     (a) Engineers shall at all times recognize that their primary obligation is to protect the safety, health, property, and welfare of the public. If their professional judgment is overruled under circumstances where the safety, health, property, or welfare of the public are endangered, they shall notify their employer or client and such other authority as may be appropriate.

     (b) Engineers shall approve only those engineering documents that are safe for public health, property, and welfare in conformity with accepted standards.

     (c) Engineers shall not reveal facts, data, or information obtained in a professional capacity without the prior consent of the client or employer except as authorized or required by law or this Code.

     (d) Engineers shall not permit the use of their name or firm name nor associate in business ventures with any person or firm that they have reason to believe is engaging in fraudulent or dishonest business or professional practices.

     (e) Engineers having knowledge of any alleged violation of this Code shall cooperate with the proper authorities in furnishing such information or assistance as may be required.

  2. Engineers shall perform services only in the areas of their competence.

     (a) Engineers shall undertake assignments only when qualified by education or experience in the specific technical fields involved.

     (b) Engineers shall not affix their signatures to any plans or documents dealing with subject matter in which they lack competence, nor to any plan or document not prepared under their direction and control.

     (c) Engineers may accept an assignment outside of their fields of competence to the extent that their services are restricted to those phases of the project in which they are qualified, and to the extent that they are satisfied that all other phases of such project will be performed by registered or otherwise qualified associates, consultants, or employees, in which case they may then sign the documents for the total project.

3. Engineers shall issue public statements only in an objective and truthful manner.

    (a) Engineers shall be objective and truthful in professional reports, statements, or testimony. They shall include all relevant and pertinent information in such reports, statements, or testimony.

    (b) Engineers may express publicly a professional opinion on technical subjects only when that opinion is founded on adequate knowledge of the facts and competence in the subject matter.

    (c) Engineers shall issue no statements, criticisms, or arguments on technical matters that are inspired or paid for by interested parties unless they have prefaced their comments by explicitly identifying the interested parties on whose behalf they are speaking, and by revealing the existence of any interest the engineers may have in the matters.

4. Engineers shall act in professional matters for each employer or client as faithful agents or trustees.

    (a) Engineers shall disclose all known or potential conflicts of interest to their employers or clients by promptly informing them of any business association, interest, or other circumstances that could influence or appear to influence their judgment or the quality of their services.

    (b) Engineers shall not accept compensation, financial or otherwise, from more than one party for services on the same project, or for services pertaining to the same project, unless the circumstances are fully disclosed to, and agreed to, by all interested parties.

    (c) Engineers shall not solicit or accept financial or other valuable consideration, directly or indirectly, from contractors, their agents, or other parties in connection with work for employers or clients for which they are responsible.

    (d) Engineers in public service as members, advisors, or employees of a governmental body or department shall not participate in decisions with respect to professional services solicited or provided by them or their organizations in private or public engineering practice.

    (e) Engineers shall not solicit or accept a professional contract from a governmental body on which a principal or officer of their organization serves as a member.

5. Engineers shall avoid improper solicitation of professional employment.

    (a) Engineers shall not falsify or permit misrepresentation of their, or their associates', academic or professional qualifications. They shall not misrepresent or exaggerate their degree of responsibility in or for the subject matter of prior assignments. Brochures or other presentations incident to the solicitation of employment shall not misrepresent pertinent facts concerning employers, employees, associates, joint venturers, or past accomplishments with the intent and purpose of enhancing their qualifications and their work.

(b)   Engineers shall not offer, give, solicit, or receive, either directly or indirectly, any political contribution in an amount intended to influence the award of a contract by public authority, or that may be reasonably construed by the public of having the effect or intent to influence the award of a contract. They shall not offer any gift, or other valuable consideration in order to secure work. They shall not pay a commission, percentage, or brokerage fee in order to secure work except to a bona fide employee or bona fide established commercial or marketing agencies retained by them.

- III.   Professional obligations

1.   Engineers shall be guided in all their professional relations by the highest standards of integrity.

   (a)   Engineers shall admit and accept their own errors when proven wrong and refrain from distorting or altering the facts in an attempt to justify their decisions.

   (b)   Engineers shall advise their clients or employers when they believe a project will not be successful.

   (c)   Engineers shall not accept outside employment to the detriment of their regular work or interest. Before accepting any outside employment they will notify their employers.

   (d)   Engineers shall not attempt to attract an engineer from another employer by false or misleading pretenses.

   (e)   Engineers shall not actively participate in strikes, picket lines, or other collective coercive action.

   (f)   Engineers shall avoid any act tending to promote their own interest at the expense of the dignity and integrity of the profession.

2.   Engineers shall at all times strive to serve the public interest.

   (a)   Engineers shall seek opportunities to be of constructive service in civic affairs and work for the advancement of the safety, health, and well-being of their community.

   (b)   Engineers shall not complete, sign, or seal plans and/or specifications that are not of a design safe to the public health and welfare and in conformity with accepted engineering standards. If the client or employer insists on such unprofessional conduct, they shall notify the proper authorities and withdraw from further service on the project.

   (c)   Engineers shall endeavor to extend public knowledge and appreciation of engineering and its achievement and to protect the engineering profession from misrepresentation and misunderstanding.

3.   Engineers shall avoid all conduct or practice that is likely to discredit the profession or deceive the public.

   (a)   Engineers shall avoid the use of statements containing a material misrepresentation of fact or omitting a material fact necessary to keep statements from being misleading; statements intended or likely to create an unjustified expectation; statements containing prediction of future success; statements containing an opinion as to the quality of the Engineers' services; or statements intended or likely to attract clients by the use of showmanship, puffery, or self-laudation, including the use of slogans, jingles, or sensational language or format.

(b) Consistent with the foregoing, Engineers may advertise for recruitment of personnel.

(c) Consistent with the foregoing, Engineers may prepare articles for the lay or technical press, but such articles shall not imply credit to the author for work performed by others.

4. Engineers shall not disclose confidential information concerning the business affairs or technical processes of any present or former client or employer without his consent.

(a) Engineers in the employ of others shall not, without the consent of all interested parties, enter promotional efforts or negotiations for work or make arrangements for other employment as a principal or to practice in connection with a specific project for which the Engineer has gained particular and specialized knowledge.

(b) Engineers shall not, without the consent of all interested parties, participate in or represent an adversary interest in connection with a specific project or proceeding in which the Engineer has gained particular specialized knowledge on behalf of a former client or employer.

5. Engineers shall not be influenced in their professional duties by conflicting interests.

(a) Engineers shall not accept financial or other considerations, including free engineering designs, from material or equipment suppliers for specifying their product.

(b) Engineers shall not accept commissions or allowances, directly or indirectly, from contractors or other parties dealing with clients or employers of the Engineer in connecton with work for which the Engineer is responsible.

6. Engineers shall uphold the principle of appropriate and adequate compensation for those engaged in engineering work.

(a) Engineers shall not accept remuneration from either an employee or employment agency for giving employment.

(b) Engineers, when employing other engineers, shall offer a salary according to professional qualifications and the recognized standards in the particular geographical area.

7. Engineers shall not compete unfairly with other engineers by attempting to obtain employment or advancement of professional engagements by taking advantage of a salaried position, by criticizing other engineers, or by other improper or questionable methods.

(a) Engineers shall not request, propose, or accept a professional commission on a contingent basis under circumstances in which their professional judgment may be compromised.

(b) Engineers in salaried positions shall accept part-time engineering work only at salaries not less than that recognized as standard in the area.

(c) Engineers shall not use equipment, supplies, laboratory, or office facilities of an employer to carry on outside private practice without consent.

8. Engineers shall not attempt to injure, maliciously or falsely, directly or indirectly, the professional reputation, prospects, practice or employment of other engineers, nor indiscriminately criticize other engineers' work. Engineers who believe others are guilty of unethical or illegal practice shall present such information to the proper authority for action.

(a) Engineers in private practice shall not review the work of another engineer for the same client, except with the knowledge of such engineer, or unless the connection of such engineer with the work has been terminated.

(b) Engineers in governmental, industrial or educational employ are entitled to review and evaluate the work of other engineers when so required by their employment duties.

(c) Engineers in sales or industrial employ are entitled to make engineering comparisons of represented products with products of other suppliers.

9. Engineers shall accept personal responsibility for all professional activities.

(a) Engineers shall conform with state registration laws in the practice of engineering.

(b) Engineers shall not use association with a nonengineer, a corporation, or partnership, as a "cloak" for unethical acts but must accept personal responsibility for all professional acts.

10. Engineers shall give credit for engineering work to those to whom credit is due, and will recognize the proprietary interests of others.

(a) Engineers shall, whenever possible, name the person or persons who may be individually responsible for designs, inventions, writings, or other accomplishments.

(b) Engineers using designs supplied by a client recognize that the designs remain the property of the client and may not be duplicated by the Engineer for others without express permission.

(c) Engineers, before undertaking work for others in connection with which the Engineer may make improvements, plans, designs, inventions, or other records that may justify copyrights or patents, should enter into a positive agreement regarding ownership.

(d) Engineers' designs, data, records, and notes referring exclusively to an employer's work are the employer's property.

11. Engineers shall cooperate in extending the effectiveness of the profession by interchanging information and experience with other engineers and students, and will endeavor to provide opportunity for the

professional development and advancement of engineers under their supervision.

(a) Engineers shall encourage engineering employees' efforts to improve their education.

(b) Engineers shall encourage engineering employees to attend and present papers at professional and technical society meetings.

(c) Engineers shall urge engineering employees to become registered at the earliest possible date.

(d) Engineers shall assign a professional engineer duties of a nature to utilize full training and experience, insofar as possible, and delegate lesser functions to subprofessionals or to technicians.

(e) Engineers shall provide a prospective engineering employee with complete information on working conditions and proposed status of employment, and after employment will keep employees informed of any changes.

Chapter 11

Communications in Engineering

*Written communication in the Department suffers seriously from arthritis complicated by delegation .... Most sentences are long. Three or four ideas are loaded in to increase the odds one will be impressive and that if another is wrong it will get lost. Paragraphs are built like sandwiches, the meat in the middle. The purpose is less to say what is right than to avoid saying anything wrong ...The Department's effectiveness would be doubled if prose were cut in half, if those who initiated documents read them, and if those who signed them wrote some of them. In none of this is the Department of Labor in any way exceptional.***W. Willard Wirtz, Secretary of Labor**

11.1 EFFECTIVE COMMUNICATIONS

Engineers are often accused of being poor writers and communicators. Industry has pushed hard for increased emphasis on written and oral communications in the engineering curriculum to better prepare graduates in this area. Universities have responded to this concern by adding courses in engineering report writing and oral communications to the curriculum. Engineering departments have also increased the writing and oral presentation requirements included in their own courses to better equip graduates with the communication skills needed to be effective as engineers. In spite of all this effort to improve the communications capabilities of graduates, the prevailing impression seems to be that engineers are ineffective communicators. If this is a factual criticism, part of the problem may be due to a lack of interest in English and language courses in junior high and high school which continues through university level courses. Many, if not most, engineering students have a natural inclination for, and interest in, math and science and little of either for writing and oral communications. Without a serious interest in becoming an effective writer and convincing speaker, it is doubtful that a student's capabilities in this area can be significantly improved.

A comment often heard that has some applicability here is that to be an effective writer, a person needs to be an avid reader. This homily is undoubtedly based on the reading of reasonably well-written books and not some of the reading material available in bookstores today. The educational system in the United States seems to do very little toward generating enthusiasm for reading on the part of students, and this may be a contributing factor in the poor writing capability and vocabulary of its graduates. If engineering students were required to have a good dictionary and thesaurus at their desk and would get in the habit of using them, the quality of their writing could be improved immensely. There is an inherent relationship between certain specific words and knowledge. Until an understanding of the meaning of these words is grasped, the related thought cannot be experienced. Thus, improvement in writing ability, to some extent, implies increased knowledge, certainly a worthy goal for all engineers, beginners and journeymen alike. Finally, some comment on the quality of the communications courses that are offered is warranted. Many secondary school and university English courses emphasize other aspects of writing and not writing itself. Engineering students often have little understanding of grammar and basic sentence and paragraph structure. Thus, they have considerable difficulty in writing on technical subjects clearly in easy-to-read context. To assist these students, some universities have established communication centers within their colleges of engineering to provide guidance for and critique of written reports, theses, and other manuscripts.

In today's environment, information flow is critical to business and industry. Engineers in industry often comment on the large amount of their time that is committed to writing and other forms of communication. Most business and industry communications are verbal, in the form of face-to-face discussions, meetings, and telephone conversations. Important communications are transmitted in writing so that the meaning can be precisely stated and a record can be established for future reference. Employees that are incapable of preparing clear and understandable written communications tend to be relegated to passive roles in this process. They become information receivers and not information generators and thus gradually find themselves out of the mainstream, out of touch with what is going on and out of mind when raises and promotions come.[1] To a great extent, an employee's value is measured by the ability to communicate clearly and effectively. In effect, engineers market their skills through the ability to communicate. It is much better to be dealing with a *bull market* rather than a *bear market* in this context.

11.2   THE FORMAL ENGINEERING REPORT

The engineering report probably constitutes the greatest writing challenge that a typical engineer faces. Engineering reports document a significant portion of work and, as such, represent an important commitment on the part of the firm in which the engineer is employed. The engineering report is commonly the only document that describes how and why the work was accomplished and what the results, recommendations, and conclusions were. It is often the only document related to the work that is maintained on file for future reference. If customers or other

[1]H. Upton, "Power Up Your Business Writing," *Spirit,* March 1991.

personnel doing the work outside the organization are involved, the report may be the principal (or only) measure of the quality of the work. Within the organization doing the work, the capability of the engineer and, to some extent, his or her department, are judged by the quality of the report.

Engineering reports are prepared for many different purposes. A few of these are listed below:

1. Test programs
2. Experiments
3. Studies
4. Investigations
5. Failure analyses
6. Evaluations

It should be evident from the variety of purposes for which engineering reports are prepared that the format must be flexible. The format must be capable of being adapted to best describe the work accomplished. A report on a test program will undoubtedly require a different format than that for a study or failure analysis. A report on an engine test will probably require a different format than that for a computer component. Some organizations will have a basic report format that can be adapted to meet the particular need. The general format outlined below includes all of the elements that should normally be considered for inclusion in a typical report. However, few reports prepared by engineers will require all of the elements described.

An important point to remember when preparing an engineering report is that the report may be read by individuals that are not intimately familiar with the work or the author. Their impressions are thus formulated by the content of the report and their ability to understand what was accomplished as well as its meaning. In some cases, engineering reports are prepared for nontechnical customers, which only exacerbates this problem. The challenge is to discuss the work accomplished in terms that are understandable to a reader not directly involved in the effort, possibly not even an engineer or technical person. It is very important to know the audience and to prepare the report so that it is completely understandable to that group of people.

A problem that many inexperienced writers have in preparing reports is establishing a logical content flow. If the reader is to understand the report, the presentation of information must follow some thread of logic. The material presented should be introduced in chronological sequence within each division of the report, and periodic *signposts* should be provided so that the reader always knows where he has been and where he is going on his journey through the report. This goal can be achieved by forecasting what is coming next in the report in each major subdivision and by beginning each paragraph with a topic statement.[2]

Formal engineering reports are normally written in third person, past tense, so personal pronouns should very rarely be used. The report is prepared in a purely objective, impersonal manner, reflecting the writer's relationship to the material presented. Results are judged on the basis of existing, applicable theory, and previous experiments. Opinions are introduced only when existing knowledge fails.

---

[2]S. A. Harbinger, A. B. Whitmer, and R. Price, *English for Engineers*, McGraw-Hill, New York, 1951.

11.2.1   The Abstract

The abstract is a brief (normally 200 words or less) statement of the essential contents of the report. It is placed before the body of the report so that the reader may decide from it whether to read the full report or not. In practice, the abstract often exists apart from the report in library collections, so that readers can review a wide range of literature relevant to their own investigations and can make decisions as to what reports are needed to complete their research. A summary (often called an executive summary) is sometimes used in place of an abstract when attached to the report. A summary will never exist separately from the report, however.

   The abstract should include brief statements of the objective(s), the method used to satisfy the objective(s), significant results (normally one or two), and conclusions. Figures and tables should not be referred to in the abstract, and equations normally should not be used. It is usually best to write the abstract last, scanning the report body for essential information and presenting this information in as few words as possible. An example of a typical abstract with the essential elements identified is given below.

Students and faculty of the Mechanical Engineering Department of Texas Tech University have developed two add-on devices that significantly reduce pickup drag without limiting vehicle utility. One device is a wing structure mounted behind the top rear of the cab; the other device is a cover over the rear portion of the pickup bed. Both devices have demonstrated 5 percent to 6 percent reduction in vehicle drag. Furthermore, the rear bed cover reduces vehicle lift by 30 percent. Most significant, scale model tests have produced drag reductions on the order of 17 percent and lift reductions of 30 percent when the wing and bed cover are used together.

   Work to date has involved scale model tests in a low-speed wind tunnel and full-scale tests in the Lockheed wind tunnel in Marietta, Georgia. In addition, computational fluid dynamics studies have been performed to aid in designing the test program and in interpreting the test results.

Objective: Development and testing of add-on devices that significantly reduce drag without limiting vehicle utility.

Significant results: Both devices have demonstrated 5 percent to 6 percent reduction in vehicle drag. Further, the rear bed cover reduces vehicle lift by 30 percent. Most significantly. Scale model tests have produced drag reductions of 17 percent and lift reductions of 30 percent when the wing and bed cover are used together.

Method: Work to date has involved scale model tests in a low-speed wind tunnel and full-scale tests in the Lockheed wind tunnel in Marietta, Georgia. In addition, computational fluid dynamics studies have been performed to aid in designing the test program and interpreting the test results.

11.2.2   The Introduction

The primary purpose of the introduction is to provide the necessary background on the subject, to describe the objective(s) of the work accomplished, and to define the scope of the investigation. The introduction should identify the sponsor and include the reason for doing the work. Any relationship that exists between the work accomplished and other related effort should also be identified. Figures and tables should not be used in the introduction. Equations and calculations should not be used unless they are absolutely necessary.

The introduction should be approximately one typewritten page, but the length can be either longer or shorter depending on the subject being introduced. An example introduction is given below.

> Radiation heat transfer in participating media is important in many applications such as the design of industrial furnaces, rocket combustion chambers, and novel high-temperature heat exchangers. A good numerical method that involves thermal radiation problems must be flexible enough to deal with complex geometries and real radiation properties, be able to deal with combined mode heat transfer problems, and be efficient so that not much computer time nor storage is required. Unfortunately, almost no existing method satisfies all of these criteria. For example, the Monte Carlo method is excellent in flexibility and requires little computer memory, but it can be very time consuming and inaccurate. The zonal method and the finite element method are both time and storage consuming although they are very accurate and can be modified to deal with complex problems such as anisotropic scattering. The product integration method (PIM), while faster than the zonal and finite element methods, does not reduce the storage requirements....
>
> In this paper, a new numerical method, the **YIX** method, is presented to solve radiation problems in multidimensional emitting, absorbing, and anisotropic scattering media. The exact integral formulation for radiation transport is used but in an alternative distance-angular integral form. Here, the method is described for a simple one-dimensional problem, extensions to a more complex case are considered, and example problems in two-dimensional participating media are solved to test the accuracy and efficiency of the method in one dimension.
>
> ***Background:***   Radiation heat transfer ... reduce the storage requirements.
>
> ***Objective:***   Presentation of a new numerical method for solving radiation problems in multidimensional emitting, absorbing, and anisotropic scattering media.
>
> ***Scope:***   Description of the method of solution for a one-dimensional problem, extensions to more complex cases, and solution of example problems in two-dimensional participating media.

11.2.3   Technical Approach (Theory)

The technical approach/theory should include an identification of the theoretical principles involved and the equations used in making calculations from the experimental data. The manner in which the various principles and equations were used in accomplishing the work should be described. Equations should be presented in the order in which they were used in making the calculations included in the report. The use of supporting sketches, diagrams, and curves is recommended to assist in clarifying and amplifying the text in this section. Whenever figures, tables, and the like are used in a report they must always be referred to in the text ahead of the figure or table, and they must be in sequential order. Supporting materials included in the technical approach should not make use of actual data recorded during the experiment, since this information is included in the results and discussion section of the report. Instead, supporting sketches, figures, and the like used in this section of the report should be related to previously existing theory applicable to the work accomplished.

The technical approach should also include any discussion of experimental uncertainties that are appropriate. The required levels of accuracy and precision vary with the nature and purposes of experimental work, but the report on any project should always examine the accuracy of the measurements involved as they relate to the overall objectives and include the appropriate discussion and/or theory.

Portions of a typical technical approach section are given below:

The total heat transfer Q_{total} was the primary measurement in this experiment. It was calculated using

$$Q_{total} = mC_p\Delta T$$

where
  m = mass flowrate, kg/s
  C_p = specific heat, $Ws/kg°C$
  $\Delta T = T_{in} - T_{out}$

After the heat loss from the receiver was measured, the convection heat loss was obtained by subtracting the calculated radiation and conduction heat losses.

The receiver loses thermal energy by radiating to its surroundings, the sky and the ground. The radiation Q_{rad} is split between the ground and sky. The ground is assumed to be at the ambient air temperature T_∞ and the sky temperature T_{sky} is estimated using a method wherein the relationship between sky temperature and ambient temperature is defined in terms of sky emissivity.

The conduction loss is a small percentage of the total heat loss from the cylinder, usually ranging from 3 to 6 percent. Nevertheless, a conduction calculation is used in the data analysis. The conduction heat transfer Q_{cond} is calculated using

$$Q_{cond} = kA(T_{wall} - T_{core})$$

where
    k = conduction heat transfer coefficient
    A = area of heat transfer surface
  T_{wall} = receiver wall temperature
  T_{core} = receiver core temperature

The back of the receiver is insulated with Kaowool insulation, with a conduction heat transfer coefficient k of 1.136 W/m² °C. The receiver front wall temperature is used as T_{wall} for this calculation and the ambient temperature serves as T_{core}, the receiver core temperature.

Uncertainty: The primary sources of error were identified as the solar flux calculation, the receiver mass flowrate measurement, and the receiver fluid temperature measurement. These potential problem areas were addressed in the experimental design in the following ways:

1. All tests were conducted with no solar flux on the receiver to eliminate the uncertainty associated with flux calculations. Except for one test set, which was conducted on an overcast day, all tests were conducted at night, which eliminated the need to account for solar input.

2. Individual boiler panel flowrates were limited to a minimum of 0.13 kg/s. Flowrates less than this are subject to unacceptably large measurement uncertainties.

3. All receiver fluid thermocouples used in the data analysis were calibrated. Thermocouples that could not be adjusted to read correctly were eliminated from analysis.

11.2.4   Test Setup

A complete identification of the component or system to be tested and of all significant test equipment should be provided when a test setup section is included. A neat sketch/schematic of the test setup should be prepared with the components and instruments used during the test or experiment identified. Identification can be provided directly on the sketch or schematic or by including a list with numerical identification corresponding to that on the schematic. A narrative should be provided that describes the function of the major components in the test setup. The reader should be able to interrelate the narrative and the schematic to gain full understanding of how the equipment was used in the test or experiment.

As a general rule only reports involving a test, an experiment, or an investigation will include a test setup section. For failure investigations, it is often better to forego the use of a test setup section and to include sketches and/or photographs of the failed hardware in the results and discussion section integrated with the text discussion of the failure.

In performing critical tests wherein data may be questioned by other contractors or the customer after the test, it is important to identify all instrumentation completely, including serial numbers. With this information available, the status of the equipment during the test can be ascertained, including instrument calibration data. For this type of report, it is important to include adequate information on the test setup so that the test can be repeated, if required. An example test setup report section is given below:

The test apparatus used in this experiment included a 1/10th-scale model pickup truck, a one-dimensional force table, strain and proximity gages, a balance scale, a velometer, and wind tunnel facility. The scale model pickup was modeled using the design of a Ford F 150 pickup. All dimensions were constructed to be 1/10th scale, and angular surfaces were duplicated to the extent possible. Figure 1 is a detail drawing of the model showing all dimensions.

The model was fabricated of wood, sanded, and painted to obtain a smooth surface to simulate the polished surface of a full-scale vehicle. The model did not include external components such as mirrors, bumpers, wheels or windshield wipers. The underbody was flat to eliminate drag due to the chassis and exhaust system. The front end was closed to eliminate internal drag due to flow through the radiator and around the engine. Two bed sizes were used during testing, a long and short bed. A movable tailgate was included so that tests could be performed with the tailgate opened or closed. The bed covers were constructed of aluminum and were 1/16th–in. thick, varying in length from 2.0 to 4.6 in.

The force table was used to determine the force on the model due to the drag. Figure 2 shows a drawing of the force table. The force table was designed to measure force in only one direction, that due to drag on the vehicle due to movement through the air mass. The model was attached to the force table by means of a 1-in. diameter cylindrical steel rod. This rod was adjustable to facilitate raising and lowering the truck and to allow rotation to simulate various yaw angles. A dashpot was connected to the force table to dampen the large vibrations induced by the wind moving over the model. To determine the force on the model the mechanism shown in Figure 3 was used to restore the loaded force table to the unloaded position. This mechanism used a pulley and cannister attached to the force table by means of a tension wire. The balance table was used to determine the weight of the cannister after each test.

A strain gage was attached to one plate of the force table to indicate plate deflection. This gage was used as a secondary check to ensure that the force table was returned to the unloaded position. The primary method used to check the force table position involved the use of a proximity gage attached to the base. Figure 4 shows the position of these gages on the table.

A pitot tube attached to the wall of the wind tunnel and connected to an oil manometer was used to measure the difference in pressure in the wind tunnel and the exterior atmospheric pressure. A type K thermocouple was used to measure the free stream air temperature, which was displayed on a digital indicator in degrees Fahrenheit. A hand-held velometer was used to indicate the approximate free stream wind speed within the tunnel. Tufts made of thread and a smoke generator were used to help visualize the flow patterns over the model.

The test was performed in the Texas Tech University Mechanical Engineering Low Speed Wind Tunnel Facility in Lubbock, Texas. The wind tunnel has a test section that is 4 ft wide by 3 ft high.

11.2.5 Procedure

This section should include a relatively brief narrative description of the test procedure. The detailed step-by-step procedure used to perform the test or experiment should not be used here but should normally be included in the appendix. The purpose of this section of the report is to inform the reader of items like the following:

1. The type and reason for individual test runs made.
2. Preliminary tests, equalizing periods, run durations, and data reading frequencies.
3. Special precautions for accuracy and means for controlling test conditions.
4. Independent variables and reasons for their selection.
5. Conformity with or divergence from standard test codes or methods.

An example of an acceptable procedure is given below. Numbers enclosed in parenthesis following each paragraph correspond to the items listed above that should be considered for inclusion in the procedure section.

The test was performed using two different bed lengths, a short-bed and long bed. Each configuration was tested at a wind speed of 51.205 mph with varying bed covers ranging in length from 2.0 to 4.6 in. The bed covers were located at the tailgate and extended forward toward the cab. The drag was also evaluated at several different yaw angles [1].

The proximity gage was calibrated before each test run to minimize error. The strain indicator was also zeroed before the start of each test. After zeroing the strain gage, the deflection meter and the proximity gage, the wind tunnel was activated and was allowed to warm up for a period of 15 min to allow the air flow to reach equilibrium. Before model testing was initiated, a test to determine the drag induced by the model supporting rod only, was made [2,3].

For the first set of tests the short bed was attached to the pickup model. The wind tunnel was activated thus deflecting the model/force table assembly. A waiting period of 2 min was adopted to allow the dashpot fluid to reach equilibrium and for the table to stabilize. Water was then added to the cannister until the model/force table assembly was restored to the unloaded position as indicated by the strain gage and proximity indicator. A 2-min waiting period was again incorporated to allow the dashpot fluid to reach equilibrium. Free stream temperature, pressure and velocity measurements were also recorded at this time. The model/force table position was then checked again to ensure zero deflection. The water in the cannister was then weighed to determine the drag force induced on the model. This procedure was repeated for all bed cover lengths and truck yaw angles evaluated. Tests were also run on the model with an open bed (no cover) and tailgate down [1,2,3].

The wind tunnel was then deactivated and the long bed was attached to the model. The above procedure was then repeated, including the open bed/tailgate down configuration [1].

11.2.6 Results and Discussion

An objective discussion of the results and analysis to show that the conclusions are warranted is the most important requirement of any engineering report. Each major conclusion must be clearly substantiated, and any contradictory theories or results must be explained. All statements should be clear to readers not as well acquainted with the subject matter as the author. All discussion in the text should be supported with graphs and tables to the extent possible. Graphs supported by tables is the preferred approach for showing trends and comparisons, substantiated by the tabulated data and calculated values. All statements about the results and any numerical values cited must agree precisely with information in the tables and on the figures.

One way to build a good discussion is to compare each result with its predicted or expected value. Some discussion as to why the experimental data should agree with the predicted or expected data should also be included. If the data do not agree with accepted theory, an explanation should be offered, if possible. If the reason for the discrepancy cannot be determined, state that this is the case. If an error or discrepancy has occurred in the data or during the conduct of the experiment, recognize it briefly and continue with the discussion. Do not dwell on events that add little or nothing to the main thrust of the effort being described. This is the most important section of the report so be sure to develop an adequate discussion and analysis that will convince the reader that the conclusions are valid. Example paragraphs from a typical results and discussion report section are given below.

> For the long bed pickup, the bed cover length that provides the greatest reduction in drag is 3.2 ft. A drag reduction of 4.7 percent was recorded for this configuration. Table 1 summarizes the drag reduction data for all bed cover lengths for both the short and long-bed configurations. A plot showing percent drag reduction versus bed cover length is given by Figure 5. The plot indicates that drag reduction levels off at this length and increased bed cover length (greater than 3.2 ft) is of little or no benefit. This is thought to be because at this bed cover length the characteristic pickup bed airflow vortex is disrupted and increased length provides little additional vortex destruction.
>
> The drag coefficient results obtained from wind tunnel testing of the open bed are comparable to the closed front end, open bed results obtained by Ford Motor Company for full-scale vehicle testing. For the open long bed, Ford reported a drag coefficient of 0.5302 compared with the model test drag coefficient of 0.4685. The differences can be attributed to several factors including model asymmetry, lack of model detail, single-force table strut stability, and force table inaccuracies.

Note that the first paragraph (above) discusses one of the principal results of the test. The data are presented in both tabular and plotted curve format, and the results are explained by resorting to known theory regarding the circulation vortex that develops in the bed of pickups. The second paragraph compares the test results with similar results obtained by a recognized and authoritative source (Ford Motor Company), and identifies the reasons for differences.

11.2.7 Conclusions

Conclusions reached in the results and discussion section should be restated here in a more general and less specific manner. Conclusions are normally listed in the order of their importance, preferably in itemized form, although a narrative presentation is completely satisfactory. Already known facts should not be included in this section, and conclusions should not be confused with factual test or experimental results. If conclusions are qualified, give the assumptions and limitations that apply.

No new information should be introduced in this section. The conclusions provide a quick reference for the reader with limited time. Often a reader will only read the abstract and the conclusions sections of the report. Thus, it is imperative that conclusions be stated concisely and accurately, with whatever qualifications that apply. Do not discuss the conclusion but use the discussion in the previous section to extract the significant thought(s). An example conclusions section is included below.[3]

---

[3]S. A. Harbarger, A. B. Whitmer, and R. Price, *English for Engineers,* McGraw-Hill, New York, 1951.

The conclusions from this study can be summarized as follows:

1. The quantity of heat lost from the charge during combustion and expansion were 14 percent and 30 percent, respectively, of the total heat of combustion.

2. The heat loss during inflammation was greater than the loss during expansion even though (a) the expansion period was nearly twice as long as the inflammation period, (b) the combustion space was entirely filled with the inflamed mixture during expansion, and (c) the wall area increased as the mixture expanded.

3. The heat lost during combustion and expansion decreased to about 70 percent of its original value when the speed was increased from 1000 to 2000 rpm.

4. The heat during combustion and expansion increased slightly when the temperature of the cast-iron and aluminum heads was increased by stopping the flow of coolant through them. The causes of this phenomenon are not fully understood.

5. The heat losses during combustion and expansion do not appear to be related directly to the thermal conductivity of the head material.

This conclusion section is clearly and succinctly written, the conclusions are presented in itemized format and can be quickly read and understood, no previously known facts are included as conclusions and no new facts are presented. In (4), the conclusion is given, and it is clearly stated that the reasons are not understood.

11.2.8 References

There are several acceptable systems for listing references in an engineering report. The author-date system of documentation as found in the *Chicago Manual of Style* is one that is commonly used.[4] In this format, quoted or paraphrased material is followed by the author's last name and the year of publication in parentheses (Weintraub, 1988). At the end of the report in the references section, the sources are listed alphabetically by the author's last name with the title in quotation marks or italicized.

Weintraub, Herman, *Baroque: A New Way of Looking at Old Things*, Johnson and Sons, 1988.

Another acceptable method is the numerical system in which the first reference used is referred to as number 1 in square brackets, [1], followed by [2], and so forth. This identification stays with the reference throughout the document. The references are usually listed at the bottom of the text page on which they were cited and in numerical order in the reference section of the report. Only those works that were used in prepearing the report should be cited in references.

Although engineering reports are not copyrighted and sold like books, it is important to get in the habit of properly recognizing work from which material is used. It is especially important that material not be plagiarized and claimed as

[4] *The Chicago Manual of Style,* 13th ed., University of Chicago Press, 1982.

one's own. Such acts are unethical and can lead to the destruction of a professional reputation and career. Plagiarism is a form of theft and should be shunned just as the theft of a physical item would be.

11.2.9 Appendix

Appendices are used to provide supporting data that are not significant enough to be included in the body of the report. In this manner, the body of the report can focus on material essential to the discussion and conclusions, and readers will not be distracted by supporting information of little importance in the main body of the report. Less important information can be included in the appendix without cluttering up the more significant portions of the report. Because only serious readers are usually interested in the data included in the appendix, the distribution of the full report (including the appendix), is sometimes restricted to only a few readers. This decision is, of course, also a function of the relative size of the appendix. Appendices normally include items such as original data sheets, sample calculations, calibration data, uncertainty analyses, instrument charts, detailed test procedures, test codes, and bulletins. Using data in the appendix, the seriously interested reader should be able to validate most, if not all, the graphs, plots, results, and the like, included in the report.

Some information associated with the test effort is of so little interest and importance that it should not normally be included in the report or in the appendix. An example of this is a description of the data reduction process. The fact that recorded parameters were originally obtained from voltage signals and subsequently converted to units appropriate for the particular test parameter is of no interest to the reader and should not be included in the report.[5] Each separate item should be included as a separate appendix (e.g., Appendix A—Data Sheets, Appendix B—Calibration Data, etc.).

11.3 PROPOSAL PREPARATION

The preparation of proposals is a task that most engineers encounter at one time or another during their professional careers. Preparation of an engineering proposal can result from a formal request for proposal (RFP), from an informal suggestion from a potential customer that a proposal in an area of interest would be favorably received, in response to a request from in-house management, or from an idea or need that is thought to exist but for which there has been no formal or informal request.

The purpose of submitting a proposal is to put (poser means to place or put) forth (pro means forth) an idea or solution to a need for which the potential customer is known or thought to be interested. Since the purpose is to sell an idea or solution, the proposal must first establish that the need exists and then provide information adequate to convince the potential customer that the solution proposed is the right one and that the proposer is capable of accomplishing the effort described within the planned schedule and budget. Several suggestions that should be kept in mind while preparing a proposal are as follows.

---

[5]J. F. Brown, *A student Guide to Engineering Report Writing*, United Western Press, Solana Beach, CA, 1985.

1. In responding to a formal RFP the format specified in the RFP should be followed verbatim and should be deviated from only in the most extreme situations. This procedure will at times cause some frustration in trying to fit the information considered essential to describing the idea into the format suggested but, nevertheless, is the prudent thing to do. Proposal evaluators are inclined to use the format specified in the RFP to check and evaluate the material provided in the proposal. If it is difficult to find or to relate certain elements required by the RFP to ones included in the proposal, evaluators are likely to reflect this difficulty in the evaluation. Page limitations and other restrictions should also be complied with or the proposal may be rejected without further consideration.

2. For an unsolicited proposal, or one in which the format is not specified, organization and document format are extremely important and should be given commensurate attention. The goal should be to make the evaluator's job as easy as possible. To achieve this goal the material submitted should be organized so that all essential information is included within the proposal and is located in a logical and ordered fashion. Information provided in the proposal should flow from a logical beginning to a suitable conclusion and be structured so that the evaluator does not have to hunt throughout the document to find all of the information on any subject.

3. It is important to be specific in regard to the product of the proposed effort. If the product is a final report, it should be stated clearly what will be included in this document. If the product is some item of hardware, it should be described adequately. The product is what the customer is going to pay for, and it thus plays an important role in determining whether or not the project will be funded.

As with formal engineering reports, the format for proposals needs to be adapted to the effort being described. However, the sections outlined below are typically included in most proposals in one form or another.

11.3.1 Background/Problem Statement

All proposals normally include information that establishes the background for the problem for which a solution is being proposed. Even when responding to an RFP, it is wise to include background material to ensure that the potential customer recognizes that the proposer understands the problem and is capable of developing a solution. It is essential for unsolicited proposals to include background material, since it is in this section of the proposal that the problem is defined and the potential customer is (hopefully) convinced that the problem exists and that finding a solution is important enough to consider funding the effort.

The background should be concise but nevertheless complete enough to provide an adequate description of the sequence of events leading up to problem recognition. The problem may be something that has not been done before, has been done poorly or incorrectly, or has been done only partially. The background/problem statement should address questions such as: What is the problem? Why is this problem of significance? What previous work has been done in this area, and what was discovered?[6] The background description should conclude

---

[6]A. Eisenberg, *Effective Technical Communication,* McGraw-Hill, New York, 1982.

with a clear (one or two sentence) statement of the problem so that the potential customer's attention is focused on the principal thrust of the proposal, that a real problem exists for which a viable solution is proposed.

11.3.2 Objective(s)

After the problem has been described, a clear and concise statement of the objective(s) of the proposal should be given. The objective of the effort is sometimes included in the background/problem statement portion of the proposal, but if this is the case it should be located toward the end of this section. If the objective(s) is included as a separate section of the proposal, it can be brief, often only a few sentences. If the effort is to be accomplished in several phases, it is sometimes helpful to identify objectives for each separate phase.

11.3.3 Technical Approach

This section of the proposal describes how the effort will be conducted. In many ways this section is the most important section of the proposal, since it spells out how the proposer will go about finding a solution to the problem that is identified. The technical approach portion of the proposal must convince the potential customer that the proposer understands the problem and that an approach has been devised for finding a solution that has a high probability of success. The technical approach should address each of the objectives and should provide an adequate level of detail so that the potential customer can understand exactly how each of the objectives would be satisfied. Individual tasks should be identified and discussed and should be shown on an easily understood schedule. Use of a bar chart/milestone schedule is recommended for this purpose, since it provides for clear delineation of the tasks and shows the associated schedule and important milestones. Typical tasks that should be discussed and scheduled include literature search activities, feasibility assessment, preliminary design, detail design, preparation of component specifications, component delivery dates, development test activity, fabrication, final assembly, qualification testing, acceptance testing, end item delivery, and final report preparation.

For projects involving delivery of an item of hardware, the technical approach section of the proposal should also include a discussion as to how the final product will be evaluated to determine whether it meets the goals of the project. This is important to the customer, since the product evaluation plan assures the customer that the end item will perform the functions intended.

11.3.4 Budget

The total proposed cost is the most important element in the proposal. It is the total cost that will determine whether the customer will fund the effort. Thus, the budget must follow logically from the tasks identified, and the allocations made must agree with the manner in which the effort is to be accomplished, as described in the technical approach. For proposals that respond to RFPs, the budget breakdown will often be required to follow a set format that may specify salaries, fringe benefits, overhead, materials and supplies, equipment, and travel.

11.3.5 Organization and Capabilities

This section of the proposal describes how the proposer will organize to accomplish the effort. The responsibilities of key personnel should be described here, and qualifications, as related to the tasks for which they are responsible, should be identified. The organizational support to be provided and the resources available for accomplishing the work also should be included in this section of the proposal.

Key Personnel   The responsibilities of only key personnel should be should be described and the qualifications, as related to these tasks, should be briefly outlined. The narrative description of qualifications provided in this section should be supported by vitae on each key individual located in the appendix. It is very helpful for at least some of the key personnel to be recognized performers in the research area and to have reputations as productive and innovative workers.

Organizational and Other Support   This section of the proposal should substantiate the enthusiastic support of company management and, when appropriate, any labor, materials, or supplies that will be provided to the project but not charged to the customer should be identified. If organizations other than the proposing entity will be involved in the project, they should be identified here, and their responsibilities should be clearly described. If letters of support from leaders in the research area outside the proposer's company can be obtained, they should be referred to here and be included in the appendix.

Facility Capabilities   Describe the facilities (laboratories, test facilities, office space, shops, computer capabilities, etc.) available to support the project. Emphasize the characteristics of the facilities that make them especially useful in accomplishing the work.

An example of a brief proposal that includes many of the features described in this section is included in Appendix E.

11.4   ORAL COMMUNICATIONS

Webster's dictionary defines the word "communicate" to mean imparting, sharing, or *to make common.* This definition implies that communication is a sharing of ideas and thoughts in a way that is understood in the same way by all the participants in the discussion. This definition places considerable responsibility on the person who is doing the speaking to not only *send* his or her message, but to ensure that it is *received* and *understood.* To accomplish this goal a speaker must know his or her audience, use language that is common to them, and elicit feedback to ensure that the message is received and understood. Although many settings do not allow all of these objectives to be satisfied, these goals are ones that speakers should strive to achieve.

There are four basic modes of delivery for oral communications that are of interest to engineers. These are described as follows:[7]

Manuscript   A manuscript is meant to be read word for word to ensure that what is communicated is precisely what is intended. This is a seldom used mode of delivery for an engineer but, nevertheless, is appropriate for certain occasions when it is imperative that the information presented be thoroughly thought out and reviewed before presentation and not be embellished upon when given. This is the type of presentation that is appropriate when the information presented will go into some significant permanent record, such as with congressional testimony, trial depositions, etc. It is also appropriate when called on to present an important technical paper in substitution for another person who did the work. The disadvantages of this type of presentation

[7]R. H. Arthur, *The Engineer's Guide to Better Communication,* Scott, Foresman, 1984.

are (1) most engineers are not skilled in reading in a natural and interesting fashion, (2) the speaker is forced to focus attention on the manuscript and thus is isolated from the audience, and (3) in reading a manuscript the reader sometimes appears to be detached from and disinterested in the material being presented.

Memorization   There appears to be little or no use for this mode of delivery for engineers. Only the best speakers can make a presentation by memory without coming across as dull, inflexible, and distant. For technical material, such as that presented by engineers, memorization has little or no application.

Extemporaneous Speaking   One meaning for the word extemporaneous is *with preparation but not read or memorized.* This is the mode of delivery that most engineers use when giving oral reports, technical papers, and other less formal presentations. The degree of preparation varies widely depending on the material being presented and the audience, but should normally consist of either committing the entire presentation to writing or preparing an outline. Practicing by using audio or video recording or having someone listen and critique the presentation is helpful. For important occasions, it may be advantageous for the inexperienced speaker to write the presentation material completely and practice using the script initially, relying on it less and less with each practice session, and finally giving the presentation with little or no reference to the written version. Committing the presentation to writing sharpens the focus on the ideas presented, helps to improve the wording used in the presentation, and assists in the elimination of poor enunciation and time-stalling phrases such as "er ... a" and "ah ... a" Using an outline assists in organizing the presentation, and in cases when the technical material is well understood by an experienced speaker, this may be very adequate. The important point to remember when using this mode of delivery is to practice, several times if possible.

Impromptu Speaking   Engineers are often required to deliver remarks, provide input to a discussion, answer a question, or give a brief summary of a project they are working on without any advanced warning or preparation. This is what is known as impromptu speaking. In this mode of delivery it is important (given the limited amount of time) that thought be given to how the remarks will be organized and structured:

1. Organize remarks to include beginnings, middles, and conclusions.
2. Clarify the purpose of the discussion at the outset for both yourself and the audience.
3. Think of the points that need to be made and start with the most important. Start the discussion on each of these points with a single thesis statement that summarizes it.
4. Conclude remarks with a concise summary and seek responses to determine whether the information presented has been understood.

11.5   ORAL PRESENTATIONS

Giving an oral presentation is possibly one of the most trying experiences a newly graduated engineer has to face. Fortunately, as experience is gained and confidence increases, giving an oral presentation becomes less formidable. However, the apprehension associated with standing before an audience never totally disappears for most people and thus some means of controlling fear and turning it to an advantage is needed. Possibly the most effective way to control fear is preparation. If the speaker is well prepared and thoroughly understands the material to be presented, the likelihood of experiencing uncontrolled fear is greatly diminished. Familiarity with the environment in which the presentation is to be given also helps to build confidence. Becoming familiar with the size and layout of the room, whether there is an audio system and/or podium, the location of light switches, position of projectors, and the like, before the presentation is to be given is very helpful in reducing anxiety and minimizing difficulties during the actual event. Wearing the proper attire also helps to establish confidence and to minimize feelings of inadequacy. Some feeling of fear can actually be an asset in that it generates concern for the importance of the event, which often results in diligent preparation and practice. Fear can also enhance alertness and may contribute to developing the "up" or "on" feeling so essential for outstanding presentations.

Uncontrollable fear and outright panic are difficult to control, since they are totally due to emotional factors not related to any correctable situation. One technique that seems to be beneficial in combating uncontrollable fear is to penetrate the barrier between the speaker and the audience by getting some sort of dialogue going. If the speaker has a co-worker in the audience, the co-worker can ask a question or make some comment that requires an answer or action from the speaker. When no one is in the audience to perform this service the speaker can ask some question of the audience that requires an answer. The question or comment can be as simple as "can you focus that projector better?" or "can everyone see that?" Another ice breaker is to get everyone in the audience to introduce themselves, to tell where they are from or what they hope to get out of the presentation. The point is to break down the wall built up in the speaker's mind between him or herself and the audience and, in so doing, relieve the tension.

A speaker should always come across to the audience as being sincere and enthusiastic. The speaker needs to be responsive to the audience's interests by anticipating why they are there and what they will derive from the presentation. Injecting a little humor into the presentation is helpful in keeping the audience's attention, but the speaker should make sure that he or she has the knack of telling a joke successfully. Relating personal experiences that support the presentation theme is also a good technique. Tying the various parts of the presentation together so that there is a logical flow is essential. The audience should always know where the speaker is and where he or she is going in relation to the subject of the talk at all times. Most importantly, the speaker must always stay in character and be him or herself. If the audience detects any falseness on the part of the speaker, any credibility that has been established will be destroyed. Finally, a strong closing should always be made. If the presentation ends by just trailing off with "that's all I have" or "thank you," the audience will probably not be persuaded by the speaker and will certainly not feel a strong impetus to support the thesis of the presentation. The presentation should end with a strong summary of what has been discussed, emphasizing the points of greatest importance.

11.5.1 Organizing the Oral Presentation

Oral presentations should be organized into three parts: an introduction, body, and conclusion. Before the presentation is actually started, there are several things to be taken care of. If visual aids are being used, which is strongly encouraged, a viewgraph or slide identifying the subject of the presentation, to whom the presentation is being given, the presenter(s) name and affiliation, and the date should be projected on the screen. This makes it clear for the audience exactly what is going to be discussed and who the presenters are. It also sets the stage for the presentation and eliminates the problem of people getting in the wrong room. It also provides an appropriate cover sheet for the copy of the presentation that is maintained in the speaker's files for future reference. As the speaker begins the presentation, appropriate introductions should be made. The speaker should introduce the other personnel in his/her party by having each individual stand up as their name is mentioned. The presentation is now ready to begin.

Introduction   An outline of what is to be discussed by what individual (if several people are involved) should be initially presented. This outline should be accompanied by an appropriate visual aid. This gives the audience an overview as to what is to be discussed and follows the military's recommendation for presentations *to tell them what you are going to tell them, tell them, and then tell them what you told them.* The background for the subject should then be presented, supported by appropriate visual aids. It is important to capture the interest and attention of the audience at this point. Fortunately, the background for most subjects is usually rich enough to provide ample material for this. The introduction should also include a clear statement and explanation of the problem as well as the objective(s) of the work being reported on. Finally, the introduction should outline the methods used to accomplish the work. The use of visual schematics and sketches of the apparatus is usually very helpful in giving the audience an understanding of the experimental method.

If theory must be included in the presentation, it can be given as section of the introduction or integrated in with the discussion in the body of the presentation. The decision as to where the theory is included depends on the degree of difficulty, the magnitude, and whether it can logically be separated from the detailed discussion in the body of the presentation. Wherever it is located, theory requires a special emphasis to make it understandable to an audience that has little time in which to absorb difficult concepts.

Body   The body is the *meat* of the presentation and corresponds generally to the results and discussion section of the written report. As is true of the written report, the body of the presentation should be supported with an adequate number of visual aids. Figures showing curves, drawings, or sketches are preferred over tables, which often contain too much information for the audience to grasp during the 30 to 60 seconds that a visual aid is normally projected. The body of the presentation should include the information necessary to substantiate the conclusions of the effort being reported on. The flow of this part of the presentation should be such that the audience can easily follow the evolution of the effort from the first steps and findings to the final conclusion.

For proposal presentations individual tasks should be described and a bar chart showing the planned schedule for completion of each task should be used. A budget depicting an appropriate cost breakdown such as that shown in Appendix E for the written proposal should also be presented.

Conclusion   The conclusion of an oral report or proposal should include a summary of what has been presented. Conclusions can be listed individually

and discussed briefly emphasizing the importance of major findings. If recommendations are appropriate this is the place to include them, as well. The
important thing to keep in mind here is that statements made at the conclusion
of the presentation are the ones most likely to be remembered by the audience. The presentation should be concluded dynamically leaving the audience
with the most important conclusion, suggestion, and recommendation.

11.5.2 Use of Visual Aids

Visual aids only need to be used when the speaker wants the audience to remember
what has been presented. If you question this statement, consider the information in
Table 11.1. Also, remember Confucius' saying that *one picture is worth a thousand
words*. The only negative factors to consider in the use of visual aids is the
preparation time, cost, and whether the appropriate equipment will be available
for the presentation. Of course, the speaker can provide the necessary equipment,
but this arrangement may be prohibitive if the presentation is to be given some
distance from the speaker's place of business. Other considerations when one opts
to use visual aids include mechanical or electrical failures, keeping the slides or
transparencies simple and uncomplicated, selecting the correct number of slides
and/or transparencies, and making sure that they are all directly related to the
subject matter.

11.5.3 Types of Visual Aids

The various types of visual aids include: [8]

1. The device itself or a model thereof
2. Blackboard
3. Flannel board
4. Flip charts
5. Motion pictures (with or without sound)
6. 35 mm slides
7. Video and sound tape recordings
8. Closed circuit television
9. Overhead projector
10. Pointer (manual or projected)

Table 11.1 Retention of Data Presented Orally
and Visually (in percent)

Data Given to Us	Amount Retained After	
	3 Hours	3 Days
Orally	25	10
Visually	72	20
Orally and visually	85	65

* Table 11.1 and the text of sections 11.5.3 and 11.5.4 are
derived in part from *Communication* by Henry B. Keiser
(© Federal Publications Inc. 1977). Further reproduction
without the author's and publisher's permission is
prohibited.

[8] H. B. Keiser, *Communications,* Federal Publications, Washington, D.C. 1977.

Probably the two most commonly used aids are the overhead projector and the 35 mm projector, although the use of video recordings is increasing rapidly. The 35 mm projector is appropriate for large audiences in more formal settings. The disadvantages in the use of this technique are that the room needs to be darkened, which may be difficult if not impossible in some settings. The darkened room also adds to the speaker's problem of keeping the audience awake, alert, and involved. With this technique, the speaker may also appear to be distant and aloof from the presentation, almost like a recording.

The overhead projector probably enjoys the greatest popularity of all the visual aid techniques. It can be located at the podium and be operated by the speaker giving an impression of dynamic involvement. The house lights do not usually need to be dimmed, so the audience is not thus encouraged to fall asleep. With equipment presently available, the overhead projector can be used to project computer output directly on the screen. Transparencies are easy to use and can be modified in real time with appropriate pens. They are also easy to prepare using a laser printer or using low-cost equipment especially made to produce transparencies.

11.5.4 Preparation of Slides and Transparencies

Slides and transparencies used in oral presentations should be simple and uncluttered. It should be remembered that a typical slide or transparency is only projected for a period of 30 to 60 seconds, and the audience does not have time to assimilate difficult or complicated material. A general rule of thumb is to make only one point for each slide or transparency and to use no more than seven words per line and no more than seven lines on each slide or transparency. Printing needs to be big and bold, and abbrevations and legends need to be avoided. Titles need to be appropriate and informative. The use of color is recommended, but its use should be limited to basic red, blue, green, yellow, and black. Overlays can be used with transparencies, which are useful in slowly revealing the information contained thereon.[8]

11.6 A FINAL WORD ON COMMUNICATIONS

Like it or not, communication is an essential and significant element in the professional life of an engineer. For those engineers who refuse to become good communicators and depend on others to do their communicating, the future probably holds an outcome similar to that of Miles Standish in Longfellow's famous poem, "The Courtship of Miles Standish." Standish sent his friend, John Alden, to carry his proposal of marriage to the maiden he loved. The maiden refused, feeling that Standish should have spoken for himself if she were worth wooing. Furthermore, she asked John why he did not speak for himself. In the same manner, engineers cannot depend on others to do their talking unless they are prepared to have others get the credit for their work. Engineers must speak for themselves and be prepared to present their work in written or oral form, as appropriate, whenever the opportunity arises. The best opportunity that an engineer has to improve his or her communications skills is while still a student, surrounded by people interested in assisting in this endeavor. The principal requirement for making significant improvement in communication skills is having a strong desire to do so. Those who have this drive will have little difficulty in becoming effective communicators.

Appendix A

Problems

A.1 CHAPTER 1

1.1 Assume that you are an employee of a pressure vessel manufacturer that has a contract to deliver the pressure vessel system shown in Fig. B.1, Design Project D.2, Appendix B. Prepare a plan that identifies all drawings and the installation procedure for the project. Identify all drawings (by title) that are required to completely describe the vessel and the associated system. Identify the next assembly on all drawings and select the drawing sizes from Table A.1. For components and parts that will be purchased from vendors, select an A-size specification drawing.

Table A.1 Drawing Sizes

Designation	Size (in.)
A	$8\frac{1}{2}\times11$
B	11×17
C	17×22
D	22×34
E	34×44

1.2 Although many improvements have been made in the manufacture of automobiles during the past 10 years, the windshield wiper is one component that has not changed dramatically. Windshield wipers are limited in effectiveness during heavy rain, and they pose a problem to further improvement in vehicle aerodynamics. Organize into groups and use brainstorming techniques to develop other solutions for keeping the windshield clear during inclement weather.

1.3 Another vehicle problem that has not been satisfactorily solved is that of engine valve actuation. The need to vary the timing of engine valves as engine speed changes has been addressed for many years, but no single system has been developed that solves all of the problems. Valve overlap (both intake and exhaust valve open at the same time) needs to be minimized to control emissions, but power output is reduced as a result. Overlap allows the cylinders to start the intake process before the combusted gas is completely expelled. In fact, with overlap the intake gas helps to expel the exhaust gas and results in more fresh fuel/air mixture for the next cycle. The valve must open and close very rapidly and provision must be made to vary the time that the valve stays open (pulse width). Develop several design concepts that could be considered as alternates to the present valve/cam actuation method.

1.4 Analyze the feasibility of using of natural gas and methanol for fueling yard and garden maintenance equipment. Consider the problems associated with refueling, equipment conversion or purchase, time between fuel tank refuelings, fuel economy, equipment maintenance, and the like. Recommend which fuel you think is best for this application and justify your conclusion.

1.5 Objectives provide the only valid criteria for assessing the success or failure of development projects. Briefly analyze the U.S. space program malaise that set in after the first lunar landing in 1969 relative to the abrupt change in program objectives thereafter.

1.6 At what points in the design process should input be made by (a) top management, (b) marketing, (c) production? What should be the nature of this input and who should take the responsibility for making sure that it happens?

1.7 Discuss the terms design, development, and engineering as related to the evolution of a product.

1.8 Describe the difference between development, qualification, and acceptance testing. Would it be possible to integrate these efforts to some extent and reduce the overall project cost?

1.9 One of the problems facing U.S. industry is how to shorten the development time for products. Using Fig. 1.1 as a basis for identifying the different tasks involved in a typical design process, develop a plan to shorten the development time to bring a new product to the marketplace.

1.10 An underground fuel tank leak is suspected at the vehicle maintenance area on the university campus where you are employed as the facilities maintenance engineer. It is your job to recommend a plan that will confirm the leakage and remediate the site, if required. Using Design Project D.18, Appendix B as a reference, develop a chronological task breakdown that will satisfy this objective.

A.2   CHAPTER 2

2.1 Develop a Gantt/milestone chart schedule showing the tasks and events identified in Exercise 2.2. Use an arbitrary time scale. Note that some of the items identified in Exercise 2.2 are neither tasks or events.

2.2 Develop an ADM network for the tasks and events identified in Exercise 2.2.

2.3 Prepare a *what* versus *how* orthogonal array for your desires relative to your first job after graduation. Be sure to consider factors such as location, type of company, department within the company, people you will work with and for, salary level, promotion possibilities, family considerations, housing availability, and cost. Characterize the *what* versus *how* relationships using the rating system from Exercise 2.1.

2.4 Complete the table (T_E, T_L, and Slack) for the network shown in Fig. A.1 and identify the critical path. Because of a critical time constraint, the overall project completion schedule must be reduced to 60 weeks by adding resources. Determine the task to which minimum resources can be added to reduce the overall schedule to 60 weeks and identify the new critical path.

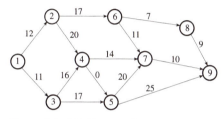

Figure A.1 PERT network.

2.5 Prepare a decision matrix comparing the use of the following alternative transportation fuels in automobiles and light trucks in the United States. Some of the factors against which the alternatives should be evaluated include conversion cost, infrastructure cost, safety, convenience, environmental effect, effect on vehicle maintenance, U.S. trade balance, and long-term feed stock resource availability.

(a) Compressed natural gas (CNG).
(b) Liquid natural gas (LNG).
(c) Neat (pure) methanol (M100).
(d) A blend of 85% methanol and 15% hydrocarbon (M85).
(e) Neat (pure) ethanol (E100).
(f) A blend of 85% ethanol and 15% hydrocarbon (E85).
(g) Liquified Petroleum Gas (LPG).

2.6 Perform a functional cost analysis on the check valve shown in Fig. A.2 using the cost data shown. Use the technique described in paragraph 2.8.1, Functional Cost Analysis, and Fig. 2.13 *a* and *b*. Identify as many functions for each part as possible.

2.7 Using the information provided in Fig. A.2 on the pieceparts that are used in the check valve assembly shown, perform a component level FMEA. Establish a piecepart functional weighting scheme that represents the relative importance of the piecepart to the overall check valve function to determine the gross and percentage weighting as shown in Fig. 2.14.

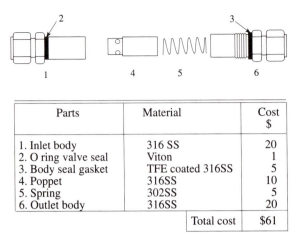

Parts	Material	Cost $
1. Inlet body	316 SS	20
2. O ring valve seal	Viton	1
3. Body seal gasket	TFE coated 316SS	5
4. Poppet	316SS	10
5. Spring	302SS	5
6. Outlet body	316SS	20
	Total cost	$61

Figure A.2   Check valve assembly.

2.8 A new automobile is purchased that has an antilock brake system (ABS). Drivers are sensitive to ABS performance and when the system cycles at a greater or lower rate than is appropriate, complaints will be registered requiring system adjustment under warranty. This adjustment is rather involved and, if performed by the dealer, costs the manufacturer $150 under the warranty agreement. If the adjustment is made at the factory the effort is considerably reduced and only costs $15. Experience indicates that customers will complain if the setting is more than plus or minus 10 cps from the nominal setting. Find the loss function k and the limits $(Y - m)$ that define when an adjustment should be made at the factory rather than at the dealer's repair shop.

A.3   CHAPTER 3

3.1 Develop a 1/10 scale model of your room with all the furniture, using cardboard, wood, glue, and the like. Furniture should be capable of being relocated so that various configurations can be considered. Decide on the best layout and justify why you think it is optimum. Compare the optimum to the present furniture arrangement.

3.2 If an automobile shown in Fig A.3 is excited by a harmonic force $F_1 \sin \omega t$, model the system using

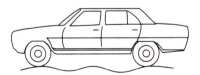

Figure A.3   Automobile on the road.

(a)   A single degree-of-freedom.

(b)   A two degree-of-freedom.

and write the mathematical expressions that describe the behavior of the automobile.

3.3 Determine the π_1, π_2, π_3, and π_5 terms for Example 3.1.

3.4 Find an expression for the distance traveled by a freely falling mass M in time t under the influence of gravity g if the initial velocity is zero.

3.5 An aluminum beam of rectangular cross section ($E = 10.5 \times 10^6$ psi) is 8 in. wide and 14 in. in height. A concentrated load of 6000 lb at 6 ft is applied from the left end of the 14-ft beam. Design a model with a 10-in. span using a steel beam to accurately predict deflection at any location along the aluminum prototype. Ignore the weight and deflection of the beam.

3.6 A bending moment of 3000 lb$_f$-in. is required to bend a model aluminum bar into an L shape. A prototype bar is made of the same material as the model with a 50 percent larger diameter. Determine the bending moment required to bend the prototype into an L shape.

3.7 Write a computer program for the example given by Fig. 3.9 which evaluates the temperatures at the grid points of Fig. 3.9 by iteration.

3.8 Write a computer program and evaluate the temperatures at the grid points for the steady-state heat conduction problem given by Fig. A.4. Compare the result with the analytical solution at $a = 5$ in. and $b = 5$ in. The heat transfer for this problem follows the equation

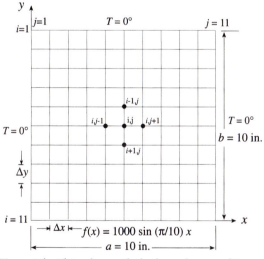

Figure A.4 Thin plate with the boundary conditions.

$$\frac{\partial^2 T}{\partial x^2} + \frac{\partial^2 T}{\partial y^2} = 0$$

The boundary conditions are:

(a) $T(0,y) = 0°$
(b) $T(x, b) = 0°$
(c) $T(a, y) = 0°$
(d) $T(x,0) = f(x)$

3.9 Consider the shape P of available area on a circuit board shown in Fig. 3.11. Using the second digits of the respective random number pair in Table 3.3 as the coordinates of the random points, estimate the area of P.

A.4   CHAPTER 4

4.1 Derive the expression for the bending moment and deflection for the drill pipe given in Example 4.2.

4.2 Construct an operation chart (variation of fatigue damage with respect to ball joint angle) for the drilling system used for drilling offshore wells from floating vessels by the use of the SN curve given in Example 4.2. Use the same design data as in the example with $\theta = 1°, 2°, 3°, 4°$, $T = 100$ kips and $T = 200$ kips.

4.3 Construct an operation chart (variation of fatigue damage with respect to ball joint angle) for the drilling system in Example 4.2 using the theory of fracture mechanics. Use the design data given in the above problem and compare the results of the two methods.

4.4 The starting valves of a two-stroke marine engine are opened by means of compressed air from an air pressure vessel. For safety, assume that the pressure of the compressed air in the pressure vessel should not exceed 16 MPa and the minimum air pressure should be 12 MPa to be able to start the engine. If the inside diameter of the pressure vessel is 1.0 m and the length is 3 m, using the principles of fracture mechanics, select the most suitable materials from Table A.2 materials. Assume an initial surface flaw of 8 mm depth and a constant $a/2c$ ratio of 0.25. Perform the analysis based on a constant mean pressure of 14 Mpa.

Table A.2   Materials for Pressure Vessel Design

Material	S_y (MPa)	K_{IC} (MPa $\sqrt{m}$)	ρ ton/m^3	E (GPa)	Cost $/ton
Steel #1	1750	87	7.7	210	650
Steel #2	1450	115	7.7	210	650
Steel #3	1200	148	7.7	210	450
Steel #4	1500	105	7.7	210	550
Steel #5	1600	58	7.7	210	500
Steel #6	1400	77	7.7	210	350

4.5 `As shown in Fig. A.5, assume that a drill pipe is subjected to a constant axial tension load of 100,000 lb, and a bending stress amplitude of 22,600 psi. Assume the initial crack depth, $a_i = 1/32$ in., initial length $c_i = 1/8$ in., and a constant a/c of 0.5. Select a suitable steel from Table A.3 for this application. Consider strength, cost, service life, and leak-before-break as the desired criteria.

Table A.3   Drill Pipe Material Properties

Steel	S_y (ksi)	K_{IC} (ksi $\sqrt{in.}$)	Cost $/lb
A	90	56	1.5
B	80	38	1.00

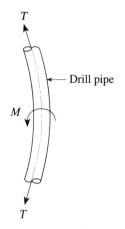

Figure A.5   Drill pipe
under bending.

Design Data

Density of the steel $= 0.283$ lb/in.^{3}
Drill pipe outside diameter $= 5.0$ in.
Drill pipe inside diameter $= 4.276$ in.
Assume that $M_k = 1$.
For both steels use $da/dN = 0.614 \times 10^{-10} \ (\Delta K)^{3.16}$.

4.6   A car radiator fan shown in Fig. A.6 ($a = 0.3$ in.) through crack introduced
during the manufacturing process. The yield strength of the fan material is 90
kpsi and the critical stress intensity factor is 120 kpsi $\sqrt{\text{in}}$. Due to the change in
acceleration in the radial direction, the force acting on the blade varies between
17,280 lb and 23,040 lb. Using a scaling factor, A $= 0.66 \times 10^{-8}$ and the slope,
m $= 2.25$ for the expression of the crack-growth rate, investigate whether the design
is adequate for 500,000 cycles. If not, what changes would you recommend so that

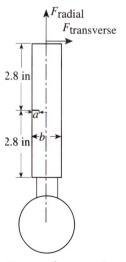

Figure A.6   Car radiator fan.

the design will be adequate for 500,000 cycles. Assume that the radial component of the force is causing the crack growth. Use $b = 2.4$ in. and thickness $= 0.12$ inch.

4.7 The landing gear on passenger airliners is operated by a hydraulic pump. When the gear is cycled the hydraulic pressure reaches to 30 kpsi. When it is locked up, the hydraulic pressure is a constant 10 kpsi. The hydraulic lines are made of 1 inch. inside diameter tubing with a wall thickness of 1/16 in. The yield strength of the steel is 255 kpsi with a K_{IC} of 79 kpsi $\sqrt{\text{in}}$.. With an initial flaw of $a_i = \frac{1}{64}$ in. in the tubing and an $a/2c$ ratio of 0.25, find the number of cycles before failure. The density of steel is 0.283 lb/in³. Use a scaling factor, $A = 0.66 \times 10^{-8}$ and the slope, $m = 2.25$ for the expression of the crack-growth rate. Assume $M_k = 1$.

4.8 Consider a thin laminate with the material properties: $E_1 = 20 \times 10^6$ 5pt psi, $E_2 = 1.5 \times 10^6$ psi, $G_{12} = 1.0 \times 10^6$ psi, $\nu_{12} = 0.2900$, $\nu_{21} = 0.02176$, $\theta = +30°$, and let the strain in the laminar be $\varepsilon_x = 3.13 \times 10^{-4}$ kpsi, $\varepsilon_y = 0.91 \times 10^{-4}$ kpsi, $\gamma_{xy} = 0.0$. Determine the stresses along the principal material axes of symmetry (σ_1, σ_2, and γ_{12}).

4.9 Consider a thin laminate with the material properties $E_1 = 20 \times 10^6$ psi, $E_2 = 1.5 \times 10^6$ psi, $G_{12} = 1.0 \times 10^6$ psi, $\nu_{12} = 0.2900$, $\nu_{21} = 0.02176$, $\theta = +45°$, and let the stresses in the laminate be $\sigma_x = 63.396 \times 10^2$ kpsi, $\sigma_y = 2.744 \times 10^2$ kpsi, $\tau_{xy} = 0.0$. Calculate the strains in the laminate.

4.10 Consider an orthotropic laminate under simple uniaxial tension as shown in Fig. A.7, with the following properties: $\sigma_1^T = 200$ kpsi, $\sigma_2^T = 100$ kpsi, $\tau_{12}^* = -20$ kpsi.

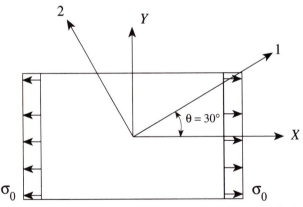

Figure A.7   Material axes of a laminate oriented at an angle θ.

For $\theta = 30°$, using the maximum stress criteria, determine the smallest σ_o at which the laminate will fail.

4.11 An aircraft propeller shown in Fig. A.8 is operated at 2500 rpm with a 5 percent variance. Select a proper material for the possibility of introducing a through crack (single edge) during the manufacturing process with a width of $2a = 0.25$ in. Pick three different materials and select the best one through analysis.

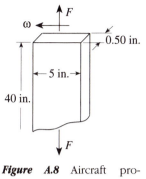

Figure A.8 Aircraft propeller.

A.5   CHAPTER 5

5.1   The parents of a first-year university student are in the process of buying a car for the student to use. Two used cars have been found. One car can be purchased for $6000 with a first-year projected maintenance cost of $300. Maintenance costs are expected to increase by $100 per year thereafter. Operational costs for this vehicle are anticipated to amount to $1000 per year, whereas the salvage value decreases at a rate of 15 percent per year. The second car can be purchased for $10,000 and the first-year maintenance cost is expected to be $200. Maintenance costs for this vehicle are expected to increase by $200 per year thereafter. Operational costs are also expected to be $1000 per year and the salvage value for this second vehicle is projected to decrease at a rate of 18 percent per year.

(a)   Find the minimum-cost life and minimum annual life-cycle cost of the two vehicles if the time value of money is ignored.

(b)   Prepare cash flow diagrams for both vehicles and calculate the minimum annual life-cycle cost using the tabular approach. Assume that the interest rate is 12 percent.

(c)   Which is the best purchase and why?

5.2   How much would the owner of a pickup truck be justified in paying for the vehicle to be converted to operate on compressed natural gas if 20,000 miles are accumulated per annum, the pickup averages 12 mpg, the savings in fuel cost amounts to $0.50/gallon, the life of the conversion system is 10 years, and the salvage value is 15 percent of its initial cost? Assume that the time value of money is 12 percent.

5.3   The annual rental income from a duplex is $9600 and annual expenses are $2000. If the potential buyer anticipates that the duplex could be sold for $70,000 at the end of 10 years, what is the purchasing price that could be justified? Assume that the interest rate is 15 percent.

5.4   Calculate the ROR for the cash flow shown in Table A.4.

Table A.4 Cash Flow
with Respect to Years

Year	Cash flow, $
1	−15,000
2	−8,000
3	10,000
4	12,000
5	15,000
6	15,000

5.5 For the three alternatives shown in Table 5.7, determine the payback period for an interest rate of 15 percent.

5.6 A fleet manager is considering the construction of a compressed natural gas refueling facility. The cost of site preparation, dispensing equipment, storage tanks, and the like has been determined, but the decision as to the compressor selection, which is the greatest single expense item, has yet to be made. The two options include (a) purchasing a 50,000-SCFM compressor at a cost of $40,000 with an interest rate of 12 percent, or (b) renting a compressor at a cost of $1000 per month, including maintenance. If a compressor is purchased, it is anticipated that annual maintenance costs will be $1000. The life of the compressor is estimated to be 10 years. Assume that the compressor is depreciated using the straight-line method and determine which is the better option for the fleet manager if the income tax rate is 30 percent.

5.7 Automobile dealers are increasingly advertising the leasing of vehicles in lieu of purchasing. In one case, an $18,000 automobile can be leased for $375 per month for 36 months, after which it is returned to the dealer. If the automobile is purchased, it could be financed for 3 years at an interest rate of 10 percent annual percentage rate with a down payment of 5 percent and 36 equal monthly payments. If at the end of the 36-month period the vehicle is estimated to be worth $8000, which would be the preferred alternative? What reasons other than financial might make leasing the vehicle advantageous? What is the lessee's minimum before-tax interest rate that makes leasing the more desireable alternative?

5.8 Find the RORs for alternatives A1, A2, and A1 - A2, in Table 5.11.

5.9 The five mutually exclusive alternatives listed below have 15-year useful lives. If the MARR is 12 percent, which alternative should be selected?

- Alternative 1—This alternative requires an initial expenditure of $10,000 and has an annual benefit (receipts less disbursements) of $1424.
- Alternative 2—This alternative requires an initial expenditure of $6000 and has an annual benefit of $1026.
- Alternative 3—This alternative requires an initial expenditure of $4000 and has an annual benefit of $742.

- Alternative 4—This alternative requires an initial expenditure of $2000 and has an annual benefit of $356.

- Alternative 5—This alternative requires an initial expenditure of $1000 and has an annual benefit of $256.

5.10 Determine the expected value of bets on Texas Tech, Texas University, and Houston winning the 1992 Southwest Conference football championship based on the data provided in Example 5.1.

5.11 A certain company has several projects under consideration and is trying to decide which should be pursued. It has been decided that the projects will be ranked according to their net present worth divided by their present worth of cost. If the company's MARR is 15 percent, rank the projects in accordance with this criterion and determine which projects from Table A.5 should be funded.

Table A.5 List of Properties

Project	Cost ($)	Uniform Annual Benefit, ($)	Useful Life (years)	Salvage ($) (%)	ROR
1	1200	267	10	0	18
2	2000	521.20	8	0	20
3	500	185.90	5	0	25
4	2000	477	10	0	20
5	2000	438.5	10	1000	20
6	1500	225	6	1500	15

5.12 An investment of $50,000 for a new air conditioning system is being considered by a company. The expected life of the new system is estimated to be 15 years, and the salvage value of system equipment at this point is thought to be $7000. The new system is expected to increase business such that annual gross receipts will grow by $8000. Expenses are expected to increase by $2700 per annum. If the time value of money is 18 percent, determine the following measures of investment worth and recommend the appropriate action.

(a) Internal rate of return.

(b) Present worth.

(c) Annual worth.

(d) Future worth.

(e) Net present worth/present worth of cost.

5.13 For the new liquid desiccant drying system as discussed, liquid desiccant can also be generated by using an open solar collector in conjunction with an auxiliary heater as shown in Fig. A.9. If the total heat energy that will be required from the auxiliary heater is $Q_T = 429,748$ Btu/container, develop an economical model and investigate if this system is feasible. Use the following assumptions:

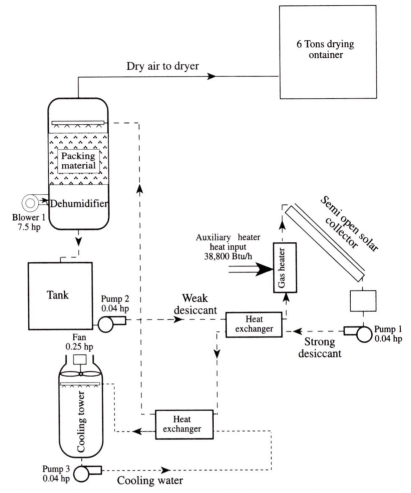

Figure A.9   Proposed new drying system with solar regeneration.

(a)   58 drying times in a drying season.

(b)   Cost of drying system with solar regenerator is $6340.

(c)   Annual interest rate, $i = 11$ percent.

(d)   Cost of natural gas is $3.84/10^6 \times$ Btu.

(e)   Cost of electricity is $0.04/kWh.

A.6   CHAPTER 6

6.1   Using differential calculus, maximize the volume of a box made of cardboard as shown in Fig. A.10, subject to the following constraints:

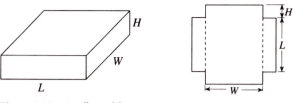

Figure A.10   Cardboard box.

(a)   Allowable area of the cardboard is equal to 40 in².

(b)   The length of the box L is equal to its width W.

6.2   Resolve question A.6.1 by using the Lagrange multiplier for the allowable area of 80 in.²

6.3   Using the graphical method, find the maximum of the objective function $U = x_1 + 4x_2$ subject to

$$x_1 \geq 0$$
$$x_2 \geq 0$$
$$3x_1 + 2x_2 \geq 0$$
$$x_1 + 2x_2 \geq 0$$

6.4   Write a computer program for the equal interval search method of Fig. 6.6. Solve Example 6.4 with different starting points and step sizes. Compare the results.

6.5   Write a computer program for the Golden section method of Fig. 6.8. Solve Example 6.5 in the interval of $2 \leq x \leq 14$ based on a convergence criterion $\varepsilon = 0.002$.

A.7   CHAPTER 7

7.1   Find the following values for a t distribution:

(**a.**) $t_{0.025}$ with $\nu=14$

(**b.**) $t_{0.05}$ with $\nu=29$

(**c.**) $t_{\alpha/2}$ so that $p(-t_{\alpha/2} < t < t_{\alpha/2})=0.90$ with $\nu=8$

7.2   Find the following values for a chi-square distribution:

(**a**) $\chi^2_{0.025}$ with $\nu = 12$

(**b**) $\chi^2_{0.05}$ with $\nu = 19$

(**c**) χ^2_α so that $p(\chi^2 < \chi^2_\alpha)=0.995$ with $\nu = 14$

7.3 Find the following values for an F distribution:

    (a) $F_{0.1}$ with $\nu = 7$, $\nu = 14$

    (b) $F_{0.05}$ with $\nu = 3$, $\nu = 28$

    (c) $F_{0.025}$ with $\nu = 40$, $\nu = 60$

7.4 The following are the spark plug gaps, in mm, of eight spark plugs in a package:

        2.50, 2.52, 2.48, 2.53, 2.47, 2.51, 2.50, and 2.55

Find a 90 percent confidence interval for the variance of all spark plug gaps.

7.5 Increasing the stiffness of a circuit board results in decreasing the deflection and, hence, a decrease in electrical lead wire and circuit board stresses during vibration. Assume that for a minimum deflection, it is desirable that the standard deviation σ of the stiffness be 3 kips/in. with a 95 percent confidence level. Assume that the circuit boards manufacturing process for a specified stiffness is normally distributed. If the stiffness of a random sample of 28 circuit boards results in a sample variance $S^2 = 2.9$ kips/in. does the given data support the claim?

7.6 Consider eight manufacturing welding processes for the drill pipe described in Example 7.8. Eight drill pipes assemblies were manufactured using both old and new welding processes. The failure strength test results of eight drill pipes at the tool joint welding location are given in Table A.6. Considering a 95 percent two-sided confidence interval, determine if there is a difference between the two manufacturing processes.

Table A.6   Comparison of Drill Pipe Welding Process

Old Method (failure strength, kpsi)	New Method (failure strength, kpsi)
120	121
118	125
105	118
107	122
101	116
115	130
113	129
121	135

7.7 A six-cylinder, 4000 hp marine diesel engine has six fuel injectors. Because of heavy fuel oil, the injectors get dirty after approximately 1000 hours. Consequently, pressure adjustment is lost, resulting in poor combustion and possible component failure. Assume that the chance of an injector failing is 1 in 200 after the 1000 hours of operation. Assume that the occurrences of failure of the injectors are independent of each other and that the engine can run if more than three injectors keep operating. What is the probability of the engine failing after 1000 hours of operation (use binomial distribution).

7.8 If the sample size of Example 7.9 had been reduced to 10 electronic components, and the sample mean and the standard deviation are found to be 4.9×10^6 cycles and 4.1×10^5 cycles, respectively, determine whether the sample data give enough evidence that the fatigue life of the materials has been improved. Assume that the life of the components is normally distributed.

7.9 The lubricating oil pressure of an engine that is coupled with a pump that delivers crude oil to shore from an offshore well should remain almost constant during the operation. Low lubricating oil pressure may cause engine failure. This is especially true for high-speed engines. However, because of vibrations, correct readings of the oil pressure using a gauge is almost impossible. Suppose, during the operation of the engine, the operator takes a reading of approximately 65 psi. Assume that the engine can operate with no trouble if the mean value of normally distributed lubricating oil pressure is 68 psi and that it may fail if the mean value of the lubricating pressure is 64 psi. In both cases, assume that the standard deviation is 2.5 psi. If the lubricating oil pressure is low, close to 64 psi, two possible decisions can be made by the operator:

(a) Stop the pumping operation and start troubleshooting.

(b) Continue pumping.

The operator should know the relative cost of a wrong decision and the failure probability of the engine before making the above-cited decision. Assume that, from experience, the failure probability of the engine when it is running under low pressure is 70 percent. Hence, the expected abnormal engine operation is 0.70 and the operator's null hypothesis is that the engine operation would not be normal.

If the operator does not stop the engine when it is not operating safely, he or she is making a Type I error. Making this error results in a loss and, in some instances, replacement of the engine for not detecting the problem in time. Assume that the relative cost of this is 1 unit, and it is designated regret r for making a Type I error.

If the operator stops the engine for troubleshooting when the engine can still operate, at least until the pumping operation finishes, he or she is making a Type II error. Consequently, this wrong decision results in loss to the company because of the down time. Assume that the relative cost of this is 5 units, and it is designated Regret R for making a Type II error.

At the observed operating oil pressure of 65 psi (average of five readings), what is the best decision the operator can make? Calculate Type I and Type II errors, show them on a tree diagram, calculate the expected loss, and discuss the results.

7.10 To evaluate the sensitivity of an explosive, 18 specimens were tested by dropping them from a certain height. The result of the experiment is shown in Table A.7, where the x's represent explosions (successes) and the 0's represents nonexplosions (failures). By using the up-and-down method, estimate the mean and standard deviation of the height for the explosion experiment.

7.11 If the result of the experiment in Problem 7.10 is as shown in Table A.8, calculate the mean and standard deviation of the height for the explosion experiment.

450   APPENDIX A/PROBLEMS

Table A.7 Up-and-Down Test for Sensitivity of Explosive

Test No.	1	2	3	4	5	6	7	8	9	10	11	12	13	14	15	16	17	18
Normalized Height 3.0							X		X		X							
2.6						0		0		0		X						
2.2			X		0								X		X		X	
1.8		0		0										0		0		0
1.4	0																	

X indicates explosion.
0 indicates nonexplosion.

Table A.8 Up-and-Down Test for Sensitivity of Explosive

Test No.	1	2	3	4	5	6	7	8	9	10	11	12	13	14	15	16	17	18
Normalized Height 3.0	X																	
2.6		X		X										X				
2.2			0		X		X		X		X		0		X			
1.8						0		0		0		0				X		0
1.4																	0	

X indicates explosion.
0 indicates nonexplosion.

7.12 Referring to Example 7.11, using the experimental data given in Table A.9 design a two-level experiment with two factors to determine the relative effects of roughness and speed on the friction coefficient between the box and pin. From the table, it can be seen that the main factors are acting independently; thus, the error is assigned to the third column instead of to interaction. Discuss the result of the ANOVA table.

Table A.9 Two-Factor L_4 Orthogonal Array

| | **Roughness** | **Speed** | **Error** | |
Col. →	1	2	3	**Observations,** Y_i
No. ↓				
1	1	1	1	0.050
2	1	2	2	0.054
3	2	1	2	0.061
4	2	2	1	0.069

7.13 Referring to Example 7.12, determine the minimum confidence level and risk α associated with the factor speed.

7.14 Electrical components have changed dramatically in the last 20 years. Instead of vacuum tubes, microcircuits are now used. These microcircuits incorporate many hundreds or thousands of electrical components; they are referred to as very-large-scale Integrated circuits (VLSIC). These electrical devices, because of their size, are very susceptible to contamination; a particle of dust can bridge a gap between very small wires called leads and cause reliability problems. Because contamination is such a serious problem, much engineering time and money is spent on limiting the number of particles that can get on a wafer. The higher the contamination, the larger the yield and, subsequently, the lower the profit.

This problem deals with a wafer undergoing one of two annealing processes, whether or not inspection under a microscope introduces particles, and whether the wafer cleaning itself introduces particles. Different manufacturing processes generate different contaminants; thus, it is necessary to use clean processes. Inspection during the various processing steps is accomplished under microscopes but, since operators are looking down at the wafer under a scope, skin particles can fall on the wafer. Cleaning the wafer is usually done in a bath, and then the wafers are placed in a machine that rotates the wafers to spin off the solvent. This is referred to as spin rinse drying. This can also introduce particles onto the surface from loosening of the photoresist.

The task is to find the largest generator of profit-robbing particles so that the problem can be quickly and adequately addressed. Using the experimental data given in Table A.10, for particle count Y_i, design a three-level experiment to determine the relative effects of process, inspection, spin rinse drying, and the interaction of the major components on contamination of VLSIC wafers. (This problem is adapted from Experimental Design and Taguchi Method, course notes, Jones Reilly & Associates, Inc., 1988).

7.15 To simplify the analysis given in Problem 7.14, assume that the interactions are minor and can be placed in a catagory called residual (error). Design a three-level lumped experiment to determine the relative effects of process, inspection, and spin rinse drying on contamination of VLSI wafers.

Table A.10   Orthogonal Array L_8 for Spin Rinse Drying Experiment

Col. →	1	2	3	4	5	6	7	Y_i
No. ↓								
1	1	1	1	1	1	1	1	40.0
2	1	1	1	2	2	2	2	335.0
3	1	2	2	1	1	2	2	25.0
4	1	2	2	2	2	1	1	80.0
5	2	1	2	1	2	1	2	23.0
6	2	1	2	2	1	2	1	248.0
7	2	2	1	1	2	2	1	45.0
8	2	2	1	2	1	1	2	196.0

Column 1 = process, column 2 = inspection, column 3 = process×
inspection, column 4 = spin rinse dry, column 5 = process×spin rinse
dry, column 6 = inspection×spin, column 7 = error.

7.16 A machine known as an ion implanter has made the fabrication of tiny
microcircuits possible. This machine is a plasma generator and extractor that dopes
wafers preferentially. The plasma field from which the dopants are extracted
depends mainly on three parameters—arc current, extraction current, and pressure
inside the plasma. The arc current is the current from the filament to the arc chamber, and the extraction current is the electromotive force applied to pull dopant ions
from the arc chamber, as shown in Fig. A.11. Using the experimental data given in
Table A.11, design a two parameter experiment to find the relationship between the

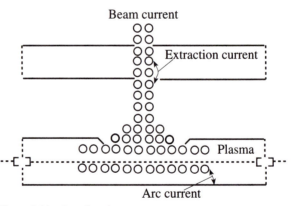

Figure A.11   Arc chamber.

Table A.11   Orthogonal Array L_8 for Arc Chamber Experiment

Col. →	Arc Current	Extraction Current	Pressure	Beam Current Y_i
No. ↓				
1	1	1	1	0.80
2	1	1	2	0.88
3	1	2	1	0.95
4	1	2	2	1.045
5	2	1	1	1.050
6	2	1	2	1.155
7	2	2	1	1.400
8	2	2	2	1.540

arc current, extraction current, and the output, which is beam current. Optimizing the beam current is essential for the most efficient use of the ion implanter. Assume that the interactions are very small.

7.17 The relationship between the viscosity VIS and temperature T of a CELD solution (50% $CaCl_2$ + 50%LiCl) can be formulated as

$$VIS = \alpha * \exp(\beta * T)$$

Referring to Table A.12 and using the SAS software package, obtain estimates of α and β.

Table A.12   Viscosity of CELD Solution

Temperature °C	Viscosity c.p.
15.55	15.713
21.11	12.786
26.66	10.548
32.22	8.424
37.00	6.871
43.33	5.439
48.88	4.515
54.44	3.900
60.00	3.494

A.8   CHAPTER 8

8.1 Suppose that $P(A) = 0.6$ is the probability of an event A, $P(B) = 0.2$ is the probability of an event B, and $P(A \text{ and } B) = 0.15$ is the joint probability of A and B. Determine if these two events are dependent or independent events.

8.2 Assume that four red pens, seven black pens, and nine blue pens are in a bag and we wish to draw them randomly. What is the probability that the first pen drawn is red and the second is blue?

8.3 Suppose a spinner game wheel has two spinning balls as shown in Fig. A.12. If the two balls are spun, what is the probability of one ball settling in the yellow region?

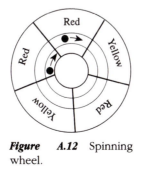

Figure **A.12** Spinning wheel.

8.4 A common procedure in testing the sensitivity of explosive materials to shock is to drop a weight on the explosive from a critical height. Assume that we wish to estimate the reliability of an explosive material that, on being tested, either explodes (success) or does not explode (failure). Suppose we have two lots of explosive materials, one of which has a reliability of 0.80. That is, 8 of 10 randomly selected explosive materials will explode. The other lot contains 50 percent reliable explosive materials. Assume that it is not known which lot has better reliability. If a single explosive material randomly selected by one of the lots has failed the test (did not explode), what is the probability that the tested explosive material was taken from

(a)   The lot having a reliability of 80 percent.

(b)   The lot having a reliability of 50 percent.

8.5   Assume that the failure of rolling bearings is described by a Weibull distribution. If the estimated values are $\theta = 1200$ h and $b = 1.25$, determine:

(a)   The reliability for a 400 h operational period.

(b)   The mean time to failure (MTTF) of the rolling bearing.

(c)   The hazard rate.

8.6   The purpose of a condenser is to convert the exhaust steam of an engine into water so that it can be reused in the boiler. Both the vacuum pump, which removes the water–air mixture from the condenser, and the cooling water circulating pump must work simultaneously for system success. If the overall reliability must not be less than 0.985, what should be the minimum reliability of each pump?

8.7   Assume a system consists of two components having reliabilities 0.85 and 0.89. For mission success, at least, one of the components must be in operation. What is the reliability of the system?

8.8   A water cooling tower operates with two independent and identical circulating pumps, both of which must work for cooling to be successful. What is the cooling tower reliability if each pump has reliability of 0.96?

8.9   Assume that 150 psi of compressed air is provided by two identical compressors to a pressure vessel. This system will fail if both compressors fail. If the constant failure rate of both compressors is 0.00015 failure/hour, calculate the system reliability and MTTF for 1500 hours operation.

8.10   Solve Problem 8.7 by using the parametric approach.

8.11   Determine the error in reliability by using the classical and parametric method for the multistage parallel redundant system given in Table A.13.

Table A.13   Reliabilities of a Four-Stage Redundant System

Stage i	Number of Elements C_i	Reliability R_i
1	2	0.90
2	3	0.85
3	4	0.80
4	5	0.70

8.12 Consider the problem given in Example 8.9. Suppose the student wants to increase the radio station's reliability from 0.78 to at least 0.95. Find the number of components required for the three-stage redundant system given in Table A.14.

Table A.14   Three-Stage Redundant System

Stage	Cost per Unit	Reliability
1	2	0.63
2	5	0.72
3	7	0.86

8.13 Consider Problem 12. If the student has a limited space of $V_L \leq 40$ unit volume in the station and a cost constraint $C_L \leq \$40$, find the number of components for optimum reliability (refer to Table A.15).

Table A.15   Three-Stage Two-Constraint Redundant System

Stage	Cost per Unit	Volume	Reliability
1	2	3	0.63
2	5	2	0.72
3	7	4	0.86

8.14 Find the number of components to minimize the cost of the four-stage redundant system given in Table A.16. Assume a reliability constraint of $R_L \geq 0.99$.

Table A.16   Four-Stage Redundant System

Stage	Cost per Unit	Reliability
1	4	0.95
2	2	0.92
3	1.5	0.80
4	1.0	0.70

8.15 Find the number of components to optimize the reliability of the four stage redundant system given in Table A.17. Assume the cost constraint $C_L \leq 35$.

Table A.17   Four-Stage Redundant System

Stage	Cost per Unit	Reliability
1	4.5	0.90
2	3.0	0.86
3	2.0	0.80
4	1.5	0.70

8.16 Find the number of components to optimize the reliability of the five-stage redundant system given in Table A.18. Assume a cost constraint $C_L \leq 90$ and weight constraint $W_L \leq 100$.

Table A.18   Five-Stage Two-constraint Redundant System

Stage	Cost per Unit	Weight per Unit	Reliability
1	4	7	0.90
2	4	9	0.75
3	8	7	0.60
4	7	7	0.80
5	5	8	0.80

8.17 A schematic of a power supply system is given in Fig. A.13, which is made up of a steam turbine, condenser, boiler, and diesel generator. Fuel, lubrication, water pumps, and valves are the other complementary devices that are used in the system. The objective is to calculate the optimum number of devices that should be included in the system to increase the system reliability. The system can be analyzed in four stages. Stage 1 is formed by the diesel engine and devices that support its operation (cooling water pump, fuel pump, generator, and lubrication pump). Stage 2 includes the boiler and other devices (feed pump, fuel pump, boiler,

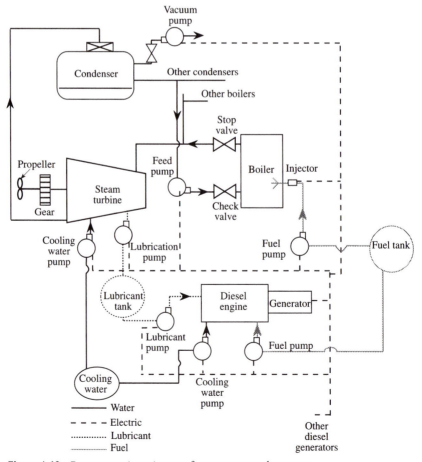

Figure A.13   Representative picture of a power supply system.

burner, stop valve, and check valve). Stage 3 includes the steam turbine and its components (lubrication pump, cooling pump, reduction gear, and pinion gear). Stage 4 includes the condenser and its devices (vacuum pump, vacuum gauges, and main control valve). The cost and reliability of these devices are given in Table A.19.

Table A.19 Cost and Reliability of the Stages

Stage	Component	Cost $	Reliability
1. Diesel generator	Cooling pump	500	0.985
	Fuel pump	700	0.980
	Diesel engine	10,000	0.995
	Generator	4,000	0.995
	Lubrication pump	800	0.980
2. Boiler generator	Feed pump	500	0.980
	Fuel pump	500	0.980
	Boiler	20,000	0.999
	Burner	400	0.999
	Stop valve	350	0.980
	Check valve	250	.980
3. Steam turbine	Lubrication pump	600	0.980
	Cooling pump	500	0.980
	Reduction gear	50,000	0.999
	Turbine	1,000,000	0.999
	Pinion gear	900	0.999
4. Condenser	Condenser	6,000	0.990
	Vacuum pump	3,500	0.985
	Vacuum gauge	250	0.980
	Main control valve	750	0.999

(a) Draw block diagrams of each stage and the whole system.

(b) Optimize the system reliability for the cost limit of $C_L = \$2,400,000$. A ship should have how many diesel generators, boilers, steam turbines, and condensers for a maximum reliablity?

(c) Minimize the cost for a given reliabilitiy limit of $R_L = 0.998$.

8.18 Develop a maintainability prediction chart as a function of time and $M(t)$ for the two constant repair rates, $\mu_1 = 0.5$ repair/hour, and $\mu_2 = 1.0$ repair/hour. Discuss the design charts in detail.

8.19 Obtain the maintainability function $M(t)$ expression when a Weilbull probability density function is used.

8.20. In turbocharged two-stroke diesel engines, the axial blower is connected directly to a gas turbine driven by the engine exhaust gases. In this instance the turbine and blower are designed as a single turboblower unit. Because of the high speed, wear of the turboblower's thrust bearings is the main concern. Assume a constant failure rate of $\lambda = 0.017$ failure/hour and a constant repair rate of $\mu = 0.08$ repair/hour. For 20 hours of operation, determine the steady state and instantaneous availability.

8.21 A pressurized water tank shown in Figure A.14 distributes water to the stations A, B, C,...N. Water is first transferred from the storage tank to the pressurized tank by the use of a transfer pump. The pressurized tank provides enough head to transfer the water to the stations. The maximum allowable water level and air pressure are 5 ft and 100 psi, respectively. Two control schemes are used with the system:

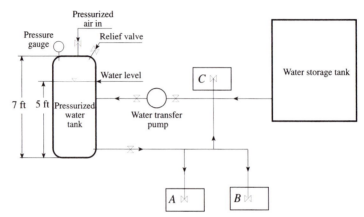

Figure A.14   Schematic diagram for a water distribution system.

- The water transfer pump stops when the maximum height is reached and starts when the minimum water level is reached.
- The water transfer pump stops when the maximum pressure is reached and starts when the minimum pressure is reached.

Develop a fault tree network for this sytem if no water is provided to the stations.

Appendix B

Design Projects

B.1 In remote areas, low-power engines are used to pump water from shallow wells for livestock and irrigation on small farms. Because farmlands requiring irrigation are often located in areas that have abundant solar insolation, and without irrigation many of these farms would have little agricultural value, the application of solar technology to water pumping for irrigation is of interest. There is a need in poor and underdeveloped countries for such low-power engines for irrigation; therefore, the cost of manufacturing these systems needs to be minimized. Migrant populations could use a source of low power, but the system must be easily transportable. The system must be capable of being left unattended for days to pump water for livestock. For this reason it must be maintenance free, not requiring the use of hazardous operations or chemicals, and must be self-starting. This task has typically been addressed by windmills and small internal combustion engines but, for reasons of maintainability, portability, and safety, an alternate system is needed.

A system has been proposed that has been identified as a thermal engine. This engine consists of a rotating wheel of diametrically opposing pods interconnected by tubes that transport a volatile liquid between the pods. As the lower pod enters a water bath that is heated by solar power, pressure is developed that forces the liquid into the upper pod, offsetting the center of gravity, which causes the wheel to rotate.

Develop a concept for this device, select the working fluid, and determine the amount of irrigation water that could be pumped with your device for water table depths of from 0 to 20 ft.

B.2 A 55 actual cubic feet (ACF) multiwall high-pressure vessel (Figure B.1) is to be designed for use in a missile launch complex . The vessel is to contain gaseous nitrogen at 6000 psi, and the system cleanliness requirement is stringent and requires special cleanliness procedures during manufacture and field assembly.

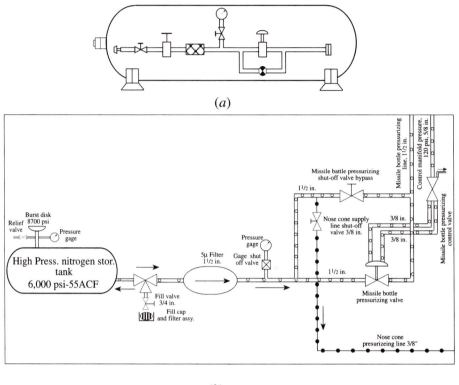

(a)

(b)

Figure B.1   (a) Pressure vessel (b) System layout.

The vessel is to be designed in accordance with ASME code requirements for unfired pressure vessels. All design features (size, thickness, etc.) must be determined in the analysis, and materials and other components, such as pressure relief devices, must be selected. It is important to control cost, and an overall cost estimate must be prepared. An installation drawing and procedure are required showing all of the details necessary to install the vessel at the launch complex site. This completed design package should include a well-organized and presented analysis with all of the necessary calculations and cost estimate, an installation drawing that identifies all of the field assembly requirements, and an installation procedure that includes all the necessary field assembly instructions.

B.3.   Industrial steam boilers consume a large portion of the energy used in this country. Even small improvements in the energy efficiency of a boiler can provide substantial energy and cost savings. For example, a 1 percent improvement in efficiency for a typical industrial boiler that produces 100,000 lbm/h of steam (medium size industrial boiler) can approach $50,000 per year. Fuel viscosity affects boiler efficiency; if the viscosity is too high or too low, the fuel will not atomize properly as it passes through the burner and, hence, will not burn efficiently.

Heavy fuel oils such as No. 5 oil, No. 6 oil, or Bunker C oil are very viscous at room temperature; thus, they will not flow easily or atomize properly when injected through the burner nozzles into the combustion chamber. Fuel preheating is required to reduce the viscosity before a heavy oil is supplied to the burner. The preheating is most effective if it can be implemented so as to provide a constant viscosity fuel to the burner. A feedback system is required to control the preheating process. The heavy oils vary in properties from batch to batch, and they even vary in properties from point to point within a batch. Thus, simply controlling the fuel delivery temperature is not sufficient to control the fuel viscosity. A means of measuring the fuel viscosity just before it enters the burner and using this information to control the preheater is needed.

A design concept for a fuel oil viscosity measuring apparatus (FOVMA) for use in an industrial boiler fuel preheat control feedback loop is to be developed. The design concept should include a description of the FOVMA (with information on the output signal), drawings showing necessary details, a cost estimate for production units, and an analysis showing the anticipated accuracy of the FOVMA.

The FOVMA must determine if the viscosity of the fuel oil is less than a specified viscosity setpoint value and must provide a proportional signal to either increase or decrease the amount of fuel preheat. The setpoint must be user selectable within the viscosity range of 50 to 200 saybolt universal seconds. The FOVMA concept must be capable of operating properly with boilers in the range of 2,000–400,000 lbm/h (fuel flow rates of 10 gal/h to 3000 gal/h); therefore, the FOVMA should continually (or periodically) sample the oil on the main supply line.

B.4.   One of the alternate fuels that has considerable potential as a future substitute for gasoline in motor vehicles is natural gas. Natural gas can be stored on the vehicle in either liquid or gaseous form but, because of its low boiling point, it is used by the engine as compressed natural gas (CNG). The advantages of using natural gas as a transportation fuel include:

1. Tailpipe emissions are lower than that of gasoline or diesel, and with CNG evaporative emissions are eliminated.
2. Natural gas engines have proved to be efficient and inexpensive to maintain.
3. Underground fuel storage problems are eliminated.
4. Natural gas reserves are abundant in the United States and worldwide.
5. Pricing is not influenced by Middle East fuel cartels.
6. National security is enhanced.

The objective of this project is to determine the feasibility of establishing a CNG refueling facility in your metropolitan area to supply natural gas for vehicles in the surrounding area. Specific items that must be considered include:

1. Design of the facility and natural gas supply system, including the selection of critical components.
2. Cost of the facility and land as well as the operating costs.
3. Source of funding and method of repayment.
4. Whether to include vehicle modification service (conversion from gasoline to CNG) with the refueling operation.

5. The number of vehicles using the refueling service required to justify building the facility. Tactics that can be used to increase the number of vehicles. Whether it is reasonable to include government owned vehicles as potential customers.

6. Code requirements (state, National Fire Protection Association, etc.).

7. Management structure and type of ownership of the facility.

8. Slow fill, fast fill, or both, including the number of vehicles that can be serviced per unit time.

9. Location of facility relative to CNG supply.

10. Business relationship to natural gas supplier. Will the supplier also be part owner of the facility? Will the supplier provide some of the capital and recover the investment with a surcharge on the gas supplied?

The end product of this project is a comprehensive report that addresses the above considerations and includes the following sections:

1. A neat schematic showing the design concept and identifying all components.

2. An economic analysis section that supports your conclusion as to the feasibility of building the facility.

3. Technical analysis to support the design.

4. A narrative description of the design philosophy, organizational approach, conclusions, and marketing techniques.

B.5. Figure B.2 represents the top and end views of a simple speed reducer using two spur gears. The gears, shafts, bearing housings, and lubrication are to be

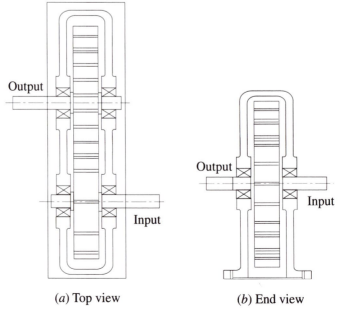

(*a*) Top view   (*b*) End view

Figure B.2   A speed reducer.

designed, and the bearings are to be selected. The speed reducer enclosure will consist of a two-part steel casting. All of the bearing housing diameters will be line-bored, and faced on a horizontal boring mill.

Individual designs are to be developed with the specifications based on the initial letter of each student's last name as follows:

$$rpm = (n)(30) + 970$$

where

n = 1 for A
n = 2 for B
n = 3 for C
⋮
n = 26 for Z

Last Name Initial	Horsepower	Speed Ratio
A–H	5	2:1
I–Q	8	2.3:1
R–Z	10	2.6:1

The components to be designed are as follows:

Gears   The gears are to be full-depth involute spur gears with a $20°$ pressure angle. Select diametral pitch, numbers of teeth, pitch diameter, face, and width; the face width should be approximately $\frac{8}{P_d}$ to $\frac{12.5}{P_d}$. Check to see that there will be no interference in the gears and, since the teeth will be hobbed, check to see that there will be no undercutting of the pinion teeth. For the tooth stress analysis, use the load applied near the middle (at the pitch circle). When making load calculations for the teeth, show free bodies of the gears.

Shafts   Shafts should be sized and designed for mounting and retaining the bearings and gears. Bearings and gears are located by accurately machined shoulders on the shaft. For this design, use a single shoulder and Woodruff key providing the torque between the gear and shaft; an appropriate interference fit will keep the gear on the shaft, since there will be no axial loading on the gear. Provide appropriate fillet radii at the bottom of the bearing-retaining shoulders, and be sure that the shoulders are of the proper outside diameters. Provide a 2-in. length of shaft between the bearing housing and the end of each keyway.

Housings   The bores in the bearing housings have shoulders similar to the ones used on the shaft, with fillets and proper shoulder diameters. Some of the bearing retainers (shoulders) will need to be removable so that the parts can be assembled and disassembled.

Bearings   Select single-row radial-ball bearings of the open type. If one-bearing will not provide sufficient load rating, use two of similar size and type. Since the housing will be filled with oil to an appropriate level, caps with gaskets or oil seals will need to be utilized. Use an interference fit of 0.0003 in. on the shaft and bearing housing at each bearing.

The Report   The report should contain the following to be complete:

1. All calculations for gear design, stresses, key and keyway selection, and shaft design.

2. Necessary sketches so a draftsman can make working drawings.

3. A detailed, dimensioned drawing of each shaft.

4. Detailed cross-sectional drawings of the bearing housing areas giving all the dimensions.

5. A half- or full-section assembly drawing of the shaft, bearing, and housing area to ascertain whether or not the speed reducer can be assembled and disassembled.

B.6   Design a steam engine as shown schematically in Figure B.3 that will run a pump that will move 1500 cfm water at 60°F over an increase in elevation of 75 ft. Assume that 5000 lb/h steam at 400°F and 247.1 psi is available to drive the steam engine. Choose the proper materials for the cylinder, gaskets, and seals and an appropriate valve for the steam inlet and exhaust. Submit a complete design package including a well-organized and presented analysis with all of the necessary calculations with an assembly drawing showing the piston, cylinder, and valve assembly.

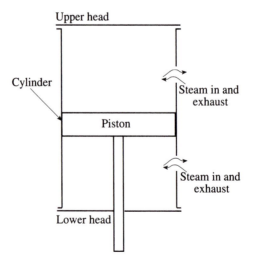

Figure B.3   A schematic of a steam engine.

B.7   Aerodynamic testing has played a significant role in improving automotive performance by allowing designers and engineers to model new body shapes without going to production units. The benefits of this testing have resulted in reducing the drag and improving in airflow for cooling systems in automobiles. The two commonly used aerodynamic testing facilities are water tunnels and wind tunnels.

The objective of this project is to design a closed-loop water tunnel for aerodynamic model testing. The design requirements are as follows:

1. The test section in which the model will be placed for testing and viewing is to have a cross-sectional area of 1 ft³ with an aspect ratio of 1.0.

2. To attain the required test section stream velocity of 1 ft/s use a construction ratio of 5. The construction ratio is the ratio of the cross-sectional area of the entrance cone to the cross-sectional area of the test section.

Include all of the analyses and provide detail drawings (sketches) to describe the individual elements.

B.8 Design a closed-loop wind tunnel for aerodynamic model testing for the following design parameters:

1. The test section in which the model will be placed for testing and viewing is to have a cross-sectional area of 160 in.² with an aspect ratio of 1.6.

2. To attain the required test section stream velocity of 100 ft/s, use a construction ratio of 3.

Include all of the analyses and provide detail drawings (sketches) to describe the individual elements.

B.9 Tanks have been used for water storage since the dawn of civilization. Because of increasing population and decreasing water supplies, designers have been asked to design larger and more sophisticated storage vessels. The demand for larger water tanks has required engineers to develop designs that result in safe and cost-effective large structures.

The versatility of material strengths enables engineers to take advantage of the *family-of-steels* concept for designing a water tank.[1] The underlying concept in the family-of-steels design technique is to vary the strength of the material composing the tank wall while maintaining a constant wall thickness. Conversely, the traditional design varies the wall thickness while maintaining the use of a single grade of steel throughout the tank wall. Theoretically, the family-of-steels design simplifies construction and saves material and shipping costs because of a reduction in the total weight of material.

The objective of this project is to design a storage tank by using the family-of-steels design concept for the following design parameters:

1. The maximum tank height is 32 ft and capacity is 6 million gal.

2. To simplify the analysis, only the tank shell will be examined. The roof, foundation, support column, and girders will not be considered. Ignore the effects of wind and earthquakes. Hence, the only load the tank shell will be subjected to is the pressure of the stored water. The specific gravity of the stored water is 1.0.

---

[1]E. D. Verink, Jr., *Methods of Materials Selection*, vol. 40, Gordon and Breach, Science Publishers, New York, 1966, pp. 34–39.

Evaluate two design options (1) the family-of-steels design concept, and (2) the traditional single steel design concept. The analysis of these options should consider cost, ease of construction, and availability of materials.

B.10   Penstock design is another application of the family-of-steels concept. The penstock, or pipeline, is usually very long and transports water down a mountainside. Because of the large changes in elevation, significant pressures are created in the lower sections of the penstock. These large pressures need to be accounted for in the material selection process.

The objective of this project is to design a penstock to transport water down a mountainside as shown in Figure B.4. The penstock is in three sections, each 5000 ft long with an inside diameter of 48 in.

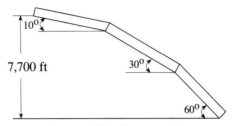

Figure B.4   A profile of a penstock.

Use two design approaches: (1) the family-of-steels design concept, and (2) the traditional single-steel design concept.

B.11   A school for the mentally and physically handicapped has a program for clients (students) to earn a small income by stuffing toys in plastic eggs. The eggs are then delivered to the customer, who sells them in vending machines located in grocery stores, cafeterias, and the like. There are nine different toys that are stuffed into the eggs, which are then stored in nine different cardboard box containers. The students can stuff approximately 200,000 eggs per week, but packaging the eggs for delivery must be accomplished by one of the workshop supervisors, since each box contains a different distribution of eggs, depending on the customer's requirements. At present, the eggs are being hand sorted and packed into boxes containing 250 eggs. This requires an individual counting of the eggs for packaging. At present, only 80,000 eggs can be packaged per week. A current backlog of 168,000 stuffed but unpackaged eggs exists.

The purpose of this project is to devise a method of automating the packaging process so that overall production and delivery can be increased to 250,000 eggs per week, minimum. The system must be automated so that severely handicapped students can perform this operation.

The product of this effort should consist of a design report including the necessary schematics, drawings, analysis, and discussion to adequately characterize the design. Special consideration should be given to the student/system interface to ensure that severely handicapped students can communicate with the system.

B.12   Another task performed by handicapped students at a state school involves tearing paper from rolls into small pieces that are acceptable for recycling. The paper available to the school is in the form of different size rolls that cannot be recycled. Before the paper can be recycled it must be torn into small pieces and shredded. The procedure that is in use at the present time involves inserting a simple wire clothes hanger through the roll of paper and attaching it to the student's wheelchair. With this arrangement the students have difficulty pulling the paper down and tearing it from the roll. The roll of paper is held in position by using masking tape, thus, each time the roll is changed the masking tape must be removed and replaced.

The purpose of this project is to develop a simple design concept that will provide an improvement in this process. The product of this effort should be a neat sketch showing your design concept, and a brief description of how it would be used in the paper processing project.

B.13 A school provides education for physically and mentally handicapped students. The school utilizes various pieces of adaptive equipment and tools to aid in the student's education. Currently, projects are in progress using learning devices to teach children with higher aptitude levels about work, responsibility, and achievement. Administrators and teachers at the school are interested in determining whether a soft-drink can crusher can be developed capable of operation by the students. Such a program could be adapted to aid children with both higher and lower aptitude levels. The collecting and sacking of the cans requires no additional equipment and could be accomplished by students at all aptitude levels. The challenge is to design and develop a can crusher that can be operated safely by students at all aptitude levels.

The recycling of aluminum cans is a profitable endeavor for the school, although making money is not the primary purpose of the project. The can crusher should accommodate children with limited mobility and motor skills and serve as a learning tool for a wide range of aptitudes. The machine will demonstrate the act of association by enabling students to recognize the reaction to their input; when a student pushes the button to activate the crusher, a can will be crushed. The repetitive action will also represent a task easily learned. Through observation of the crusher action the student will learn when to push the button to initiate another cycle and eventually, will learn to recognize that a worthwhile task is being accomplished through their action.

The design criteria must be driven by the limited capabilities of the students for whom the device is developed. Some of the students have such limited mobility that they are only capable of moving their elbow. This lack of mobility dictates that the machine be capable of activation from a single, easily contacted button. The machine must also be capable of operation without the continuous aid of a teacher or supervisor. The can crusher must be safe, effective, and economical in performing the following actions:

1. Loading the can automatically.
2. Crushing the can on command.
3. Discharging the crushed can.
4. Stopping the cycle after the can is crushed.

For the machine to be safe it must be constructed so that students cannot get a hand or finger in the mechanism. Finally, the crusher must be structurally sound, reliable, and easy to maintain. Design a can crusher to meet these requirements.

Provide an assembly drawing showing the overall crusher and enough detail drawings (sketches) to describe the individual elements. Provide an operational step-by-step procedure describing how to use the crusher.

B.14   Hearing loss is a problem for our society. There are numerous causes of hearing loss ranging from birth defects and accidents to extended exposure to noise levels above 90 dBA. The major cause of hearing loss is due to damage to the cochlea caused by stimulation at certain frequencies and high intensities. Several hearing aids are currently available; however, most are expensive and considerable time and testing is required to adjust the device to each individual's hearing loss. Implants are also available but, again, they are very expensive and entail a surgical process.

A device has been proposed that does not enhance hearing but increases the level of communications interaction and understanding for the hearing impaired. This system is composed of a transmitter and vibrotactile receiver that allows a person to attract the attention of the hearing impaired over some distance. For example, using this device a teacher could attract the attention of hearing-impaired children on a school playground when a recess period is over. The cost of this device needs to be held to under $100 to make it attractive to potential customers. The transmitter is to send out a signal with a frequency of 1 MHz with a range of, at least, 100 yds. The receiver is to respond to this signal by vibrating for 5s. The transmitter and receiver are to be portable, durable, aesthetically pleasing and incorporate rechargeable batteries. Design a device that will accomplish these objectives. Select appropriate components including the vibrotactile device and design the circuitry and packaging for the receiver and transmitter.

B.15   The automobile transport business faces many challenges in safely moving certain types of vehicles. Automobiles differ in size, weight, and shape, and equipment must be capable of transporting all of them, regardless of the design or condition of the vehicle. According to the *AAA Domestic and Imported Vehicle Towing Manual*, every automobile on the road today must be towed according to the procedures recommended by the manufacturer to avoid damage. With the current trend to more aerodynamic designs, the need for wheel-lift and flatbed-hauling equipment in the towing business is increasingly apparent. Of all models of automobiles currently produced in the world, a total of 186 classifications are identified. Thirty-one percent of these cannot be towed with the conventional sling-type towing equipment. In addition, 21 percent require that the automobile be entered and the gear shift be placed in neutral before towing. Obviously, there exists a need for an alternative towing system that can accommodate these vehicles.

Numerous problems are encountered by the towing industry because of auto-mobile designs. Low-hanging components are subject to damage caused by dragging during towing or as a result of being crushed by the towing equipment. Many automobiles are equipped with bumpers designed to carry only a horizontal load rather than a vertical load, such as that experienced with some towing operations. To prevent transmission and differential damage, the drive wheels of an automobile should be raised from the ground (a rear-wheel drive auto should be towed from the rear). If an automobile is towed on its drive wheels it must be placed in neutral, and speed and towing distance must be minimized because of lubrication needs. Any car being towed from the rear should have its steering column unlocked to avoid damaging the antitheft mechanism. This mechanism is located within the steering column; therefore, it is necessary for the towing operator to have the ignition key for the vehicle. Finally, an automobile that has been damaged in a collision is a challenge to tow. A damaged automobile may be unable to roll; therefore, a quick and simple method for transporting it must be available without causing further damage.

An alternative to conventional towing methods is to be devised. This method should not require that the vehicle be rolled on its drive wheels or lifted at vulnerable points. The vehicle should be capable of being towed without the tow operator having to enter the vehicle. The product of this effort should be a model that can be used to demonstrate the alternative towing method.

B.16 *Correction of Groundwater Contamination.* An aerospace engineering facility is located approximately 20 miles from a city with an estimated population of 60,000. A number of organic solvents are routinely used in various design and testing activities. Waste solvents have been discharged in the past to unlined, earthen evaporation ponds for disposal. A review of the test facility's past and current waste management operations by the state environmental agency indicates a high potential for threats to human health and the environment. This has prompted regulatory action and forced the facility to determine if releases of hazardous wastes to the environment have occurred.

Groundwater monitoring wells were subsequently installed downgradient from each evaporation pond. Groundwater samples were collected from each well and analyzed for volatile organic constituents. Results indicated that releases have occurred to groundwater beneath the facility. The following compounds were detected: freon-11 (MF), trichloroethylene (TCE), and tetrachloroethylene (PCE).

The facility then conducted a contamination assessment for the site to determine the lateral and vertical extent of groundwater contamination resulting from past activities. Two hundred monitoring wells were drilled over a three-year period in an effort to define the extent of groundwater contamination. The results indicated that groundwater contaminants would impact on public drinking water supplies (located approximately four miles downgradient) within one year unless a remediation program was initiated.

The facility plans to drill a series of interception wells to prevent potential contamination of the public water supply. Pumping of these wells will prevent continued migration of contaminants and will recover contaminated groundwater for subsequent treatment and restoration.

A corrective measures study was undertaken to determine the most efficient and cost effective method for treating groundwater containing volatile organic compounds (VOCs). Results of the study indicated that air stripping the volatile organic constituents was the most appropriate method for this site. Accordingly, project hydrogeologists have installed the interception well network designed to capture the contaminant plume. Air stripper design has not been completed to date.

Given the site-specific information below, design an air-stripper system to treat contaminated groundwater to acceptable standards. The design should include the diameter and height of individual stripper tower(s), the number of towers required, and the size of pipe and airblower requirements. The type of packing material should also be included in the design. Design and size the system based on the most difficult compound to strip. Assume that no additional pumps are required to transfer the water from the well pump to the air stripping tower. Design the air-stripping system to provide the most efficient and cost-effective approach.

The system is scheduled to be in operation continuously in a climatic environment similar to that found in southern New Mexico. The following information has been provided:

1. Total flow rate from production wells = 150 gpm.
2. Atmospheric pressure = 0.85 atm.
3. Water temperature = 23°C.
4. Influent concentration of TCE = 220 μg/liter.
5. Influent concentration of PCE = 20 μg/liter.
6. Influent concentration of MF = 160 μg/liter.
7. Required effluent concentration of each contaminant = 0.1 μg/liter.
8. Henry's constant for MF = 3239 atm.

The design package should include a schematic of the tower showing significant components and other elements, analysis justifying the tower design and components selected, theory involved in air stripping of VOCs, and an economic analysis supporting the cost effectiveness of the design. Since this is an iterative design process, computer printouts or other data should be included to show the number of tower design options considered and to validate the option selected.

B.17 Environmental Design Concept for Disposal and Release of Propellants. A design concept that will provide for the safe disposal and release of earth storable hypergolic propellants at an aerospace test complex located within a large research, development, and flight operations facility is required.

Background   The test complex consists of five individual test facilities located within a fenced and access controlled area of approximately 100 acres. This test complex was established to test low-thrust rocket engines, fluid components, and fuel cells and batteries, and to provide the capability for hard vacuum testing of propellant system components and subsystems. The complex includes a central office building and supporting laboratories, a warehouse, and the five test facilities listed below:

1. Thermal vacuum
2. Pyrotechnics
3. Propulsion
4. Power systems
5. Fluid systems

Figure. B.5 depicts the layout of the facilities and their general relationship to one another.

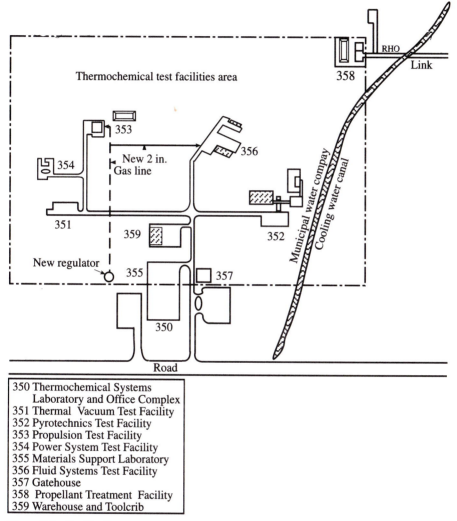

350 Thermochemical Systems
    Laboratory and Office Complex
351 Thermal Vacuum Test Facility
352 Pyrotechnics Test Facility
353 Propulsion Test Facility
354 Power System Test Facility
355 Materials Support Laboratory
356 Fluid Systems Test Facility
357 Gatehouse
358 Propellant Treatment Facility
359 Warehouse and Toolcrib

Figure B.5   Facility layout.

The problems in the handling of propellants in these facilities are the potential exposure of off-site personnel to toxic vapor from propellant releases, both intentional and unintentional, and the potential contamination of ground and surface water from oxidizers, fuel, or other fluid seepage or flow. Personnel assigned to work in the test area are knowledgeable as to the hazards of propellants and as to the safety procedures necessary to ensure that area personnel are not exposed to toxic vapors in excess of that allowed. The primary concern is for off-site personnel exposure to air-transported vapor and to contaminated water as a result of propellant seeping into the ground and water table or as a result of surface flow into a municipal power company cooling water canal that flows through the area (see Figure B.5).

Only two of the facilities, the Fluid Systems and Propulsion Test Facilities, require the use of propellants that are of concern [nitrogen tetroxide, N_2O_4 and the family of hydrazines including monomethylhydrazine (MMH), unsymmetrical dimethylhydrazine (UDMH), Aerozine-50, A-50, and neat hydrazine, N_2H_4]. These facilities include several test cells in which testing is accomplished and in which small amounts of propellant (less than one gallon) must necessarily be disposed of routinely. These cells are constructed of reinforced concrete and are open at one end. Testing can be accomplished remotely from the control room, but test cell preparations and pretest operations require that personnel work in the cell when the systems are charged with propellant. These personnel are well trained and equipped with personal protective equipment. They are also under the control of the test conductor or test engineer when the cell systems are charged. Personnel are also required to work around charged systems when filling run tanks, performing system maintenance or modification, or transferring propellant from the receiver tank.

Oxidizer and fuel storage equipment at these two facilities are separated by location on opposite sides of the facility. Storage tanks are approximately 400 gal. in size, and propellant transfer is accomplished by pressurization of the run (storage) tank with gaseous nitrogen. Propellant flows from the run tank to the test cell through stainless steel lines. Normally, the propellant lines are filled during pretest operations up to a valve located in the test cell. This valve, which could be the rocket engine propellant valve, depending on the particular component under test, is controlled during test from the control room.

Following testing, propellant is removed from the system by flowing into a receiver tank, and the system is cleared of propellant by blowdown with hot, dry nitrogen. The fuel system infrequently requires flushing with de-ionized water prior to drying. Since receiver tank propellant can normally be reused, this contaminated water is not transferred to the receiver tank but must be disposed of. Small leaks and spills within the test cell can be decontaminated by flushing with water. After drying the system is normally maintained under a "blanket" pressure to ensure that leakage flows from the system to the environment.

Toxic Vapor Exposure Limits TWA and ceiling limits for the propellants used in this facility are shown in Table B.1 and are based on the following assumptions:

Table B.1 Propellant Spill Criteria

Propellant	TWA (ppm)	EEL (ppm)	Emergency Source Strength (lb/min ft²)	Routine Source Strength (lb/min ft²)	Wind Speed (mph)
Oxidizer N_2O_4	3.0[a]	0.44	30	30	30
MMH	0.2	10	0.089	0.021	30
Hydrazine (N_2H_4)	0.1	30	0.025	0.005	30
UDMH	0.5	100	0.250	0.065	30

[a] N_2O_4 emergency source strength is 30 percent of the total quantity spilled in 10 min but not less than 12 lb/min.

1. Exposures at these levels will be accidental and not the result of engineering controls designed to yield exposures at these levels.

2. Normally prevailing values of airborne contamination will be below TLV/s.

3. These accidental exposures will be single events, that is, if a person is exposed at these levels further exposure will be prevented until normal resistance is regained.

4. Persons who could be exposed under these conditions are not idiosyncratic, hypertensive, or otherwise predisposed to disease from the specific contaminate.

5. Persons who are exposed under these conditions are under medical surveillance.

6. The probable severity of injury due to secondary accidents, including those resulting from impairment of vision, judgment, and coordination, must be considered in applying these values.

In addition to the exposure limits for these test facilities, Table B.1 gives the source strengths based on operational procedures. The design concept must be based on allowable contaminant level values for both air and water. The research, development, and flight operations facility is located adjacent to a recreational lake, and effluent from these test facilities could well end up in this lake. The water quality criteria for recreational lakes in this locale states that sewage effluents shall not exhibit either acute or chronic toxicity to human, animal, or aquatic life to such an extent that lake use is interfered with.

Required Product Provide the information necessary to convey an adequate understanding of the proposed design concept, including the necessary discussion, calculations, and sketches organized in a neat and orderly fashion. The two specific considerations that must be accounted for in the design concept are (1) how to ensure that personnel outside the perimeter fence are not affected in any way by activities (spills or tank venting) within the test complex, and (2) how to ensure that routine and emergency propellant spills are disposed of in a manner that will not result in ground or surface water contamination and will not be a prohibitive operational burden on facility personnel.

B.18 *Leaking Petroleum Tank Site Remediation*. The discovery of a leaking underground storage tank was made in November, 1989 while a university was in the process of removing four underground storage tanks at the physical plant. The tanks that were removed consisted of a 2000-gal regular gasoline steel tank, a 1000-gal regular gasoline steel tank, a 500-gal regular gasoline steel tank, and a 500-gal diesel steel tank. A 2000-gal unleaded gasoline fiberglass tank remains at the site. As the tanks were removed, it was discovered that the 500-gal regular gasoline steel tank had been leaking an undetermined length of time. On inspection of the tank, several small holes, the size of buckshot, were found. The small holes were in an area of approximately 1.5 ft^2. The holes were located approximately at or below the centerline of the tank. The cause of the holes is not known.

The four tanks that were removed were installed in 1960, and were used to service university vehicles. The remaining 2000 gal fiberglass tank, which was installed in 1976, is currently still in use and shows no evidence of leaking.

Table B.2 Soil and Water Analysis

Well ID	Sample Type	Sample Depth	Sample Date	TRPHC (ppb)	MTBE (ppb)	Benzene (ppb)	Toluene (ppb)	Ethyl-benzene (ppb)	m, p & o Xylene (ppb)
1	Soil	30–32	1/24/90	5,364.0	45.3	6.5	14.0	<0.2	20.2
	Water		3/06/90		70.3	46.8	82.6	<0.2	163.2
			4/14/90	519.0	<0.2	238.4	62.6	<0.2	144.7
2	Soil	20	4/07/90	<500	<0.2	<0.2	<0.2	<0.2	<0.2
		25	4/07/90	5,692.0	23.4	1.7	4.4	0.2	1.4
		30	4/07/90	33,843.0	65.7	7.3	108.6	11.9	710.9
		33	4/07/90	295,100.0	41.4	8.6	52.9	34.9	183.2
	Water		4/14/90	767.0	54.5	12.2	8.7	<0.2	13.0
3	Soil	20	4/07/90	10,733.0	<0.2	<0.2	<0.2	<0.2	<0.2
		25	4/07/90	6,849.0	26.2	23.7	32.6	6.4	49.7
		30	4/07/90	3,682.0	6.5	22.3	21.1	15.9	56.8
		33	4/07/90	262,859.0	245.0	196.2	197.7	196.5	603.6
	Water		4/14/90	<200	<0.2	0.2	0.9	<0.2	<0.2
B1	Soil	20	4/07/90	15,139.0	<0.2	<0.2	<0.2	<0.2	<0.2
		25	4/07/90	4,500	<0.2	<0.2	<0.2	<0.2	<0.2
		30	4/07/90	7,189.0	13.5	3.3	7.4	0.5	4.3
		33	4/07/90	225,334.0	41.8	43.8	688.4	698.6	2,729.0

Emergency abatement measures were subsequently taken, and an engineering firm was retained to assess the extent of the contamination. Some soil tests were performed and a monitoring well was installed before the consulting firm became involved in the project. Soil and water samples were taken during the drilling of the monitoring well and were analyzed by a certified laboratory. In addition, organic vapor meter readings were taken at various depths during the drilling of the monitoring well.

After studying the results of the laboratory work performed on the soil and water samples, the consulting engineer recommended additional monitoring wells and one soil boring to better define the extent of the contamination. Soil samples were taken from these wells and were analyzed. Results of this analysis are shown by Table B.2.

From the data provided by the soil and water analysis, the contaminant was determined to have migrated vertically to a depth of from 27 to 32 ft, and horizontally to points just outside of the monitoring wells. Migration was facilitated by the sandy soil comprising much of the lower unsaturated strata. A cross-section of the soil structure is shown in Figure B.6. Figures B.7. and B.8 depict the site layout and location of storage tanks and wells.

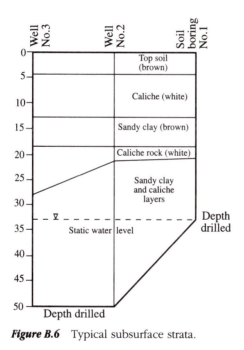

Figure B.6   Typical subsurface strata.

A site remediation plan contemplating the use of soil venting is required. The plan must provide answers to the following questions:

1. What contaminant vapor concentrations are likely to be obtained?
2. Under ideal vapor flow conditions (100–1000 scfm vapor flowrates) is this concentration great enough to yield acceptable removal rates?
3. What range of vapor flowrates can be achieved?

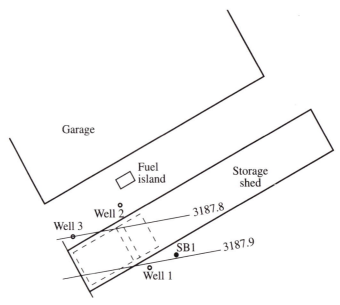

Figure B.7 The site layout.

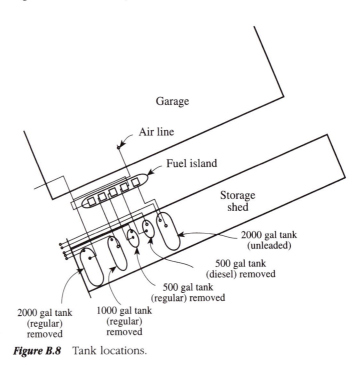

Figure B.8 Tank locations.

4. Will the contaminant concentrations and realistic vapor flowrates produce acceptable removal rates?

5. What residual, if any, will be left in the soil? What vapor composition and concentration changes will occur with time? How do these values relate to regulatory requirements?

6. Are there likely to be any negative effects of soil venting?

Appendix C

Statistical Tables

Table C.1   Area under the Normal Curve[a]

Area

z	0.00	0.01	0.02	0.03	0.04	0.05	0.06	0.07	0.08	0.09
−3.4	0.0003	0.0003	0.0003	0.0003	0.0003	0.0003	0.0003	0.0003	0.0003	0.0002
−3.3	0.0005	0.0005	0.0005	0.0004	0.0004	0.0004	0.0004	0.0004	0.0004	0.0003
−3.2	0.0007	0.0007	0.0006	0.0006	0.0006	0.0006	0.0006	0.0005	0.0005	0.0005
−3.1	0.0010	0.0009	0.0009	0.0009	0.0008	0.0008	0.0008	0.0008	0.0007	0.0007
−3.0	0.0013	0.0013	0.0013	0.0012	0.0012	0.0011	0.0011	0.0011	0.0010	0.0010
−2.9	0.0019	0.0018	0.0017	0.0017	0.0016	0.0016	0.0015	0.0015	0.0014	0.0014
−2.8	0.0026	0.0025	0.0024	0.0023	0.0023	0.0022	0.0021	0.0021	0.0020	0.0019
−2.7	0.0035	0.0034	0.0033	0.0032	0.0031	0.0030	0.0029	0.0028	0.0027	0.0026
−2.6	0.0047	0.0045	0.0044	0.0043	0.0041	0.0040	0.0039	0.0038	0.0037	0.0036
−2.5	0.0062	0.0060	0.0059	0.0057	0.0055	0.0054	0.0052	0.0051	0.0049	0.0048
−2.4	0.0082	0.0080	0.0078	0.0075	0.0073	0.0071	0.0069	0.0068	0.0066	0.0064
−2.3	0.0107	0.0104	0.0102	0.0099	0.0096	0.0094	0.0091	0.0089	0.0087	0.0084
−2.2	0.0139	0.0136	0.0132	0.0129	0.0125	0.0122	0.0119	0.0116	0.0113	0.0110
−2.1	0.0179	0.0174	0.0170	0.0166	0.0162	0.0158	0.0154	0.0150	0.0146	0.0143
−2.0	0.0228	0.0222	0.0217	0.0212	0.0207	0.0202	0.0197	0.0192	0.0188	0.0183
−1.9	0.0287	0.0281	0.0274	0.0268	0.0262	0.0256	0.0250	0.0244	0.0239	0.0233
−1.8	0.0359	0.0352	0.0344	0.0336	0.0329	0.0322	0.0314	0.0307	0.0301	0.0294
−1.7	0.0446	0.0436	0.0427	0.0418	0.0409	0.0401	0.0392	0.0384	0.0375	0.0367
−1.6	0.0548	0.0537	0.0526	0.0516	0.0505	0.0495	0.0485	0.0475	0.0465	0.0455
−1.5	0.0668	0.0655	0.0643	0.0630	0.0618	0.0606	0.0594	0.0582	0.0571	0.0559
−1.4	0.0808	0.0793	0.0778	0.0764	0.0749	0.0735	0.0722	0.0708	0.0694	0.0681
−1.3	0.0968	0.0951	0.0934	0.0918	0.0901	0.0885	0.0869	0.0853	0.0838	0.0823

Reprinted with the permission of Macmillan Publishing Company, Inc. from *Probability and Statistics for Engineers and Scientists, Fourth Edition* by Ronald E. Walpole and Raymond H. M yers. Copyright 1989 by Macmillan Publishing Company.

Table C.1   (Continued)

z	0.00	0.01	0.02	0.03	0.04	0.05	0.06	0.07	0.08	0.09
-1.2	0.1151	0.1131	0.1112	0.1093	0.1075	0.1056	0.1038	0.1020	0.1003	0.0985
-1.1	0.1357	0.1335	0.1314	0.1292	0.1271	0.1251	0.1230	0.1210	0.1190	0.1170
-1.0	0.1587	0.1562	0.1539	0.1515	0.1492	0.1469	0.1446	0.1423	0.1401	0.1379
-0.9	0.1841	0.1814	0.1788	0.1762	0.1736	0.1711	0.1685	0.1660	0.1635	0.1611
-0.8	0.2119	0.2090	0.2061	0.2033	0.2005	0.1977	0.1949	0.1922	0.1894	0.1867
-0.7	0.2420	0.2389	0.2358	0.2327	0.2296	0.2266	0.2236	0.2206	0.2177	0.2148
-0.6	0.2743	0.2709	0.2676	0.2643	0.2611	0.2578	0.2546	0.2514	0.2483	0.2451
-0.5	0.3085	0.3050	0.3015	0.2981	0.2946	0.2912	0.2877	0.2843	0.2810	0.2776
-0.4	0.3446	0.3409	0.3372	0.3336	0.3300	0.3264	0.3228	0.3192	0.3156	0.3121
-0.3	0.3821	0.3783	0.3745	0.3707	0.3669	0.3632	0.3594	0.3557	0.3520	0.3483
-0.2	0.4207	0.4168	0.4129	0.4090	0.4052	0.4013	0.3974	0.3936	0.3897	0.3859
-0.1	0.4602	0.4562	0.4522	0.4483	0.4443	0.4404	0.4364	0.4325	0.4286	0.4247
-0.0	0.5000	0.4960	0.4920	0.4880	0.4840	0.4801	0.4761	0.4721	0.4681	0.4641
0.0	0.5000	0.5040	0.5080	0.5120	0.5160	0.5199	0.5239	0.5279	0.5319	0.5359
0.1	0.5398	0.5438	0.5478	0.5517	0.5557	0.5596	0.5636	0.5675	0.5714	0.5753
0.2	0.5793	0.5832	0.5871	0.5910	0.5948	0.5987	0.6026	0.6064	0.6103	0.6141
0.3	0.6179	0.6217	0.6255	0.6293	0.6331	0.6368	0.6406	0.6443	0.6480	0.6517
0.4	0.6554	0.6591	0.6628	0.6664	0.6700	0.6736	0.6772	0.6808	0.6844	0.6879
0.5	0.6915	0.6950	0.6985	0.7019	0.7054	0.7088	0.7123	0.7157	0.7190	0.7224
0.6	0.7257	0.7291	0.7324	0.7357	0.7389	0.7422	0.7454	0.7486	0.7517	0.7549
0.7	0.7580	0.7611	0.7642	0.7637	0.7704	0.7734	0.7764	0.7794	0.7823	0.7852
0.8	0.7881	0.7910	0.7939	0.7967	0.7995	0.8023	0.8051	0.8078	0.8106	0.8133
0.9	0.8159	0.8186	0.8212	0.8238	0.8264	0.8289	0.8315	0.8340	0.8365	0.8389
1.0	0.8413	0.8438	0.8461	0.8485	0.8508	0.8531	0.8554	0.8577	0.8599	0.8621
1.1	0.8643	0.8665	0.8686	0.8708	0.8729	0.8749	0.8770	0.8790	0.8810	0.8830
1.2	0.8849	0.8869	0.8888	0.8907	0.8925	0.8944	0.8962	0.8980	0.8997	0.9015
1.3	0.9032	0.9049	0.9066	0.9082	0.9099	0.9115	0.9131	0.9147	0.9162	0.9177
1.4	0.9192	0.9207	0.9222	0.9236	0.9251	0.9265	0.9278	0.9292	0.9306	0.9319
1.5	0.9332	0.9345	0.9357	0.9370	0.9382	0.9394	0.9406	0.9418	0.9429	0.9441

Table C.1 (Continued)

z	0.00	0.01	0.02	0.03	0.04	0.05	0.06	0.07	0.08	0.09
1.6	0.9452	0.9463	0.9474	0.9484	0.9495	0.9505	0.9515	0.9525	0.9535	0.9545
1.7	0.9554	0.9564	0.9573	0.9582	0.9591	0.9599	0.9608	0.9616	0.9625	0.9633
1.8	0.9641	0.9649	0.9656	0.9664	0.9671	0.9678	0.9686	0.9693	0.9699	0.9706
1.9	0.9713	0.9719	0.9726	0.9732	0.9738	0.9744	0.9750	0.9756	0.9761	0.9767
2.0	0.9772	0.9778	0.9783	0.9788	0.9793	0.9798	0.9803	0.9808	0.9812	0.9817
2.1	0.9821	0.9826	0.9830	0.9834	0.9838	0.9842	0.9846	0.9850	0.9854	0.9857
2.2	0.9861	0.9864	0.9868	0.9871	0.9875	0.9878	0.9881	0.9884	0.9887	0.9890
2.3	0.9893	0.9896	0.9898	0.9901	0.9904	0.9906	0.9909	0.9911	0.9913	0.9916
2.4	0.9918	0.9920	0.9922	0.9925	0.9927	0.9929	0.9931	0.9932	0.9934	0.9936
2.5	0.9938	0.9940	0.9941	0.9943	0.9945	0.9946	0.9948	0.9949	0.9951	0.9952
2.6	0.9953	0.9955	0.9956	0.9957	0.9959	0.9960	0.9961	0.9962	0.9963	0.9964
2.7	0.9965	0.9966	0.9967	0.9968	0.9969	0.9970	0.9971	0.9972	0.9973	0.9974
2.8	0.9974	0.9975	0.9976	0.9977	0.9977	0.9978	0.9979	0.9979	0.9980	0.9981
2.9	0.9981	0.9982	0.9982	0.9983	0.9984	0.9984	0.9985	0.9985	0.9986	0.9986
3.0	0.9987	0.9987	0.9987	0.9988	0.9988	0.9989	0.9989	0.9989	0.9990	0.9990
3.1	0.9990	0.9991	0.9991	0.9991	0.9992	0.9992	0.9992	0.9992	0.9993	0.9993
3.2	0.9993	0.9993	0.9994	0.9994	0.9994	0.9994	0.9994	0.9995	0.9995	0.9995
3.3	0.9995	0.9995	0.9995	0.9996	0.9996	0.9996	0.9996	0.9996	0.9996	0.9997
3.4	0.9997	0.9997	0.9997	0.9997	0.9997	0.9997	0.9997	0.9997	0.9997	0.9998

Table C.2 Critical Values of the *t* Distribution[a]

ν	α				
	0.10	0.05	0.025	0.01	0.005
1	3.078	6.314	12.706	31.821	63.657
2	1.886	2.920	4.303	6.965	9.925
3	1.638	2.353	3.182	4.541	5.841
4	1.533	2.132	2.776	3.747	4.604
5	1.476	2.015	2.571	3.365	4.032
6	1.440	1.943	2.447	3.143	3.707
7	1.415	1.895	2.365	2.998	3.499
8	1.397	1.860	2.306	2.896	3.355
9	1.383	1.833	2.262	2.821	3.250
10	1.372	1.812	2.228	2.764	3.169
11	1.363	1.796	2.201	2.718	3.106
12	1.356	1.782	2.179	2.681	3.055
13	1.350	1.771	2.160	2.650	3.012
14	1.345	1.761	2.145	2.624	2.977
15	1.341	1.753	2.131	2.602	2.947
16	1.337	1.746	2.120	2.583	2.921
17	1.333	1.740	2.110	2.567	2.898
18	1.330	1.734	2.101	2.552	2.878
19	1.328	1.729	2.093	2.539	2.861
20	1.325	1.725	2.086	2.528	2.854
21	1.323	1.721	2.080	2.518	2.831
22	1.321	1.717	2.074	2.508	2.819
23	1.319	1.714	2.069	2.500	2.807
24	1.318	1.711	2.064	2.492	2.797
25	1.316	1.708	2.060	2.485	2.787
26	1.315	1.706	2.056	2.479	2.779
27	1.314	1.703	2.052	2.473	2.771
28	1.313	1.701	2.048	2.467	2.763
29	1.311	1.699	2.045	2.462	2.756
inf.	1.282	1.645	1.960	2.326	2.576

From R. A. Fisher, "Statistical Methods for Research Workers," published by Oliver & Boyd, Edinburgh. Copyright © 1970 University of Adelaide.

Table C.3 Critical Values of the Chi-Square Distribution[a]

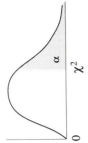

ν	0.995	0.99	0.975	0.95	0.05	0.025	0.01	0.005
1	.000039	.00016	.00098	.0039	3.84	5.02	6.63	7.88
2	.0100	.0201	.0506	.1026	5.99	7.38	9.21	10.60
3	.0717	.115	.216	.352	7.81	9.35	11.34	12.84
4	.207	.297	.484	.711	9.49	11.14	13.28	14.86
5	.412	.554	.831	1.15	11.07	12.83	15.09	16.75
6	.676	.872	1.24	1.64	12.59	14.45	16.81	18.55
7	.989	1.24	1.69	2.17	14.07	16.01	18.48	20.28
8	1.34	1.65	2.18	2.73	15.51	17.53	20.09	21.96
9	1.73	2.09	2.70	3.33	16.92	19.02	21.67	23.59
10	2.16	2.56	3.25	3.94	18.31	20.48	23.21	25.19
11	2.60	3.05	3.82	4.57	19.68	21.92	24.73	26.76
12	3.07	3.57	4.40	5.23	21.03	23.34	26.22	28.30
13	3.57	4.11	5.01	5.89	22.36	24.74	27.69	29.82
14	4.07	4.66	5.63	6.57	23.68	26.12	29.14	31.32
15	4.60	5.23	6.26	7.26	24.00	27.49	30.58	32.80
16	5.14	5.81	6.91	7.96	26.30	28.85	32.00	35.27

[a] Reproduced with permission of McGraw-Hill, from Table A-6a Of W. J. Dixon and F. J. Masset Jr., *Introduction to Statistical Analysis*, McGraw-Hill, New York, 1957.

Table C.3 (Continued)

					α					
ν	0.995	0.99	0.975	0.95	0.05	0.025	0.01	0.005		
17	5.70	6.41	7.56	8.67	27.59	30.19	33.41	35.72
18	6.26	7.02	8.23	9.39	28.87	31.53	34.81	37.16
19	6.84	7.63	8.91	10.12	30.14	32.85	36.19	38.58
20	7.43	8.26	9.59	10.85	31.41	34.17	37.57	40.00
21	8.03	8.90	10.28	11.59	32.67	35.49	38.93	41.40
22	8.64	9.54	10.98	12.34	33.92	36.78	40.29	42.80
23	9.26	10.20	11.69	13.09	35.17	38.08	41.64	44.18
24	9.89	10.86	12.40	13.85	36.42	39.36	42.98	45.56
25	10.52	11.52	13.12	14.61	37.65	40.65	44.31	46.93
26	11.16	12.20	13.84	15.38	38.89	41.92	45.64	48.29
27	11.81	12.88	14.57	16.15	40.12	43.19	46.96	49.65
28	12.46	13.57	15.31	16.93	41.34	44.46	48.28	50.99
29	13.12	14.26	16.05	17.71	42.56	45.72	49.59	52.34
30	13.79	14.95	16.79	18.49	43.77	46.98	50.89	53.67
40	20.71	22.16	24.43	26.51	55.76	59.34	63.69	66.77
60	35.53	37.48	40.48	43.19	79.08	83.30	88.38	91.95
120	83.85	86.92	91.58	95.70	146.57	152.21	158.95	163.64

Table C.4 Percentage Points of the F Distribution[a]

$$F^o_{.25, \nu_1, \nu_2}$$

Degrees of Freedom for the Numerator (ν_1)

ν_2 \ ν_1	1	2	3	4	5	6	7	8	9	10	12	15	20	24	30	40	60	120	∞
1	5.83	7.50	8.20	8.58	8.82	8.98	9.10	9.19	9.26	9.32	9.41	9.49	9.58	9.63	9.67	9.71	9.76	9.80	9.85
2	2.57	3.00	3.15	3.23	3.28	3.31	3.34	3.35	3.37	3.38	3.39	3.41	3.43	3.43	3.44	3.45	3.46	3.47	3.48
3	2.02	2.28	2.36	2.39	2.41	2.42	2.43	2.44	2.44	2.44	2.45	2.46	2.46	2.46	2.47	2.47	2.47	2.47	2.47
4	1.81	2.00	2.05	2.06	2.07	2.08	2.08	2.08	2.08	2.08	2.08	2.08	2.08	2.08	2.08	2.08	2.08	2.08	2.08
5	1.69	1.85	1.88	1.89	1.89	1.89	1.89	1.89	1.89	1.89	1.89	1.89	1.88	1.88	1.88	1.88	1.87	1.87	1.87
6	1.62	1.76	1.78	1.79	1.79	1.78	1.78	1.78	1.77	1.77	1.77	1.76	1.76	1.75	1.75	1.75	1.74	1.74	1.74
7	1.57	1.70	1.72	1.72	1.71	1.71	1.70	1.70	1.70	1.69	1.68	1.68	1.67	1.67	1.66	1.66	1.65	1.65	1.65
8	1.54	1.66	1.67	1.66	1.66	1.65	1.64	1.64	1.63	1.63	1.62	1.62	1.61	1.60	1.60	1.59	1.59	1.58	1.58
9	1.51	1.62	1.63	1.63	1.62	1.61	1.60	1.60	1.59	1.59	1.58	1.57	1.56	1.56	1.55	1.54	1.54	1.53	1.53
10	1.49	1.60	1.60	1.59	1.59	1.58	1.57	1.56	1.56	1.55	1.54	1.53	1.52	1.52	1.51	1.51	1.50	1.49	1.48
11	1.47	1.58	1.58	1.57	1.56	1.55	1.54	1.53	1.53	1.52	1.51	1.50	1.49	1.49	1.48	1.47	1.47	1.46	1.45
12	1.46	1.56	1.56	1.55	1.54	1.53	1.52	1.51	1.51	1.50	1.49	1.48	1.47	1.46	1.45	1.45	1.44	1.43	1.42
13	1.45	1.55	1.55	1.53	1.52	1.51	1.50	1.49	1.49	1.48	1.47	1.46	1.45	1.44	1.43	1.42	1.42	1.41	1.40
14	1.44	1.53	1.53	1.52	1.51	1.50	1.49	1.48	1.47	1.46	1.45	1.44	1.43	1.42	1.41	1.41	1.40	1.39	1.38
15	1.43	1.52	1.52	1.51	1.49	1.48	1.47	1.46	1.46	1.45	1.44	1.43	1.41	1.41	1.40	1.39	1.38	1.37	1.36
16	1.42	1.51	1.51	1.50	1.48	1.47	1.46	1.45	1.44	1.44	1.43	1.41	1.40	1.39	1.38	1.37	1.36	1.35	1.34
17	1.42	1.51	1.50	1.49	1.47	1.46	1.45	1.44	1.43	1.43	1.41	1.40	1.39	1.38	1.37	1.36	1.35	1.34	1.33
18	1.41	1.50	1.49	1.48	1.46	1.45	1.44	1.43	1.42	1.42	1.40	1.39	1.38	1.37	1.36	1.35	1.34	1.33	1.32
19	1.41	1.49	1.49	1.47	1.46	1.44	1.43	1.42	1.41	1.41	1.40	1.38	1.37	1.36	1.35	1.34	1.33	1.32	1.30

[a] Reprinted by permission of John Wiley & Sons, Inc. From Table V of W. W. Hines and D. C. Montgomery, *Probability and Statistics in Engineering and Management Sciences*, Wiley, New York, Copyright © 1990.

Table C.4 (Continued)

$$F^o_{.25, \nu_1, \nu_2}$$

ν_2 \ ν_1	1	2	3	4	5	6	7	8	9	10	12	15	20	24	30	40	60	120	∞
							Degrees of Freedom for the Numerator(ν_1)												
20	1.40	1.49	1.48	1.47	1.45	1.44	1.43	1.42	1.41	1.40	1.39	1.37	1.36	1.35	1.34	1.33	1.32	1.31	1.29
21	1.40	1.48	1.48	1.46	1.44	1.43	1.42	1.41	1.40	1.39	1.38	1.37	1.35	1.34	1.33	1.32	1.31	1.30	1.28
22	1.40	1.48	1.47	1.45	1.44	1.42	1.41	1.40	1.39	1.39	1.37	1.36	1.34	1.33	1.32	1.31	1.30	1.29	1.28
23	1.39	1.47	1.47	1.45	1.43	1.42	1.41	1.40	1.39	1.38	1.37	1.35	1.34	1.33	1.32	1.31	1.30	1.28	1.27
24	1.39	1.47	1.46	1.44	1.43	1.41	1.40	1.39	1.38	1.38	1.36	1.35	1.33	1.32	1.31	1.30	1.29	1.28	1.26
25	1.39	1.47	1.46	1.44	1.42	1.41	1.40	1.39	1.38	1.37	1.36	1.34	1.33	1.32	1.31	1.29	1.28	1.27	1.25
26	1.38	1.46	1.45	1.44	1.42	1.41	1.39	1.38	1.37	1.37	1.35	1.34	1.32	1.31	1.30	1.29	1.28	1.26	1.25
27	1.38	1.46	1.45	1.43	1.42	1.40	1.39	1.38	1.37	1.36	1.35	1.33	1.32	1.31	1.30	1.28	1.27	1.26	1.24
28	1.38	1.46	1.45	1.43	1.41	1.40	1.39	1.38	1.37	1.36	1.34	1.33	1.31	1.30	1.29	1.28	1.27	1.25	1.24
29	1.38	1.45	1.45	1.43	1.41	1.40	1.38	1.37	1.36	1.35	1.34	1.32	1.31	1.30	1.29	1.27	1.26	1.25	1.23
30	1.38	1.45	1.44	1.42	1.41	1.39	1.38	1.37	1.36	1.35	1.34	1.32	1.30	1.29	1.28	1.27	1.26	1.24	1.23
40	1.36	1.44	1.42	1.40	1.39	1.37	1.36	1.35	1.34	1.33	1.31	1.30	1.28	1.26	1.25	1.24	1.22	1.21	1.19
60	1.35	1.42	1.41	1.38	1.37	1.35	1.33	1.32	1.31	1.30	1.29	1.27	1.25	1.24	1.22	1.21	1.19	1.17	1.15
120	1.34	1.40	1.39	1.37	1.35	1.33	1.31	1.30	1.29	1.28	1.26	1.24	1.22	1.21	1.19	1.18	1.16	1.13	1.10
∞	1.32	1.39	1.37	1.35	1.33	1.31	1.29	1.28	1.27	1.25	1.24	1.22	1.19	1.18	1.16	1.14	1.12	1.08	1.00

Table C.4 (Continued)

$$F^{\circ}_{.10, v_1, v_2}$$

v_1									Degrees of Freedom for the Numerator(v_1)										
v_2	1	2	3	4	5	6	7	8	9	10	12	15	20	24	30	40	60	120	∞
1	39.86	49.50	53.59	55.83	57.24	58.20	58.91	59.44	59.86	60.19	60.71	61.22	61.74	62.00	62.26	62.53	62.79	63.06	63.33
2	8.53	9.00	9.16	9.24	9.29	9.33	9.35	9.37	9.38	9.39	9.41	9.42	9.44	9.45	9.46	9.47	9.47	9.48	9.49
3	5.54	5.46	5.39	5.34	5.31	5.28	5.27	5.25	5.24	5.23	5.22	5.20	5.18	5.18	5.17	5.16	5.15	5.14	5.13
4	4.54	4.32	4.19	4.11	4.05	4.01	3.98	3.95	3.94	3.92	3.90	3.87	3.84	3.83	3.82	3.80	3.79	3.78	3.76
5	4.06	3.78	3.62	3.52	3.45	3.40	3.37	3.34	3.32	3.30	3.27	3.24	3.21	3.19	3.17	3.16	3.14	3.12	3.10
6	3.78	3.46	3.29	3.18	3.11	3.05	3.01	2.98	2.96	2.94	2.90	2.87	2.84	2.82	2.80	2.78	2.76	2.74	2.72
7	3.59	3.26	3.07	2.96	2.88	2.83	2.78	2.75	2.72	2.70	2.67	2.63	2.59	2.58	2.56	2.54	2.51	2.49	2.47
8	3.46	3.11	2.92	2.81	2.73	2.67	2.62	2.59	2.56	2.54	2.50	2.46	2.42	2.40	2.38	2.36	2.34	2.32	2.29
9	3.36	3.01	2.81	2.69	2.61	2.55	2.51	2.47	2.44	2.42	2.38	2.34	2.30	2.28	2.25	2.23	2.21	2.18	2.16
10	3.29	2.92	2.73	2.61	2.52	2.46	2.41	2.38	2.35	2.32	2.28	2.24	2.20	2.18	2.16	2.13	2.11	2.08	2.06
11	3.23	2.86	2.66	2.54	2.45	2.39	2.34	2.30	2.27	2.25	2.21	2.17	2.12	2.10	2.08	2.05	2.03	2.00	1.97
12	3.18	2.81	2.61	2.48	2.39	2.33	2.28	2.24	2.21	2.19	2.15	2.10	2.06	2.04	2.01	1.99	1.96	1.93	1.90
13	3.14	2.76	2.56	2.43	2.35	2.28	2.23	2.20	2.16	2.14	2.10	2.05	2.01	1.98	1.96	1.93	1.90	1.88	1.85
14	3.10	2.73	2.52	2.39	2.31	2.24	2.19	2.15	2.12	2.10	2.05	2.01	1.96	1.94	1.91	1.89	1.86	1.83	1.80
15	3.07	2.70	2.49	2.36	2.27	2.21	2.16	2.12	2.09	2.06	2.02	1.97	1.92	1.90	1.87	1.85	1.82	1.79	1.76
16	3.05	2.67	2.46	2.33	2.24	2.18	2.13	2.09	2.06	2.03	1.99	1.94	1.89	1.87	1.84	1.81	1.78	1.75	1.72
17	3.03	2.64	2.44	2.31	2.22	2.15	2.10	2.06	2.03	2.00	1.96	1.91	1.86	1.84	1.81	1.78	1.75	1.72	1.69
18	3.01	2.62	2.42	2.29	2.20	2.13	2.08	2.04	2.00	1.98	1.93	1.89	1.84	1.81	1.78	1.75	1.72	1.69	1.66
19	2.99	2.61	2.40	2.27	2.18	2.11	2.06	2.02	1.98	1.96	1.91	1.86	1.81	1.79	1.76	1.73	1.70	1.67	1.63

Table C.4 (Continued)

$$F^{\circ}_{.10, \nu_1, \nu_2}$$

Degrees of Freedom for the Numerator(ν_1)

ν_2 \ ν_1	1	2	3	4	5	6	7	8	9	10	12	15	20	24	30	40	60	120	∞
20	2.97	2.59	2.38	2.25	2.16	2.09	2.04	2.00	1.96	1.94	1.89	1.84	1.79	1.77	1.74	1.71	1.68	1.64	1.61
21	2.96	2.57	2.36	2.23	2.14	2.08	2.02	1.98	1.95	1.92	1.87	1.83	1.78	1.75	1.72	1.69	1.66	1.62	1.59
22	2.95	2.56	2.35	2.22	2.13	2.06	2.01	1.97	1.93	1.90	1.86	1.81	1.76	1.73	1.70	1.67	1.64	1.60	1.57
23	2.94	2.55	2.34	2.21	2.11	2.05	1.99	1.95	1.92	1.89	1.84	1.80	1.74	1.72	1.69	1.66	1.62	1.59	1.55
24	2.93	2.54	2.33	2.19	2.10	2.04	1.98	1.94	1.91	1.88	1.83	1.78	1.73	1.70	1.67	1.64	1.61	1.57	1.53
25	2.92	2.53	2.32	2.18	2.09	2.02	1.97	1.93	1.89	1.87	1.82	1.77	1.72	1.69	1.66	1.63	1.59	1.56	1.52
26	2.91	2.52	2.31	2.17	2.08	2.01	1.96	1.92	1.88	1.86	1.81	1.76	1.71	1.68	1.65	1.61	1.58	1.54	1.50
27	2.90	2.51	2.30	2.17	2.07	2.00	1.95	1.91	1.87	1.85	1.80	1.75	1.70	1.67	1.64	1.60	1.57	1.53	1.49
28	2.89	2.50	2.29	2.16	2.06	2.00	1.94	1.90	1.87	1.84	1.79	1.74	1.69	1.66	1.63	1.59	1.56	1.52	1.48
29	2.89	2.50	2.28	2.15	2.06	1.99	1.93	1.89	1.86	1.83	1.78	1.73	1.68	1.65	1.62	1.58	1.55	1.51	1.47
30	2.88	2.49	2.28	2.14	2.03	1.98	1.93	1.88	1.85	1.82	1.77	1.72	1.67	1.64	1.61	1.57	1.54	1.50	1.46
40	2.84	2.44	2.23	2.09	2.00	1.93	1.87	1.83	1.79	1.76	1.71	1.66	1.61	1.57	1.54	1.51	1.47	1.42	1.38
60	2.79	2.39	2.18	2.04	1.95	1.87	1.82	1.77	1.74	1.71	1.66	1.60	1.54	1.51	1.48	1.44	1.40	1.35	1.29
120	2.75	2.35	2.13	1.99	1.90	1.82	1.77	1.72	1.68	1.65	1.60	1.55	1.48	1.45	1.41	1.37	1.32	1.26	1.19
∞	2.71	2.30	2.08	1.94	1.85	1.77	1.72	1.67	1.63	1.60	1.55	1.49	1.42	1.38	1.34	1.30	1.24	1.17	1.00

Table C.4 (Continued)

$$F^o_{.05, \nu_1, \nu_2}$$

ν_2 \ ν_1	Degrees of Freedom for the Numerator(ν_1)																		
	1	2	3	4	5	6	7	8	9	10	12	15	20	24	30	40	60	120	∞
1	161.4	199.5	215.7	224.6	230.2	234.0	236.8	238.9	240.5	241.9	243.9	245.9	248.0	249.1	250.1	251.1	252.2	253.3	254.3
2	18.51	19.00	19.16	19.25	19.30	19.33	19.35	19.37	19.38	19.40	19.41	19.43	19.45	19.45	19.46	19.47	19.48	19.49	19.50
3	10.13	9.55	9.28	9.12	9.01	8.94	8.89	8.85	8.81	8.79	8.74	8.70	8.66	8.64	8.62	8.59	8.57	8.55	8.53
4	7.71	6.94	6.59	6.39	6.26	6.16	6.09	6.04	6.00	5.96	5.91	5.86	5.80	5.77	5.75	5.72	5.69	5.66	5.63
5	6.61	5.79	5.41	5.19	5.05	4.95	4.88	4.82	4.77	4.74	4.68	4.62	4.56	4.53	4.50	4.46	4.43	4.40	4.36
6	5.99	5.14	4.76	4.53	4.39	4.28	4.21	4.15	4.10	4.06	4.00	3.94	3.87	3.84	3.81	3.77	3.74	3.70	3.67
7	5.59	4.74	4.35	4.12	3.97	3.87	3.79	3.73	3.68	3.64	3.57	3.51	3.44	3.41	3.38	3.34	3.30	3.27	3.23
8	5.32	4.46	4.07	3.84	3.69	3.58	3.50	3.44	3.39	3.35	3.28	3.22	3.15	3.12	3.08	3.04	3.01	2.97	2.93
9	5.12	4.26	3.86	3.63	3.48	3.37	3.29	3.23	3.18	3.14	3.07	3.01	2.94	2.90	2.86	2.83	2.79	2.75	2.71
10	4.96	4.10	3.71	3.48	3.33	3.22	3.14	3.07	3.02	2.98	2.91	2.85	2.77	2.74	2.70	2.66	2.62	2.58	2.54
11	4.84	3.98	3.59	3.36	3.20	3.09	3.01	2.95	2.90	2.85	2.79	2.72	2.65	2.61	2.57	2.53	2.49	2.45	2.40
12	4.75	3.89	3.49	3.26	3.11	3.00	2.91	2.85	2.80	2.75	2.69	2.62	2.54	2.51	2.47	2.43	2.38	2.34	2.30
13	4.67	3.81	3.41	3.18	3.03	2.92	2.83	2.77	2.71	2.67	2.60	2.53	2.46	2.42	2.38	2.34	2.30	2.25	2.21
14	4.60	3.74	3.34	3.11	2.96	2.85	2.76	2.70	2.65	2.60	2.53	2.46	2.39	2.35	2.31	2.27	2.22	2.18	2.13
15	4.54	3.68	3.29	3.06	2.90	2.79	2.71	2.64	2.59	2.54	2.48	2.40	2.33	2.29	2.25	2.20	2.16	2.11	2.07
16	4.49	3.63	3.24	3.01	2.85	2.74	2.66	2.59	2.54	2.49	2.42	2.35	2.28	2.24	2.19	2.15	2.11	2.06	2.01
17	4.45	3.59	3.20	2.96	2.81	2.70	2.61	2.55	2.49	2.45	2.38	2.31	2.23	2.19	2.15	2.10	2.06	2.01	1.96
18	4.41	3.55	3.16	2.93	2.77	2.66	2.58	2.51	2.46	2.41	2.34	2.27	2.19	2.15	2.11	2.06	2.02	1.97	1.92
19	4.38	3.52	3.13	2.90	2.74	2.63	2.54	2.48	2.42	2.38	2.31	2.23	2.16	2.11	2.07	2.03	1.98	1.93	1.88

Table C.4 (Continued)

$$F^\circ_{.05, \nu_1, \nu_2}$$

ν_2 \ ν_1	Degrees of Freedom for the Numerator(ν_1)																		
	1	2	3	4	5	6	7	8	9	10	12	15	20	24	30	40	60	120	∞
20	4.35	3.49	3.10	2.87	2.71	2.60	2.51	2.45	2.39	2.35	2.28	2.20	2.12	2.08	2.04	1.99	1.95	1.90	1.84
21	4.32	3.47	3.07	2.84	2.68	2.57	2.49	2.42	2.37	2.32	2.25	2.18	2.10	2.05	2.01	1.96	1.92	1.87	1.81
22	4.30	3.44	3.05	2.82	2.66	2.55	2.46	2.40	2.34	2.30	2.23	2.15	2.07	2.03	1.98	1.94	1.89	1.84	1.78
23	4.28	3.42	3.03	2.80	2.64	2.53	2.44	2.37	2.32	2.27	2.20	2.13	2.05	2.01	1.96	1.91	1.86	1.81	1.76
24	4.26	3.40	3.01	2.78	2.62	2.51	2.42	2.36	2.30	2.25	2.18	2.11	2.03	1.98	1.94	1.89	1.84	1.79	1.73
25	4.24	3.39	2.99	2.76	2.60	2.49	2.40	2.34	2.28	2.24	2.16	2.09	2.01	1.96	1.92	1.87	1.82	1.77	1.71
26	4.23	3.37	2.98	2.74	2.59	2.47	2.39	2.32	2.27	2.22	2.15	2.07	1.99	1.95	1.90	1.85	1.80	1.75	1.69
27	4.21	3.35	2.96	2.73	2.57	2.46	2.37	2.31	2.25	2.20	2.13	2.06	1.97	1.93	1.88	1.84	1.79	1.73	1.67
28	4.20	3.34	2.95	2.71	2.56	2.45	2.36	2.29	2.24	2.19	2.12	2.04	1.96	1.91	1.87	1.82	1.77	1.71	1.65
29	4.18	3.33	2.93	2.70	2.55	2.43	2.35	2.28	2.22	2.18	2.10	2.03	1.94	1.90	1.85	1.81	1.75	1.70	1.64
30	4.17	3.32	2.92	2.69	2.53	2.42	2.33	2.27	2.21	2.16	2.09	2.01	1.93	1.89	1.84	1.79	1.74	1.68	1.62
40	4.08	3.23	2.84	2.61	2.45	2.34	2.25	2.18	2.12	2.08	2.00	1.92	1.84	1.79	1.74	1.69	1.64	1.58	1.51
60	4.00	3.15	2.76	2.53	2.37	2.25	2.17	2.10	2.04	1.99	1.92	1.84	1.75	1.70	1.65	1.59	1.53	1.47	1.39
120	3.92	3.07	2.68	2.45	2.29	2.17	2.09	2.02	1.96	1.91	1.83	1.75	1.66	1.61	1.55	1.50	1.43	1.35	1.25
∞	3.84	3.00	2.60	2.37	2.21	2.10	2.01	1.94	1.88	1.83	1.75	1.67	1.57	1.52	1.46	1.39	1.32	1.22	1.00

Table C.4 (Continued)

$$F^\circ_{.025, \nu_1, \nu_2}$$

ν_2 \ ν_1	1	2	3	4	5	6	7	8	9	10	12	15	20	24	30	40	60	120	∞
									Degrees of Freedom for the Numerator(ν_1)										
1	647.8	799.5	864.2	899.6	921.8	937.1	948.2	956.7	963.3	968.6	976.7	984.9	993.1	997.2	1001	1006	1010	1014	1018
2	38.51	39.00	39.17	39.25	39.30	39.33	39.36	39.37	39.39	39.40	39.41	39.43	39.45	39.46	39.46	39.47	39.48	39.49	39.50
3	17.44	16.04	15.44	15.10	14.88	14.73	14.62	14.54	14.47	14.42	14.34	14.25	14.17	14.12	14.08	14.04	13.99	13.95	13.90
4	12.22	10.65	9.98	9.60	9.36	9.20	9.07	8.98	8.90	8.84	8.75	8.66	8.56	8.51	8.46	8.41	8.36	8.31	8.26
5	10.01	8.43	7.76	7.39	7.15	6.98	6.85	6.76	6.68	6.62	6.52	6.43	6.33	6.28	6.23	6.18	6.12	6.07	6.02
6	8.81	7.26	6.60	6.23	5.99	5.82	5.70	5.60	5.52	5.46	5.37	5.27	5.17	5.12	5.07	5.01	4.96	4.90	4.85
7	8.07	6.54	5.89	5.52	5.29	5.12	4.99	4.90	4.82	4.76	4.67	4.57	4.47	4.42	4.36	4.31	4.25	4.20	4.14
8	7.57	6.06	5.42	5.05	4.82	4.65	4.53	4.43	4.36	4.30	4.20	4.10	4.00	3.95	3.89	3.84	3.78	3.73	3.67
9	7.21	5.71	5.08	4.72	4.48	4.32	4.20	4.10	4.03	3.96	3.87	3.77	3.67	3.61	3.56	3.51	3.45	3.39	3.33
10	6.94	5.46	4.83	4.47	4.24	4.07	3.95	3.85	3.78	3.72	3.62	3.52	3.42	3.37	3.31	3.26	3.20	3.14	3.08
11	6.72	5.26	4.63	4.28	4.04	3.88	3.76	3.66	3.59	3.53	3.43	3.33	3.23	3.17	3.12	3.06	3.00	2.94	2.88
12	6.55	5.10	4.47	4.12	3.89	3.73	3.61	3.51	3.44	3.37	3.28	3.18	3.07	3.02	2.96	2.91	2.85	2.79	2.72
13	6.41	4.97	4.35	4.00	3.77	3.60	3.48	3.39	3.31	3.25	3.15	3.05	2.95	2.89	2.84	2.78	2.72	2.66	2.60
14	6.30	4.86	4.24	3.89	3.66	3.50	3.38	3.29	3.21	3.15	3.05	2.95	2.84	2.79	2.73	2.67	2.61	2.55	2.49
15	6.20	4.77	4.15	3.80	3.58	3.41	3.29	3.20	3.12	3.06	2.96	2.86	2.76	2.70	2.64	2.59	2.52	2.46	2.40
16	6.12	4.69	4.08	3.73	3.50	3.34	3.22	3.12	3.05	2.99	2.89	2.79	2.68	2.63	2.57	2.51	2.45	2.38	2.32
17	6.04	4.62	4.01	3.66	3.44	3.28	3.16	3.06	2.98	2.92	2.82	2.72	2.62	2.56	2.50	2.44	2.38	2.32	2.25
18	5.98	4.56	3.95	3.61	3.38	3.22	3.10	3.01	2.93	2.87	2.77	2.67	2.56	2.50	2.44	2.38	2.32	2.26	2.19
19	5.92	4.51	3.90	3.56	3.33	3.17	3.05	2.96	2.88	2.82	2.72	2.62	2.51	2.45	2.39	2.33	2.27	2.20	2.13

Table C.4 (Continued)

$$F^\circ_{.025,\nu_1,\nu_2}$$

ν_2 \ ν_1	Degrees of Freedom for the Numerator(ν_1)																		
	1	2	3	4	5	6	7	8	9	10	12	15	20	24	30	40	60	120	∞
20	5.87	4.46	3.86	3.51	3.29	3.13	3.01	2.91	2.84	2.77	2.68	2.57	2.46	2.41	2.35	2.29	2.22	2.16	2.09
21	5.83	4.42	3.82	3.48	3.25	3.09	2.97	2.87	2.80	2.73	2.64	2.53	2.42	2.37	2.31	2.25	2.18	2.11	2.04
22	5.79	4.38	3.78	3.44	3.22	3.05	2.93	2.84	2.76	2.70	2.60	2.50	2.39	2.33	2.27	2.21	2.14	2.08	2.00
23	5.75	4.35	3.75	3.41	3.18	3.02	2.90	2.81	2.73	2.67	2.57	2.47	2.36	2.30	2.24	2.18	2.11	2.04	1.97
24	5.72	4.32	3.72	3.38	3.15	2.99	2.87	2.78	2.70	2.64	2.54	2.44	2.33	2.27	2.21	2.15	2.08	2.01	1.94
25	5.69	4.29	3.69	3.35	3.13	2.97	2.85	2.75	2.68	2.61	2.51	2.41	2.30	2.24	2.18	2.12	2.05	1.98	1.91
26	5.66	4.27	3.67	3.33	3.10	2.94	2.82	2.73	2.65	2.59	2.49	2.39	2.28	2.22	2.16	2.09	2.03	1.95	1.88
27	5.63	4.24	3.65	3.31	3.08	2.92	2.80	2.71	2.63	2.57	2.47	2.36	2.25	2.19	2.13	2.07	2.00	1.93	1.85
28	5.61	4.22	3.63	3.29	3.06	2.90	2.78	2.69	2.61	2.55	2.45	2.34	2.23	2.17	2.11	2.05	1.98	1.91	1.83
29	5.59	4.20	3.61	3.27	3.04	2.88	2.76	2.67	2.59	2.53	2.43	2.32	2.21	2.15	2.09	2.03	1.96	1.89	1.81
30	5.57	4.18	3.59	3.25	3.03	2.87	2.75	2.65	2.57	2.51	2.41	2.31	2.20	2.14	2.07	2.01	1.94	1.87	1.79
40	5.42	4.05	3.46	3.13	2.90	2.74	2.62	2.53	2.45	2.39	2.29	2.18	2.07	2.01	1.94	1.88	1.80	1.72	1.64
60	5.29	3.93	3.34	3.01	2.79	2.63	2.51	2.41	2.33	2.27	2.17	2.06	1.94	1.88	1.82	1.74	1.67	1.58	1.48
120	5.15	3.80	3.23	2.89	2.67	2.52	2.39	2.30	2.22	2.16	2.05	1.94	1.82	1.76	1.69	1.61	1.53	1.43	1.31
∞	5.02	3.69	3.12	2.79	2.57	2.41	2.29	2.19	2.11	2.05	1.94	1.83	1.71	1.64	1.57	1.48	1.39	1.27	1.00

Table C.4 (Continued)

$$F^{\circ}_{.01, v_1, v_2}$$

v_2 \ v_1	1	2	3	4	5	6	7	8	9	10	12	15	20	24	30	40	60	120	∞
1	4052	4999.5	5403	5625	5764	5859	5928	5982	6022	6056	6106	6157	6209	6235	6261	6287	6313	6339	6366
2	98.50	99.00	99.17	99.25	99.30	99.33	99.36	99.37	99.39	99.40	99.42	99.43	99.45	99.46	99.47	99.47	99.48	99.49	99.50
3	34.12	30.82	29.46	28.71	28.24	27.91	27.67	27.49	27.35	27.23	27.05	26.87	26.69	26.00	26.50	26.41	26.32	26.22	26.13
4	21.20	18.00	16.69	15.98	15.52	15.21	14.98	14.80	14.66	14.55	14.37	14.20	14.02	13.93	13.84	13.75	13.65	13.56	13.46
5	16.26	13.27	12.06	11.39	10.97	10.67	10.46	10.29	10.16	10.05	9.89	9.72	9.55	9.47	9.38	9.29	9.20	9.11	9.02
6	13.75	10.92	9.78	9.15	8.75	8.47	8.26	8.10	7.98	7.87	7.72	7.56	7.40	7.31	7.23	7.14	7.06	6.97	6.88
7	12.25	9.55	8.45	7.85	7.46	7.19	6.99	6.84	6.72	6.62	6.47	6.31	6.16	6.07	5.99	5.91	5.82	5.74	5.65
8	11.26	8.65	7.59	7.01	6.63	6.37	6.18	6.03	5.91	5.81	5.67	5.52	5.36	5.28	5.20	5.12	5.03	4.95	4.86
9	10.56	8.02	6.99	6.42	6.06	5.80	5.61	5.47	5.35	5.26	5.11	4.96	4.81	4.73	4.65	4.57	4.48	4.40	4.31
10	10.04	7.56	6.55	5.99	5.64	5.39	5.20	5.06	4.94	4.85	4.71	4.56	4.41	4.33	4.25	4.17	4.08	4.00	3.91
11	9.65	7.21	6.22	5.67	5.32	5.07	4.89	4.74	4.63	4.54	4.40	4.25	4.10	4.02	3.94	3.86	3.78	3.69	3.60
12	9.33	6.93	5.95	5.41	5.06	4.82	4.64	4.50	4.39	4.30	4.16	4.01	3.86	3.78	3.70	3.62	3.54	3.45	3.36
13	9.07	6.70	5.74	5.21	4.86	4.62	4.44	4.30	4.19	4.10	3.96	3.82	3.66	3.59	3.51	3.43	3.34	3.25	3.17
14	8.86	6.51	5.56	5.04	4.69	4.46	4.28	4.14	4.03	3.94	3.80	3.66	3.51	3.43	3.34	3.27	3.18	3.09	3.00
15	8.68	6.36	5.42	4.89	4.36	4.32	4.14	4.00	3.89	3.80	3.67	3.52	3.37	3.29	3.21	3.13	3.05	2.96	2.87
16	8.53	6.23	5.29	4.77	4.44	4.20	4.03	3.89	3.78	3.69	3.55	3.41	3.26	3.18	3.10	3.02	2.93	2.84	2.75
17	8.40	6.11	5.18	4.67	4.34	4.10	3.93	3.79	3.68	3.59	3.46	3.31	3.16	3.08	3.00	2.92	2.83	2.75	2.65
18	8.29	6.01	5.09	4.58	4.25	4.01	3.84	3.71	3.60	3.51	3.37	3.23	3.08	3.00	2.92	2.84	2.75	2.66	2.57
19	8.18	5.93	5.01	4.50	4.17	3.94	3.77	3.63	3.52	3.43	3.30	3.15	3.00	2.92	2.84	2.76	2.67	2.58	2.49

Degrees of Freedom for the Numerator(v_1)

Table C.4 (Continued)

$$F^\circ_{.01,\nu_1,\nu_2}$$

ν_2 \\ ν_1	Degrees of Freedom for the Numerator (ν_1)																		
	1	2	3	4	5	6	7	8	9	10	12	15	20	24	30	40	60	120	∞
20	8.10	5.85	4.94	4.43	4.10	3.87	3.70	3.56	3.46	3.37	3.23	3.09	2.94	2.86	2.78	2.69	2.61	2.52	2.42
21	8.02	5.78	4.87	4.37	4.04	3.81	3.64	3.51	3.40	3.31	3.17	3.03	2.88	2.80	2.72	2.64	2.55	2.46	2.36
22	7.95	5.72	4.82	4.31	3.99	3.76	3.59	3.45	3.35	3.26	3.12	2.98	2.83	2.75	2.67	2.58	2.50	2.40	2.31
23	7.88	5.66	4.76	4.26	3.94	3.71	3.54	3.41	3.30	3.21	3.07	2.93	2.78	2.70	2.62	2.54	2.45	2.35	2.26
24	7.82	5.61	4.72	4.22	3.90	3.67	3.50	3.36	3.26	3.17	3.03	2.89	2.74	2.66	2.58	2.49	2.40	2.31	2.21
25	7.77	5.57	4.68	4.18	3.85	3.63	3.46	3.32	3.22	3.13	2.99	2.85	2.70	2.62	2.54	2.45	2.36	2.27	2.17
26	7.72	5.53	4.64	4.14	3.82	3.59	3.42	3.29	3.18	3.09	2.96	2.81	2.66	2.58	2.50	2.42	2.33	2.23	2.13
27	7.68	5.49	4.60	4.11	3.78	3.56	3.39	3.26	3.15	3.06	2.93	2.78	2.63	2.55	2.47	2.38	2.29	2.20	2.10
28	7.64	5.45	4.57	4.07	3.75	3.53	3.36	3.23	3.12	3.03	2.90	2.75	2.60	2.52	2.44	2.35	2.26	2.17	2.06
29	7.60	5.42	5.45	4.04	3.73	3.50	3.33	3.20	3.09	3.00	2.87	2.73	2.57	2.49	2.41	2.33	2.23	2.14	2.03
30	7.56	5.39	4.51	4.02	3.70	3.47	3.30	3.17	3.07	2.98	2.84	2.70	2.55	2.47	2.39	2.30	2.21	2.11	2.01
40	7.31	5.18	4.31	3.83	3.51	3.29	3.12	2.99	2.89	2.80	2.66	2.52	2.37	2.29	2.20	2.11	2.02	1.92	1.80
60	7.08	4.98	4.13	3.65	3.34	3.12	2.95	2.82	2.72	2.63	2.50	2.35	2.20	2.12	2.03	1.94	1.84	1.73	1.60
120	6.85	4.79	3.95	3.48	3.17	2.96	2.79	2.66	2.56	2.47	2.34	2.19	2.03	1.95	1.89	1.76	1.66	1.53	1.38
∞	6.63	4.61	3.78	3.32	3.02	2.80	2.64	2.51	2.41	2.32	2.18	2.04	1.88	1.79	1.70	1.59	1.47	1.32	1.00

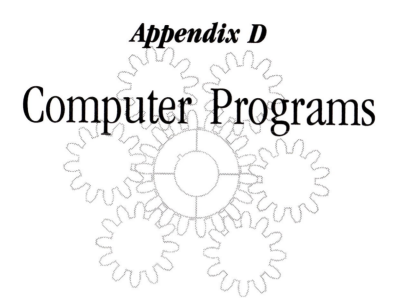

Appendix D
Computer Programs

D.1 PROGRAM LISTING FOR MAPLE FOR EVALUATING THE EIGENVALUES

```
# Define the solution of the beam problem as a MAPLE
  function

y:=
   proc(x)
   a*cosh(beta*x) + b*cos(beta*x) + c*sinh(beta*x)
   + d*sin(beta*x) end;

y := proc(x) a*cosh(beta*x)+b*cos(beta*x)+c*sinh(beta*x)
     +d*sin(beta*x) end

# Check to see if we have the correct   function
 y(x);
         a cosh(beta x) + b cos(beta x) + c sinh(beta x)
         + d sin(beta x)

# Apply the displacement condition at x=0
# Solve for b, and substitute b=-a in y(x)
 solve(y(0),b);
                                   - a
 subs(b=-a,y(x));
         a cosh(beta x) - a cos(beta x) + c sinh(beta x)
         + d sin(beta x)

# Now apply the slope condition at x=0
# Solve for d, and substitute d=-c in y(x)
 diff(y(x),x);
         a sinh(beta x) beta - b sin(beta x) beta
         + c cosh(beta x) beta + d cos(beta x) beta
 subs(x=0,");
```

```
        a sinh(0) beta - b sin(0) beta + c cosh(0) beta
        + d cos(0) beta
 evalf(");
                              c beta + d beta
 solve(",d);

# Now define another MAPLE function v(x), which does not
# contain b and d
 v:=proc(x) subs(b=-a,d=-c,y(x)) end;
 v:= proc(x) subs(b = -a,d = -c,y(x)) end

# Define the 2nd and 3rd derivatives as functions
 d2v:=proc(x) diff(v(x),x$2) end;
 d2v:= proc(x) diff(v(x),x $ 2) end
 d3v:=proc(x) diff(v(x),x$3) end;
# Now apply the bending moment and the shear force condtions
# at x=L
# Determine the coefficients of "a" and "c" in the bending
# moment and the shear force boundary conditions
#c11:= coefficient of "a" in the bending moment condition at
 x=L
 c11:=coeff(d2v(L),a);
                               2                        2
        c11 := cosh(beta L) beta  + cos(beta L) beta
#c12:= coefficient of "c" in the bending moment condition at
 x=L
 c12:=coeff(d2v(L),c);
                               2                      2
        c12 := sinh(beta L) beta  + sin(beta L) beta
#c21:= coefficient of "a" in the shear force condition at
 x=L
 c21:=coeff(d3v(L),a);
                               3                        3
        c21 := sinh(beta L) beta  - sin(beta L) beta
#c22:= coefficient of "c" in the shear force condition at
 x=L
 c22:=coeff(d3v(L),c);
                               3                      3
        c22 := cosh(beta L) beta  + cos(beta L) beta
# Now define a Wranskian Matrix, W=c(i,j) of the above
 calculated    coefficients as
 MAPLE array
 W:=array([[c11,c12],[c21,c22]]);
 W:=                       2                      2
      [cosh(beta L) beta  + cos(beta L) beta ,
                       2                      2
        sinh(beta L) beta  + sin(beta L) beta ]
                 3                      3
```

```
[sinh(beta L)  beta    - sin(beta L)  beta ,
                   3                        3
       cosh(beta L)  beta   + cos(beta L)  beta ]
```

```
# Load the Linear Algebra Package
  with(linalg);
# Define the Characteristic polynomial as the determinant of
# Wranskian
  p:=proc(beta) simplify(det(W)) end;
  p:= proc(beta) simplify(det(W)) end
```

```
# Check the characteristic polynomial
  p(beta);
```
$$2 \cosh(beta\ L)\ beta^5\ \cos(beta\ L) + 2\ beta^5$$

```
# Factor out  p(beta) and simplify
  factor(p(beta));
```
$$2\ beta^5\ (\cosh(beta\ L)\ \cos(beta\ L) + 1)$$

```
#Assign p(beta)->q is the simplified characteristic poly-
  nomial
  q:= "/(2*beta^5);
```
$$q := \cosh(beta\ L)\ \cos(beta\ L) + 1$$
```
# For the length L=10, plot q, for 0<beta<1.2,-10000<q<2000,
# the roots are the eigenvalues
  L:=10;
```
$$L := 10$$
```
  plot(q,beta=0.0..1.2,-10000..2000);
```

```
# Now solve for the roots in the range
# beta=[0,0.25],[0.2,0.5],[0.5,0.8],[0.8,1.2]
  beta1:=fsolve(q,beta,0.0..0.25);
```
$$beta1 := .1875104069$$
```
  beta2:=fsolve(q,beta,0.2..0.50);
```
$$beta2 := .469409113$$
```
  beta3:=fsolve(q,beta,0.5..0.80);
```
$$beta3 := .7854757438$$
```
  beta4:=fsolve(q,beta,0.8..1.20);
```
$$beta4 := 1.099554073$$
```
# Now substitute a=1,c=-c12/c11 ,beta=beta1,beta2,beta3,
  beta4 into v(x) and plot the mode-shapes
  z1:=proc(x) subs(a=1,c=-c11/c12,beta=beta1,v(x)) end;
  z1:= proc(x) subs(a = 1,c = -c11/c12,beta = beta1,v(x))
  end
  z2:=proc(x) subs(a=1,c=-c11/c12,beta=beta2,v(x)) end;
  z2:= proc(x) subs(a = 1,c = -c11/c12,beta = beta2,v(x))
  end
```

```
  z3:=proc(x) subs(a=1,c=-c11/c12,beta=beta3,v(x)) end;
 z3:= proc(x) subs(a = 1,c = -c11/c12,beta = beta3,v(x)) end
  z4:=proc(x) subs(a=1,c=-c11/c12,beta=beta4,v(x)) end;
 z4:= proc(x) subs(a = 1,c = -c11/c12,beta = beta4,v(x)) end

  plot({z1(x),z2(x),z3(x),z4(x)},x=0..L);
```

D.2 MAXIMIZATION OF A RELIABILITY FOR A GIVEN COST CONSTRAINT

*** .... *******

MAXIMIZATION OF A RELIABILITY FOR A GIVEN COST CONSTRAINT:

THIS PROGRAM CALCULATES THE OPTIMUM NUMBER OF REDUNDANCY IN
AN N STAGE SERIES SYSTEM FOR A GIVEN COST LIMIT AND CONSTANT RE-
LIABILITY. THE LAGRANGIAN METHOD IS USED FOR THE OPTIMIZATION.
PROGRAM CALCULATES THE SYSTEM RELIABILITY, SYSTEM TOTAL COST AND
ALSO PRINTS OUT THE LAGRANGE COEFFICIENT LAMBDA. DECREMENTS OF
LAMBDA SHOULD BE CAREFULLY CHOSEN TO BE ABLE TO OBTAIN SUFFICIENT
NUMBER OF STEPS BEFORE PROGRAM QUITS.

*** .... *******

```
C.... VARIABLES:
C.... A : COEFFICIENT FOR LAMBDA
C.... B : COEFFICIENT FOR LAMBDA
C.... C : COST OF EACH COMPONENT
C.... CL : COST CONSTRAINT
C.... DELLAM : CHANGE IN THE LAMBDA
C.... N : TOTAL NUMBER OF STAGES
C.... NK  : COEFFICIENT OF LAMBDA
C.... PHI  : PARAMETER
C.... R : RELIABILITY OF EACH COMPONENT
C.... REL  : RELIABILITY OF THE SYSTEM
C.... SUM1 : DUMMY VARIABLES FOR THE CALCULATION OF LAMBDA
C.... SUM2 : DUMMY VARIABLES FOR THE CALCULATION OF LAMBDA
C.... SUMC : TOTAL COST OF THE ELEMENT
C.... SUMP : TOTAL RELIABILITY OF THE SYSTEM
C.... XLAM : LAGRANGE MULTIPLIER

      REAL*8 R(10),C(10),PHI(100),A(10),NK(10),B(10),XLAM
      REAL*8 H(100),SUM1,SUM2,SUMC,SUMP,CL,AL,REL,PROF,PRO
      REAL*8 RELL(100),RELLL(100)
      DIMENSION NY(10)
      OPEN(UNIT=7,FILE='OUTH1.DAT',STATUS='NEW')
C....
C.... READ IN THE DATA
C....
```

```
      WRITE(*,*)'WHAT IS THE NUMBER OF STAGES?'
      READ(*,*) N
      WRITE(*,*)'WHAT IS THE COST LIMIT?'
      READ(*,*) CL
      DO 1 I=1,N
      WRITE(*,*) 'ENTER THE COST AND THE RELIABILITY OF
                 STAGE'
      WRITE(*,*) I
      READ(*,*) C(I),R(I)
   1 CONTINUE
C....
C.... CALCULATE PARAMETERS AND CONSTANTS FOR THE CALCULATION
C.... OF LAMBDA
C....
      DO 4 I=1,N
      PHI(I)=(1-R(I))/R(I)
      A(I)=1/DLOG(PHI(I))
      NK(I)=-C(I)/DLOG(PHI(I))
      B(I)=DLOG(NK(I))/DLOG(PHI(I))
   4 CONTINUE
C....
C.... CALCULATE LAMBDA
C.... SUM1=0.0
      SUM2=0.0
      DO 3 I=1,N
      SUM1=SUM1+C(I)*B(I)
      SUM2=SUM2+C(I)*A(I)
   3 CONTINUE
      S=(CL-SUM1)/SUM2
      XLAM=EXP(S)
      WRITE(*,*)'LAMBDA IS CALCULATED AS='
      WRITE(*,*) XLAM
      WRITE(*,*)'CHOOSE DELTA LAMBDA'
      READ(*,*)DELLAM
   9 CONTINUE
C....
C.... CALCULATE THE SMALLEST NUMBER OF ELEMENTS THAT
C.... WILL SATISFY THE UNDOMINATED SEQUENCE INEQUALITY
C.... FOR EVERY STAGE.
C....
      WRITE(7,89)
  89 FORMAT(///,2X,'NUMBER OF ELEMENTS',2X,'STAGES')
      DO 5 I=1,N
      NY(I)=0.0
   7 H(I)=(PHI(I)**NY(I))/((1+PHI(I))**(NY(I)+1))
      AL=XLAM*C(I)
      IF (H(I).LT.AL) GO TO 6
      NY(I)=NY(I)+1
      GO TO 7
   6 WRITE(7,90) NY(I),I
  90 FORMAT(2I,5X,2I)
```

```
      5 CONTINUE
C....
C.... CALCULATE THE COST AND RELIABILITY OF THE SYSTEM
C....
        SUMC=0.0
        SUMP=1.0
        sumpp=0.0
     10 CONTINUE
        DO 11 NI=1,N
        SUMC=SUMC+C(NI)*NY(NI)
        RELL(NI)=1/(1+PHI(NI))
        RELLL(NI)=1-(1-RELL(NI))**NY(NI)
        SUMP=SUMP*RELLL(NI)
     11 CONTINUE
        REL=SUMP
C.... RE=1/(1+SUMPp)
        PROF=PRO*REL-SUMC
C....
C.... PRINT OUT THE RESULTS IN OUT.DAT FILE
C....
        WRITE(7,91) SUMC, REL,XLAM
     91 FORMAT(/,'COST=',F16.13,2X,/,'RELIABILITY=',F16.13,
        $/,'LAMBDA=',F16.13,///)
        XLAM=XLAM-DELLAM
        IF(XLAM.LE.0.0) GO TO 98
        GO TO 101
     98 WRITE(*,*) 'CALCULATED LAMBDA IS LESS THAN 0,'
        WRITE(*,*)'RECHOOSE THE DELTA LAMBDA'
        WRITE(*,*)'RECOMMENDED DELTA LAMDA IS 20
        WRITE(*,*)'THE FIRST GUESSED DELTA LAMBDA,'
        WRITE(*,*)'KEEP YOUR FINGERS CROSSED'
        GO TO 8
C....
C.... TERMINATE THE CALCULATION WHEN COST CONSTRAINT NO
C.... MORE SATISFIED
C....
    101 IF(SUMC.GT.CL) GO TO 8
        GO TO 9
      8 CONTINUE

        END
```

D.3   MAXIMIZATION OF A COST FOR A GIVEN RELIABILITY CONSTRAINT

***..**

MINIMIZATION OF COST TO OBTAIN THE GIVEN SYSTEM RELIABILITY:

THIS PROGRAM CALCULATES THE OPTIMUM NUMBER OF REDUNDANCY IN AN
N STAGE SERIES SYSTEM FOR A GIVEN RELIABILITY LIMIT AND COST CON-

STRAINT. THE LAGRANGIAN METHOD IS USED FOR THE OPTIMIZATION. PRO-
GRAM CALCULATES THE SYSTEM RELIABILITY, SYSTEM TOTAL COST AND ALSO
PRINTS OUT THE LAGRANGE COEFFICIENT LAMBDA. DECREMENTS OF LAMBDA
SHOULD BE CAREFULLY CHOSEN TO BE ABLE TO OBTAIN SUFFICIENT NUMBER
OF STEPS BEFORE PROGRAM QUITS.

```
***  ..............................................................................  ***
C.... VARIABLES:
C.... A : COEFFICIENT FOR LAMBDA
C.... B : COEFFICIENT FOR LAMBDA
C.... C : COST OF EACH COMPONENT
C.... DELLAM : CHANGE IN THE LAMBDA
C.... N : TOTAL NUMBER OF STAGES
C.... NK : COEFFICIENT OF LAMBDA
C.... PHI : PARAMETER
C.... R : RELIABILITY OF EACH COMPONENT
C.... REL : RELIABILITY OF THE SYSTEM
C.... RL: RELIABILITY CONSTRAINT
C.... SUM1 : DUMMY VARIABLES FOR THE CALCULATION OF LAMBDA
C.... SUM2 : DUMMY VARIABLES FOR THE CALCULATION OF LAMBDA
C.... SUMC : TOTAL COST OF THE ELEMENT
C.... SUMP : TOTAL RELIABILITY OF THE SYSTEM
C.... XLAM : LAGRANGE MULTIPLIER
      REAL*8 R(10),C(10),PHI(100),A(10),K(10),B(10),XLAM
      REAL*8 H(100),Y(100),PH,RL,S
      REAL*8 RELL(100),RELLL(100),SUM1,SUMC,SUMP,REL
      OPEN(UNIT=7,FILE='OUTH2.DAT',STATUS='NEW')
C....
C.... READ IN THE DATA
C....
      WRITE(*,*)'WHAT IS THE NUMBER OF STAGES?'
      READ(*,*) N
      WRITE(*,*)'WHAT IS THE RELIABILITY CONS?'
      READ(*,*) RL
      RS=RL-0.0001*RL
      DO 1 I=1,N
      WRITE(*,*) 'ENTER THE COST AND THE RELIABILITY OF
                 STAGE'
      WRITE(*,*) I
      READ(*,*) C(I),R(I)
    1 CONTINUE
C....
C.... CALCULATE PARAMETERS AND CONSTANTS FOR THE CALCULA-
C.... TION OF LAMBDA
C....
      DO 4 I=1,N
      PHI(I)=(1-R(I))/R(I)
      K(I)=(C(I)*DLOG(PHI(1)))/(C(1)*DLOG(PHI(I)))
    4 CONTINUE
      PH=(1-RL)/RL
```

```
      SUM1=0.0
      DO 3 I=1,N
      SUM1=SUM1+K(I)
    3 CONTINUE
C....
C.... CALCULATE LAMBDA
C....
      S=PH/SUM1
      XLAM=-C(1)/(S*DLOG(PHI(1)))
      WRITE(*,*)'LAMBDA IS CALCULATED AS='
      WRITE(*,*) XLAM
      WRITE(*,*)'CHOOSE DELTA LAMBDA'
      READ(*,*)DELLAM
C....
C.... CALCULATE THE SMALLEST NUMBER OF ELEMENTS THAT WILL
C.... SATISFY THE UNDOMINATED SEQUENCE INEQUALITY
C.... FOR EVERY STAGE.
C....
    9 CONTINUE
      WRITE(7,89)
   89 FORMAT(///,2X,'NUMBER OF ELEMENTS',2X,'STAGES')
      DO 5 I=1,N
      Y(I)=0.0
    7 H(I)=(PHI(I)**Y(I))/((1+PHI(I))**(Y(I)+1))*XLAM
      IF (H(I).LT.C(I)) GO TO 6
      Y(I)=Y(I)+1
      GO TO 7
    6 WRITE(7,90) Y(I),I
   90 FORMAT(2I,5X,2I)
    5 CONTINUE
C....
C.... CALCULATE THE COST AND RELIABILITY OF THE SYSTEM
C....
      SUMC=0.0
      SUMP=1.0
      DO 11 I=1,N
      SUMC=SUMC+C(I)*Y(I)
      RELL(I)=1/(1+PHI(I))
      RELLL(I)=1-(1-RELL(I))**Y(I)
      SUMP=SUMP*RELLL(I)
   11 CONTINUE
      REL=SUMP
C.... REL=1./(SUMP+1.)
C....
C.... PRINT OUT THE RESULTS IN A.DAT FILE
C....
      WRITE(7,91) SUMC, REL,XLAM
   91 FORMAT(/,'COST=',F16.13,2X,/,'RELIABILITY=',F16.13
     $,/,'LAMBDA=',F16.10,///)
      XLAM=XLAM-DELLAM
      IF(XLAM.LE.0.0) GO TO 98
```

```
      GO TO 101
   98 WRITE(*,*) 'CALCULATED LAMBDA IS LESS THAN 0,'
      WRITE(*,*)'RECHOOSE THE DELTA LAMBDA'
      WRITE(*,*)'RECOMMENDED DELTA LAMDA IS 20
      WRITE(*,*)'THE FIRST GUESSED DELTA LAMBDA,'
      WRITE(*,*)'KEEP YOUR FINGERS CROSSED'
      GO TO 8
C....
C.... TERMINATE THE CALCULATION WHEN COST CONSTRAINT NO
C.... MORE SATISFIED
C....
  101 IF(REL.LT.RS) GO TO 8
      GO TO 9
    8 CONTINUE

      END
```

D.4   RELIABILITY MAXIMIZATION WITH MULTIPLE CONSTRAINT

*** .... **

RELIABILITY MAXIMIZATION WITH MULTIPLE CONSTRAINTS:

THIS PROGRAM CALCULATES THE OPTIMUM NUMBER OF REDUNDANCY IN AN
N STAGE SERIES SYSTEM FOR A GIVEN COST AND RELIABILITY CONTRAINT. THE
LAGRANGIAN METHOD IS USED FOR THE OPTIMIZATION. PROGRAM CALCU-
LATES THE SYSTEM RELIABILITY, SYSTEM TOTAL COST AND ALSO PRINTS OUT
THE LAGRANGE COEFFICIENT LAMBDA. FOR THE CALCULATIONS IT IS NECCE-
SARY TO GUESS TWO LAMBDA VALUES. DECREMENTS OF LAMBDA SHOULD BE
CAREFULLY CHOSEN TO BE ABLE TO OBTAIN SUFFICIENT NUMBER OF STEPS
BEFORE PROGRAM QUITS.

*** .... ***

```
C.... VARIABLES:
C.... A : COEFFICIENT FOR LAMBDA
C.... B : COEFFICIENT FOR LAMBDA
C.... C : COST OF EACH COMPONENT
C.... CL : COST CONSTRAINED
C.... DELLAM  : CHANGE IN THE LAMBDA
C.... G1 : FRACTION OF LAMBDA1
C.... G2 : FRACTION OF LAMBDA2
C.... N : TOTAL NUMBER OF STAGES
C.... NK  : COEFFICIENT OF LAMBDA
C.... PHI  : PARAMETER
C.... R : RELIABILITY OF EACH COMPONENT
C.... REL : RELIABILITY OF THE SYSTEM
C.... SUM1 : DUMMY VARIABLES FOR THE CALCULATION OF LAMBDA
C.... SUM2 : DUMMY VARIABLES FOR THE CALCULATION OF LAMBDA
C.... SUMC : TOTAL COST OF THE ELEMENT
```

```
C.... SUMP : TOTAL RELIABILITY OF THE SYSTEM
C.... SUMW : TOTAL WEIGHT OF THE SYSTEM
C.... W : WEIGHT OF EACH COMPONENT
C.... WL : WEIGHT CONSTRAINT
C.... XLAM : LAGRANGE MULTIPLIER
C.... XLAM1  : LAGRANGE MULTIPLIER GUESS
C.... XLAM2  : LAGRANGE MULTIPLIER GUESS

      REAL*8 R(10),C(10),PHI(100),A(10),NK(10),B(10),XLAM1,
           W(10)
      REAL*8 T(100),U(100),FC,FW,WL,D1FC,D1FW,D2FC,D2FW,G1,
           G2,XLAM,SUMW
      REAL*8 H(100),SUM1,SUM2,SUMC,SUMP,CL,AL,REL,XLAM2
      REAL*8 RELL(100),RELLL(100)
      DIMENSION NY(10)
      OPEN(UNIT=7,FILE='OUTH3.DAT',STATUS='NEW')
C....
C.... READ IN THE DATA
C....
      WRITE(*,*)'WHAT IS THE NUMBER OF STAGES?'
      READ(*,*) N
      WRITE(*,*)'WHAT IS THE COST LIMIT?'
      READ(*,*) CL
      WRITE(*,*)'WHAT IS THE WEIGHT LIMIT?'
      READ(*,*) WL
      WRITE(*,*) CL,WL
      DO 1 I=1,N
      WRITE(*,*) 'ENTER THE COST, WEIGHT, AND THE RELIABIL-
                 ITY OF STAGE'
      WRITE(*,*) I
      READ(*,*) C(I),W(I),R(I)
    1 CONTINUE
      WRITE(*,*) 'GIVE TWO GUESSES OF LAMBDA'
      READ(*,*) XLAM1,XLAM2
C....
C.... CALCULATE PARAMETERS AND CONSTANTS FOR THE CALCULATION
C.... of LAMBDA
C....
      DO 4 I=1,N
      PHI(I)=(1-R(I))/R(I)
      A(I)=-C(I)/DLOG(PHI(I))
      B(I)=-W(I)/DLOG(PHI(I))
    4 CONTINUE
C....
C.... CALCULATE  OF LAMBDA BY NEWTON-RAPHSON METHOD
C....
      T(1)=XLAM1
      U(1)=XLAM2
      J=1
   55 FC=CL
      FW=WL
      D1FC=0.0
      D2FC=0.0
      D1FW=0.0
```

```
      D2FW=0.0
      DO 46 I=1,N
      FC=FC+A(I)*DLOG(T(J)*A(I)+U(J)*B(I))
      FW=FW+B(I)*DLOG(T(J)*A(I)+U(J)*B(I))
      D1FC=D1FC+ A(I)*A(I)/(T(J)*A(I)+U(J)*B(I))
      D2FC= D2FC+A(I)*B(I)/(T(J)*A(I)+U(J)*B(I))
      D1FW=D1FW+ A(I)*B(I)/(T(J)*A(I)+U(J)*B(I))
      D2FW=D2FW+ B(I)*B(I)/(T(J)*A(I)+U(J)*B(I))
   46 CONTINUE
      T(J+1)=T(J)-(FC*D2FW-FW*D2FC)/(D1FC*D2FW-D1FW*D2FC)
      U(J+1)=U(J)+(FC*D1FW-FW*D1FC)/(D1FC*D2FW-D1FW*D2FC)
      ERR=ABS((T(J+1)-T(J))/T(J))
      ER=ABS((U(J+1)-U(J))/U(J))
      IF(ERR.LT.0.01.AND.ER.LT.0.01) GO TO 56
      J=J+1
      IF(J.GT.30) GO TO 19
      GO TO 55
   56 XLAM1=T(J)
      XLAM2=U(J)
      XLAM=XLAM1+XLAM2
      G1=XLAM1/XLAM
      G2=XLAM2/XLAM
      WRITE(*,*)'LAMBDA IS CALCULATED AS='
      WRITE(*,*) XLAM
      WRITE(*,*)'CHOOSE DELTA LAMBDA'
      READ(*,*)DELLAM
    9 CONTINUE
C....
C.... CALCULATE THE SMALLEST NUMBER OF ELEMENTS THAT WILL
C.... SATISFY THE UNDOMINATED SEQUENCE INEQUALITY
C.... FOR EVERY STAGE.
C....
      WRITE(7,89)
   89 FORMAT(///,2X,'NUMBER OF ELEMENTS',2X,'STAGES')
      DO 5 I=1,N
      NY(I)=0.0
    7 H(I)=(PHI(I)**NY(I))/((1+PHI(I))**(NY(I)+1))
      AL=XLAM*(G1*C(I)+G2*W(I))
      IF (H(I).LT.AL) GO TO 6
      NY(I)=NY(I)+1
      GO TO 7
    6 WRITE(7,90) NY(I),I
   90 FORMAT(2I,5X,2I)
    5 CONTINUE
C....
C.... CALCULATE THE COST AND RELIABILITY OF THE SYSTEM
C....
      SUMC=0.0
      SUMP=1.0
      SUMW=0.0
   10 CONTINUE
```

```
      DO 11 NI=1,N
      SUMW=SUMW+W(NI)*NY(NI)
      SUMC=SUMC+C(NI)*NY(NI)
      RELL(NI)=1/(1+PHI(NI))
      RELLL(NI)=1-(1-RELL(NI))**NY(NI)
      SUMP=SUMP*RELLL(NI)
   11 CONTINUE
      REL=SUMP
C.... C REL=1/(1+SUMP)
C....
C.... PRINT OUT THE RESULTS IN AA.DAT FILE
C....
      WRITE(7,91) SUMC,SUMW, REL,XLAM
   91 FORMAT(/,'COST=',F16.11,2X,/,'WEIGHT=',F16.11,2X,/
  ,'RELIABILITY=' $,F16.11,/,'LAMBDA=',F16.14,///)
      IF(SUMC.GT.CL) GO TO 8
      IF(SUMW.GT.WL) GO TO 8
      XLAM=XLAM-DELLAM
      IF(XLAM.LE.0.0) GO TO 98
      GO TO 9
   98 WRITE(*,*) 'CALCULATED LAMBDA IS LESS THAN 0,'
      WRITE(*,*)'RECHOOSE THE DELTA LAMBDA'
      WRITE(*,*)'RECOMMENDED DELTA LAMDA IS 20
      WRITE(*,*)'THE FIRST GUESSED DELTA LAMBDA,'
      WRITE(*,*)'KEEP YOUR FINGERS CROSSED'

      GO TO 8
      GO TO 9
   19 WRITE(*,*) 'LAMBDA NOT CONVERGE'
      WRITE(*,*) 'CHOOSE A NEW INITIAL GUESS LAMBDA'
    8 CONTINUE

      END
```

Table D.1 Management Activities (adapted from Project Management Short Course, TU Electric)

	Step 1 Individual Ranking	Step 2 Group Ranking	Step 3 Planning Experts Ranking	Step 4 Difference Steps 1 and 3	Step 5 Difference Steps 2 and 3
A. Find quality people to fill positions.			12		
B. Measure progress toward and/or deviation from the project's goals.			17		
C. Identify and analyze the various tasks necessary to implement the project.			8		
D. Develop strategies (priorities, sequence, timing of major steps).			6		
E. Develop possible alternative courses of action.			3		
F. Establish appropriate policies for recognizing individual performance.			20		
G. Assign responsibility/ accountability/authority.			15		
H. Establish project objectives.			2		
I. Train and develop personnel for new responsibilities/authority.			13		
J. Gather and analyze the facts of the current project situation.			1		
K. Establish qualifications for positions.			10		
L. Take corrective action on project (recycle project plans).			19		
M. Coordinate ongoing activities.			16		
N. Determine the allocation of resources (budget, facilities, etc.).			11		
O. Measure individual performance against objectives and standards.			18		
P. Identify the negative consequences of each course of action.			4		
Q. Develop individual performance objectives that are mutually agreeable to the individual and supervisor.			14		
R. Define scope of relationships, responsibilities, and authority of new positions.			5		
S. Decide on a basic course of action.			5		
T. Determine measurable checkpoints for the project and variations expected.			7		
Totals (the lower the ⟶ score the better)					
				Ind. Score Step 4	Team Score Step 5

Appendix E

A Proposal: The Effect of the Long-Term Use of Methanol

E.1 BACKGROUND

Methanol, one of the leading alternatives to gasoline as a motor vehicle fuel, has been highlighted in national competitions like as the SAE Methanol Marathon in 1989 and the SAE Methanol Challenge in 1990, but little has been done in the area of long-term testing of methanol as a motor vehicle fuel. The 1988 Chevrolet Corsica modified by Texas Tech University, which finished fifth in the 1989 competition and second in the 1990 competition, is an ideal test bed to determine the long-term effects of methanol on engine and emission systems performance.

The Texas Tech Corsica was optimized to operate on M85 for the SAE competitions; however, it has recently been modified to use M100. A methanol compatible fuel system was installed for the SAE competitions. The engine has been modified by increasing the stroke to take advantage of methanol's increased energy availability and by decreasing the bore to maintain economy. The resulting displacement is 2.8 liters, which is the same as the original gasoline engine. Because methanol has a higher octane rating than gasoline, the compression ratio was increased to 11.7:1 by installing custom flat-top pistons. A custom camshaft was employed to compensate for the slow-burn characteristics of methanol. Allied Signal, Inc., Tulsa, Oklahoma, provided the specially designed catalysts to control exhaust emissions.

E.2 OBJECTIVE

The objective of this project is to determine the effects of methanol fuel on engine performance and exhaust emissions during long-term use. Engine wear and tear, gasket performance, fuel system performance, emissions level, and overall vehicle performance will be monitored over 25,000 to 30,000 miles of vehicle operation.

E.3   TECHNICAL APPROACH

A vehicle performance baseline will be established initially and will be used for comparative purposes during the program. The engine will be removed from the vehicle and will be disassembled to record all bearing and ring clearances and cam profiles to determine any preexisting wear. Any needed repairs will be made to the engine at this time. After reassembling the engine, a Super-Flow dynamometer will be used to determine the engine performance at peak and road loads. Performance parameters to be measured will include brake torque, brake power, brake-specific fuel consumption, and engine out levels of hydrocarbons, carbon monoxide, carbon dioxide, and oxygen. The engine will then be installed in the vehicle.

Once the engine is installed in the Corsica, chassis dynamometer testing will be accomplished for engine/vehicle final calibration and performance evaluation. Track testing will determine 0–60 mph and quarter-mile elapsed times. On-road tests will measure both city and highway fuel economy. The vehicle will then be driven to Southwest Research Institute in San Antonio where it will undergo a full Environmental Protection Agency (EPA) Federal Test Procedure (FTP) emissions test to determine a basis for future emission comparisons.

The vehicle will then be driven approximately 25,000 miles. To accumulate this mileage, the vehicle will be displayed at various automobile, emissions, and energy seminars in Texas and across the nation and the vehicle will be driven daily in the Lubbock area. The mileage amassed on the vehicle will consist of approximately one-third highway and two-thirds city miles. Two oil samples will be taken every 3000 miles when the oil is changed. One sample will be analyzed locally and the other sample will be sent to the Lubrizol Corp. for analysis. Emissions testing will be accomplished periodically during the program using a four-gas analyzer.

When the required mileage has been accumulated on the Corsica, the vehicle will be driven back to Southwest Research Institute for a second round of emissions tests. These results will indicate any emission degradation caused by the long-term use of methanol in the Corsica. The engine will then be removed from the vehicle and retested on the dynamometer to identify any performance loss. After the dynamometer test, the engine will be disassembled and all fits and clearances will be remeasured. The gaskets will be removed and sent to FelPro, Inc. for analysis. The data from the initial tests and measurements will be compared with the data from the final tests, and a detailed report will be prepared and disseminated to all participants in the project. It is also anticipated that a paper describing the project and the results will be submitted to the Society of Automotive Engineers (SAE).

E.4   PROJECT PERSONNEL

The principal investigators will be Dr. Timothy T. Maxwell and Jesse C. Jones. Dr. Maxwell's interests include automotive engineering, internal combustion engines, aerodynamics, computational fluid dynamics, combustion, heat transfer, and energy utilization. Dr. Maxwell is the Technical Program Manager for a Ford Motor Company-funded aerodynamics research program that is currently in its fifth year. Mr. Jones has 25 years experience in the aerospace industry. He was site manager of the NASA White Sands Test Facility with responsibility for management, administration, engineering, technical support, and operation. Mr. Jones' interests include aerodynamics, automotive engineering, and design. Mr. Jones is coordinator for the cooperative industry/university design project program.

Both Dr. Maxwell and Mr. Jones are extensively involved in alternative fuels research programs. They are coprincipal investigators for a project funded through the Texas Advanced Technology Program to investigate the starting of methanol fueled engines in cold weather, and they were coadvisors for the 1989–90 Methanol Marathon/Challenge and the 1991 Natural Gas Vehicle Challenge. Mr. Jones and Dr. Maxwell have several other alternate fuels programs in progress. One of these programs is a survey of the Lubbock, Texas, area to determine what vehicle fleets could be converted to burn alternate fuels. Another project is concerned with the development of a direct injection capability for natural gas fueled engines. A series of workshops was recently completed across the State of Texas on Managing Vehicle Fleets for Fuel Economy.

Mr. Jones and Dr. Maxwell will work together to manage the project. Mr. Jones will be responsible for all engine and vehicle testing, and Dr. Maxwell will direct the engine disassembly, characterization, and reassembly activities. They will be assisted in these tasks by three undergraduate students who will compile the data, demonstrate the vehicle, and assist in the test and engine disassembly efforts.

E.5 SCHEDULE

The project was scheduled to start in June 1992 and will continue over a two-year period. Table E.1 presents the work schedule by quarters.

E.6 BUDGET

Several organizations have agreed to provide either funding or in-kind support. FelPPro, Inc., will provide $10,000 and all gaskets and seals needed for the project. Air Products and Chemicals, Inc. will provide 2000 gallons of methanol fuel and the Lubrizol Corp. will provide all of the oil required for the project. Allied Signal will provide the exhaust system catalysts. The Texas Tech University Center for Energy Research has agreed to provide $10,000 for the project, and the university will charge only 10 percent indirect costs as partial cost sharing.

Table E.2 presents the details of the budget request for this project. The total budget is $43,263.

Table E.1 Project Schedule

Tasks	Quarters Starting in June 1992							
	1	2	3	4	5	6	7	8
Remove engine from car, disassemble, inspect, measure clearance, and reassemble	••							
Engine and chassis dyno tests	•							
Performance and fuel economy tests	••							
Emissions tests	•							
Mileage accumulation		•••	•••	•••	•••	•••		
Emissions retest							•	
Performance and fuel economy retest							••	
Engine and chassis dyno retests								•
Remove engine from car, disassemble, inspect, measure clearance, and reassemble								•
Prepare report								•

Table E.2   Project Budget

A. Salaries and Wages	
1. Undergraduate students (500 hours @ $5/h)	$ 2,500
Total Salaries and Wages	**$ 2,500**
B. Fringe Benefits (18% of A.1)	$ 450
C. Parts and Supplies	
1. Six fuel injectors	300
2. Fuel pump	100
3. Custom piston rings	120
4. Engine bearings	40
5. Gaskets	150
6. Filters (oil and air)	80
7. Spark plugs	15
Total Parts and Supplies	**$ 805**
D. Engine Work	
1. Remove engine from vehicle and replace (2)	900
2. Disassemble and clean engine (2)	200
3. Inspect, measure, record data, and reassemble engine (2)	3,000
Total Engine Work	**$ 4,100**
E. Engine and Vehicle Testing	
1. Dynamometer testing of engine (2)	3,400
2. FTP emissions tests (2)	5,000
Total Engine and Vehicle Testing	**$ 8,500**
F. Fuel and Oil	
1. 2000 gallons of methanol	3,000
2. 60 quarts of oil	75
Total Fuel and Oil	**$ 3,075**
G. Travel (20 trips)	$20,000
H. Total Direct	$39,330
I. Indirect Costs (10% of H)	$ 3,933
J. Total Project Costs	**$43,263**

Index

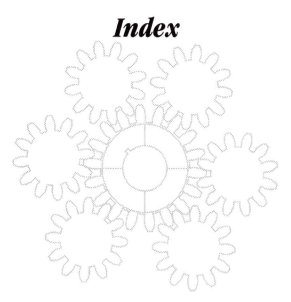